AF719340

FUZZY RELATIONAL SYSTEMS

International Federation for Systems Research
International Series on Systems Science and Engineering

Series Editor: George J. Klir
State University of New York at Binghamton

Editorial Board

Gerrit Broekstra
Erasmus University, Rotterdam, The Netherlands

John L. Casti
Santa Fe Institute, New Mexico

Brian Gaines
University of Calgary, Canada

Ivan M. Havel
Charles University, Prague, Czech Republic

Manfred Peschel
Academy of Sciences, Berlin, Germany

Franz Pichler
University of Linz, Austria

Volume 14 *INTELLIGENT ROBOTIC SYSTEMS: Design, Planning, and Control*
Witold Jacak

Volume 15 *FACETS OF SYSTEMS SCIENCE, Second Edition*
George J. Klir

Volume 16 *PROBABILISTIC ANALYSIS OF BELIEF FUNCTIONS*
Ivan Kramosil

Volume 17 *UNDERSTANDING SYSTEMS: Conversations on Epistemology and Ethics*
Heinz von Foerster and Bernhard Poerksen

Volume 18 *CREATING INTERNET INTELLIGENCE: Wild Computing, Distributed Digital Consciousness, and the Emerging Global Brain*
Ben Goertzel

Volume 19 *FLEXIBLE ROBOT DYNAMICS AND CONTROLS*
Rush D. Robinett, III, Clark R. Dohrmann, G. Richard Eisler, John T. Feddema, Gordon G. Parker, David G. Wilson, and Dennis Stokes

Volume 20 *FUZZY RELATIONAL SYSTEMS: Foundations and Principles*
Radim Bělohlávek

IFSR was established "to stimulate all activities associated with the scientific study of systems and to coordinate such activities at international level." The aim of this series is to stimulate publication of high-quality monographs and textbooks on various topics of systems science and engineering. This series complements the Federation's other publications.

A Continuation Order Plan is available for this series. A continuation order will bring delivery of each new volume immediately upon publication. Volumes are billed only upon actual shipment. For further information please contact the publisher.

Volumes 1–6 were published by Pergamon Press.

FUZZY RELATIONAL SYSTEMS

FOUNDATIONS AND PRINCIPLES

Radim Bělohlávek

Kluwer Academic/Plenum Publishers
New York, Boston, Dordrecht, London, Moscow

Library of Congress Cataloging-in-Publication Data

Belohlávek, Radim.
Fuzzy relational systems: foundations and principles/Radim Belohlávek.
p. cm. — (International Federation for Systems Research international series on systems science and engineering; v. 20)
Includes bibliographical references and index.
ISBN 0-306-46777-1
1. Fuzzy sets. 2. Fuzzy logic. 3. Fuzzy systems. I. Title. II. IFSR international series on systems science and engineering; v. 20.

QA248.5 .B45 2002
511.3′22—dc21

2002022111

ISBN 0-306-46777-1

©2002 Kluwer Academic/Plenum Publishers, New York
233 Spring Street, New York, N.Y. 10013

http://www.wkap.com

10 9 8 7 6 5 4 3 2 1

A C.I.P. record for this book is available from the Library of Congress

All rights reserved

No part of this book may be reproduced, stored in a retrieval system, or transmitted in any form or by any means, electronic, mechanical, photocopying, microfilming, recording, or otherwise, without written permission from the Publisher

Printed in the United States of America

To my wife Jana and
our sons Ivan and Jan
with love.

Preface

Since their inception, fuzzy sets and fuzzy logic became popular. The reason is that the very idea of fuzzy sets and fuzzy logic attacks an old tradition in science, namely bivalent (black-or-white, all-or-none) judgment and reasoning and the thus resulting approach to formation of scientific theories and models of reality. The idea of fuzzy logic, briefly speaking, is just the opposite of this tradition: instead of full truth and falsity, our judgment and reasoning also involve intermediate truth values. Application of this idea to various fields has become known under the term fuzzy approach (or graded truth approach). Both practice (many successful engineering applications) and theory (interesting nontrivial contributions and broad interest of mathematicians, logicians, and engineers) have proven the usefulness of fuzzy approach.

One of the most successful areas of fuzzy methods is the application of fuzzy relational modeling. Fuzzy relations represent formal means for modeling of rather nontrivial phenomena (reasoning, decision, control, knowledge extraction, systems analysis and design, etc.) in the presence of a particular kind of indeterminacy called vagueness. Models and methods based on fuzzy relations are often described by logical formulas (or by natural language statements that can be translated into logical formulas). Therefore, in order to approach these models and methods in an appropriate formal way, it is desirable to have a general theory of fuzzy relational systems with basic connections to (formal) language which enables us to describe relationships in these systems. From the engineering point of view, such a theory should provide us with a useful conceptual framework and results that can be directly applied when designing relational models of systems using fuzzy approach. From the mathematical point of view, such a theory is a kind of a model theory in fuzzy setting and should therefore reflect the recent development in fuzzy logic.

The present book attempts to present a general theory of fuzzy relational systems with respect to the above-sketched view. We aim to show that fuzzy relational systems not only provide us with a reasonable and nontrivial generalization of results of ordinary (i.e. bivalent) relational systems but, importantly, they offer us means for modeling of various phenomena that are degenerate and thus hidden in the ordinary case but naturally appear in the context of relational modeling in the presence of vagueness.

The book is written in a mathematical style (we present definitions and formulate theorems). Perhaps more than usual, we pay attention to the extramathematical meaning of the presented notions and results. This is mainly because there are basically two intended groups of readers. In addition to mathematicians interested in "fuzzy", we hope that (parts of) the book will be interesting for engineers working with fuzzy relational models who want to learn more about mathematical foundations and also about issues which, however useful, are not so frequently discussed. The book provides a presentation of several general aspects of fuzzy relational modeling (structures of truth values and their properties, basics of syntax (enabling us to translate natural language statements into logical formulas) and semantics (enabling us to evaluate the truth degrees of the statements), results on sensitivity of fuzzy relational models, results on reducing the complexity, and applications of these results to selected models) as well as a survey of several particular fuzzy relational models (models based on similarity relations, models based on analysis of information contained in a fuzzy relation between objects and attributes, composition of fuzzy relations, methods for solving of fuzzy relational equations). From the mathematical point of view, the book develops foundations of fuzzy relational systems considered as structures of first-order fuzzy logic. The primary attention is paid to phenomena which are hidden in classical model theory (graded similarity in models, preservation of similarity by logical formulas, morphism of structures as a pair of a mapping between universes and a mapping between sets of truth values, representation of fuzzy structures by bivalent structures, "almost" morphisms, etc.). However, we also present several classical results in their generalized form. The structures of truth values we use are complete residuated lattices. Although the most important residuated lattices are those with support set being (possibly a subset of) the interval $[0, 1]$, the use of abstract residuated lattices makes explicit the algebraic character of fuzzy relational systems (or, at least, of the parts of fuzzy relational systems we deal with). The book *is not* about deductive systems of fuzzy logic (with one exception being a part of Section 3.6). However, a deductive system of a first-order fuzzy logic (including references) is briefly presented in Section 7.3.

The book is organized as follows. The necessary mathematical concepts are surveyed in Chapter 1. Chapter 2 contains a general discussion on fuzzy approach, structures of truth values, and a section devoted to structures of truth values and their properties. Fundamental notions and results on fuzzy relational systems are contained in Chapter 3. Investigated are important structural notions (e.g. subsystems, morphisms, products); special attention is paid to topics which are degenerate in the ordinary case (introducing gradedness into classical notions, similarity relations, etc.). Several of these notions and results are directly connected to fuzzy relational modeling. Chapter 4 focuses on binary fuzzy relations, especially on operations and special types of binary relations and their properties. Binary fuzzy relations interpreted as relations between objects and attributes and related topics are studied in Chapter 5. Attention is paid to conceptual interpretation of object–attribute knowledge. Chapter 6 presents the problem of composition and decomposition of fuzzy relations (fuzzy relational equations), an important topic in the design of relational systems. Surveyed are the solutions and approximate solutions of relational equations, systems of equations, sensitivity of relational equations, and the structure of solutions. Chapter 7 contains miscellaneous topics: fuzzy closure operators and similarity spaces; further applications of selected results of Chapter 3; a deductive system of first-order fuzzy logic.

The book contains an index and a bibliography. At the close of each of Chapters 2–7 is a section "Bibliographical remarks" which contains references to bibliographical items (most of all references in the book are cited in these sections) and also description of the development and/or references to alternative approaches.

Generally, we focus more on issues and results that are new, not so frequently discussed, or not generally known. Thus, for instance, although there is a vast literature on fuzzy relational equations, we have only a relatively short section on this topic where we bring up the basic types of equations as well as the general aspects and principles. However, this section is accompanied by references. On the other hand, we have a relatively long section on binary fuzzy relations interpreted as relations between objects and attributes.

The book can be read chapter by chapter, each chapter section by section, from the beginning to the end. This is the recommended style of reading (of course one may skip proofs and even lemmas and concentrate on theorems). However, the chapters can be read relatively separately—having in mind especially engineers who may find interesting information here. Thus, the other style of reading is the following: Read Section 2.1 for motivations and general ideas behind fuzzy approach; read the definition of a complete residuated lattice in Section 2.3 and skip just to (left-continuous) t-norms (definition and examples are in Section 2.3; further information is in Section 2.4). For the rest of the book you can assume that the structure **L** of truth values is the real interval $[0, 1]$ equipped with min, max, a left-continuous t-norm $\otimes$, and the corresponding residuum (implication) $\rightarrow$. Then read Section 3.1 on fuzzy sets and fuzzy relations and read the definitions and examples from Section 3.2. You can then read any of Chapters 4–7 and go back to Chapters 2, 3 if necessary.

My thanks go to my colleague Vilém Vychodil for preparing some figures and symbols, and for comments to some chapters. More than I can say I thank my wife Jana for her patience and support, and also for preparation of some figures.

Christmas 2001 Radim Bělohlávek

Contents

Chapter 1

PRELIMINARIES

Just as everybody must strive to learn language and writing before he can use them freely for the expression of his thoughts, here too there is only one way to escape the weight of formulae. It is to acquire such power over the tool ... that, unhampered by formal technique, one can turn to the true problems ...

—Herman Weyl (Raum, Zeit, Materie, 5th ed. Berlin, 1923)

The aim of this chapter is to give an overview of some basic notions and results (with almost no proofs).

1.1 SETS, RELATIONS, MAPPINGS

We assume familiarity with the notion of a *set* as used in the intuitive set theory, i.e. as a (sharply delineated) collection of objects (that is, each object either belongs or does not belong to a given set). Objects which belong to a given set are also called its elements.

We write $x = y$ to denote that x and y are identical.

The fact that the object x belongs to the set A (A contains x) is denoted by $x \in A$, the opposite by $x \notin A$. Given a property φ, $\{x \mid \varphi(x)\}$ denotes the set of all objects having φ. The empty set (a set containing no elements) is denoted by $\emptyset$.

A set A is a *subset* of a set B (denoted by $A \subseteq B$) if each element of A belongs to B. We write $A \subset B$ if $A \subseteq B$ but $A \neq B$. Given sets A and B, the *intersection* of A and B (denoted by $A \cap B$) is the set containing all elements which belong to both A and B, the *union* of A and B (denoted by $A \cup B$) is the set containing all elements which belong to A or to B (or both), the *difference* between A and B (denoted by $A - B$) is the set containing the elements belonging to A but not to B. In some situations, only elements of some given set are taken into account. Such a set is then called a *universe of discourse* (universal set, universe). If X is a universe of discourse and $A \subseteq X$, then a *complement* of A (w.r.t. X, denoted by $\overline{A}$) is the set $X - A$.

The *ordered n-tuple* of objects $x_1, \ldots, x_n$ is denoted by $\langle x_1, \ldots, x_n \rangle$ (this notion may be defined set-theoretically: $\langle x_1, x_2 \rangle = \{\{x_1\}, \{x_1, x_2\}\}$; and by induction, $\langle x_1 \rangle = x_1$, $\langle x_1, \ldots, x_n, x_{n+1} \rangle = \langle \langle x_1, \ldots, x_n \rangle, x_{n+1} \rangle$). We have that $\langle x_1, \ldots, x_n \rangle = \langle y_1, \ldots, y_m \rangle$ iff $n = m$ and $x_1 = y_1, \ldots, x_n = y_n$ (this fact can be proved if tuples are defined set-theoretically as above). The *direct product* (Cartesian product) of sets $X_1, \ldots, X_n$ is the set $X_1 \times \cdots \times X_n = \{\langle x_1, \ldots, x_n \rangle \mid x_1 \in X_1, \ldots, x_n \in X_n\}$, i.e. the set of all n-tuples with x_i taken from X_i. If $X_1 = \cdots = X_n = X$, we denote the direct product by X^n.

A *relation* between sets $X_1, \ldots, X_n$ is any subset R of the direct product of these sets, i.e. $R \subseteq X_1 \times \cdots \times X_n$. The number n is called the *arity* of R, R is said to be n-ary. For $n = 1, 2, 3, \ldots$ we use the terms unary, binary, ternary,.... If $X_1 = \cdots = X_n = X$, R is said to be a relation in the set X. The *inverse relation* of a binary relation $R \subseteq X \times Y$ is the relation $R^{-1} \subseteq Y \times X$ defined by $R^{-1} = \{\langle y, x \rangle \mid \langle x, y \rangle \in R\}$. If $R \subseteq X \times Y$, $A \subset X$, we put $R(A) = \{y \mid \langle x, y \rangle \in R \text{ for some } x \in A\}$. $R(\{x\})$ is also denoted by $[x]_R$.

A *mapping* (function, map) of a set X into a set Y is a binary relation f between X and Y such that (i) for each $x \in X$ there is some $y \in Y$ such that $\langle x, y \rangle \in f$, and (ii) if $\langle x, y_1 \rangle \in f$ and $\langle x, y_2 \rangle \in f$ then $y_1 = y_2$. In that case, f may be thought of as representing an assignment of elements of Y to elements of X. The fact that f is a mapping of X to Y is denoted by $f : X \to Y$, the set of all mappings of X to Y is denoted by Y^X. If $X = Y^n$, f is also called an n-ary *operation* in Y. If $f : X \to Y$, the element y such that $\langle x, y \rangle \in f$ is denoted by $f(x)$, i.e. $f(x) = y$ means $\langle x, y \rangle \in f$. A mapping f is called *injective* (injection, one-one mapping) if $x_1 \neq x_2$ implies $f(x_1) \neq f(x_2)$, *surjective* (surjection, onto mapping) if for each $y \in Y$ there is some $x \in X$ such that $f(x) = y$, *bijective* (bijection, one-one onto mapping) if it is both injective and surjective. If $f : X \to Y$, $g : Y \to Z$, then the *composition* of f and g is the mapping $f \circ g$ of X to Z defined by $(f \circ g)(x) = f(g(x))$.

Put $2 = \{0, 1\}$ (there will be no danger of confusion with the integer 2). Then 2^X denotes the set of all mappings $\chi : X \to \{0, 1\}$. There is a natural bijection between the set of all subsets of X and the set 2^X: each $\chi \in 2^X$ induces a subset $A_\chi = \{x \in X \mid \chi(x) = 1\}$ of X; each subset A of X induces a mapping $\chi_A : X \to \{0, 1\}$ by $\chi_A(x) = 1$ if $x \in A$ and $\chi_A(x) = 0$ if $x \notin A$; χ is called a characteristic function of the corresponding subset (i.e. of A_χ). For convenience, one sometimes does not distinguish between subsets of X and their characteristic functions (modulo the above bijection) and thus denotes the set of all subsets of X also by 2^X.

If $I \neq \emptyset$ is a set and for each $i \in I$ there is a set X_i, we say that $\{X_i \mid i \in I\}$ is a system of sets indexed by I. In a natural way, intersection and union is extended to a system of sets by putting $\bigcap_{i \in I} X_i = \{x \mid x \in X_j \text{ for all } j \in I\}$, $\bigcup_{i \in I} X_i = \{x \mid x \in X_j \text{ for some } j \in I\}$. The direct product of a system $\{X_i \mid i \in I\}$ is the set $\times_{i \in I} X_i = \{f : I \to \bigcup_{i \in I} X_i \mid f(j) \in X_j \text{ for all } j \in I\}$. For $I = \{1, \ldots, n\}$, the notion coincides in an obvious way with the notion of a direct product of $X_1, \ldots, X_n$ introduced above. The j-th projection ($j \in I$) of the direct product $\times_{i \in I} X_i$ is a mapping $\mathrm{pr}_j : \times_{i \in I} X_i \to X_j$ defined for any $a \in \times_{i \in I} X_i$ by $\mathrm{pr}_j(a) = a(j)$ (i.e. pr_j selects the j-th component).

A binary relation R in a set X is called *reflexive* if for each $x \in X$ we have $\langle x, x \rangle \in R$; *symmetric* if for each $x, y \in X$, $\langle x, y \rangle \in R$ implies $\langle y, x \rangle \in R$; *antisymmetric* if for each $x, y \in X$, $\langle x, y \rangle \in R$ and $\langle y, x \rangle \in R$ imply $x = y$; *transitive* if for each $x, y, z \in X$, $\langle x, y \rangle \in R$ and $\langle y, z \rangle \in R$ imply $\langle x, z \rangle \in R$.

$\mathbf{N}$, $\mathbf{Z}$, $\mathbf{Q}$, and $\mathbf{R}$ denote the set of all nonnegative integers, the set of all integers, the set of all rationals, and the set of all reals, respectively.

1.2 ORDERED SETS, LATTICES

A binary relation in a set is called a *partial order* (or simply order) if it is reflexive, antisymmetric, and transitive. If $\leq$ is a partial order, we write $x < y$ if $x \leq y$ and $x \neq y$. If $\leq$ is a partial order on X, the pair $\mathbf{X} = \langle X, \leq \rangle$ is called a (partially) *ordered set*.

Let $\langle X, \leq \rangle$ be an ordered set and $A \subseteq X$. An element $x \in A$ is called a *minimal* element of A if for each $y \in A$ such that $y \leq x$ we have $x = y$; a *maximal* element of A if for each $y \in A$ such that $x \leq y$ we have $x = y$; the *least* element of A if for each $y \in A$ we have $x \leq y$; the *greatest* element of A if for each $y \in A$ we have $y \leq x$. If there is a least (greatest) element of X, it is denoted by 0 (1). We define the *lower cone* of A by $\mathcal{L}(A) = \{x \in X \mid x \leq y \text{ for all } y \in A\}$, and the *upper cone* of A by $\mathcal{U}(A) = \{x \in X \mid y \leq x \text{ for all } y \in A\}$. If $\mathcal{L}(A)$ has a greatest element x, then x is called the *infimum* of A in $\langle X, \leq \rangle$. Dually, if $\mathcal{U}(A)$ has a least element x, then x is called the *supremum* of A in $\langle X, \leq \rangle$. The infimum (supremum) of A is denoted by $\bigwedge A$ or $\inf A$ ($\bigvee A$ or $\sup A$). Note that if 0 (1) exists then $\bigwedge X = 0 = \bigvee \emptyset$ ($\bigwedge \emptyset = 1 = \bigvee X$), by definition.

A partially ordered set $\langle X, \leq \rangle$ is said to be *lattice ordered* (or a lattice) if infimum and supremum exist for any two-element (and hence for any nonempty finite) subset of X; it is said to be *completely lattice ordered* (or a complete lattice) if infimum and supremum exist for any subset of X.

If $\langle X, \leq \rangle$ is a complete lattice, a subset $K \subseteq X$ is called *infimally dense* (supremally dense) if each $x \in X$ is the infimum (supremum) of some subset of K. For example, $\mathbf{Q}$ is both infimally and supremally dense in $\langle \mathbf{R}, \leq \rangle$.

A partially ordered set $\langle X, \leq \rangle$ is said to be *linearly ordered* (or a chain) if for every $x, y \in X$ we have $x \leq y$ or $y \leq x$, i.e. every two elements are comparable; in this case, $\leq$ is called a linear order. Each chain is a lattice with $\inf\{x, y\} = \min\{x, y\}$ and $\sup\{x, y\} = \max\{x, y\}$ (where $\min\{x, y\}$ and $\max\{x, y\}$ denote the smaller and the bigger one of x and y, respectively).

Lattices can be alternatively described as algebras (for a while, we will use the term lattice in two ways; Proposition 1.1 shows why one can afford it). A lattice is an algebra $\mathbf{L} = \langle L, \wedge, \vee \rangle$ (i.e. L is a nonempty set) with two binary operations which satisfy

$$\begin{array}{lll} x \vee y = y \vee x & x \wedge y = y \wedge x & \text{(commutativity)}, \\ x \vee (y \vee z) = (x \vee y) \vee z & x \wedge (y \wedge z) = (x \wedge y) \wedge z & \text{(associativity)}, \\ x \vee (x \wedge y) = x & x \wedge (x \vee y) = x & \text{(absorption)}. \end{array}$$

Proposition 1.1 *(1) If $\mathbf{X} = \langle X, \leq \rangle$ is a lattice ordered set, then $\langle X, \wedge, \vee \rangle$ where $x \wedge y = \inf\{x, y\}$, $x \vee y = \sup\{x, y\}$ is a lattice.*

(2) If $\mathbf{L} = \langle L, \wedge, \vee \rangle$ is a lattice, then $\langle L, \leq \rangle$ where $x \leq y$ iff $x \wedge y = x$ (or, which is equivalent, iff $x \vee y = y$) is a lattice ordered set such that $x \wedge y = \inf\{x, y\}$, and $x \vee y = \sup\{x, y\}$.

(3) The constructions described in (1) and (2) are mutually inverse.

If a lattice $\mathbf{L}$ contains both the least and the greatest element 0 and 1, respectively, it is called a *bounded lattice*. In this case we write $\mathbf{L} = \langle L, \wedge, \vee, 0, 1 \rangle$ or $\mathbf{L} = \langle L, \leq, 0, 1 \rangle$.

We assume so-called Zorn lemma which says that if any chain of an ordered set $\mathbf{X}$ (i.e. a subset $A \subseteq X$ with pairwise comparable elements) has an upper bound (i.e. for some $x \in X$ we have $a \leq x$ for each $a \in A$) then each element of X is contained in some maximal element of $\mathbf{X}$ (i.e. for each $x \in X$ there is some y maximal in $\mathbf{X}$ with $x \leq y$).

Example 1.2 (1) The natural order $\leq$ on the set $\mathbf{R}$ of all real numbers is a partial order. Moreover, $\langle \mathbf{R}, \leq \rangle$ is a chain and thus a lattice (which is not complete).

(2) For any set X, $\langle 2^X, \subseteq \rangle$ is a complete lattice (which is not linear iff X has at least two elements) with $\inf\{A_i \subseteq X \mid i \in I\} = \bigcap_{i \in I} A_i$ and $\sup\{A_i \subseteq X \mid i \in I\} = \bigcup_{i \in I} A_i$.

(3) More generally, if $\mathcal{S} \subseteq 2^X$ is a nonempty system of subsets of X that is closed w.r.t. arbitrary intersections (i.e. $\bigcap_{i\in I} A_i \in \mathcal{S}$ whenever $A_i \in \mathcal{S}$) then $\langle \mathcal{S}, \subseteq\rangle$ is a complete lattice where infima coincide with intersections and suprema are given by $\sup_{i\in I} A_i = \bigcap\{A \in \mathcal{S} \mid \bigcup_{i\in I} A_i \subseteq A\}$. A system $\mathcal{S}$ of subsets of X that is closed w.r.t. arbitrary intersections is called a *closure system* in X. For a closure system $\mathcal{S}$ and a subset $A \subseteq X$, the set $[A]_{\mathcal{S}} = \bigcap\{B \in \mathcal{S} \mid A \subseteq B\}$ is called the closure of A; it is the least (w.r.t. $\subseteq$) subset of X that belongs to A.

Let $\mathbf{U} = \langle U, \leq_U\rangle$ and $\mathbf{V} = \langle V, \leq_V\rangle$ be complete lattices. A map $c : U \to U$ is called a *closure operator* in $\mathbf{U}$ if (i) $u \leq_U c(u)$; (ii) $u_1 \leq_U u_2$ implies $c(u_1) \leq_U c(u_2)$; (iii) $c(u) = c(c(u))$ (for any $u, u_i \in U$). A pair $\langle {}^\uparrow, {}^\downarrow\rangle$ of maps ${}^\uparrow : U \to V$ and ${}^\downarrow : V \to U$ is said to form a Galois connection between $\mathbf{U}$ and $\mathbf{V}$ if (i) $u_1 \leq_U u_2$ implies $u_2^\uparrow \leq_V u_1^\uparrow$; (ii) $v_1 \leq_V v_2$ implies $v_2^\downarrow \leq_U v_1^\downarrow$; (iii) $u \leq_U u^{\uparrow\downarrow}$ and $v \leq_V v^{\downarrow\uparrow}$ (for $u, u_i \in U$, $v, v_i \in V$).

Proposition 1.3 *The set $\{u \in U \mid u = c(u)\}$ of all fixed points of c equipped with the restriction of $\leq_U$ is a complete lattice where infima coincide with infima in $\mathbf{U}$ and a supremum of a_i's ($i \in I$) is $c(\sup_{i\in I} a_i)$. The set $\{\langle u, v\rangle \in U \times V \mid u^\uparrow = v, v^\downarrow = u\}$ of all fixed points of $\langle {}^\uparrow, {}^\downarrow\rangle$ is a complete lattice w.r.t. a partial order $\leq$ given by $\langle u_1, v_1\rangle \leq \langle u_2, v_2\rangle$ iff $u_1 \leq_U u_2$ (or, equivalently, iff $v_2 \leq_V v_1$); infima and suprema are given by $\inf_{i\in I}\langle u_i, v_i\rangle = \langle \inf_{i\in I} u_i, (\sup_{i\in I} v_i)^{\downarrow\uparrow}\rangle$ and $\sup_{i\in I}\langle u_i, v_i\rangle = \langle (\sup_{i\in I} u_i)^{\uparrow\downarrow}, \inf_{i\in I} v_i\rangle$.*

For a Galois connection $\langle {}^\uparrow, {}^\downarrow\rangle$, the composite maps ${}^{\uparrow\downarrow}$ and ${}^{\downarrow\uparrow}$ are closure operators in $\mathbf{U}$ and $\mathbf{V}$, respectively; furthermore, $u^\uparrow = u^{\uparrow\downarrow\uparrow}$ and $v^\downarrow = v^{\downarrow\uparrow\downarrow}$. Particularly, if one takes $U = 2^X$, $V = 2^Y$ and the subsethood relations as the partial orders $\leq_U$ and $\leq_V$, one gets the notion of a closure operator in a set X and that of a Galois connection between sets X and Y.

If $\mathcal{S}$ is a closure system in X (cf. Example 1.2 (3)), then $c : 2^X \to 2^X$ defined by $c(A) = [A]_{\mathcal{S}}$ is a closure operator in X.

1.3 TOLERANCES, EQUIVALENCES

A binary relation in a set is called a *tolerance* if it is reflexive and symmetric; it is called an *equivalence* if it is reflexive, symmetric, and transitive.

Let R be a tolerance in the set X. A set $A \subseteq X$ is called a *block* of R if $\langle x, y\rangle \in R$ holds for all $x, y \in A$. A block A of R is called a *maximal block* of R if it is maximal w.r.t. set inclusion, i.e. if there is no block

B of R such that $A \subset B$. The set of all maximal blocks of a tolerance R in X will be denoted by X/R and called a *factor set* of X by R. The set $[x]_R$ is called the *class* of R determined by x.

Proposition 1.4 *Let R be a tolerance in X. Each block A of R is contained in some maximal block of R. Each $x \in X$ is contained in some maximal block of R.*

Proof. Clearly, a union of any chain $A = A_1 \subseteq A_2 \subseteq \cdots$ of blocks of R is again a block of R. By Zorn lemma, A is contained in some maximal block of R. The second assertion follows from the first one by the fact that $\{x\}$ is a block of R for any $x \in X$. □

A *covering* of a set X is a set Π of nonempty subsets of X such that each $x \in X$ belongs to some $A \in \Pi$. A *partition* of a set X is a covering Π of X such that $A \neq B$ implies $A \cap B = \emptyset$ for all $A, B \in \Pi$.

Remark 1.5 By Proposition 1.4, the set of all maximal blocks of a tolerance relation forms a covering.

Proposition 1.6 *Let R be an equivalence in X, Π be a partition of X. Then $\Pi_R = \{[x]_R \mid x \in X\}$ is a partition of X, $R_\Pi = \{\langle x, y\rangle \mid x, y \in A$ for some $A \in \Pi\}$ is an equivalence in X, and we have $R = R_{\Pi_R}$ and $\Pi = \Pi_{R_\Pi}$.*

Remark 1.7 (1) Hence, there is one-to-one correspondence between equivalences and partitions on a given set.

(2) While maximal blocks and classes $[x]_R$ coincide if R is an equivalence relation (i.e. $\Pi_R = X/R$), $[x]_R$ may not be a block if R is a tolerance relation.

For a mapping $h : A \to B$ we denote by θ_h a binary relation on A given by $\langle a_1, a_2\rangle \in \theta_h$ iff $h(a_1) = h(a_2)$; one easily verifies that θ_h is an equivalence relation. If θ is an equivalence on A, we denote by h_θ a mapping of A into A/θ given by $h_\theta(a) = [a]_\theta$; clearly, h_θ is surjective.

1.4 ALGEBRAS

Algebras are sets equipped with operations with prescribed arities. Arities of operations of an algebra are given by a so-called type. Formally, a *type* is a pair $\langle F, \sigma\rangle$ of a set F (of symbols of operations) and a mapping σ of F into a set $\mathbf{N}_0$ of nonnegative integers. For $f \in F$, $\sigma(f)$ is called the *arity* of f. An *algebra of type σ* is a pair

$\mathbf{A} = \langle A, F^{\mathbf{A}} \rangle$ where A is a nonempty set (called a *universe* of $\mathbf{A}$), and $F^{\mathbf{A}} = \{f^{\mathbf{A}} : A^{\sigma(f)} \to A \mid f \in F\}$ is the set of the operations of $\mathbf{A}$ ($f^{\mathbf{A}}$ is $\sigma(f)$-ary).

Following common usage, we adopt the following conventions: If there is no danger of confusion, we omit the superscript $^{\mathbf{A}}$, i.e. we write F or f instead of $F^{\mathbf{A}}$ or $f^{\mathbf{A}}$, respectively. If $\langle A, F^{\mathbf{A}} \rangle$ is an algebra and $F = \{f_1, \ldots, f_n\}$, we write only $\langle A, f_1^{\mathbf{A}}, \ldots, f_n^{\mathbf{A}} \rangle$ instead of $\langle A, \{f_1^{\mathbf{A}}, \ldots, f_n^{\mathbf{A}}\} \rangle$. Also the phrase "of type $\langle F, \sigma \rangle$" will often be omitted.

If $\mathbf{A} = \langle A, F \rangle$ is an algebra, a nonempty subset B of A is a *subuniverse* of $\mathbf{A}$ if it is closed under any $f \in F$, i.e. if for any $f \in F$, $n = \sigma(f)$, and every $a_1, \ldots, a_n \in A$, we have $f(a_1, \ldots, a_n) \in B$ whenever $a_1, \ldots, a_n \in B$. The set of all subuniverses of an algebra $\mathbf{A}$ is denoted by $\operatorname{Sub} \mathbf{A}$. Any subuniverse $B \neq \emptyset$ equipped with restrictions of operations of $\mathbf{A}$ to B is again an algebra of the same type as $\mathbf{A}$; it is called a *subalgebra* of $\mathbf{A}$.

If $\mathbf{A} = \langle A, F \rangle$ is an algebra, a binary relation θ on A is said to be *compatible* (with the operations of $\mathbf{A}$) if for each $f \in F$ s.t. $\sigma(f) = n$ and every $a_1, b_1, \ldots, a_n, b_n \in A$, we have that if $\langle a_1, b_1 \rangle \in \theta, \ldots, \langle a_n, b_n \rangle \in \theta$ then $\langle f(a_1, \ldots, a_n), f(b_1, \ldots, b_n) \rangle \in \theta$. A compatible equivalence on A is called a *congruence*. $\operatorname{Con} \mathbf{A}$ denotes the set of all congruence relations on the algebra $\mathbf{A}$.

Given an algebra $\mathbf{A} = \langle A, F^{\mathbf{A}} \rangle$ and $\theta \in \operatorname{Con} \mathbf{A}$, the *factor algebra* $\mathbf{A}/\theta = \langle A/\theta, F^{\mathbf{A}/\theta} \rangle$ is defined by $f^{\mathbf{A}/\theta}([a_1]_\theta, \ldots, [a_n]_\theta) = [f^{\mathbf{A}}(a_1, \ldots, a_n)]_\theta$.

Let $\mathbf{A}$ and $\mathbf{B}$ be algebras of the same type $\langle F, \sigma \rangle$. A mapping $h : A \to B$ is called a *morphism* (or homomorphism) if for each $f \in F$ s.t. $\sigma(f) = n$ and every $a_1, \ldots, a_n \in A$, we have $h(f^{\mathbf{A}}(a_1, \ldots, a_n)) = f^{\mathbf{B}}(h(a_1), \ldots, h(a_n))$; in such a case we write $h : \mathbf{A} \to \mathbf{B}$. A morphism h is called monomorphism (epimorphism, isomorphism) if h is injective (surjective, bijective); if there is an isomorphism of $\mathbf{A}$ to $\mathbf{B}$, we write $\mathbf{A} \cong \mathbf{B}$.

If θ and ϕ are congruences on $\mathbf{A}$ such that $\theta \subseteq \phi$, we denote by ϕ/θ a binary relation on A/θ defined by $\langle [a]_\theta, [a']_\theta \rangle$ iff $\langle a, a' \rangle \in \phi$.

Proposition 1.8 *(1) For $h : \mathbf{A} \to \mathbf{B}$ and $\theta \in \operatorname{Con} \mathbf{A}$, θ_h is a congruence on $\mathbf{A}$; h_θ is a surjective morphism of $\mathbf{A}$ to $\mathbf{A}/\theta$; $\theta = \theta_{h_\theta}$ and if h is surjective then there is an isomorphism $g : \mathbf{A}/\theta \to \mathbf{B}$ such that $h = h_{\theta_h} \circ g$.*

(2) If $h : \mathbf{A} \to \mathbf{B}$ is a morphism then there is an algebra $\mathbf{C}$, an epimorphism $g_1 : \mathbf{A} \to \mathbf{C}$, and a monomorphism $g_2 : \mathbf{C} \to \mathbf{B}$ such that $h = g_1 \circ g_2$ (in fact, one may take $\mathbf{C} = \mathbf{A}/\theta_h$, $g_1(a) = [a]_{\theta_h}$, and $g_2([a]_{\theta_h}) = h(a)$).

(3) For congruences $\theta \subseteq \phi$ on $\mathbf{A}$*, ϕ/θ is a congruence on* $\mathbf{A}/\theta$*; $(\mathbf{A}/\theta)/(\phi/\theta)$ is isomorphic to* $\mathbf{A}/\phi$*.*

(4) For a congruence θ on $\mathbf{A}$ *and a subalgebra* $\mathbf{B}$ *of* $\mathbf{A}$*, $B^\theta = \{a \in A \mid \langle a, b\rangle \in \theta$ for some $b \in B\}$ is a subuniverse of A and the mapping sending $[a]_{\theta\cap B^2}$ to $[a]_{\theta\cap(B^\theta)^2}$ is an isomorphism between* $\mathbf{B}/(\theta \cap B^2)$ *and* $\mathbf{B}^\theta/(\theta \cap (B^\theta)^2)$.

Remark 1.9 Proposition 1.8 shows that congruences and surjective morphisms are in natural correspondence (1) and that each morphism can be decomposed into a surjective and an injective one (2).

A *direct product* of a system $\{\mathbf{A}_i \mid i \in I\}$ of algebras of the same type $\langle F, \sigma\rangle$ is the algebra $\times_{i\in I}\mathbf{A}_i = \langle \times_{i\in I}A_i, F^{\times_{i\in I}\mathbf{A}_i}\rangle$ where the operations $f^{\times_{i\in I}\mathbf{A}_i}$ on $\times_{i\in I}A_i$ are defined by $f^{\times_{i\in I}\mathbf{A}_i}(a_1, \ldots, a_n)(i) = f^{\mathbf{A}_i}(a_1(i), \ldots, a_n(i))$.

A *term* of type $\langle F, \sigma\rangle$ over a set X (of elements called *variables*) is inductively defined as follows: (i) each $x \in X$ is a term; (ii) if $f \in F$, $\sigma(f) = n$, and $t_1, \ldots, t_n$ are terms, then $f(t_1, \ldots, t_n)$ is a term. As usual, we do not define the notions of an occurrence of a variable in a term etc. Rather, we proceed on an intuitive level. To stress the fact that all the variables which occur in a term t are among $x_1, \ldots, x_n$, one writes $t(x_1, \ldots, x_n)$. A phrase like "given a term $t(x_1, \ldots, x_n)$" then translates into "given a term t whose variables are among $x_1, \ldots, x_n$" etc. Given an algebra $\mathbf{A}$ of type $\langle F, \sigma\rangle$, each term $t(x_1, \ldots, x_n)$ of type $\langle F, \sigma\rangle$ over X in an obvious way induces a function $t^{\mathbf{A}} : A^n \to A$, the so-called *term function* (the definition: for $a_1, \ldots, a_n \in A$ we have (i) $t^{\mathbf{A}}(a_1, \ldots, a_n) = a_i$ if $t = x_i$, and (ii) $t^{\mathbf{A}}(a_1, \ldots, a_n) = f^{\mathbf{A}}(t_1^{\mathbf{A}}(a_1, \ldots, a_n), \ldots, t_n^{\mathbf{A}}(a_1, \ldots, a_n))$ if $t = f(t_1, \ldots, t_n)$). If $f \in F$ is binary, we write also $t_1 f t_2$ instead of $f(t_1, t_2)$, i.e. $3 + x$ rather than $+(3, x)$.

An *identity* (of a type $\langle F, \sigma\rangle$) is a pair $\langle t, s\rangle$ (written also simply $t = s$) where t, s are terms of type $\langle F, \sigma\rangle$. Examples of identities (of type $\langle\{\otimes, 1\}, \sigma\rangle$ with $\sigma(\otimes) = 2$, $\sigma(1) = 0$) are $x \otimes (y \otimes z) = (x \otimes y) \otimes z$ (associativity), $x \otimes y = y \otimes x$ (commutativity), $x \otimes 1 = x$, $1 \otimes x = x$ (neutrality of 1 from right and left, respectively). We say that an identity $t(x_1, \ldots, x_n) = s(x_1, \ldots, x_n)$ is satisfied in an algebra $\mathbf{A}$ if for every $a_1, \ldots, a_n \in A$ we have $t^{\mathbf{A}}(a_1, \ldots, a_n) = s^{\mathbf{A}}(a_1, \ldots, a_n)$; $t = s$ is satisfied in a class $\mathcal{K}$ of algebras if it is satisfied in any $\mathbf{A} \in \mathcal{K}$.

A class $\mathcal{V}$ of algebras of the same type is called *variety* if it is the class of all algebras satisfying a given set of identities. Equivalently, a variety is a class of algebras which is closed under taking homomorphic images, subalgebras, and direct products of algebras of this class (that is, for any $\mathbf{A} \in \mathcal{V}$ and $\mathbf{A}_i \in \mathcal{V}$ $(i \in I)$: if $h : A \to B$ is a surjective homomorphism of $\mathbf{A}$ to $\mathbf{B}$ then $\mathbf{B} \in \mathcal{V}$; any subalgebra of $\mathbf{A}$ belongs to $\mathcal{V}$; the direct product $\times_{i \in I} \mathbf{A}_i$ belongs to $\mathcal{V}$). An example of a variety is the class of all lattices, see Section 1.2 (it is the class of all algebras of type $\langle \{\wedge, \vee\}, \sigma \rangle$, $\sigma(\wedge) = \sigma(\vee) = 2$ satisfying associativity (for $\wedge$ and $\vee$), commutativity (for $\wedge$ and $\vee$), and absorption). An algebra of type $\langle \{\otimes, 1\}, \sigma \rangle$, $\sigma(\otimes) = 2$, $\sigma(1) = 0$, satisfying associativity, neutrality of 1 from right and from left is called a *monoid*; a commutative monoid is a monoid satisfying commutativity. Clearly, $\langle [0,1], \cdot, 1 \rangle$ and $\langle [0,1], \min, 1 \rangle$ are monoids. The class of all monoids as well as the class of all commutative monoids are further examples of varieties. On the other hand, linearly ordered sets do not form a variety—they are not closed under direct product.

Chapter 2

FUZZY APPROACH, GRADED TRUTH, AND STRUCTURES OF TRUTH VALUES

Think of arm chairs and reading chairs and dining-room chairs, and kitchen chairs, chairs that pass into benches, chairs that cross the boundary and become settees, dentist's chairs, thrones, opera stalls, seats of all sorts, those miraculous fungoid growths that cumber the floor of arts and crafts exhibitions, and you will perceive what a lax bundle in fact is this simple straightforward term. In cooperation with an intelligent joiner I would undertake to defeat any definition of chair or chairishness that you gave me.

— H. G. Wells (First and Last Things, London, 1908)

It is perfectly obvious, since colours form a continuum, that there are shades of colour concerning which we shall be in doubt whether to call them red or not, not because we are ignorant of the meaning of the word "red", but because it is a word the extent of whose application is essentially doubtful.

— B. Russell (Vagueness, 1923)

A symbol's vagueness is held to consist in the existence of objects concerning which it is intrinsically impossible to say either that the symbol in question does, or does not, apply. ...
Reserving the terms of logic and mathematics for separate consideration, we can say that all "material" terms, all whose application requires the recognition of the presence of sensible qualities, are vague in the sense described.

— M. Black (Vagueness: an exercise in logical analysis, 1937)

2.1 FUZZY APPROACH: FROM RESOLVING PARADOXES TO ENGINEERING APPLICATIONS

Paradoxes of vague terms. Human description of the world provides us with remarkable situations of a special kind. Consider the following form of the ancient *sorites* paradox (paradox of heap). "A heap consisting of just one grain of sand is small. If you add one grain to a small heap, it remains small. Therefore, each heap is small." Note that various forms of *sorites* paradoxes (e.g. the *phalakros* paradox, i.e. a paradox of a bald man that will be mentioned shortly) were known in antiquity and their formulation is attributed to Eubulides of Miletus (fourth century B.C.).

To see clearly that there is nothing wrong in the inference of the paradox (a classically minded logician says), let us formalize: let $S(x)$ stand for "a heap with x grains of sand is small". Then the assumptions (axioms) read: $S(1)$, and $S(x) \Rightarrow S(x+1)$ (if $S(x)$ then $S(x+1)$). Now, let there be a heap consisting of n grains. We want to prove that the heap is small, i.e. we want to prove $S(n)$. This can be done be repeated application of *modus ponens* (an inference rule which is evident; recall that *modus ponens* says: from φ and $\varphi \Rightarrow \psi$ infer ψ). Indeed, from $S(1)$ and $S(1) \Rightarrow S(2)$ we get $S(2)$ which together with $S(2) \Rightarrow S(3)$ gives $S(3)$ which ... together with $S(n-1) \Rightarrow S(n)$ gives $S(n)$.

Another example. Let $A \approx B$ denote that the colors A and B are not much different. Intuitively clear is that if A is not much different from B and B is not much different from C then A is not much different from C, i.e. $A \approx B$ and $B \approx C$ implies $A \approx C$ (transitivity of $\approx$). Now, take any two colors A and B. There is no doubt that we can find a series $B_1, \ldots, B_n$ of colors such that $A \approx B_1$, $B_1 \approx B_2$, ..., $B_{n-1} \approx B_n$, $B_n \approx C$, i.e. a chain of colors starting with A and ending with C such that the successive colors are not much different. By transitivity, we have $A \approx C$, i.e. any two colors are not much different. This paradox is closely related to Henri Poincaré's considerations about physical and mathematical continuum. Here is an excerpt from [166, p. 240]:

> It happens that we are capable of distinguishing two impressions one from the other, while each is indistinguishable from a third. ...Such a statement, translated into symbols, may be written:
>
> $$A = B, \ B = C, \ A < C.$$
>
> ..., if A and B are two distinguishable elements of a continuum C, a series of elements may be found, $E_1, E_2, \ldots, E_n$, all belonging to this same continuum and such that each of them is indistinguishable from the pre-

ceding, that E_1 is indistinguishable from A, E_n is indistinguishable from B.

Notice the term "paradox" used in these situations. Situations referred to as paradoxical obey two attributes. First, they are contradictory (in some obvious sense). Second, analyzing the arguments which lead to the contradiction "step by step", they are all correct. Usually, such paradoxes are "resolved" by depreciation of some standpoint. Such a standpoint, held before for self-evident, is considered as a dogma afterwards.

Vagueness, fuzzy sets, and fuzzy approach. Looking closer at the above paradoxes we find a common feature: they are formulated in natural language and contain terms which are referred to as vague ("small", "not much different"). When analyzing paradoxes, terms used to formulate them should have exact meaning. And there is a problem with vague terms because they have no exact meaning.

The discussion leads us to an old tradition in science, namely to the *ideal of precision* and to the *principle of bivalence*. Statements that are to be useful and scientific statements in particular have been traditionally required to be precise (the more, the better). Thus, data and their characteristics, measurements, expert knowledge, scientific laws, conclusions, etc. were required to be as precise as possible in order to be considered serious, valuable, and amenable to scientific analysis. On the other hand, if the data, knowledge, laws, etc. were not precise, they were considered unreliable, not amenable to serious analysis, and somewhat suspicious. This view, called *ideal of precision*, prevailed in philosophy and in science for centuries (and still has a strong influence up to now). The principle of bivalence says that any statement is either true or false [4]. Thus, given a number x, either "x is odd" is true or "x is odd" is false; given a heap h, either "h is small" is true or "h is small" is false; etc.

It was the formulation of Heisenberg's principle of uncertainty and the developments in quantum mechanics [46] in the first half of the twentieth century that led science to reconsider the ideal of precision. Basically, the principle of uncertainty places a limit on the accuracy with which a pair of incompatible observables (i.e. measurable quantities like position or momentum) of a particle can be measured simultaneously: the more precise a measurement of an observable (a position) the less precise a simultaneous measurement of an incompatible observable (a momentum); that is, we cannot know (by measurement) both of the incompatible observables of a particle at a given moment. Therefore, the ideal of precision cannot be reached in principle due to the laws of nature (provided the principle is right). It is to be noted that probability calculus (that is used in quantum mechanics as a fundamental tool) was available at the time quantum mechanics was born (the birth of probability calculus is associated with Jakob Bernoulli's treatise *Ars Conjectandi*, 1713).

The principle of uncertainty (if accepted) was the first clear example showing that, in principle, imprecision cannot be removed from scientific description of (physical) reality. The indeterminacy concerning quantum mechanics is the uncertainty about the exact value of an observable (there is some but we do not know it). Since the beginning of twentieth century, philosophers began to address another kind of indeterminacy, namely *vagueness*. We only give some quotations from two important papers devoted to vagueness that will approach the nature of vagueness. Perhaps the first paper systematically devoted to vagueness is that of Bertrand Russell [176]:

> ...Let us consider the various ways in which common words are vague, and let us begin with such a word as "red". It is perfectly obvious, since colours form a continuum, that there are shades of colour concerning which we shall be in doubt whether to call them red or not, not because we are ignorant of the meaning of the word "red", but because it is a word the extent of whose application is essentially doubtful. This, of course, is the answer to the old puzzle about the man who went bald. It is supposed that at first he was not bald, that he lost his hairs one by one, and that in the end he was bald; therefore, it is argued, there must have been one hair the loss of which converted him into a bald man. This, of course, is absurd. Baldness is a vague conception; some men are certainly bald, some are certainly not bald, while between them, there are men of whom it is not true to say they must either be bald or not bald. The law of excluded middle is true when precise symbols are employed, but it is not true when symbols are vague, as, in fact, all symbols are. All words describing sensible qualities have the same kind of vagueness which belongs to the word "red".

A further important paper on vagueness was by Max Black [42] who considered vagueness as one of the fundamental issues of (philosophy of) science. His paper begins:

> It is a paradox, whose importance familiarity fails to diminish, that the most highly developed and useful scientific theories are ostensibly expressed in terms of objects never encountered in experience. The line traced by a draftsman, no matter how accurate, is seen beneath the microscope as a kind of corrugated trench, far removed from the ideal line of pure geometry. And the "point-planet" of astronomy, the perfect glass of thermodynamics, or the "pure species" of genetics are equally remote from exact realization. Indeed the unintelligibility at the atomic or subatomic level of the notion of a rigidly demarcated boundary shows that such objects not merely are not but could not be encountered. While the

> mathematician constructs a theory in terms of "perfect" objects, the experimental scientist observes of which the properties demanded by theory are and can, in the very nature of measurement, be only approximately true. As Duhem remarks, mathematical deduction is not useful to the physicist if interpreted rigorously. It is necessary to know that its validity is unaltered when the premise and conclusion are only "approximately true". But the indeterminacy thus introduced, it is necessary to add in criticism, will invalidate the deduction unless the permissible limits of variation are specified. To do so, however, replaces the original mathematical deduction by a more complicated mathematical theory in respect of whose the same problem arises, and whose exact nature is in any case unknown. This lack of exact correlation between a scientific theory and its empirical interpretation can be blamed either upon the world or upon the theory. We can regard the shape of an orange or a tennis ball as imperfect copies of an ideal form of which perfect knowledge is to be had in pure geometry, or we can regard the geometry of spheres as a simplified and imperfect version of the spatial relation between the members of a certain class of physical objects. On either view there remains a gap between scientific theory and its application which ought to be, but is not, bridged. To say that all language (symbolism, or thought) is vague is a favorite method for evading the problems involved and lack of analysis has the disadvantage of tempting even the most eminent thinkers into appearance of absurdity. Duhem claims that "for the strict logician", a physical law is neither true nor false. For Einstein mathematics is either uncertain or inapplicable, and Russell cheerfully sacrifices logic as well.
>
> The aim of this essay is to avoid such whole destruction of the formal sciences by supplying in greater detail than has hitherto been attempted an analysis and symbolism for the "vagueness" or "lack of precision" of a language.
>
> We shall not assume that "laws" of logic or mathematics prescribe models of existence to which intelligible discourse must necessarily conform. It will be argued, on the contrary, that deviations from the logical or mathematical standards of precision are all pervasive in symbolism; that to label them as subjective aberrations sets an impassable gulf between formal laws and experience and leaves the *usefulness* of the formal sciences an insoluble mystery.

Black then comments on the example of H. G. Wells (see the first quotation of this chapter):

> One can imagine an exhibition in some unlikely museum of applied logic of a series of "chairs" differing in quality by least noticeable amounts. At one end of a long line, containing perhaps thousands of exhibits, might be a Chippendale chair; at the other, a small nondescript lump of wood. Any "normal" observer inspecting the series finds extreme difficulty in "drawing the line" between chair and not-chair. Indeed the demand to perform this operation is felt to be inappropriate *in principle*: "*chair* is not the kind of word which admits of this sharp distinction" is the kind of reply which is made "and if it were, if we were forbidden to use it for any object which varied in the slightest way from the limiting term, it would not be as useful to us as it is".

After analyzing several aspects of vagueness, Black devised a formalism (so-called consistency profiles) for dealing with vague terms.

However penetrating Black's analysis was, it is now generally agreed that the final turn in recognizing vagueness as immanent to human description of the world and in utilizing vagueness came with Lotfi A. Zadeh who devised a natural formalism, so-called fuzzy sets, for dealing with vagueness. Zadeh's motivations came from systems engineering. In the early 1960s, Zadeh emphasized a need for a "different kind of mathematics" for dealing with certain aspects of complex systems [215]:

> There are some ... who feel the fundamental inadequacy of the conventional mathematics—the mathematics of precisely-defined points, functions, sets, probability measures, etc.—for coping with the analysis of biological systems, and that to deal efficiently with such systems, we need a radically different kind of mathematics, the mathematics of fuzzy or cloudy quantities which are not describable in terms of probability distributions. Indeed the need for such mathematics is becoming increasingly apparent even in the realm of inanimate systems, for in most practical cases the a priori data as well as the criteria by which the performance of a man-made system is judged are far from being precisely specified ...

Finally, in his 1965 paper [216] he summarized the arguments:

> More often than not, the classes of objects encountered in the real physical world do not have precisely defined criteria of membership. For example, the class of animals clearly includes dogs, horses, birds, etc. as its members, and clearly excludes such objects as rocks, fluids, plants etc. However, such objects as starfish, bacteria, etc. have an ambiguous status with respect to the class of animals. The same kind of ambiguity arises in the case of a number such as 10 in relation to the "class" of all real numbers which are much greater than 1.
>
> Clearly, the "class of all real numbers which are much greater than 1", or "the class of beautiful women", or "the class of tall men", do not constitute classes or sets in the usual mathematical sense of these terms. Yet, the fact remains that such imprecisely defined "classes" play an important role in human thinking, particularly in the domains of pattern recognition, communication of information, and abstraction.

and came up with the concept of a *fuzzy set*:

> A *fuzzy set* (*class*) A in X is characterized by a *membership* (*characteristic*) *function* $f_A(x)$ which associates with each point[2] in X a real number in the interval $[0,1]$,[3] with the value of $f_A(x)$ at x representing the "grade of membership" of x in A. Thus, the nearer the value of $f_A(x)$ to unity, the higher the grade of membership of x in A. When A is a set in the ordinary sense of the term, its membership function can take on only two values 0 and 1, with $f_A(x) = 1$ or 0 according as x does or does not belong to A. Thus, in this case $f_A(x)$ reduces to the familiar characteristic function of a set A.
>
> [footnote 2 refers to:] More generally, the domain of definition of $f_A(x)$ may be restricted to a subset of X.
>
> [footnote 3 refers to:] In a more general setting, the range of the membership function can be taken to be a suitable partially ordered set P. For

> our purposes, it is convenient and sufficient to restrict the range of f to the unit interval. If the values of $f_A(x)$ are interpreted as truth values, the latter case corresponds to a multivalued logic with a continuum of truth values in the interval [0,1].

Zadeh's message points out two important facts. First, collections of objects encountered in human description of the world are, more often than not, nonsharp in that there is a gradual transition between being a member and not being a member of the collection. Second, classical mathematics does not have the means for description of such collections in a natural way. To overcome this difficulty, Zadeh proposed the concept of a fuzzy set as a formal model of a collection with gradual transition from nonmembership to membership. The main idea is that, in addition to "fully belong" (truth value 1) and "fully not belong" (truth value 0), there can also be other cases (other truth values) of belonging of an element to a fuzzy set; being a member of a fuzzy set is thus a graded property. This is the contrary to the principle of bivalence stating that a property either applies to an element or not. A fuzzy set is specified by a rule assigning to each element of a given universe set its membership degree, i.e. the degree to which the element belongs to the fuzzy set. Formally, fuzzy sets are just those rules: a fuzzy set A in a given universe set X is a mapping of X into L where L is a suitable set of membership degrees; $A(x)$, the element of L assigned to x by A, is called the membership degree of x in A. $A(x)$ can be thought of as the truth degree (truth value) of the proposition "x belongs to A". Note that in the above-quoted paper, Zadeh distinguishes a fuzzy set A from its membership function f_A (the membership degree of x in A is then $f_A(x)$, not $A(x)$); some authors proceed this way, we will not. Assigning (possibly intermediate) truth degrees to (elementary) propositions like "an element belongs to a fuzzy set" is a particular case of a more general one. In general, propositions may be composed out of elementary propositions using logical connectives (e.g. "and", "or", "if ... then ...") and quantifiers (e.g. "for all x ...", "there is some x ...") and possibly other constructions. This way one obtains propositions like "for all x and y, if x is big and if x and y do not differ much then y is also big" containing elementary propositions that concern possibly nonsharp collections or relationships. Intuitively, truth degrees can be compared, and 0 (degree corresponding to (full) falsity) and 1 (degree corresponding to (full) truth) are the least and the greatest truth degrees, respectively (that is, L is partially ordered and bounded by 0 from below and by 1 from above). The most appealing set of truth degrees is the interval $[0, 1]$ (or some of its subsets containing 0 and 1) with its natural ordering.

The approach to modeling that acknowledges gradual transition from nonbeing a member to being a member of a collection, from not having to having a relationship, from not being true to being true, etc., and that uses the concept of a truth degree with the understanding that there are intermediate truth degrees (between 0 and 1) is called *fuzzy approach* or, perhaps more succinctly, *graded truth approach.*

Fuzzy approach seems to be most natural in modeling of human reasoning and human description of the outer world. For example, consider the term *tall man.* Using fuzzy approach, the collection of tall men can be modeled by a fuzzy set A to which a man 5 feet tall belongs with membership degree 0 (is not tall at all), a man 6 feet tall belongs with membership degree 0.8, and a man 6 feet 5 inches tall belongs with membership degree 1 (is absolutely tall). If we were to model the collection of tall men by an ordinary set B, one would have to pick up a particular height x (say 6 feet) and define B to be the set of those men whose height is at least x. This yields problems: First, why just x? Second and more important, a man whose height is just a little bit under 6 feet is not considered tall while a man who is 6 feet tall is—this is very counterintuitive.

Let us add a technical but still important comment. Literature on fuzzy sets contains statements like "classical mathematics is not capable of modeling nonsharp collections" which, although heading to the point, are somewhat misleading and need clarification. The theory of fuzzy sets as pursued by Zadeh and as it is pursued almost exclusively up to now is developed "in classical mathematics" using classical set theory (cf. the above-quoted paper by Zadeh: a membership function of a fuzzy set is an ordinary function) in an intuitive manner. This is, however, not the point. The point and the breakthrough of Zadeh's proposal is that the concept of a set, i.e. the very fundamental mathematical concept used for modeling of collections of objects, is not appropriate for collections with gradual membership (most of the collections employed in human reasoning). Therefore, in order to have appropriate means, we need a new concept that is powerful enough to describe collections with gradual membership, yet simple enough to serve as a fundamental one. Zadeh proposed the concept of a fuzzy set. Whether this new concept will be introduced using classical set theory or independently of it (e.g. axiomatically) is a question which is possibly very interesting from the foundational point of view, but may be only of technical importance from the point of view of modeling (and applications).

Resolving the paradoxes. Using graded truth approach, the *sorites* paradox can be resolved. Being a small heap is clearly a graded property and thus, $S(x)$ can take more truth values than just 0 or 1, heaps with less grains are "more small" than heaps with more grains. Take the real interval [0,1] for the set of truth values. Consider the axioms ($S(1)$ and $S(x) \Rightarrow S(x+1)$). There is no doubt that a heap consisting of one grain is small, i.e. we can accept $S(1)$ with the truth degree 1. On the other hand, it is intuitively appealing that $S(x) \Rightarrow S(x+1)$ should be accepted with a high truth degree which is, however, (strictly) less than 1, say 0.999 (the exact value doesn't matter for the argument). A direct generalization of *modus ponens* gives the following: if the truth degree of φ is (at least) a and the truth degree of $\varphi \Rightarrow \psi$ is (at least) b then the truth degree of ψ is (at least) the truth degree of $\varphi \& (\varphi \Rightarrow \psi)$. Then if $\otimes$ defined by $a \otimes b = \max(0, a+b-1)$ is the operation corresponding to the connective & (we will see later that this is a reasonable choice, but we will also see that there are other reasonable choices), the truth degree of $\varphi \& (\varphi \Rightarrow \psi)$ is (at least) $a \otimes b$.

But now, there is no paradox any more. Indeed, let $||\varphi||$ denote the truth degree of φ. Then from $||S(1)|| = 1$ and $||S(1) \Rightarrow S(2)|| = 0.999$ we get $||S(2)|| \geq 0.999$ which together with $||S(2) \Rightarrow S(3)|| = 0.999$ gives $||S(3)|| \geq 0.999^2$ which ... together with $||S(n-1) \Rightarrow S(n)|| \geq 0.999$ gives $||S(n)|| \geq 0.999^{n-1}$ (the power is taken w.r.t. $\otimes$, so 0.999^3 is $0.999 \otimes 0.999 \otimes 0.999$). So, for example, we can deduce that the truth degree of "a heap with 10 grains of sand is small" is (at least) $0.999^{10} = 0.99$, that the truth degree of "a heap with 200 grains of sand is small" is (at least) $0.999^{200} = 0.8$, etc. (one may change 0.999 to get intuitively more acceptable results).

The graded truth approach also helps to resolve the paradox "any two colors are not much different". The point to be rejected is the assumption that the relation "not differ much" is a bivalent one (i.e. two colors either are much different or are not much different). We should rather construe the relation $\approx$ as admitting graded truth which is quite a natural assumption. For example, take colors A, B, and C with wave-lengths x, y, and z, respectively, and suppose, moreover, $x < y < z$ and $|x-y| < |y-z|$. Then A is closer to B than B is to C, and thus the truth value of $A \approx B$ is greater than the truth value $B \approx C$. Starting from the (intuitively acceptable) assumption that the formula "if $A \approx B$ and $B \approx C$ then $A \approx C$" (transitivity of $\approx$) should be true and taking again [0,1] and product as the set of truth grades and the operation corresponding to conjunction (we stress again that it is only one particular choice of many others possible), respectively, graded approach to $\approx$ translates the assumption into the condition $||A \approx B|| \cdot ||B \approx C|| \leq ||A \approx C||$ ($||X||$ denotes the truth value of X). Thus, if, for instance, the truth values of "A and B are not much different" and "B and C are not much different" are 0.8 and 0.5, respectively, then the restriction imposed on A and C by "if $A \approx B$ and $B \approx C$ then $A \approx C$" is that the truth degree of "A and C are not much different" is at least $||A \approx B|| \cdot ||B \approx C|| = 0.8 \cdot 0.5 = 0.4$ which is intuitively satisfactory.

From fuzzy approach to engineering applications. The choice of paradoxes for introducing fuzzy logic has not been made to present fuzzy logic merely as a tool for resolving "strange paradoxical situations". Resolving a paradox with a long-termed tradition usually means to reveal a kind of a dogma—an uncritically accepted tenet limiting possibly the scope of applications. In our case, the dogma is bivalence and the scope of applications is the world of bivalent reasoning. However beautiful bivalence and the world of bivalent reasoning are, humanlike description of the outer world and humanlike reasoning lie beyond their borders. Bivalent logic, i.e. the formalism originated from bivalence, is thus not appropriate for modeling of humanlike description of the outer world and humanlike reasoning.

As demonstrated, replacing bivalence with graded truth approach extends the scope of modeling capabilities of logic beyond the original limits forced by bivalence. Ignoring fuzzy approach in applications does not necessarily have to lead to a "disaster" like in the case of the paradoxes (like ignoring the theory of relativity and using Newtonian physics in applications does not usually lead to disasters). It may, however, imply a significant loss compared to the situation when fuzzy approach is used. In this sense, the paradoxes are extreme examples showing the usefulness of fuzzy approach. As repeatedly succinctly expressed by Zadeh, the need for replacing bivalent approach by fuzzy approach is particularly appealing in the area of so-called humanistic systems. From [219]:

> Given the deeply entrenched tradition of scientific thinking which equates the understanding of a phenomenon with the ability to analyze it in quantitative terms, one is certain to strike a dissonant note by questioning the growing tendency to analyze the behavior of humanistic systems as if they were mechanistic systems governed by difference, differential, or integral equations.
>
> Essentially, our contention is that the conventional quantitative techniques of system analysis are intrinsically unsuited for dealing with humanistic systems or, for that matter, any system whose complexity is comparable to that of humanistic systems. The basis for this contention rests on what might be called the *principle of incompatibility*. Stated informally, the essence of this principle is that as the complexity of a system increases, our ability to make precise and yet significant statements about its behavior diminishes until a threshold is reached beyond which precision and significance (or relevance) become almost mutually exclusive characteris-

tics. It is in this sense that precise analyses of the behaviour of humanistic systems are not likely to have much relevance to the real-world societal, political, economic, and other types of problems which involve humans either as individuals or in groups.

An alternative approach ... is based on the premise that the key elements in human thinking are not numbers, but labels of fuzzy sets, that is, classes of objects in which the transition from membership to nonmembership is gradual rather than abrupt. Indeed, pervasiveness of fuzziness in human thought processes suggests that much of the logic behind human reasoning is not the traditional two-valued or even multivalued logic, but a logic with fuzzy truths, fuzzy connectives, and fuzzy rules of inference. In our view, it is this fuzzy, and as yet not well-understood, logic that plays a basic role in what may well be one of the most important facets of human thinking, namely, the ability to *summarize* information—to extract from the collection of masses of data impinging upon the human brain those and only those subcollections which are relevant to the performance of the task at hand.

By this nature, a summary is an approximation to what it summarizes. For many purposes, a very approximate characterization of a collection of data is sufficient because most of the basic tasks performed by humans do not require a high degree of precision in their execution. The human brain takes advantage of this tolerance for imprecision by encoding the "task-relevant" (or "decision-relevant") information into labels of fuzzy sets which bear an approximate relation to the primary data. In this way, the stream of information reaching the brain via the visual, auditory, tactile, and other senses is eventually reduced to the trickle that is needed to perform a specific task with a minimal degree of precision. Thus, the ability to manipulate fuzzy sets and the consequent summarizing capability constitute one of the most important assets of the human mind as well as a fundamental characteristic that distinguishes human intelligence from the type of machine intelligence that is embodied in present-day digital computers.

Viewed in this perspective, the traditional techniques of system analysis are not well suited for dealing with humanistic systems because they fail to come to grisp with the reality of the fuzziness of human thinking and behavior. Thus to deal with such systems radically, we need approaches which do not make a fetish of precision, rigor, and mathematical formalism, and which employ instead a methodological framework which is tolerant of imprecision and partial truths.

Up to now, fuzzy approach has proven to be useful in several areas. There are many real-world applications, several of them successful commercially. Perhaps the main breakthrough in considering fuzzy approach a useful tool for real-world applications was the development of a fuzzy controller by E. Mamdani and S. Assilian [138]. The main idea behind a fuzzy controller is to utilize expert's knowledge of a control strategy that is formulated in natural language in the form of so-called if-then rules (like "if the temperature is high then turn the cap to the right")—the methods of fuzzy control enable us to transform the natural language description of a control strategy into a control algorithm. Nowadays, fuzzy controllers (and if-then rule-based systems in general) are the most successful applications of fuzzy approach. Their use in consumer electronics led to a boom of "fuzzy products" in Japan in the 1990s. We can mention fuzzy washing machines, vacuum cleaners, dishwashers, cameras, microwave ovens, etc. However, fuzzy controllers have also been applied in industry and other areas (subway in Sendai (Japan) where fuzzy control leads to very smooth slowing-down and acceleration, helicopter control [191], car-braking systems, kiln control); further information can be found in [127, 129]. Except for fuzzy control, fuzzy approach was applied to systems analysis and design, decision making, data analysis, pattern recognition, prediction, database and information systems, and others [127].

Fuzzy logic: formal foundations of fuzzy approach. These days, the term *fuzzy logic* is used basically in two ways. In the broad sense, fuzzy logic is understood to cover any kind of methods and applications inspired by fuzzy approach. In the narrow sense, fuzzy logic refers to logical calculi that admit propositions to also take intermediate truth values (between 0 and 1) which are interpreted as truth degrees; these calculi aim at formalization of reasoning in the presence of vagueness. The main concern of fuzzy logic in the broad sense is to provide tools enabling us to deal with and utilize fuzziness. Inspiration for this is the kind of approximate reasoning people perform when reasoning with "fuzzy data" in everyday life. Formal rules of reasoning are the subject of logic. If we allow for fuzziness, i.e. more truth degrees, we get to fuzzy logic (in narrow sense). This is the way fuzzy logic in the broad and narrow senses are connected: fuzzy logic in the narrow sense offers foundations for fuzzy logic in the broad sense. Among other things, fuzzy logic in the narrow sense tells us: how to express formally propositions formulated in natural language (mathematically: what is the language of fuzzy logic, what are formulas), how to formally describe the (part of a) real world the propositions refer to (mathematically: what are structures in which formulas are to be evaluated); how to interpret natural language connectives (like "...or ...", "...and ...", "if ...then ...") and quantifiers (like "for all ...", "for some ...") that apply to propositions with possibly intermediate truth values (mathematically: what is the semantics of connectives and quantifiers in fuzzy setting); how to evaluate truth degrees of propositions; how propositions follow from other propositions that are possibly valid to only a certain degree and how to formalize approximate inference (that is: what is deduction, provability, etc. in fuzzy logic); what are the properties of approximate inference (that is: what about the classical properties like syntactico-semantical completeness, what about the complexity of proving, etc.). Fuzzy logic in the narrow sense has been substantially developed in recent years. Good overviews are [85, 99, 105, 154] where one can find further references.

Fuzzy relational systems. Fuzzy relational systems are systems of fuzzy relations and (ordinary) functions defined on a set (or on sets of possibly different sorts). We meet fuzzy relational systems when formalizing what we see around us. A simple example: Suppose we are interested in what patients have what diseases. The medical evidence may yield the following situation. The first step is to identify relevant sets of elements. In our case, we may have a set P of patients (that we are concerned with), a set D of diseases, a set S of symptoms of diseases, and a set R of hospital rooms. The next step is to identify relevant fuzzy relations and functions. In our case, we may have a binary fuzzy relation h between P and S representing what patients have what symptoms; a binary fuzzy relation i between S and D representing what symptoms are characteristic for what diseases; a binary fuzzy relation has between P and D representing what patients have what diseases; and a unary function f of P to R assigning hospital rooms to patients. Thus, for a patient $p \in P$, a disease $d \in D$, and a symptom $s \in S$, we have that $has(p,d)$ is the truth degree to which patient p has disease d, $h(p,s)$ is the truth degree to which patient p has symptom s, $i(s,d)$ is the truth degree to which symptom s is characteristic for disease d, and $f(p)$ is the hospital room patient p is staying in. In this way we came to a fuzzy relational system with four sets P, D, S, and R, three fuzzy relations h, i, and has, and one function f.

Fuzzy relational systems are closely connected to fuzzy logic in the narrow sense: they are the structures in which formulas of fuzzy logic are interpreted. Having a formula that represents a proposition concerning elements of some universe of discourse, one can evaluate the truth degree of this formula once we specify the semantic counterparts of the symbols appearing in the formula. We assume first-order formulas, i.e. formulas containing variables for elements of the universe, symbols of fuzzy relations and ordinary functions defined on the universe, logical connectives, and quantifiers. Then a fuzzy relational system is just the structure that consists of the semantic counterparts of variables, and relation and function symbols. Thus, fuzzy logic enables us to formulate propositions about elements of a given universe of discourse on which there are defined fuzzy relations and functions, and enables us to evaluate truth degrees of these propositions.

This is useful and by no means trivial—consider the following situation. Suppose that in the previous example, we have evidence about h (medical examination tells) and about i (medical books tell) but we have no evidence about *has*. Then we may want to obtain *has* from h and i: intuitively, a patient p has a disease d iff p has exactly those symptoms that are characteristic for d. What is going on here is that we define a fuzzy relation using a logical formula: we have a fuzzy relational system having h and i (but not *has*) and a formula φ_{has} that corresponds to the proposition "for each x_S: x_P has x_S if and only if x_S is characteristic for x_D" (variables x_P, x_S, and x_D refer to patients, symptoms, and diseases, respectively) and define a fuzzy relation *has* by saying that the truth degree $has(p, d)$ to which a patient p has a disease d equals the truth degree of φ_{has} when x_P refers to p and x_D refers to d. Fuzzy relation *has* gives us a nontrivial information obtained from "input" fuzzy relations h and i using evaluation of a truth degree of a formula. Suppose, furthermore, that the set of truth values is the real interval $[0, 1]$. A natural question arises as to what extent *has* will change if h and i change. For instance, what happens with *has* if we take, say, 0.75 for $h(p, s)$ (for some p and s) instead of, say, 0.76? Needless to say, questions like this one about the sensitivity of the output fuzzy relation *has* to the input fuzzy relations h and i are very important from the application point of view—they tell us where to take care. Clearly, the above question translates into a logical question of what can be said about truth degrees of a formula that result in two similar but different interpretations. It is to be noted that these questions are degenerate and thus not interesting in the bivalent case.

These considerations show the following. First, when formalizing knowledge about some universe of discourse we often arrive at a system of fuzzy relations and functions plus some formulas which express further properties of elements of the universe, define further relationships, etc. Second, the problems that arise this way belong to fuzzy logic (in the narrow sense). However, rather than deduction, the main concern here is the systems of fuzzy relations and functions themselves, i.e. those parts that traditionally belong to model theory rather than proof theory. Except for the general relational systems, one may be interested in special fuzzy relations, namely in binary fuzzy relations defined on a set (fuzzy equivalences, fuzzy orders, etc.). The conclusion that follows is that fuzzy logic in the narrow sense and particularly the theory of fuzzy relational systems play a crucial role for fuzzy relational modeling, i.e. for modeling of the outer world using fuzzy relations. It offers not only a "language" but also (as we will see) a couple of useful results describing important properties of models built out of fuzzy relations.

The study of general fuzzy relational systems (i.e. systems of fuzzy relations and (ordinary) functions) and properties of elements expressible by logical formulas is our main concern here. We do not attempt to develop a model theory for fuzzy logic—this still needs to be done. Rather, we concentrate on natural problems that are directly met in fuzzy relational modeling. Some of them are direct generalizations from the crisp (i.e. ordinary bivalent) case, some of them are new since they are degenerate in the crisp case.

2.2 GRADED TRUTH AND STRUCTURES OF TRUTH VALUES

From the previous discussion it is clear that the structure of the set of truth values deserves special attention. The aim of this section is to show how certain natural logical assumptions reflect themselves in corresponding properties of the structure of truth values.

The graded truth approach directly leads to the assumption that the set L of truth values is partially ordered (we denote the partial order by $\leq$) with 0 and 1 being the least and the greatest element, respectively. For every two truth values a and b there is a truth value greater than both a and b (one can take 1). Moreover, one may require that there is the least truth value which is greater than both a and b. In this way we come to the requirement of existence of suprema (and dually of infima) of two-element subsets in L. Let $\{\varphi_i \mid i \in I\}$ be a set of propositions. A generalization from the classical case of two-valued logic leads to the assumption that the truth value of "there exists $i \in I$ such that φ_i" is the supremum of the truth values of φ_i, i.e. $||$"there exists $i \in I$ such that φ_i"$|| = \bigvee_{i\in I} ||\varphi_i||$ ($||\varphi||$ is the truth value of φ). Therefore, if one wants to evaluate such existential (and dually, universal) propositions, suprema (and infima) of arbitrary subsets of L should exist. In this way one comes to the assumption that L should be a complete lattice.

We now get to the question of operations on L which model logical connectives. The general principle to which we adhere says that they should extend the classical operations in that restricting the operations to "classical truth values" 0 and 1, they coincide with classical operations. Furthermore, as in the classical case we want the logic to be truth functional, i.e. the truth value of a compound formula depends only on the truth values of its parts. We start by conjunction (we denote it by &). Denote the operation which corresponds to & by $\otimes$, i.e. $\otimes$ is a binary operation on L. Since $\otimes$ should extend the operation corresponding to classical conjunction, we require $1 \otimes 1 = 1$, $1 \otimes 0 = 0 \otimes 1 = 0 \otimes 0 = 0$. If we want the truth value of $\varphi\&\psi$ to be the same as the truth value of $\psi\&\varphi$, the truth functionality leads to the requirement of commutativity of $\otimes$ (indeed, by truth functionality we have $||\varphi\&\psi|| = ||\varphi|| \otimes ||\psi||$ and $||\psi\&\varphi|| = ||\psi|| \otimes ||\varphi||$, therefore $||\varphi\&\psi|| = ||\varphi\&\psi||$ leads to $||\varphi|| \otimes ||\psi|| = ||\psi|| \otimes ||\varphi||$). Similarly, if the truth values of $\varphi\&(\psi\&\chi)$ and $(\varphi\&\psi)\&\chi$ are to be the same, truth functionality leads to associativity of $\otimes$. In this way we came to the assumption that $\langle L, \otimes, 1\rangle$ is a commutative monoid. Furthermore, it is intuitively appealing to require that $\otimes$ is nondecreasing, i.e. that $a_1 \leq a_2$ and $b_1 \leq b_2$ imply $a_1 \otimes b_1 \leq a_2 \otimes b_2$ (the bigger the truth degrees of two propositions the bigger the truth degree of their conjunction).

Let us turn to implication. In classical logic, conjunction and implication play an important role in the formulation of an inference rule called *modus ponens*. *Modus ponens* says that if φ is valid and $\varphi \Rightarrow \psi$ is valid ($\Rightarrow$ denotes "implies") then (we may infer that) ψ is valid. Let us now reformulate a little bit using validity degree. Let us say that φ is valid in degree 1 if φ is valid and that φ is valid in degree 0 if φ is not valid (we have no other degrees than 0 and 1 in bivalent logic). A moment's reflection shows that an equivalent formulation of *modus ponens* is the following one: If φ is valid in degree at least a, and $\varphi \Rightarrow \psi$ is valid in degree at least b ($a, b \in \{0, 1\}$), then ψ is valid in degree at least $a \otimes_2 b$ ($\otimes_2$ denotes the operation on $\{0, 1\}$ that corresponds to classical conjunction). This rule is sound in the following sense: If φ and ψ are formulas such that $a \leq ||\varphi||$ and $b \leq ||\varphi \Rightarrow \psi||$ then $a \otimes_2 b \leq ||\psi||$. This means that when using *modus ponens* it cannot happen that "we infer more than is actually true", i.e. it cannot happen that the actual truth degree of ψ is less than the degree inferred by *modus ponens*. Therefore, the role of $\otimes_2$ in *modus ponens* is to get the lower estimation of the validity of the inferred formula ψ from the lower estimations of validities of φ and $\varphi \Rightarrow \psi$. It is easy to see that $\otimes_2$ gives the highest possible estimation under the condition that the inference rule still be sound. This way of looking at *modus ponens* is suitable for generalization to graded-truth case. Knowing that $a \leq ||\varphi||$ and $b \leq ||\varphi \Rightarrow \psi||$ ($a, b \in L$) we want to use $\otimes$ to (1) obtain the lower estimation of validity of ψ. Moreover, we want (2) to obtain the highest possible lower estimation such that the rule still be sound. Denote by $\rightarrow$ the binary operation on L which corresponds to $\Rightarrow$ (implication connective). Condition (1) translates then to: $a \leq ||\varphi||$ and $b \leq ||\varphi \Rightarrow \psi||$ implies $a \otimes b \leq ||\psi||$, i.e., by truth-functionality, $a \leq ||\varphi||$ and $b \leq ||\varphi|| \rightarrow ||\psi||$ implies $a \otimes b \leq ||\psi||$. Putting $a = ||\varphi||$ and denoting $c = ||\psi||$ we obtain a special case of this implication, i.e. $b \leq a \rightarrow c$ implies $a \otimes b \leq c$. Condition (2): We want the inference by *modus ponens* to be as powerful as possible. From $a = ||\varphi||$ and $||\varphi \Rightarrow \psi||$ we get a lower estimation $a \otimes ||\varphi \Rightarrow \psi||$ of

$c = ||\psi||$, i.e. $a \otimes ||\varphi \Rightarrow \psi|| \leq c$. The rule is the more powerful the bigger the estimated value $a \otimes ||\varphi \Rightarrow \psi||$ (of course, under the condition that the rule is still sound, i.e. it really gets the lower estimation of $||\psi||$). Since a is given, $a \otimes ||\varphi \Rightarrow \psi||$ depends on $||\varphi \Rightarrow \psi||$. Since $\otimes$ is nondecreasing, bigger $||\varphi \Rightarrow \psi||$ leads to bigger (or at least the same) $a \otimes ||\varphi \Rightarrow \psi||$. Now, as $||\varphi \Rightarrow \psi|| = ||\varphi|| \rightarrow ||\psi|| = a \rightarrow c$, the requirement that *modus ponens* be as powerful as possible yields that $a \rightarrow c$ should be the largest possible value that leads to lower estimation of c. That is, we want that whenever $a \otimes b \leq c$ (i.e. whenever b is a possible candidate for $||\varphi|| \rightarrow ||\psi|| = a \rightarrow c$; b is considered a possible candidate since it leads to lower estimation) then $b \leq a \rightarrow c$ (i.e. then $a \rightarrow c$ is at least as good as b since it leads to at least as good lower estimation of c).

Putting the conditions which we derived from (1) and (2) together we get

$$a \otimes b \leq c \qquad \text{iff} \qquad b \leq a \rightarrow c.$$

This condition (one may check that it is true for $\otimes_2$ and $\rightarrow_2$) will be called the adjointness property and will play a very important role in our investigation. We just saw the assumptions from which it was derived.

Algebraic structures which satisfy the above conditions will be called residuated lattices and will play the role of the basic structures of truth values in this book. The logical assumptions from which the algebraic conditions have been derived are relatively simple. However, we will see that they lead to rich structures. Further logical requirements can be taken into account by adding appropriate algebraic conditions. For example, one may require idempotent conjunction (i.e. the truth value of $\varphi\&\varphi$ be the same as the truth value of φ) and this leads to an additional condition $x \otimes x = x$. In a similar way one can obtain MV-algebras (the algebras of Łukasiewicz logic) as a special case of residuated lattices etc. We will see that despite the generality of residuated lattices, their properties are strong enough to allow us to pursue important questions of fuzzy relational systems and fuzzy relational modeling.

2.3 RESIDUATED LATTICES

This section is devoted to residuated lattices—the structures of truth values of fuzzy logic. We first define residuated lattices, some derived operations, and several additional conditions which are relevant from the logical point of view. The definitions are accompanied by examples. We then present basic properties of residuated lattices. First, properties valid in all residuated lattices are established. Then we turn attention to properties of residuated lattices satisfying additional conditions. Then we consider some classical algebraic notions like morphisms, factors, subdirect products etc. of residuated lattices. We close the section by considering residuated lattices with support set being $[0,1]$ (the real unit interval); such structures are the most important for applications.

2.3.1 DEFINITIONS AND BASIC EXAMPLES

Definition 2.1 A *residuated lattice* is an algebra $\mathbf{L} = \langle L, \wedge, \vee, \otimes, \rightarrow, 0, 1\rangle$ where

(i) with the least element 0 and the greatest element 1,

(ii) a commutative monoid, i.e. $\otimes$ is associative, commutative, and the identity $x \otimes 1 = x$ holds,

(iii) *adjointness property*, i.e.

$$x \leq y \rightarrow z \quad \text{iff} \quad x \otimes y \leq z \tag{2.1}$$

holds for each $x, y, z \in L$ ($\leq$ denotes the lattice ordering).

A residuated lattice is called *complete* if $\langle L, \wedge, \vee, 0, 1\rangle$ is a complete lattice.

Remark 2.2 (1) We know from the previous section that the conditions satisfied by a complete residuated lattice may be thought of as consequences of desirable properties of logical connectives.

(2) The operations $\otimes$ and $\rightarrow$ are called *multiplication* and *residuum*, respectively, $\langle\otimes, \rightarrow\rangle$ is called an *adjoint couple*. For $a, b \in L$, $a \rightarrow b$ is called the residuum of b by a. For a given $\otimes$, there is at most one operation $\rightarrow$ satisfying (2.1) (for if $\rightarrow^*$ is another one, we get $x \leq y \rightarrow z$ iff $x \otimes y \leq z$ iff $x \leq y \rightarrow^* z$ for any x, from which it clearly follows $y \rightarrow z = y \rightarrow^* z$). Similarly, $\rightarrow$ uniquely determines $\otimes$.

Example 2.3 Consider $L = [0,1]$ (the interval of all reals between 0 and 1). The natural ordering of $[0,1]$ makes $\langle L, \wedge, \vee, 0, 1\rangle$ a lattice (where $a \wedge b = \min(a,b)$, $a \vee b = \max(a,b)$) which is moreover complete. Each of the following pairs of operations makes $\langle L, \wedge, \vee, \otimes, \rightarrow, 0, 1\rangle$ a (complete) residuated lattice:

(1) $a \otimes b = \max(a+b-1, 0)$, $a \rightarrow b = \min(1-a+b, 1)$ (Łukasiewicz structure);

(2) $a \otimes b = \min(a,b)$, $a \rightarrow b = 1$ if $a \leq b$ and $= b$ if $a > b$ (Gödel structure);

(3) $a \otimes b = a \cdot b$, $a \to b = 1$ if $a \leq b$ and $= b/a$ if $a > b$ (product structure).

The corresponding algebras are called standard Łukasiewicz algebra, s-tandard Gödel algebra, and standard product algebra (on $[0,1]$).

Example 2.4 Let $L = \{a_0, a_1, \ldots, a_n\}$ with the order given by $0 = a_0 < \cdots < a_n = 1$. We have $a_i \wedge a_j = a_{\min(i,j)}$ and $a_i \vee a_j = a_{\max(i,j)}$. The following two pairs of operations determine a (complete) residuated lattice structure on L:

(1) $a_k \otimes a_l = a_{\max(k+l-n,0)}$, $a_k \to a_l = a_{\min(n-k+l,n)}$ (Łukasiewicz structure); if $\{a_0, \ldots, a_n\} \subseteq [0,1]$ and $a_i = \frac{i}{n}$, then all the operations are restrictions of Łukasiewicz operations on $[0,1]$ to $\{a_0, \ldots, a_n\}$, i.e. $\{a_0, \ldots, a_n\}$ is a subuniverse of the standard Łukasiewicz algebra.

(2) $a_k \otimes a_l = a_{\min(k,l)}$, $a_k \to a_l = 1$ if $k \leq l$ and $= a_l$ if $k > l$ (Gödel structure); if $\{a_0, \ldots, a_n\} \subseteq [0,1]$ and $a_0 = 0$, $a_n = 1$, then all the operations are restrictions of the Gödel operations on $[0,1]$ to $\{a_0, \ldots, a_n\}$, i.e. $\{a_0, \ldots, a_n\}$ is a subuniverse of the standard Gödel algebra.

If $L = \{0,1\}$ then both of the structures (1) and (2) of Example 2.4 yield the same, namely, the two-element Boolean algebra, i.e. the structure of truth values of classical two-valued logic. We denote this algebra by **2** and write $2 = \{0,1\}$. There is no other residuated lattice structure on $\{0,1\}$ (easy to check).

Residuated lattices from Example 2.3 are special cases of residuated lattices induced by so-called t-norms.

Definition 2.5 A *t-norm* is a binary operation on $[0,1]$ which is associative, commutative, monotone, and with 1 acting as its unit element, i.e. $\otimes$ is a mapping $\otimes : [0,1] \times [0,1] \to [0,1]$ satisfying

$$(x \otimes y) \otimes z = x \otimes (y \otimes z), \tag{2.2}$$

$$x \otimes y = y \otimes z, \tag{2.3}$$

$$y_1 \leq y_2 \text{ implies } x \otimes y_1 \leq x \otimes y_2, \tag{2.4}$$

$$x \otimes 1 = x. \tag{2.5}$$

Remark 2.6 For each t-norm we have $x \otimes 0 = 0$. Indeed, $x \leq 1$ implies $x \otimes 0 \leq 1 \otimes 0 = 0$.

Example 2.7 Let $\otimes$ be a left-continuous t-norm (i.e. $\lim_{n\to\infty}(a_n \otimes b) = (\lim_{n\to\infty} a_n) \otimes b$ for any increasing sequence $\{a_n \in [0,1] \mid n = 1,2,3,\ldots\}$). Put

$$a \to b = \sup\{c \mid a \otimes c \leq b\}. \tag{2.6}$$

Then $\langle [0,1], \min, \max, \otimes, \rightarrow, 0, 1\rangle$ is a residuated lattice (a complete one). Later on, we prove this statement (Theorem 2.88) and more. It is easy to see that each of the three operations $\otimes$ from Example 2.3 is a continuous t-norm and that the corresponding residua are obtained by (2.6).

We will see examples of t-norms in Section 2.4.

Example 2.8 Let $\mathbf{R} = \langle R, +, \cdot, 0, 1\rangle$ be a commutative associative ring with unit 1 (i.e. both $+$ and $\cdot$ are commutative and associative, $x+0=x$ and $x \cdot 1 = x$ hold, for each $a \in R$ there exists $b \in R$ such that $a+b=0$, and $x \cdot (y+z) = (x \cdot y) + (x \cdot z)$ holds). Let $\mathrm{I}(R)$ denote the set of all ideals of $\mathbf{R}$ (i.e. the set of nonempty subgroups I of R such that $a \cdot i \in I$ for every $a \in R$, $i \in I$). As is well-known, $\mathrm{I}(R)$ is a complete lattice w.r.t. set inclusion (the least ideal is $\{0\}$, the greatest one is R itself). For $I, J \in \mathrm{I}(R)$, put

$$I \otimes J = \left\{ \sum_{k=1}^{n} i_k \cdot j_k \mid n \in \mathbf{N},\ i_k \in I\ j_k \in J \right\}$$

and

$$I \rightarrow J = \{k \in R \mid i \cdot k \in J \text{ for each } i \in I\}$$

(notice that $I \rightarrow J$ is an ideal). Then $\langle \mathrm{I}(R), \cap, \vee, \otimes, \rightarrow, \{0\}, R\rangle$ is a complete residuated lattice (this follows by direct verification or by Theorem 2.28 observing that $I \rightarrow J = \bigvee\{K \in \mathrm{I}(R) \mid I \otimes K \subseteq J\}$).

The following are important operations derived in terms of the basic operations of a residuated lattice.

Definition 2.9 Let $\mathbf{L} = \langle L, \wedge, \vee, \otimes, \rightarrow, 0, 1\rangle$ be a residuated lattice. A *biresiduum* in $\mathbf{L}$ is a binary operation $\rightarrow$ on L defined by

$$a \leftrightarrow b = (a \rightarrow b) \wedge (b \rightarrow a). \tag{2.7}$$

A *negation* in $\mathbf{L}$ is a unary operation $\neg$ defined by

$$\neg a = a \rightarrow 0. \tag{2.8}$$

For a nonnegative integer n, the n-th *power* of $a \in L$ is defined by

$$a^0 = 1 \qquad \text{and} \qquad a^{n+1} = a^n \otimes a. \tag{2.9}$$

Remark 2.10 In the case of the two-element Boolean algebra of classical logic, the operations $\neg$ and $\leftrightarrow$ coincide with the classical ones.

Example 2.11 One can easily check that for the biresidua corresponding to the Łukasiewicz, Gödel, and product structures on $[0, 1]$ we have

$$a \leftrightarrow b = 1 - |a - b| \quad \text{(Łukasiewicz)}, \tag{2.10}$$

$$a \leftrightarrow b = \begin{cases} 1 & \text{for } a = b \\ \min(a, b) & \text{otherwise} \end{cases} \quad \text{(Gödel)}, \tag{2.11}$$

$$a \leftrightarrow b = \frac{\min(a, b)}{\max(a, b)} \quad \text{(product)} \tag{2.12}$$

(we put $0/0 = 1$).

Example 2.12 For the negations of the Łukasiewicz, Gödel, and product structures on $[0, 1]$ we have

$$\neg a = 1 - a \quad \text{(Łukasiewicz)}, \tag{2.13}$$

$$\neg a = \begin{cases} 1 & \text{for } a = 0 \\ 0 & \text{for } a > 0 \end{cases} \quad \text{(both Gödel and product)}. \tag{2.14}$$

Example 2.13 For powers of the Łukasiewicz, Gödel, and product structures on $[0, 1]$ we have

$$a^n = \max(0, 1 - n(1 - a)) \quad \text{(Łukasiewicz)}, \tag{2.15}$$

$$a^n = a \quad \text{(Gödel)}, \tag{2.16}$$

$$a^n = \text{the usual } n\text{-th power of } a \quad \text{(product)}. \tag{2.17}$$

It is clear that $1^n = 1$, $0^n = 0$ for any n in each of the three cases. Furthermore, for a such that $0 < a < 1$ we have $a^n < a$ for any $n > 0$ in both Łukasiewicz and product structure. Moreover, w.r.t. the Łukasiewicz structure, each $a < 1$ is *nilpotent*, i.e. there is some n (the least one is $n \geq 1/(1 - a)$) such that $a^n = 0$. On the other hand, except for 0, no element is nilpotent in Gödel and product structure.

We are now going to introduce some additional conditions for residuated lattices. Each of them has a particular relevance from the logical point of view. For example, $a = (a \to 0) \to 0$ (i.e. $a = \neg\neg a$) asserts that the structure $\mathbf{L}$ of truth values satisfies the well-known law of double negation.

Definition 2.14 Let $\mathbf{L}$ be a residuated lattice. $\mathbf{L}$ satisfies the *prelinearity axiom* if it satisfies

$$(x \to y) \vee (y \to x) = 1. \tag{2.18}$$

$\mathbf{L}$ is *divisible* if it satisfies

$$x \wedge y = x \otimes (x \to y). \tag{2.19}$$

L satisfies the *law of double negation* if it satisfies

$$x = (x \to 0) \to 0. \tag{2.20}$$

L satisfies *idempotency* if it satisfies

$$x \otimes x = x. \tag{2.21}$$

Definition 2.15 An *integral commutative Girard-monoid* is a residuated lattice which satisfies the law of double negation. A *Heyting algebra* is a residuated lattice satisfying

$$x \otimes y = x \wedge y. \tag{2.22}$$

A *BL-algebra* is a residuated lattice which is divisible and satisfies the prelinearity axiom. An *MV-algebra* is a residuated lattice satisfying

$$x \vee y = (x \to y) \to y. \tag{2.23}$$

A Π-*algebra* (product algebra) is a BL-algebra satisfying

$$\begin{aligned} (z \to 0) \to 0 &\leq ((x \otimes z) \to (y \otimes z)) \to (x \to y) && (2.24)\\ x \wedge (x \to 0) &= 0. && (2.25) \end{aligned}$$

A *G-algebra* (Gödel algebra) is a BL-algebra which satisfies $x \otimes x = x$. A *Boolean algebra* is a residuated lattice which is both a Heyting algebra and an MV-algebra.

Remark 2.16 Looking at the residuated lattices from Examples 2.3 and 2.4, one can easily check that those with Łukasiewicz structure are MV-algebras, those with Gödel structure are G-algebras, and that with product structure is a Π-algebra.

2.3.2 BASIC PROPERTIES OF RESIDUATED LATTICES

We now consider several properties of residuated lattices. Some of them are (in)equalities involving general infima or suprema (e.g. $x \otimes \bigvee_{i\in I} y_i = \bigvee_{i\in I}(x \otimes y_i)$). They are to be read as follows: if both sides of the (in)equality exist then the (in)equality is true. We will call the (in)equalities containing possibly general infima and suprema *generalized (in)equalities* (or identities). That is, generalized (in)equalities are (in)equalities composed of terms that are allowed to contain general infima and suprema (terms of this kind will be called generalized terms). Note also that an inequality $t_1 \leq t_2$ may be expressed as an equality, namely, $t_1 \leq t_2$ is equivalent $t_1 \vee t_2 = t_2$.

Theorem 2.17 (basic properties) *Each residuated lattice satisfies the following conditions:*

$$x \otimes (x \to y) \leq y, \quad y \leq x \to (x \otimes y), \quad x \leq (x \to y) \to y, \tag{2.26}$$

$$x \leq y \quad \textit{iff} \quad x \to y = 1, \tag{2.27}$$

$$x \to x = 1, \quad x \to 1 = 1, \quad 0 \to x = 1, \tag{2.28}$$

$$1 \to x = x, \tag{2.29}$$

$$x \otimes 0 = 0, \tag{2.30}$$

$$x \otimes y \leq x, \quad x \leq y \to x, \tag{2.31}$$

$$x \otimes y \leq x \wedge y, \tag{2.32}$$

$$(x \otimes y) \to z = x \to (y \to z), \tag{2.33}$$

$$(x \to y) \otimes (y \to z) \leq (x \to z), \tag{2.34}$$

$$x \to y \textit{ is the greatest element of } \{z \mid x \otimes z \leq y\}, \tag{2.35}$$

$$x \otimes y \textit{ is the least element of } \{z \mid x \leq y \to z\}. \tag{2.36}$$

Proof. (2.26) follows directly by adjointness ($(x \to y) \otimes x = x \otimes (x \to y) \leq y$ iff $x \to y \leq x \to y$, the latter being true by reflexivity of $\leq$).

(2.27) By adjointness, $x \leq y$ iff $1 \otimes x \leq y$ iff $1 \leq x \to y$ iff $1 = x \to y$.

(2.28) are direct consequences of (2.27).

(2.29) by checking both inequalities. $x \leq 1 \to x$ iff $x = 1 \otimes x \leq x$ which is true. Conversely, $1 \to x \leq 1 \to x$ yields $1 \to x = 1 \otimes (1 \to x) \leq x$.

(2.30) is equivalent to $x \otimes 0 \leq 0$ which is, by adjointness, equivalent to $0 \leq x \to 0$.

(2.31) By (2.28), $x \to x = 1$, hence $y \leq x \to x$, and so $x \otimes y \leq x$ by adjointness. Applying adjointness to the last inequality once more, we get $x \leq y \to x$.

(2.32) By (2.31), $x \otimes y \leq x$ and $x \otimes y \leq y$ from which (2.32) immediately follows.

(2.33) We check both inequalities. $(x \otimes y) \to z \leq x \to (y \to z)$ iff (applying adjointness twice) $(x \otimes y) \otimes ((x \otimes y) \to z) \leq z$ which is true by (2.26). Conversely, $x \to (y \to z) \leq (x \otimes y) \to z$ iff (by adjointness) $y \otimes x \otimes (x \to (y \to z)) \leq z$ which is true (applying (2.26) yields $y \otimes x \otimes (x \to (y \to z)) \leq y \otimes (y \to z) \leq z$).

(2.34) is (by adjointness) equivalent to $x \otimes (x \to y) \otimes (y \to z) \leq z$ which holds (apply (2.26) twice).

(2.35) $x \otimes (x \to y) \leq y$ by (2.26), whence $x \to y \in \{z \mid x \otimes z \leq y\}$. If $x \otimes z \leq y$ then by adjointness, $z \leq x \to y$.

(2.36) $x \leq y \to (x \otimes y)$ by (2.26), whence $x \otimes y \in \{z \mid x \leq y \to z\}$. If $x \leq y \to z$ then by adjointness, $x \otimes y \leq z$. □

Theorem 2.18 *The class of all residuated lattices is a variety of algebras.*

Proof. We give a set of defining identities for residuated lattices. Clearly, the conditions (i) and (ii) of the definition of residuated lattices are expressible by identities. We claim that these identities plus

$$(x \otimes y) \to z = x \to (y \to z), \tag{2.37}$$
$$(x \otimes (x \to y)) \vee y = y, \tag{2.38}$$
$$x \to (x \vee y) = 1 \tag{2.39}$$

define residuated lattices. First, (2.37)–(2.39) hold in residuated lattices. Indeed, this claim directly follows from Theorem 2.17 (observe that (2.38) and (2.39) say that $x \otimes (x \to y) \leq y$ and $x \leq x \vee y$).

To complete the proof we have to show that (i), (ii), (2.37)–(2.39) imply the adjointness condition. First, notice that (2.38) and (2.39) imply (2.27). Indeed, if $x \leq y$ then (by (2.39)) $x \to y = x \to (x \vee y) = 1$. Conversely, if $x \to y = 1$ then (by (2.38)) $x \vee y = (x \otimes (x \to y)) \vee y = y$, i.e. $x \leq y$. Now, by (2.27) and (2.37) we have $x \otimes y \leq z$ iff $(x \otimes y) \to z = 1$ iff $x \to (y \to z) = 1$ iff $x \leq y \to z$. □

Remark 2.19 In a structure satisfying (i) and (ii), each of the three identities (2.37)–(2.39) is independent of the two remaining. Indeed, take any t-norm $\otimes$. First, put $a \to b = 1$ for every $a, b \in [0, 1]$. Then (2.38) does not hold while (2.37) and (2.39) do. Second, put $a \to b = 0$ for every $a, b \in [0, 1]$. Then (2.37) and (2.38) hold but (2.39) does not. Third, let $\otimes$ be the Łukasiewicz t-norm and put $a \to b = 1$ if $a \leq b$ and $a \to b = 0$ otherwise. In this case, (2.37) does not hold (e.g. for $x = 0.7$, $y = 0.6$, $z = 0.5$) but both (2.38) and (2.39) do.

Residuated lattices are defined as algebras with lattice structure, monoidal structure, and an "additional" operation $\to$ bound by adjointness. The following proposition shows that an alternative definition is possible where $\otimes$ plays the role of the "additional" operation.

Theorem 2.20 $\mathbf{L} = \langle L, \wedge, \vee, \otimes \to, 0, 1\rangle$ *is a residuated lattice iff*

(i) $\langle L, \wedge, \vee, 0, 1\rangle$ *is a bounded lattice,*

(ii') $\langle L, \to, 1\rangle$ *satisfies*

$$x = 1 \to x, \tag{2.40}$$
$$x \to (y \to z) = y \to (x \to z), \tag{2.41}$$

(iii) $\otimes$ *and* $\rightarrow$ *satisfy the adjointness property.*

Proof. By Theorem 2.17, (i), (ii'), and (iii) hold in any residuated lattice (note that (2.41) follows from (2.33) by commutativity of $\otimes$). Conversely, it suffices to show that (i), (ii'), and (iii) imply that $\langle L, \otimes, 1\rangle$ is a commutative monoid. We have $x \otimes 1 \leq t$ iff $x \leq 1 \rightarrow t$ iff (by (2.40)) $x \leq t$, which implies $x \otimes 1 = x$. Furthermore, $x \otimes y \leq t$ iff $(1 \rightarrow x =)x \leq y \rightarrow t$ iff $1 \leq x \rightarrow (y \rightarrow t)$ iff (by (2.41)) $1 \leq y \rightarrow (x \rightarrow t)$ iff $\cdots$ iff $y \otimes x \leq t$, i.e. $x \otimes y = y \otimes x$. Finally, $(x \otimes y) \otimes z \leq t$ iff $\cdots$ iff $(1 \rightarrow x =)x \leq y \rightarrow (z \rightarrow t)$ iff $1 \leq x \rightarrow (y \rightarrow (z \rightarrow t))$ iff (by (2.41)) $1 \leq z \rightarrow (y \rightarrow (x \rightarrow t))$ iff $\cdots$ iff $x \otimes (y \otimes z) \leq t$, i.e. $(x \otimes y) \otimes z = x \otimes (y \otimes z)$. Therefore, $\langle L, \otimes, 1\rangle$ is a commutative monoid.
□

We are going to show (a little bit more than) that for every set of identities of residuated lattices there is an equivalent (single) generalized identity.

Theorem 2.21 *Let $p_i = q_i$ $(i \in I)$ be identities of residuated lattices. Then there is a generalized identity $p = q$ such that for any class $\mathcal{K}$ of residuated lattices we have that*

$$\mathcal{K} \text{ satisfies } p_i = q_i \ (i \in I) \quad \text{iff} \quad \mathcal{K} \text{ satisfies } p = q.$$

If I is finite then $p = q$ can be chosen to be an identity.

Proof. Let all variables which occur in $p_i = q_i$ be $x_1, \ldots, x_n$, and let p'_i and q'_i be terms which result from p_i and q_i, respectively, by replacing x_j by x_{ij}, $j = 1, \ldots, n$ (x_{ij} are new variables). Doing so, the sets of variables of different identities $p'_i = q'_i$ are disjoint. Put $p = \bigwedge_{i \in I}(p'_i \vee x_i)$ and $q = \bigwedge_{i \in I}(q'_i \vee x_i)$. Clearly, if $\mathcal{K}$ satisfies $p_i = q_i$ $(i \in I)$ then $\mathcal{K}$ also satisfies $p = q$. Conversely, let $\mathcal{K}$ satisfy $p = q$. Take any $\mathbf{L} \in \mathcal{K}$ and take any evaluation e of variables such that $e(x_k) = 0$ for $k = i$ and $e(x_k) = 1$ for $k \neq i$. It is immediate that $p = p'_i$ and $q = q'_i$ hold in $\mathbf{L}$, therefore $p'_i = q'_i$ and also $p_i = q_i$ hold in $\mathbf{L}$ which completes the proof. □

Each of the defining identities of residuated lattices of Theorems 2.18 and 2.20 contains at most three variables. Next we prove that residuated lattices cannot be defined by identities with at most two variables.

Theorem 2.22 *Residuated lattices cannot be defined by identities with at most two variables.*

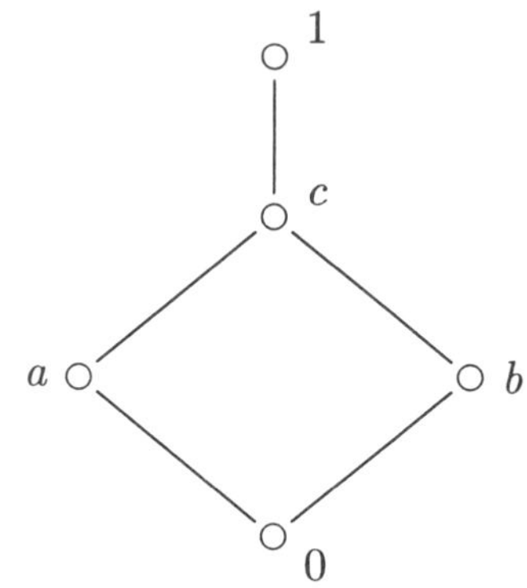

Figure 2.1. Residuated lattice from the proof of Theorem 2.22.

Proof. Consider the lattice in Fig. 2.1. Put $x \otimes y = x \wedge y$, $x \to y = 1$ for $x \leq y$, $x \to y = y$ for $x > y$, $a \to b = b$, and $b \to a = a$. Consider the operation $\vee'$ which is defined the same way as $\vee$ except for $a \vee' b = 1$ and $b \vee' a = 1$. It is easy to see that $\mathbf{L} = \langle \{0, a, b, c, 1\}, \wedge, \vee, \otimes, \to, 0, 1 \rangle$ is a residuated lattice (in fact, a Heyting algebra) but $\mathbf{L}' = \langle \{0, a, b, c, 1\}, \wedge, \vee', \otimes, \to, 0, 1 \rangle$ is not. However, every subalgebra of $\mathbf{L}'$ generated by an (at most) two-element subset is a residuated lattice. Therefore, $\mathbf{L}'$ (which is not a residuated lattice) satisfies every identity with at most two variables which is valid in all residuated lattices. □

The following proposition states that $\otimes$ is isotone in both arguments (in fact, isotony in the first argument follows by commutativity of $\otimes$) and $\to$ is isotone in the second and antitone in the first argument.

Theorem 2.23 (monotony of $\otimes$ and $\to$) *In each residuated lattice the following are true:*

$$y_1 \leq y_2 \textit{ implies } x \otimes y_1 \leq x \otimes y_2, \tag{2.42}$$

$$y_1 \leq y_2 \textit{ implies } x \to y_1 \leq x \to y_2, \tag{2.43}$$

$$x_1 \leq x_2 \textit{ implies } x_2 \to y \leq x_1 \to y. \tag{2.44}$$

Proof. (2.42) By adjointness, $x \otimes y_1 \leq x \otimes y_2$ is equivalent to $y_1 \leq x \to (x \otimes y_2)$ which follows directly from $y_1 \leq y_2$ (assumption) and $y_2 \leq x \to (x \otimes y_2)$ (2.26).

(2.43) $x \to y_1 \leq x \to y_2$ iff $x \otimes (x \to y_1) \leq y_2$, the latter follows from $y_1 \leq y_2$ (assumption) and $x \otimes (x \to y_1) \leq y_1$ (2.26).

(2.44) $x_2 \to y \leq x_1 \to y$ is equivalent to $x_1 \otimes (x_2 \to y) \leq y$ which is true. Indeed, from $x_1 \leq x_2$ (assumption) we have by (2.42) and (2.26) $x_1 \otimes (x_2 \to y) \leq x_2 \otimes (x_2 \to y) \leq y$. □

Theorem 2.24 *In each residuated lattice the following are true:*

$$x \to y \leq (x \wedge z) \to (y \wedge z), \tag{2.45}$$

$$x \to y \leq (x \vee z) \to (y \vee z), \tag{2.46}$$

$$x \to y \leq (x \otimes z) \to (y \otimes z), \tag{2.47}$$

$$x \to y \leq (z \to x) \to (z \to y), \tag{2.48}$$

$$x \to y \leq (y \to z) \to (x \to z). \tag{2.49}$$

Proof. (2.45) By adjointness, the inequality is equivalent to $(x \wedge z) \otimes (x \to y) \leq y \wedge z$ which follows from $(x \wedge z) \otimes (x \to y) \leq (x \otimes (x \to y)) \wedge (z \otimes (x \to y)) \leq y \wedge z$, by (2.26) and (2.31).

(2.46) is equivalent to $(x \vee z) \otimes (x \to y) \leq y \otimes z$ which is true since $(x \vee z) \otimes (x \to y) = (x \otimes (x \to y)) \vee (z \otimes (x \to y)) \leq y \wedge z$, by (2.26) and (2.31).

(2.47) is equivalent to $x \otimes z \otimes (x \to y) \leq y \otimes z$ which is true since $x \otimes (x \to y) \leq y$.

(2.48) is equivalent to $(z \to x) \otimes (x \to y) \leq z \to y$ which holds true by (2.34).

(2.49) by the same argument as (2.48). □

The next proposition summarizes basic properties of multiplication and residuum w.r.t. suprema and infima.

Theorem 2.25 (distributivity of $\otimes, \to$ over $\wedge, \vee$) *The following are true for each index set I. Moreover, in (2.50)–(2.52), if the left side makes sense, then so does the right side.*

$$x \otimes \bigvee_{i \in I} y_i = \bigvee_{i \in I} (x \otimes y_i), \tag{2.50}$$

$$x \to \bigwedge_{i \in I} y_i = \bigwedge_{i \in I} (x \to y_i), \tag{2.51}$$

$$\bigvee_{i \in I} x_i \to y = \bigwedge_{i \in I} (x_i \to y), \tag{2.52}$$

$$x \otimes \bigwedge_{i \in I} y_i \leq \bigwedge_{i \in I} (x \otimes y_i), \tag{2.53}$$

$$\bigvee_{i \in I} (x \to y_i) \leq x \to \bigvee_{i \in I} y_i, \tag{2.54}$$

$$\bigvee_{i \in I} (x_i \to y) \leq \bigwedge_{i \in I} x_i \to y. \tag{2.55}$$

Proof. (2.50) Let $x \otimes \bigvee_{i\in I} y_i$ make sense (i.e. $\bigvee_{i\in I} y_i$ exists). Isotony of $\otimes$ gives $x \otimes y_i \leq x \otimes \bigvee_{i\in I} y_i$ (for each $i \in I$). If t is such that $x \otimes y_i \leq t$ (for each $i \in I$) then $y_i \leq x \to t$, thus $\bigvee_{i\in I} y_i \leq x \to t$ which yields $x \otimes \bigvee_{i\in I} y_i \leq t$. Therefore, $x \otimes \bigvee_{i\in I} y_i$ is the least element s satisfying $x \otimes y_i \leq s$ for each $i \in I$. By definition of $\bigvee$, $x \otimes \bigvee_{i\in I} y_i = \bigvee_{i\in I}(x \otimes y_i)$.

(2.51) Let $x \to \bigwedge_{i\in I} y_i$ make sense. Isotony of $\to$ in the second argument yields $x \to \bigwedge_{i\in I} y_i \leq x \to y_i$ (for each $i \in I$). If for some t, $t \leq x \to y_i$, then $t \otimes x \leq y_i$, thus $t \otimes x \leq \bigwedge_{i\in I} y_i$ which yields $t \leq x \to \bigwedge_{i\in I} y_i$, i.e. $x \to \bigwedge_{i\in I} y_i = \bigwedge_{i\in I}(x \to y_i)$.

(2.52) Let $\bigvee_{i\in I} x_i \to y$ make sense. Antitony of $\to$ in the first argument yields $\bigvee_{i\in I} x_i \to y \leq x_i \to y$. Take t such that $t \leq x_i \to y$ for any $i \in I$. Then $x_i \leq t \to y$ (apply adjointness twice), therefore $\bigvee_{i\in I} x_i \leq t \to y$, i.e. $t \leq \bigvee_{i\in I} x_i \to y$, i.e. $\bigvee_{i\in I} x_i \to y = \bigwedge_{i\in I}(x_i \to y)$.

(2.53) follows from the isotony of $\otimes$ while (2.54) and (2.55) follow from antitony and isotony of $\to$, respectively. □

Remark 2.26 Let us comment on Theorem 2.25. First, if I is finite then every side of each of the (in)equalities (2.50)–(2.55) makes sense. If $|I| = 2$, Theorem 2.25 yields useful identities. For example, (2.50) becomes $x \otimes (y \vee z) = (x \otimes y) \vee (x \otimes z)$. Second, in a complete residuated lattice, the expressions of (2.50)–(2.55) clearly make sense no matter what I.

Theorem 2.27 *The following are further properties of residuated lattices:*

$$x \to y = ((x \to y) \to y) \to y, \tag{2.56}$$

$$x \otimes (x \to y) = y \quad \textit{iff} \quad (\exists z)(x \otimes z = y), \tag{2.57}$$

$$x \to (x \otimes y) = y \quad \textit{iff} \quad (\exists z)(x \to z = y), \tag{2.58}$$

$$(y \to x) \to x = y \quad \textit{iff} \quad (\exists z)(z \to x = y), \tag{2.59}$$

$$(x \wedge y) \otimes (x \vee y) \leq x \otimes y, \tag{2.60}$$

$$x \vee y \leq ((x \to y) \to y) \wedge ((y \to x) \to x), \tag{2.61}$$

$$x \wedge y \geq x \otimes (x \to y), \tag{2.62}$$

$$x \otimes (y \to z) \leq y \to (x \otimes z). \tag{2.63}$$

Proof. (2.56) By (2.26) it suffices to show $((x \to y) \to y) \to y \leq x \to y$ which equivalent to $x \otimes (((x \to y) \to y) \to y) \leq y$. This is indeed true since $x \otimes (((x \to y) \to y) \to y) \leq x \otimes (x \to y) \leq y$, by (2.26) and antitony of $\to$ in the first argument.

(2.57) If $x \otimes z = y$ then (by adjointness) $z \leq x \to y$ and hence $y = x \otimes z \leq x \otimes (x \to y) \leq y$ by (2.26) and isotony of $\otimes$, i.e. $x \otimes (x \to y) = y$; the converse implication is trivial.

(2.58) If $x \to z = y$ then (by adjointness) $x \otimes y \leq z$ and hence $y \leq x \to (x \otimes y) \leq x \to z = y$ by (2.26) and isotony of $\to$ in the second argument, i.e. $x \to (x \otimes y) = y$; the converse is again trivial.

(2.59) If $z \to x = y$ then $z \leq y \to x$ and so $y \leq (y \to x) \to x \leq z \to x = y$ by (2.26) and antitony of $\to$ in the second argument; the converse is trivial.

(2.60) By (2.50) we have $(x \wedge y) \otimes (x \vee y) = (x \wedge y) \otimes x \vee (x \wedge y) \otimes y \leq y \otimes x \vee x \otimes y = x \otimes y$.

(2.61) We have to show that both x and y are less or equal to both $(x \to y) \to y$ and $(y \to x) \to x$. By (2.26) we have $x \leq (x \to y) \to y$ and by (2.31) we have $y \leq (x \to y) \to y$; the rest is symmetric.

(2.62) We have to show both $x \otimes (x \to y) \leq x$ and $x \otimes (x \to y) \leq y$. The first inequality holds by (2.31) and the second one by (2.26).

(2.63) By adjointness (apply twice), the inequality is equivalent to $y \otimes (y \to z) \leq x \to (x \otimes z)$ which holds by (2.26). □

Theorem 2.28 *Let* $\mathbf{L} = \langle L, \wedge, \vee, \otimes, 0, 1 \rangle$ *be a structure satisfying (i) and (ii) of the definition of residuated lattices. Moreover, let* L *be a complete lattice. Then the following conditions are equivalent:*

$$\text{There exists } \to \text{ satisfying adjointness (with } \otimes). \tag{2.64}$$

$$\text{For any } a, b,\ \{c \mid a \otimes c \leq b\} \text{ has a greatest element.} \tag{2.65}$$

$$x \otimes \bigvee_{i \in I} y_i = \bigvee_{i \in I} (x \otimes y_i) \text{ holds in } \mathbf{L}. \tag{2.66}$$

$$\text{For } x \to y = \bigvee \{z \mid x \otimes z \leq y\}, \quad \otimes, \to \text{ satisfy adjointness.} \tag{2.67}$$

Proof. Trivially, (2.67) implies (2.64). Since, by Theorems 2.17 and 2.25, (2.64) implies (2.65) and (2.64) implies (2.66), it suffices to prove that (2.65) implies (2.67) and (2.66) implies (2.67).

Assume (2.65) and define $\to$ as in (2.67). Then $a \to b$ is the greatest element of $\{c \mid a \otimes c \leq b\}$. Therefore, if $a \otimes c \leq b$ then $c \leq a \to b$. If $c \leq a \to b$ then, by isotony of $\otimes$ and by the property of $a \to b$, $a \otimes c \leq a \otimes (a \to b) \leq b$. Hence, $\otimes$ and $\to$ defined in (2.67) satisfy adjointness, i.e. (2.67) holds true.

Assume (2.66) and define $\to$ as in (2.67). If $a \otimes c \leq b$ then clearly $c \leq a \to b$ ($a \to b$ is the supremum of such c's). If $c \leq a \to b$ then $a \otimes c \leq a \otimes (a \to b) = a \otimes \bigvee \{d \mid a \otimes d \leq b\} = \bigvee \{a \otimes d \mid a \otimes d \leq b\} \leq b$. Therefore, (2.67) is true. □

The following is an analogous version of the previous statement for the alternative definition from Theorem 2.20.

Theorem 2.29 *Let* $\mathbf{L} = \langle L, \wedge, \vee, \to, 0, 1\rangle$ *be a structure satisfying (i) and (ii') of Theorem 2.20. Moreover, let* L *be a complete lattice. Then the following conditions are equivalent.*

$$\text{There exists } \otimes \text{ satisfying adjointness (with } \to\text{).} \tag{2.68}$$

$$\text{For any } a, b,\ \{c \mid a \le b \to c\} \text{ has a least element.} \tag{2.69}$$

$$x \to \bigwedge_{i\in I} y_i = \bigwedge_{i\in I}(x \to y_i) \text{ is true } \mathbf{L}. \tag{2.70}$$

$$\text{For } x \otimes y = \bigwedge\{z \mid x \le y \to z\},\ \otimes, \to \text{ satisfy adjointness.} \tag{2.71}$$

Proof. Trivially, (2.71) implies (2.68). Since, by Theorems 2.17 and 2.25, (2.68) implies (2.69) and (2.68) implies (2.70), it suffices to prove that (2.69) implies (2.71) and (2.70) implies (2.71).

Assume (2.69) and define $\otimes$ as in (2.71). Then $a \otimes b$ is the least element of $\{c \mid a \le b \to c\}$. Therefore, $a \le b \to c$ implies $a \otimes b \le c$. If $a \otimes b \le c$ then $b \to c \ge b \to (a \otimes b) \ge a$. Thus, $\otimes$ and $\to$ satisfy adjointness.

Assume (2.70) and define $\otimes$ as in (2.71). If $a \le b \to c$ then clearly $a \otimes b \le c$ ($a \otimes b$ is the infimum of such c's). If $a \otimes b \le c$ then $b \to c \ge b \to (a \otimes b) = b \to \{c \mid a \le b \to c\} = \bigwedge\{b \to c \mid a \le b \to c\} \ge a$ proving (2.71). □

Theorem 2.30 (negation) *The following are properties of negation:*

$$\neg 0 = 1, \qquad \neg 1 = 0, \tag{2.72}$$

$$x \otimes \neg x = 0, \tag{2.73}$$

$$x \le \neg\neg x, \qquad \neg x = \neg\neg\neg x, \tag{2.74}$$

$$x \le y \qquad \text{implies} \qquad \neg y \le \neg x, \tag{2.75}$$

$$\neg(\bigvee_{i\in I} x_i) = \bigwedge_{i\in I} \neg x_i, \tag{2.76}$$

$$\neg(\bigwedge_{i\in I} x_i) \ge \bigvee_{i\in I} \neg x_i. \tag{2.77}$$

Proof. The assertions are immediate consequences of previous statements. Namely, (2.72) follows from (2.28) and (2.29), (2.73) from (2.26), (2.74) from (2.26) and (2.56), (2.75) from antitony of $\to$ in the first argument, (2.76) from (2.52), (2.77) from (2.55). □

Theorem 2.31 (biresiduum) *The following are properties of biresiduum:*

$$0 \leftrightarrow 1 = 1 \leftrightarrow 0 = 0, \qquad 0 \leftrightarrow 0 = 1 \leftrightarrow 1 = 1, \tag{2.78}$$
$$x \leftrightarrow x = 1, \tag{2.79}$$
$$x \leftrightarrow y = y \leftrightarrow x, \tag{2.80}$$
$$(x \leftrightarrow y) \otimes (y \leftrightarrow z) \leq x \leftrightarrow z, \tag{2.81}$$
$$x \leftrightarrow 1 = x, \qquad x \leftrightarrow 0 = \neg x, \tag{2.82}$$
$$x \leftrightarrow y = 1 \qquad \textit{iff} \qquad x = y, \tag{2.83}$$
$$(x_1 \leftrightarrow y_1) \wedge (x_2 \leftrightarrow y_2) \leq (x_1 \wedge x_2) \leftrightarrow (y_1 \wedge y_2), \tag{2.84}$$
$$(x_1 \leftrightarrow y_1) \wedge (x_2 \leftrightarrow y_2) \leq (x_1 \vee x_2) \leftrightarrow (y_1 \vee y_2), \tag{2.85}$$
$$(x_1 \leftrightarrow y_1) \otimes (x_2 \leftrightarrow y_2) \leq (x_1 \otimes x_2) \leftrightarrow (y_1 \otimes y_2), \tag{2.86}$$
$$(x_1 \leftrightarrow y_1) \otimes (x_2 \leftrightarrow y_2) \leq (x_1 \rightarrow x_2) \leftrightarrow (y_1 \rightarrow y_2), \tag{2.87}$$
$$\bigwedge_{i \in I} (x_i \leftrightarrow y_i) \leq (\bigwedge_{i \in I} x_i) \leftrightarrow (\bigwedge_{i \in I} y_i), \tag{2.88}$$
$$\bigwedge_{i \in I} (x_i \leftrightarrow y_i) \leq (\bigvee_{i \in I} x_i) \leftrightarrow (\bigvee_{i \in I} y_i), \tag{2.89}$$
$$x \leftrightarrow y = (x \vee y) \rightarrow (x \wedge y). \tag{2.90}$$

Proof. (2.78)–(2.80), (2.82), (2.83) are immediate consequences of previously stated properties of $\rightarrow$.

In the rest, we have to show inequalities of the form $\cdots \leq t \leftrightarrow s$. To do so, it is sufficient to show both $\cdots \leq t \rightarrow s$ and $\cdots \leq s \rightarrow t$. Since both of these cases are symmetric in (2.81)–(2.87), we limit ourselves to the verification of the first inequality in (2.81)–(2.87).

(2.81) We have to show $(x \leftrightarrow y) \otimes (y \leftrightarrow z) \leq x \rightarrow z$. By adjointness, this is equivalent to $x \otimes (x \leftrightarrow y) \otimes (y \leftrightarrow z) \leq z$ which follows from $x \otimes (x \leftrightarrow y) \otimes (x \leftrightarrow z) \leq x \otimes (x \rightarrow y) \otimes (y \rightarrow z) \leq y \otimes (y \rightarrow z) \leq z$.

(2.84) $(x_1 \leftrightarrow y_1) \wedge (x_2 \leftrightarrow y_2) \leq (x_1 \wedge x_2) \rightarrow (y_1 \wedge y_2)$ is true iff $(x_1 \wedge x_2) \otimes ((x_1 \leftrightarrow y_1) \wedge (x_2 \leftrightarrow y_2)) \leq (y_1 \wedge y_2)$. The last inequality holds since $(x_1 \wedge x_2) \otimes ((x_1 \leftrightarrow y_1) \wedge (x_2 \leftrightarrow y_2)) \leq ((x_1 \wedge x_2) \otimes (x_1 \leftrightarrow y_1)) \wedge ((x_1 \wedge x_2) \otimes (x_2 \leftrightarrow y_2)) \leq (x_1 \otimes (x_1 \leftrightarrow y_1)) \wedge (x_2 \otimes (x_2 \leftrightarrow y_2)) \leq (x_1 \otimes (x_1 \rightarrow y_1)) \wedge (x_2 \otimes (x_2 \rightarrow y_2)) \leq y_1 \wedge y_2$.

(2.85) $(x_1 \leftrightarrow y_1) \wedge (x_2 \leftrightarrow y_2) \leq (x_1 \vee x_2) \rightarrow (y_1 \vee y_2)$ is equivalent to $(x_1 \vee x_2) \otimes (x_1 \leftrightarrow y_1) \wedge (x_2 \leftrightarrow y_2) \leq y_1 \vee y_2$. This inequality indeed holds, since $(x_1 \vee x_2) \otimes ((x_1 \leftrightarrow y_1) \wedge (x_2 \leftrightarrow y_2)) = (x_1 \otimes ((x_1 \leftrightarrow y_1) \wedge (x_2 \leftrightarrow y_2))) \vee (x_2 \otimes ((x_1 \leftrightarrow y_1) \wedge (x_2 \leftrightarrow y_2))) \leq (x_1 \otimes (x_1 \rightarrow y_1)) \vee (x_2 \otimes (x_2 \rightarrow y_2)) \leq y_1 \vee y_2$.

(2.86) $(x_1 \leftrightarrow y_1) \otimes (x_2 \leftrightarrow y_2) \leq (x_1 \otimes x_2) \rightarrow (y_1 \otimes y_2)$ holds true iff $x_1 \otimes x_2 \otimes (x_1 \leftrightarrow y_1) \otimes (x_2 \leftrightarrow y_2) \leq y_1 \otimes y_2$ which is easy to verify.

(2.87) $(x_1 \leftrightarrow y_1) \otimes (x_2 \leftrightarrow y_2) \leq (x_1 \rightarrow x_2) \rightarrow (y_1 \rightarrow y_2)$ is equivalent to $y_1 \otimes (x_1 \rightarrow x_2) \otimes (x_1 \leftrightarrow y_1) \otimes (x_2 \leftrightarrow y_2) \leq y_2$ which holds since $y_1 \otimes (x_1 \rightarrow x_2) \otimes (x_1 \leftrightarrow y_1) \otimes (x_2 \leftrightarrow y_2) \leq y_1 \otimes (y_1 \rightarrow x_1) \otimes (x_1 \rightarrow x_2) \otimes (x_2 \rightarrow y_2) \leq y_2$ by repeated use of (2.26).

(2.88) and (2.89) can be proved in a similar way to (2.84) and (2.85), respectively.

(2.90) $x \leftrightarrow y \leq (x \vee y) \rightarrow (x \wedge y)$ iff $(x \vee y) \otimes (x \leftrightarrow y) \leq (x \wedge y)$ which is true: $(x \vee y) \otimes (x \leftrightarrow y) \leq x \otimes (x \rightarrow y) \leq y$ and $(x \vee y) \otimes (x \leftrightarrow y) \leq y \otimes (y \rightarrow x) \leq x$. On the other hand, to show $(x \vee y) \rightarrow (x \wedge y) \leq x \leftrightarrow y$ we verify $(x \vee y) \rightarrow (x \wedge y) \leq x \rightarrow y$ (due to symmetry this is sufficient) which is equivalent to $x \otimes ((x \vee y) \rightarrow (x \wedge y)) \leq y$. We have $x \otimes ((x \vee y) \rightarrow (x \wedge y)) \leq x \otimes (x \rightarrow y) \leq y$, completing the proof. □

2.3.3 PROPERTIES OF ADDITIONAL CONDITIONS FOR RESIDUATED LATTICES

Lemma 2.32 *Each of the conditions from Definitions 2.14 and 2.15 is expressible by identities.*

Proof. The assertion is almost clear since each inequality can be expressed as an equality, namely, $t_1(x_1, \ldots, x_n) \leq t_2(x_1, \ldots, x_n)$ by $t_1(x_1, \ldots, x_n) \vee t_2(x_1, \ldots, x_n) = t_2(x_1, \ldots, x_n)$. □

Theorem 2.33 *All classes of special residuated lattices from Definition 2.15 form varieties of algebras, namely subvarieties of the variety of all residuated lattices.*

Proof. The assertion follows directly from Theorem 2.18 and Lemma 2.32. □

Theorem 2.34 (prelinearity) *The following conditions are equivalent in each residuated lattice:*

prelinearity (i.e. (2.18))

$$x \rightarrow (y_1 \vee y_2) = (x \rightarrow y_1) \vee (x \rightarrow y_2), \tag{2.91}$$

$$(x_1 \wedge x_2) \rightarrow y = (x_1 \rightarrow y) \vee (x_2 \rightarrow y). \tag{2.92}$$

Proof. First we show that prelinearity implies both (2.91) and (2.92). Due to Theorem 2.25 we have to show only the "$\leq$" inequalities. By prelinearity and (2.63) we have

$$\begin{aligned}
x \to (y_1 \vee y_2) &= ((y_1 \to y_2) \vee (y_2 \to y_1)) \otimes (x \to (y_1 \vee y_2)) = \\
&= ((y_1 \to y_2) \otimes (x \to (y_1 \vee y_2))) \vee ((y_2 \to y_1) \otimes (x \to (y_1 \vee y_2))) \leq \\
&\leq (x \to ((y_1 \to y_2) \otimes (y_1 \vee y_2))) \vee (x \to ((y_2 \to y_1) \otimes (y_1 \vee y_2))) = \\
&= (x \to (y_1 \otimes (y_1 \to y_2) \vee y_2 \otimes (y_1 \to y_2))) \vee \\
&\quad \vee (x \to (y_1 \otimes (y_2 \to y_1) \vee y_2 \otimes (y_2 \vee y_2))) \leq \\
&\leq (x \to y_2) \vee (x \to y_1),
\end{aligned}$$

therefore (2.91) holds true. Furthermore,

$$\begin{aligned}
(x_1 \wedge x_2) \to y &= ((x_1 \wedge x_2) \to y) \otimes ((x_1 \to x_2) \vee (x_2 \to x_1)) = \\
&= (((x_1 \wedge x_2) \to y) \otimes (x_1 \to x_2)) \vee (((x_1 \wedge x_2) \to y) \otimes (x_2 \to x_1)) \leq \\
&\leq (((x_1 \otimes (x_1 \to x_2)) \to y) \otimes (x_1 \to x_2)) \vee \\
&\quad \vee (((x_2 \otimes (x_2 \to x_1)) \to y) \otimes (x_2 \to x_1)) = \\
&= (((x_1 \to x_2) \to (x_1 \to y)) \otimes (x_1 \to x_2)) \vee \\
&\quad \vee (((x_2 \to x_1) \to (x_2 \to y)) \otimes (x_2 \to x_1)) \leq \\
&\leq (x_1 \to y) \vee (x_2 \to y).
\end{aligned}$$

From (2.91) we infer prelinearity as follows:

$$\begin{aligned}
(x \to y) \vee (y \to x) &= ((y \to y) \wedge (x \to y)) \vee ((x \to x) \wedge (y \to x)) = \\
&= ((x \vee y) \to y) \vee ((x \vee y) \to x)) = (x \vee y) \to (x \vee y) = 1.
\end{aligned}$$

From (2.92) we infer prelinearity as follows:

$$\begin{aligned}
(x \to y) \vee (y \to x) &= ((x \to x) \wedge (x \to y)) \vee ((y \to y) \wedge (y \to x)) = \\
&= (x \to (x \wedge y)) \vee (y \to (x \wedge y)) = (x \wedge y) \to (x \wedge y) = 1.
\end{aligned}$$

The proof is complete. □

Remark 2.35 Note that prelinearity does not imply $x \to \bigvee_{i \in I} y_i = \bigvee_{i \in I}(x \to y_i)$. Namely, by Lemma 2.85 the just-mentioned equality means left-continuity of $\to$ in the second argument, the right continuity follows from (2.51). Hence, $\to$ would be continuous in the second argument. A counterexample to this is the Gödel structure on $[0, 1]$: prelinearity is valid but $\to$ is not continuous in the second argument (take $a > 0$ and any sequence $\{b_n \mid b_n < a\}$ such that $\lim_{n \to \infty} b_n = a$; then $\lim_{n \to \infty} a \to b_n = \lim_{n \to \infty} b_n = a$ while $a \to \lim_{n \to \infty} b_n = a \to a = 1$).

Theorem 2.36 (divisibility) *The following conditions are equivalent in each residuated lattice:*

$$\begin{aligned}
&\textit{divisibility (i.e. (2.19))}, \\
&x \leq y \quad \textit{implies } (\exists z)(x = y \otimes z), && (2.93)\\
&x \to (y \wedge z) = (x \to y) \otimes ((x \wedge y) \to z). && (2.94)
\end{aligned}$$

Proof. We show that (2.93) implies divisibility. Let z be such that $x \wedge y = x \otimes z$ (z exists on account of (2.93)). By adjointness we have $z \leq x \to (x \wedge y) \leq x \to y$, therefore $x \wedge y = x \otimes z \leq x \otimes (x \to y)$. By (2.62), divisibility is verified. Starting from divisibility we have $x \to (y \wedge z) = (x \to y) \wedge (x \to z) = (x \to y) \otimes ((x \to y) \to (x \to z)) = (x \to y) \otimes ((x \otimes (x \to y)) \to z) = (x \to y) \otimes ((x \wedge y) \to z)$, i.e. (2.94) holds. Finally, assume (2.94) and let $x \leq y$. Then, by (2.94), $y \otimes (y \to x) = (1 \to y) \otimes ((1 \wedge y) \to x) = 1 \to (x \wedge y) = x$, verifying (2.93). □

Theorem 2.37 (consequences of prelinearity) *If a residuated lattice satisfies prelinearity then it also satisfies the following conditions:*

$$\begin{aligned}
&(x \otimes x) \wedge (y \otimes y) \leq x \otimes y \leq (x \otimes x) \vee (y \otimes y), && (2.95)\\
&x \otimes (y_1 \wedge y_2) = (x \otimes y_1) \wedge (x \otimes y_2), && (2.96)\\
&x \wedge (y_1 \vee y_2) = (x \wedge y_1) \vee (x \wedge y_2), && (2.97)\\
&x \vee y = ((x \to y) \to y) \wedge ((y \to x) \to x). && (2.98)
\end{aligned}$$

Proof. (2.95) We verify as follows:

$$\begin{aligned}
(x \otimes x) \wedge (y \otimes y) &= ((x \otimes x) \wedge (y \otimes y)) \otimes ((x \to y) \vee (y \to x)) \leq \\
&\leq (x \otimes x \otimes (x \to y)) \wedge (y \otimes y \otimes (y \to x)) \leq x \otimes y = \\
&= (x \otimes y) \otimes ((x \to y) \vee (y \to x)) = \\
&= (y \otimes x \otimes (x \to y)) \vee (x \otimes y \otimes (y \to x)) \leq (x \otimes x) \vee (y \otimes y).
\end{aligned}$$

(2.96) By Theorem 2.25 we have to check the "$\geq$" inequality. We have

$$\begin{aligned}
&(x \otimes y_1) \wedge (x \otimes y_2) = \\
&= ((x \otimes y_1) \otimes ((y_1 \to y_2) \vee (y_2 \to y_1))) \wedge \\
&\quad \wedge ((x \otimes y_2) \otimes ((y_1 \to y_2) \vee (y_2 \to y_1))) \leq \\
&\leq (x \otimes y_1 \otimes (y_1 \to y_2)) \wedge (x \otimes y_2 \otimes (y_2 \to y_1)) \leq \\
&\leq (x \otimes y_2) \wedge (x \otimes y_1).
\end{aligned}$$

(2.97) Since $x \wedge (y_1 \vee y_2) \geq (x \wedge y_1) \vee (x \wedge y_2)$ is always the case we have to prove the converse inequality. We have

$$\begin{aligned} & x \wedge (y_1 \vee y_2) \to ((x \wedge y_1) \vee (x \wedge y_2)) = \\ = \;& ((x \wedge (y_1 \vee y_2)) \to (x \wedge y_1)) \vee ((x \wedge (y_1 \vee y_2)) \to (x \wedge y_2)) \geq \\ \geq \;& ((y_1 \vee y_2) \to y_1) \vee ((y_1 \vee y_2) \to y_2) = (y_1 \vee y_2) \to (y_1 \vee y_2) = 1, \end{aligned}$$

thus by (2.27), the converse inequality is established.

(2.98) On account of (2.61) it remains to prove the "$\geq$" part. We have

$$\begin{aligned} & ((x \to y) \to y) \wedge ((y \to x) \to x) = \\ = \;& (((x \to y) \to y) \wedge ((y \to x) \to x)) \otimes ((x \to y) \vee (y \to x)) \leq \\ \leq \;& ((x \to y) \otimes ((x \to y) \to y)) \vee ((y \to x) \otimes ((y \to x) \to x)) \leq \\ \leq \;& y \vee x = x \vee y, \end{aligned}$$

establishing the inequality. □

Remark 2.38 Prelinearity does not imply $x \otimes \bigwedge_{i \in I} y_i = \bigwedge_{i \in I}(x \otimes y_i)$. By Lemma 2.85 and Theorem 2.25 this would imply continuity of $\otimes$. However, the residuated lattice on $[0,1]$ induced by the "nilpotent minimum" t-norm from Remark 2.89 satisfies prelinearity but $\otimes$ is not continuous (it is only left-continuous).

Theorem 2.39 (consequences of divisibility) *If a residuated lattice satisfies divisibility then it also satisfies the following conditions:*

$$x \otimes y \leq (x \otimes x) \vee (y \otimes y), \tag{2.99}$$

$$x \otimes (y_1 \wedge y_2) = (x \otimes y_1) \wedge (x \otimes y_2), \tag{2.100}$$

$$x \wedge \bigvee_{i \in I} y_i = \bigvee_{i \in I} (x \wedge y_i), \tag{2.101}$$

$$x \otimes x = x \quad \textit{implies} \quad (\forall y)(x \wedge y = x \otimes y). \tag{2.102}$$

Proof. (2.99) By (2.93) which is equivalent to divisibility, there exist z_1, z_2 such that $x = (x \vee y) \otimes z_1$, $y = (x \vee y) \otimes z_2$. Then clearly $x \vee y = (z_1 \vee z_2) \otimes (x \vee y)$. We thus have

$$\begin{aligned} & x \otimes y = (x \vee y) \otimes z_1 \otimes (x \vee y) \otimes z_2 = \\ = \;& z_1 \otimes z_2 \otimes (z_1 \vee z_2) \otimes (x \vee y) \otimes (x \vee y) \leq \\ \leq \;& ((z_1 \otimes z_1) \vee (z_2 \otimes z_2)) \otimes (x \vee y) \otimes (x \vee y) = (x \otimes x) \vee (y \otimes y). \end{aligned}$$

(2.100) Due to (2.53) we need only to establish $x\otimes(y_1\wedge y_2)\geq(x\otimes y_1)\wedge(x\otimes y_2)$. We have

$$\begin{aligned}(x\otimes y_1)\wedge(x\otimes y_2) &= (x\otimes y_1)\otimes((x\otimes y_1)\to(x\otimes y_2)) = \\ &= x\otimes(y_1\otimes(y_1\to(x\to(x\otimes y_2)))) = x\otimes(y_1\wedge(x\to(x\otimes y_2))) = \\ &= x\otimes(x\to(x\otimes y_2))\otimes((x\to(x\otimes y_2))\to y_1)\leq \\ &\leq x\otimes y_2\otimes(y_2\to y_1) = x\otimes(y_1\wedge y_2).\end{aligned}$$

(2.101) We only have to verify the "$\leq$" part (the converse inequality is true in every lattice). We have

$$\begin{aligned}x\wedge\bigvee_{i\in I}y_i &= (\bigvee_{i\in I}y_i)\otimes((\bigvee_{i\in I}y_i)\to x) = \\ &= \bigvee_{i\in I}(y_i\otimes((\bigvee_{i\in I}y_i)\to x))\leq\bigvee_{i\in I}(y_i\otimes(y_i\to x)) = \\ &= \bigvee_{i\in I}(x\wedge y_i).\end{aligned}$$

(2.102) $x\wedge y = x\otimes(x\to y) = x\otimes x\otimes(x\to y)\leq x\otimes y$. Since the converse is always true the assertion follows. □

Theorem 2.40 (consequences of double negation) *If a residuated lattice satisfies the law of double negation then it also satisfies the following conditions:*

$$x\to y = \neg(x\otimes\neg y), \tag{2.103}$$

$$\neg(\bigwedge_{i\in I}x_i) = \bigvee_{i\in I}\neg x_i, \tag{2.104}$$

$$x\to y = \neg y\to\neg x, \tag{2.105}$$

prelinearity is equivalent to (2.106)

$$x\otimes(y_1\wedge y_2) = (x\otimes y_1)\wedge(x\otimes y_2).$$

Proof. (2.103) Immediately by (2.33) $x\to y = x\to((y\to 0)\to 0) = (x\otimes(y\to 0))\to 0$.

(2.104) By Theorem 2.25 we have $(\bigvee_{i\in I}(x_i\to 0))\to 0 = \bigwedge_{\in I}((x_i\to 0)\to 0) = \bigwedge_{i\in I}x_i$, hence $\bigvee_{i\in I}(x_i\to 0) = ((\bigvee_{i\in I}(x_i\to 0))\to 0)\to 0 = (\bigwedge_{i\in I}x_i)\to 0$, which proves the assertion.

(2.105) By (2.49) we only have to prove the "$\geq$" inequality. This easily follows from (2.49) and the law of double negation: $\neg y\to\neg x\leq\neg\neg x\to\neg\neg y = x\to y$.

(2.106) On account of Theorem 2.37 we have only to prove the "$\Leftarrow$" implication. Using $x \otimes (y_1 \wedge y_2) = (x \otimes y_1) \wedge (x \otimes y_2)$, (2.103), and (2.104) we have

$$\begin{aligned} x \to (y_1 \vee y_2) &= (x \otimes ((y_1 \vee y_2) \to 0)) \to 0 = \\ &= (x \otimes ((y_1 \to 0) \wedge (y_2 \to 0))) \to 0 = \\ &= ((x \otimes (y_1 \to 0)) \wedge (x \otimes (y_2 \to 0))) \to 0 = \\ &= ((x \otimes (y_1 \to 0)) \to 0) \vee ((x \otimes (y_2 \to 0)) \to 0) = \\ &= (x \to y_1) \vee (x \to y_2), \end{aligned}$$

which proves (2.91), a condition equivalent to prelinearity. □

Theorem 2.41 *If a residuated lattice satisfies prelinearity and the law of double negation then*

$$x = x \to 0 \quad \textit{and} \quad y = y \to 0 \quad \textit{implies} \quad x = y,$$

i.e. negation has at most one fixpoint.

Proof. By (2.105) we have

$$\begin{aligned} x = x \otimes ((x \to y) \vee (y \to x)) &= \\ &= (x \otimes (x \to y)) \vee (x \otimes (y \to x)) = \\ &= (x \otimes (x \to y)) \vee (x \otimes ((y \to 0) \to (x \to 0))) = \\ &= (x \otimes (x \to y)) \vee (x \otimes (x \to y)) \leq y \vee y = y, \end{aligned}$$

hence $x \leq y$. Symmetrically, one may obtain $y \leq x$ which shows $x = y$ whenever x and y are fixpoints of the negation. □

Theorem 2.42 *In any residuated lattice* **L** *the following are equivalent:*

L *is an MV-algebra.*

L *is divisible and satisfies the law of double negation.*

Proof. Let **L** be an MV-algebra, i.e. $x \vee y = (x \to y) \to y$. Since the law of double negation is immediate to see (just put $y = 0$), we need only to check divisibility, i.e. $x \wedge y = x \otimes (x \to y)$. By (2.62), "$\geq$" always holds. We therefore need to verify "$\leq$". Due to the law of double negation, the inequality is equivalent to $(x \otimes (x \to y)) \to 0 \leq (x \wedge y) \to 0$. By (2.55), $(x \to 0) \vee (y \to 0) \leq (x \wedge y) \to 0$. Therefore, it suffices to show $(x \otimes (x \to y)) \to 0 \leq (x \to 0) \vee (y \to 0)$. Using

$(x \otimes (x \to y)) \to 0 = (x \to y) \to (x \to 0)$ and $(x \to 0) \vee (y \to 0) = ((y \to 0) \to (x \to 0)) \to (x \to 0) = (x \to y) \to (x \to 0)$, the inequality becomes $(x \to y) \to (x \to 0) \leq (x \to y) \to (x \to 0)$, hence **L** is divisible.

Conversely, by (2.61) we need to show $(x \to y) \to y \leq x \vee y$. The law of double negation yields $x \vee y = ((x \vee y) \to 0) \to 0 = ((x \to 0) \wedge (y \to 0)) \to 0$. By adjointness, the inequality therefore becomes $((x \to 0) \wedge (y \to 0)) \otimes ((x \to y) \to y) \leq 0$. Employing furthermore divisibility, we get $((x \to 0) \wedge (y \to 0)) \otimes ((x \to y) \to y) = (y \to 0) \otimes ((y \to 0) \to (x \to 0)) \otimes ((x \to y) \to y) = (y \to 0) \otimes (x \to y) \otimes ((x \to y) \to y) \leq (y \to 0) \otimes y = 0$, verifying the required inequality. □

Theorem 2.43 *Any MV-algebra satisfies*

$$x \to (x \otimes y) = (x \to 0) \vee y, \tag{2.107}$$

prelinearity. (2.108)

Proof. (2.107) We have $x \to (x \otimes y) = ((x \otimes y) \to 0) \to (x \to 0) = (y \to (x \to 0)) \to (x \to 0) = (x \to 0) \vee y$.

(2.108) The assertion follows directly by (2.106), Theorem 2.39, and Theorem 2.42. □

Among the properties of a structure of truth values, linearity is very appealing (any two truth values are comparable). In the following we list some properties of linearly ordered residuated lattices.

Theorem 2.44 (consequences of linearity) *Each linearly ordered residuated lattice satisfies prelinearity; moreover, idempotency of $\otimes$ implies $x \otimes y = x \wedge y$.*

Proof. Due to linearity of $\leq$, either $x \leq y$ or $y \leq x$, i.e. either $x \to y = 1$ or $y \to x = 1$, whence $(x \to y) \vee (y \to x) = 1$.

Assume now $\otimes$ to be idempotent. If $x \leq y$ then $x = x \otimes x \leq x \otimes y \leq x \otimes 1 = x$, i.e. $x \otimes y = x = x \wedge y$. If $y < x$, one proceeds analogously. □

Theorem 2.45 (linear MV-algebras) *Under $x \oplus y = \neg(\neg x \otimes \neg y)$, the following are true in each linearly ordered MV-algebra:*

$$x \oplus y < 1 \textit{ implies } x \otimes y = 0, \tag{2.109}$$

$$x \oplus y = 1 \textit{ and } x \otimes y = 0 \textit{ implies } x = \neg y, \tag{2.110}$$

$$x \oplus y = x \oplus z < 1 \textit{ or } x \otimes y = x \otimes z > 0 \textit{ implies } y = z, \tag{2.111}$$

$$x \oplus y = x \oplus z \text{ and } x \otimes y = x \otimes z \textit{ implies } y = z, \tag{2.112}$$

$$x \oplus y = x \textit{ implies } x = 1 \textit{ or } y = 0. \tag{2.113}$$

Proof. (2.109): By (2.103) we have $\neg x \to y = \neg(\neg x \otimes \neg y) = x \oplus y < 1$, i.e. $\neg x \not\leq y$ and so, by linearity of $\leq$, $y < \neg x$. Therefore, $x \otimes y \leq x \otimes \neg x = 0$.

(2.110): By (2.103) we have $\neg x \to y = 1$, i.e. $\neg x \leq y$, and $y \to \neg x = 1$, i.e. $y \leq \neg x$, thus $\neg x = y$ and, by the law of double negation, $x = \neg y$.

(2.111): As in (2.109), $x \oplus y = x \oplus z < 1$ yields $y < \neg x$ and $z < \neg x$ whence $y = y \wedge \neg x = \neg x \otimes(\neg x \to y) = \neg x \otimes(x \oplus y) = \neg x \otimes(x \oplus z) = \cdots = z$. If $x \otimes y = x \otimes z > 0$ then $\neg x \oplus \neg y = \neg x \oplus \neg z < 1$; by the above statement, $\neg y = \neg z$ whence $y = z$.

(2.112): Directly from (2.110) and (2.111).

(2.113): If $x < 1$ then $x \oplus y = x + 0 < 1$ and thus $y = 0$ by (2.111). □

Theorem 2.46 (linear Π-algebras) *The following are true in each linearly ordered Π-algebra:*

$$0 < x \ \textit{implies}\ \neg x = 0, \tag{2.114}$$

$$\textit{if}\ 0 < x\ \textit{then}\ x \otimes y_1 = x \otimes y_2\ \textit{implies}\ y_1 = y_2, \tag{2.115}$$

$$\textit{if}\ 0 < x\ \textit{then}\ x \otimes y_1 < x \otimes y_2\ \textit{implies}\ y_1 < y_2. \tag{2.116}$$

Proof. (2.114) By (2.25) and linearity we have $\min(x, x \to 0) = x \wedge (x \to 0) = 0$. Since $x > 0$, $x \to 0 = 0$ follows.

(2.115) By (2.114), $\neg\neg x = 1$, therefore, by (2.24) $((x \otimes y_1) \to (x \otimes y_2)) \to (y_1 \to y_2) = 1$ which yields $((x \otimes y_1) \to (x \otimes y_2)) \leq (y_1 \to y_2)$. By assumption $(x \otimes y_1) \to (x \otimes y_2) = 1$, i.e. $y_1 \to y_2 = 1$, whence $y_1 \leq y_2$. By symmetry, $y_2 \leq y_1$, thus $y_1 = y_2$.

(2.116) The argument in the proof of (2.115) shows that $y_1 \leq y_2$. Clearly, $y_1 = y_2$ cannot be the case (since it implies $x \otimes y_1 = x \otimes y_2$), and so $y_1 < y_2$. □

Theorem 2.47 (linear G-algebras) *The following are true in each linearly ordered G-algebra:*

$$x < y \quad \textit{implies}\ y \to x = x, \tag{2.117}$$

$$\textit{each subset containing 0 and 1 is a subalgebra.} \tag{2.118}$$

Proof. (2.117) If $x < y$ then x is clearly the largest element such that $y \wedge x \leq x$, i.e. $y \to x = x$.

(2.118) Linearity implies closedness under $\wedge$, $\vee$, and $\otimes(= \wedge)$ while closedness under $\to$ follows from (2.117). □

2.3.4 MORPHISMS, SUBALGEBRAS, AND PRODUCTS OF RESIDUATED LATTICES

Morphisms, congruences, and filters

Since residuated lattices are particular algebras, definitions and results concerning morphisms and congruences on universal algebras (Chapter 1) apply. In addition to these, congruences (and thus also surjective morphisms) correspond to special subsets called filters in the case of residuated lattices. If, moreover, one takes into account general infima and suprema, one gets complete morphisms, congruences, and filters. The details follow.

Definition 2.48 A *morphism* of a residuated lattice $\mathbf{L}_1$ into a residuated lattice $\mathbf{L}_2$ is a mapping $h : L_1 \to L_2$ that is compatible w.r.t. all operations; h is called a *complete morphism* if it is, moreover, compatible with all existing infima and suprema, i.e. if $h(\bigwedge_{i\in I} a_i) = \bigwedge_{i\in I} h(a_i)$ and $h(\bigvee_{i\in I} a_i) = \bigvee_{i\in I} h(a_i)$ whenever the infima and suprema exist.

Remark 2.49 (1) Recall that $h : \mathbf{L}_1 \to \mathbf{L}_2$ (i.e. h is a morphism of $\mathbf{L}_1$ into $\mathbf{L}_2$) means that $h(a \wedge^{\mathbf{L}_1} b) = h(a) \wedge^{\mathbf{L}_2} h(b)$, $h(a \vee^{\mathbf{L}_1} b) = h(a) \vee^{\mathbf{L}_2} h(b)$, $h(a \otimes^{\mathbf{L}_1} b) = h(a) \otimes^{\mathbf{L}_2} h(b)$, $h(a \to^{\mathbf{L}_1} b) = h(a) \to^{\mathbf{L}_2} h(b)$, $h(0^{\mathbf{L}_1}) = 0^{\mathbf{L}_2}$, $h(1^{\mathbf{L}_1}) = 1^{\mathbf{L}_2}$, for all $a, b \in L_1$.

(2) Morphisms are isotone, i.e. $a \leq b$ implies $h(a) \leq h(b)$: if $a \leq b$ then $a \vee b = b$ and thus $h(a) \vee h(b) = h(a \vee b) = h(b)$, i.e. $h(a) \leq h(b)$;

(3) If some operation can be defined using other operations (e.g. $\wedge$ can be defined using $\otimes, \to$ in a divisible residuated lattice: $x \wedge y = x \otimes (x \to y)$), then we only need to verify compatibility of h w.r.t. the defining operations (i.e. to $\otimes$ and $\to$ in the above case). Indeed, taking our example, we get $h(a \wedge b) = h(a \otimes (a \to b)) = h(a) \otimes h(a \to b) = h(a) \otimes (h(a) \to h(b)) = h(a) \wedge h(b)$.

Definition 2.50 A *congruence* on a residuated lattice $\mathbf{L}$ is a binary relation θ on L that is compatible w.r.t. operations of $\mathbf{L}$; θ is called *complete* if it is compatible with all existing suprema and infima, i.e. if $\langle a_i, b_i\rangle \in \theta$ $(i \in I)$ imply $\langle \bigwedge_{i\in I} a_i, \bigwedge_{i\in I} b_i\rangle \in \theta$ and $\langle \bigvee_{i\in I} a_i, \bigvee_{i\in I} b_i\rangle \in \theta$ whenever the infima and suprema exist.

Remark 2.51 (1) We denote by $\mathrm{Con}\,\mathbf{L}$ the set of all congruences on $\mathbf{L}$ and by $\mathrm{Con_C}\,\mathbf{L}$ the set of all complete congruences on $\mathbf{L}$. Note that $\theta \in \mathrm{Con}\,\mathbf{L}$ means that $\langle a_1, b_1\rangle \in \theta$ and $\langle a_2, b_2\rangle \in \theta$ implies $\langle a_1 \wedge a_2, b_1 \wedge b_2\rangle \in \theta$, $\langle a_1 \vee a_2, b_1 \vee b_2\rangle \in \theta$, $\langle a_1 \otimes a_2, b_1 \otimes b_2\rangle \in \theta$, and $\langle a_1 \to a_2, b_1 \to b_2\rangle \in \theta$, for all $a_1, a_2, b_1, b_2 \in L$.

(2) It is easy to see that $\mathrm{Con}\,\mathbf{L}$ and $\mathrm{Con_C}\,\mathbf{L}$ are closed w.r.t. arbitrary intersections, thus (Example 1.2 (3)) $\langle \mathrm{Con}\,\mathbf{L}, \subseteq\rangle$ and $\langle \mathrm{Con_C}\,\mathbf{L}, \subseteq\rangle$ are

complete lattices; $L \times L$ and $\{\langle a,a\rangle \mid a \in L\}$ are the greatest and the least element, respectively.

(3) Similarly as for morphisms, if some operations of $\mathbf{L}$ are definable from others, we do not have to check (in order to verify that a binary relation is a congruence) the compatibility for the definable operations.

Given a congruence θ on a residuated lattice $\mathbf{L}$, one can form a factor algebra $\mathbf{L}/\theta$. Recall from Chapter 1 that the support of $\mathbf{L}/\theta$ is L/θ (i.e. elements of the factor algebra are congruence classes $[a]_\theta$, $a \in L$); operations on $\mathbf{L}/\theta$ are defined by $[a]_\theta \wedge^{\mathbf{L}/\theta} [b]_\theta = [a \wedge^{\mathbf{L}} b]_\theta$, $[a]_\theta \vee^{\mathbf{L}/\theta} [b]_\theta = [a \vee^{\mathbf{L}} b]_\theta$, $[a]_\theta \otimes^{\mathbf{L}/\theta} [b]_\theta = [a \otimes^{\mathbf{L}} b]_\theta$, $[a]_\theta \rightarrow^{\mathbf{L}/\theta} [b]_\theta = [a \rightarrow^{\mathbf{L}} b]_\theta$, $0^{\mathbf{L}/\theta} = [0^{\mathbf{L}}]_\theta$, $1^{\mathbf{L}/\theta} = [1^{\mathbf{L}}]_\theta$. For the sake of brevity, we usually omit the superscripts $\mathbf{L}/\theta$ and $\mathbf{L}$ that distinguish the operations in $\mathbf{L}/\theta$ and $\mathbf{L}$ (there will be no danger of confusion).

Remark 2.52 If $a' \leq b'$ for some $a' \in [a]_\theta$ and $b' \in [b]_\theta$, then $a' \vee b' = b'$ and so $[a]_\theta \vee [b]_\theta = [a']_\theta \vee [b']_\theta = [a' \vee b']_\theta = [b']_\theta = [b]_\theta$, i.e. $[a]_\theta \leq [b]_\theta$. Conversely, if $[a]_\theta \leq [b]_\theta$ then $[a \vee b]_\theta = [a]_\theta \vee [b]_\theta = [b]_\theta$, i.e. we have $a \in [a]_\theta$ and $a \vee b \in [b]_\theta$. To sum up, $[a]_\theta \leq [b]_\theta$ iff $a' \leq b'$ for some $a' \in [a]_\theta$ and $b' \in [b]_\theta$.

Lemma 2.53 *(1) If $\mathbf{L}$ is a complete residuated lattice and θ is a complete congruence on $\mathbf{L}$ then $\mathbf{L}/\theta$ is a complete residuated lattice with infima and suprema described by $\bigwedge_{i\in I}[a_i]_\theta = [\bigwedge_{i\in I} a_i]_\theta$ and $\bigvee_{i\in I}[a_i]_\theta = [\bigvee_{i\in I} a_i]_\theta$.*

(2) If θ is a congruence on a residuated lattice $\mathbf{L}$, then each identity that is valid in $\mathbf{L}$ is also valid in $\mathbf{L}/\theta$. Moreover, if both θ and $\mathbf{L}$ are complete then $\mathbf{L}/\theta$ satisfies each generalized identity that is true in $\mathbf{L}$.

Proof. (1) We show that $[\bigwedge_{i\in I} a_i]_\theta$ is the infimum of $[a_i]_\theta$'s ($i \in I$). Since $\bigwedge_{i\in I} a_i \leq a_j$, we have $[\bigwedge_{i\in I} a_i]_\theta \leq [a_i]_\theta$. If there is some $b \in L$ with $[b]_\theta \leq [a_i]_\theta$ for each $i \in I$, then, by the above remark, there are $b_i \in [b]_\theta$ and $a'_i \in [a_i]_\theta$ with $b_i \leq a'_i$. Therefore, $\bigwedge_{i\in I} b_i \leq \bigwedge_{i\in I} a'_i$. Compatibility yields $\bigwedge_{i\in I} b_i \in [b]_\theta$ and $\bigwedge_{i\in I} a'_i \in [\bigwedge_{i\in I} a_i]_\theta$, whence $[b]_\theta \leq [\bigwedge_{i\in I} a_i]_\theta$ proving that $[\bigwedge_{i\in I} a_i]_\theta$ is the infimum of $[a_i]_\theta$'s. The proof for supremum is similar.

(2) Follows from the well-known fact from universal algebra (factor algebra satisfies all identities satisfied in the original algebra, see Chapter 1); by (1) one easily shows the second part concerning generalized identities. □

The relationship between congruences and morphisms is the subject of the following assertion.

Theorem 2.54 *(1) For a morphism $h : \mathbf{L}_1 \to \mathbf{L}_2$ and a congruence $\theta \in \mathrm{Con}\,\mathbf{L}$, θ_h is a congruence on $\mathbf{L}_1$ which is complete whenever h is complete; h_θ is a surjective morphism of $\mathbf{L}$ to $\mathbf{L}/\theta$ which is complete provided θ is complete; $\theta = \theta_{h_\theta}$ and if h is surjective then there is an isomorphism $g : \mathbf{A}/\theta \to \mathbf{B}$ such that $h = h_{\theta_h} \circ g$.*

(2) If $h : \mathbf{L}_1 \to \mathbf{L}_2$ is a (complete) morphism of (complete) residuated lattices then there is a (complete) residuated lattice $\mathbf{L}$, a (complete) epimorphism $g_1 : \mathbf{L}_1 \to \mathbf{L}$, and a (complete) monomorphism $g_2 : \mathbf{L} \to \mathbf{L}_2$ such that $h = g_1 \circ g_2$ (in fact, one may take $\mathbf{L} = \mathbf{L}_1/\theta_h$, $g_1(a) = [a]_{\theta_h}$, and $g_2([a]_{\theta_h}) = h(a)$).

Proof. (1) Due to Proposition 1.8, we only have to verify that if h and θ are compatible with arbitrary infima and suprema then so are θ_h and h_θ. Let thus $\langle a_i, b_i \rangle \in \theta_h$ $(i \in I)$: by definition, $h(a_i) = h(b_i)$, thus $h(\bigwedge_{i\in I} a_i) = \bigwedge_{i\in I} h(a_i) = \bigwedge_{i\in I} h(b_i) = h(\bigwedge_{i\in I} b_i)$, i.e. $\langle \bigwedge_{i\in I} a_i, \bigwedge_{i\in I} b_i \rangle \in \theta_h$. The rest of the proof is equally easy.

(2) Follows directly from Proposition 1.8 and from (1). □

Next we show a bijective correspondence of congruences to so-called filters.

Definition 2.55 A *filter* in $\mathbf{L}$ is a nonempty subset $F \subseteq L$ such that (i) $a \otimes b \in F$ whenever $a, b \in F$ and (ii) $b \in F$ whenever $a \leq b$ for some $a \in F$. F is called *complete* if $\bigwedge_{i\in I} a_i \in F$ for any $a_i \in F$ $(i \in I)$ such that $\bigwedge_{i\in I} a_i$ exists in $\mathbf{L}$.

Clearly, $1 \in F$ for any filter. In a sense, filters are sets of truth values located around 1.

Theorem 2.56 *For any (complete) congruence θ on $\mathbf{L}$, $F_\theta = [1]_\theta$ is a (complete) filter. For any (complete) filter F in $\mathbf{L}$, θ_F, defined by $\langle a, b \rangle \in \theta_F$ iff $a \leftrightarrow b \in F$, is a (complete) congruence on $\mathbf{L}$. Moreover, $\theta = \theta_{F_\theta}$ and $F = F_{\theta_F}$.*

Proof. Let θ be a congruence. If $a, b \in [1]_\theta$ then $\langle a, 1 \rangle, \langle b, 1 \rangle \in \theta$ and so, by compatibility of θ, $\langle a \otimes b, 1 \rangle = \langle a \otimes b, 1 \otimes 1 \rangle \in \theta$, i.e. $a \otimes b \in [1]_\theta$. If $a \in [1]_\theta$ and $a \leq b$ then $\langle b, 1 \rangle \in \theta$: Indeed, by reflexivity and compatibility of θ we get $\langle b, b \rangle \in \theta$ and $\langle a, b \rangle = \langle a \wedge b, 1 \wedge b \rangle \in \theta$, thus by transitivity of θ, $\langle b, 1 \rangle \in \theta$, i.e. $b \in [1]_\theta$ proving that $F_\theta = [1]_\theta$ is a filter. If θ is complete, then $a_i \in [1]_\theta$ $(i \in I)$, i.e. $\langle a_i, 1 \rangle \in \theta$, yields $\langle \bigwedge_{i\in I} a_i, 1 \rangle \in \theta$, i.e. $\bigwedge_{i\in I} a_i \in [1]_\theta$, proving that F_θ is complete.

Let F be a filter, $\langle a_i, b_i \rangle \in \theta_F$, i.e. $a_i \leftrightarrow b_i \in F$, $i = 1, 2$. Compatibility of θ_F is now a direct consequence of $x \otimes y \leq x \wedge y$ and Theorem 2.31

((2.84)–(2.87)); if F is complete, then compatibility with arbitrary infima and suprema follows from Theorem 2.31 ((2.88) and (2.89)). For instance, (2.84) yields $(a_1 \leftrightarrow b_1)\otimes(a_2 \leftrightarrow b_2) \leq (a_1 \leftrightarrow b_1) \wedge (a_2 \leftrightarrow b_2) \leq (a_1 \wedge a_2) \leftrightarrow (b_1 \wedge b_2)$. As $a_1 \leftrightarrow b_1, a_2 \leftrightarrow b_2 \in F$, $(a_1 \leftrightarrow b_1)\otimes(a_2 \leftrightarrow b_2) \in F$ and thus also $(a_1 \wedge a_2) \leftrightarrow (b_1 \wedge b_2) \in F$, i.e. $\langle a_1 \wedge a_2, b_1 \wedge b_2\rangle \in \theta_F$.

$a \in F_{\theta_F}$ iff $\langle a, 1\rangle \in \theta_F$ iff $a = a \leftrightarrow 1 \in F$, i.e. $F = F_{\theta_F}$. $\langle a, b\rangle \in \theta_{F_\theta}$ iff $a \leftrightarrow b \in F_\theta$ iff $\langle a \leftrightarrow b, 1\rangle \in \theta$. Now, $\langle a \leftrightarrow b, 1\rangle \in \theta$ is equivalent to $\langle a, b\rangle \in \theta$. Indeed, $\langle a, b\rangle \in \theta$ implies $\langle a \leftrightarrow b, 1\rangle = \langle a \leftrightarrow b, b \leftrightarrow b\rangle \in \theta$. Conversely, let $\langle a \leftrightarrow b, 1\rangle \in \theta$. Observe first that $\langle x, y_1\rangle, \langle x, y_2\rangle \in \theta$, $y_1 \leq y \leq y_2$ implies $\langle x, y\rangle \in \theta$ (for $\langle x, y\rangle = \langle x \vee (x \wedge y), y_1 \vee (y_2 \wedge y)\rangle \in \theta$). Therefore, we have $\langle a \rightarrow b, 1\rangle, \langle b \rightarrow a, 1\rangle \in \theta$, and so $\langle b\otimes(b \rightarrow a), b\otimes 1\rangle = \langle b\otimes(b \rightarrow a), b\rangle \in \theta$ and $\langle (a \rightarrow b) \rightarrow b, 1 \rightarrow b\rangle = \langle (a \rightarrow b) \rightarrow b, b\rangle \in \theta$. Using $b\otimes(b \rightarrow a) \leq a \leq (a \rightarrow b) \rightarrow b$ we get $\langle a, b\rangle \in \theta$. □

Remark 2.57 (1) Due to a bijective correspondence between filters and congruences we may also denote by $\mathbf{L}/F$ the factor algebra $\mathbf{L}/\theta_F$ (F a filter).

(2) Note that Theorem 2.56 implies that each congruence on a residuated lattice $\mathbf{L}$ is determined by its class containing 1, i.e., if $[1]_{\theta_1} = [1]_{\theta_2}$ (for $\theta_1, \theta_2 \in \mathrm{Con}\,\mathbf{L}$) then $\theta_1 = \theta_2$. A consequence of this is that if θ is a congruence with $[1]_\theta = \{1\}$ then θ is the identity (i.e. $\langle x, y\rangle \in \theta$ iff $x = y$). It has to be noted that [18, 25] imply that if $\mathbf{L}$ satisfies the law of double negation then each congruence is determined by its class containing 0; if $\mathbf{L}$ is an MV-algebra then each congruence is determined by any of its classes.

Example 2.58 (1) Let $\mathbf{L}$ be the standard Gödel algebra on $[0, 1]$. Filters on $\mathbf{L}$ are exactly subsets $[a, 1]$ (closed intervals) or $(a, 1]$ (intervals open from left) for $a \in [0, 1]$ (except for $(1, 1]$). Indeed, if F is a filter, take $a = \bigwedge F$. Then any b such that $a < b$ belongs to F (otherwise a is not the infimum) and no b with $b < a$ belongs to F. Either $a \in F$ and so $F = [a, 1]$ (in this case, F is a complete filter), or $a \notin F$ and so $F = (a, 1]$. Clearly, $[a, 1]$ and $(a, 1]$ are filters (closed w.r.t. $\otimes$, i.e., w.r.t. min).

The corresponding congruences are $\theta_{[a,1]} = \{\langle x, y\rangle \mid a \leq x \leftrightarrow y\} = \{\langle x, x\rangle \mid x < a\} \cup [a, 1] \times [a, 1]$ and $\theta_{(a,1]} = \{\langle x, y\rangle \mid a < x \leftrightarrow y\} = \{\langle x, x\rangle \mid x \leq a\} \cup (a, 1] \times (a, 1]$.

Similarly for a subalgebra $\mathbf{L}'$ of $\mathbf{L}$: Filters are either $[a, 1] \cap L'$ or $(a, 1] \cap L'$.

(2) In the standard product algebra on $[0, 1]$, the only filters are $\{1\}$, $(0, 1]$, and $[0, 1]$. Indeed, if F is a filter and $a \in F$, $0 \neq a \neq 1$, then for

any $b > 0$ there is some n with $a^n < b$ (power w.r.t. $\otimes$), thus $F = [0, 1]$ or $F = (0, 1]$.

(3) There are only trivial filters in the standard Łukasiewicz algebra on $[0, 1]$, i.e. the only filters are $\{1\}$ and $[0, 1]$. Indeed, if F is a filter and $a \in F$, $0 \neq a \neq 1$, then there is some n with $a^n = 0$ (power w.r.t. $\otimes$), whence $F = [0, 1]$.

In the following, we concentrate on important special types of filters (prime and maximal) and their basic properties.

Definition 2.59 A filter F is called *prime* if for each $a, b \in L$, $a \vee b \in F$ implies $a \in F$ or $b \in F$.

Lemma 2.60 *In a residuated lattice* $\mathbf{L}$ *satisfying prelinearity, a filter* F *is prime iff for each* $a, b \in L$ *we have* $a \to b \in F$ *or* $b \to a \in F$.

Proof. Let F be prime. Since $(a \to b) \vee (b \to a) = 1 \in F$, primality of F implies $a \to b \in F$ or $b \to a \in F$.

Conversely, let F satisfy the condition of lemma, let $a \vee b \in F$. Then $a \to b \in F$ or $b \to a \in F$. If $a \to b \in F$ then from $a \vee b \in F$ and $a \vee b \leq (a \to b) \to b$ we have $(a \to b) \to b \in F$ and thus also $(a \to b) \otimes ((a \to b) \to b) \in F$. Using $(a \to b) \otimes ((a \to b) \to b) \leq b$ we get $b \in F$. If $b \to a \in F$, one can proceed analogously. □

Lemma 2.61 *If* $\mathbf{L}$ *satisfies prelinearity, a filter* F *in* $\mathbf{L}$ *is prime iff* $\mathbf{L}/F$ *is a chain (i.e. linearly ordered).*

Proof. Take any $a, b \in L$. F is prime iff $a \to b \in F = [1]_{\theta_F}$ or $b \to a \in F = [1]_{\theta_F}$ iff $[a \to b]_{\theta_F} = [1]_{\theta_F}$ or $[b \to a]_{\theta_F} = [1]_{\theta_F}$ iff $[a]_{\theta_F} \to [b]_{\theta_F} = [1]_{\theta_F}$ or $[b]_{\theta_F} \to [a]_{\theta_F} = [1]_{\theta_F}$ iff $[a]_{\theta_F} \leq [b]_{\theta_F}$ or $[b]_{\theta_F} \leq [a]_{\theta_F}$ iff $\mathbf{L}/F$ is a chain. □

Definition 2.62 A filter $F \neq L$ is called *maximalmaximal filter* if $F' = L$ for any filter $F' \supset F$.

That is, maximal filters are maximal elements in the set of all filters different from L ordered by set inclusion. We say that a residuated lattice is *locally finite* if each $a < 1$ is *nilpotent*, i.e. for each $a < 1$ there is a positive integer n with $a^n = 0$.

Lemma 2.63 *A filter* F *in* $\mathbf{L}$ *is maximal iff* $\mathbf{L}/F$ *is locally finite.*

Proof. Let F be maximal, take $[a]_{\theta_F} \neq [1]_{\theta_F}$ (thus $a \notin F$). Since F is maximal, there are n and $b \in F$ with $a^n \otimes b = 0$ (otherwise F is not maximal for $\{c \mid a^n \otimes b \leq c,\ b \in F, n \in \mathbf{N}\} \neq L$ would be a filter containing F). We thus have $b \leq a^n \to 0$, i.e. $a^n \leftrightarrow 0 = a^n \to 0 \in F$, i.e. $[a]^n_{\theta_F} = [a^n]_{\theta_F} = [0]_{\theta_F}$, i.e. $\mathbf{L}/F$ is locally finite.

Conversely, let $\mathbf{L}/F$ be locally finite, take a filter $G \supset F$ and $a \in G - F$. By local finiteness, $[a]^n_{\theta_F} = [0]_{\theta_F}$ for some n which means $a^n \leftrightarrow 0 = a^n \to 0 \in F \subseteq G$. By $a^n \in G$ and $a^n \otimes (a^n \to 0) \leq 0$ we get $0 \in G$, thus $G = L$. Therefore, F is maximal. □

Lemma 2.64 *(i) Each filter different from L is contained in a maximal filter. (ii) Each maximal filter is prime. (iii) For any $a \neq 1$ there exists a prime filter F with $a \notin F$.*

Proof. (i) follows by Zorn lemma: An immediate verification shows that a union of a chain of filters that are different from L is a filter different from L, thus, each filter is contained in some maximal filter.

(ii) Let F be a maximal filter. If F is not prime, there are $a, b \notin F$ with $a \vee b \in F$. Maximality of F again implies (proof of Lemma 2.63) that there are $c, d \in F$ and m, n with $a^m \otimes c = 0$ and $b^n \otimes d = 0$. But then, $(a \vee b)^{m+n} \otimes c \otimes d \leq a^m \otimes c \otimes b^n \otimes d = 0$ and $(a \vee b)^{m+n} \otimes c \otimes d \in F$ imply $0 \in F$, a contradiction with maximality of F.

(iii) The set $\mathcal{F}_a$ of all filters that do not contain a is nonempty (it contains $\{1\}$) and forms a partially ordered set w.r.t. $\subseteq$. Clearly, the union of any chain in $\mathcal{F}_a$ belongs to $\mathcal{F}_a$ and thus, Zorn lemma implies that $\mathcal{F}_a$ contains some maximal element F. We claim that F is prime. Indeed, if not, there are $b, b' \notin F$ such that $b \vee b' \in F$. Maximality of F in $\mathcal{F}_a$ yields $c, d \in F$ and m, n with $b^m \otimes c \leq a$ and $b'^n \otimes d \leq a$. Therefore, $(b \vee b')^{m+n} \otimes c \otimes d \leq b^m \otimes c \otimes b'^n \otimes d \leq a$ which implies $a \in F$, a contradiction. □

Remark 2.65 Note that one could define a maximal complete filter as a filter $F \neq L$ that is complete and satisfies $F' = L$ for any complete filter $F' \supset F$. Then, for instance, Lemma 2.64 (i) is not true: Consider the standard Gödel algebra on $[0, 1]$. Complete filters are $[a, 1]$ for $a \in [0, 1]$. For any $0 < a \leq 1$, $[\frac{a}{2}, 1]$ is a complete complete filter larger than $[a, 1]$ but different from $[0, 1]$. Thus there is no maximal complete filter. Note also that the proof of Lemma 2.64 (i) fails in that a union of a chain of complete filters need not be a complete filter; to see this, consider the chain $\{[a, 1] \mid 0 < a \leq 1\}$.

Products

Recall from Chapter 1 that the direct product of residuated lattices $\mathbf{L}_i$ $(i \in I)$ is a residuated lattice $\times_{i\in I}\mathbf{L}_i$ whose support is $\times_{i\in I}L_i$ and operations are defined componentwise (i.e. for $a = \langle \dots, a(i), \dots \rangle, b = \langle \dots, b(i), \dots \rangle \in \times_{\in I}L_i$ we have $(a \vee b)(i) = a(i) \vee b(i)$ etc.). It is easy to see that $\times_{i\in I}\mathbf{L}_i$ is complete iff each $\mathbf{L}_i$ is complete. In this case we have for $a_j \in \times_{i\in I}\mathbf{L}_i$ $(j \in J)$, $(\bigwedge_{j\in J} a_j)(i) = \bigwedge_{j\in J} a_j(i)$ and $(\bigvee_{j\in J} a_j)(i) = \bigvee_{j\in J} a_j(i)$. It is immediate that each generalized identity is true in $\times_{i\in I}\mathbf{L}_i$ iff it is true in each $\mathbf{L}_i$.

Subalgebras

Definition 2.66 Let $\mathbf{L}$ be a residuated lattice. A subset $A \subseteq L$ is called a *subuniverse* if it is closed w.r.t. each operation of $\mathbf{L}$. If, moreover, for every $a_i \in A$ $(i \in I)$, both $\bigwedge_{i\in I} a_i$ and $\bigvee_{i\in I} a_i$ belong to A whenever they exist in $\mathbf{L}$, A is called a *complete subuniverse.*

Recall from Chapter 1 that the set of all subuniverses of $\mathbf{L}$ is denoted by $\mathrm{Sub}\,\mathbf{L}$; that is, $A \in \mathrm{Sub}\,\mathbf{L}$ iff $0, 1 \in A$, and $a, b \in A$ imply $a \wedge b, a \vee b, a \otimes b, a \rightarrow b \in A$. The set of all complete subuniverses of $\mathbf{L}$ will be denoted by $\mathrm{Sub_C}\,\mathbf{L}$. Note that $\mathrm{Sub}\,\mathbf{L} \supseteq \mathrm{Sub_C}\,\mathbf{L}$ and that we always have $\{0, 1\} \in \mathrm{Sub_C}\,\mathbf{L}$. Any $\emptyset \neq A \in \mathrm{Sub}\,\mathbf{L}$ equipped with operations that are restrictions of the operations of $\mathbf{L}$ to A is called a *subalgebra* of $\mathbf{L}$; if, moreover, $A \in \mathrm{Sub_C}\,\mathbf{L}$, the subalgebra is called *complete*.

Remark 2.67 (1) It is easy to verify that both $\mathrm{Sub}\,\mathbf{L}$ and $\mathrm{Sub_C}\,\mathbf{L}$ are closed w.r.t. arbitrary intersections. Therefore, both $\langle \mathrm{Sub}\,\mathbf{L}, \subseteq \rangle$ and $\langle \mathrm{Sub_C}\,\mathbf{L}, \subseteq \rangle$ are complete lattices. We may thus speak of the least subalgebra $[A]$ containing a given subset $A \subseteq L$ (it is the intersection of all $B \in \mathrm{Sub}\,\mathbf{L}$ such that $A \subseteq B$, Example 1.2 (3)); similarly for the least complete subalgebra.

(2) Note that for (complete) morphisms $h_1 : \mathbf{L}_1 \to \mathbf{L}_2$, $h_2 : \mathbf{L}_2 \to \mathbf{L}_3$, $h_1(\mathbf{L}_1)$ (i.e. $h_1(L_1) = \{h(a) \mid a \in L_1\}$ equipped with restrictions of operations from $\mathbf{L}_2$) is a (complete) subalgebra of $\mathbf{L}_2$, $h_1^{-1}(\mathbf{L}_2)$ (i.e. $h_1^{-1}(L_2) = \{a \in L_1 \mid h(a) \in L_2\}$ equipped with restrictions of operations from $\mathbf{L}_1$) is a (complete) subalgebra of $\mathbf{L}_1$, and $h_1 \circ h_2$ (the composition of h_1 and h_2) is a (complete) morphism of $\mathbf{L}_1$ into $\mathbf{L}_3$.

Subdirect products and subdirect representation

Definition 2.68 A residuated lattice $\mathbf{L}$ is called a *subdirect product* of residuated lattices $\mathbf{L}_i$ $(i \in I)$ if $\mathbf{L}$ is a subalgebra of the direct product $\times_{i\in I}\mathbf{L}_i$ such that for each $j \in I$ we have $\mathrm{pr}_j(\times_{i\in I}L_i) = L_j$. If $\mathbf{L}$ is, moreover, a complete subalgebra of $\times_{i\in I}L_i$ then it is called a *complete subdirect product* of $\mathbf{L}_i$'s.

Linearity of a structure of truth values is a particularly appealing property (any two truth degrees comparable). A question arises whether each residuated lattice can be in some way constructed from linear ones, i.e. whether or not linearly ordered structures of truth values are in some way the basic ones. The direct product of nontrivial (i.e. having at least two elements) residuated lattices is clearly not linearly ordered; direct product should therefore play a role. In order not to restrict the cardinality to be a product of cardinalities, one should take subalgebras of direct products into consideration. Obviously, direct product of linearly ordered residuated lattices satisfies prelinearity. The next theorem shows that prelinearity is in fact a necessary and sufficient property for a residuated lattice to be representable as a subalgebra of a direct product of linearly ordered residuated lattices.

Theorem 2.69 (subdirect representation) *A residuated lattice* $\mathbf{L}$ *is isomorphic to a subdirect product of linearly ordered residuated lattices iff it satisfies prelinearity. Moreover, the linear residuated lattices can be chosen so that they satisfy all identities satisfied by* $\mathbf{L}$.

Proof. Any subdirect product of residuated lattices satisfies each identity in each of these residuated lattices. Since prelinearity holds true in linear residuated lattices, the "$\Rightarrow$" part follows.

Conversely, let $\mathbf{L}$ satisfy prelinearity. Denote by $P(\mathbf{L})$ the set of all prime filters of $\mathbf{L}$. By Lemma 2.61, $\mathbf{L}/F$ is linearly ordered for each $F \in P(\mathbf{L})$; furthermore, each identity satisfied by $\mathbf{L}$ is satisfied by any $\mathbf{L}/F$ (satisfaction of identities is preserved by formation of factor algebras). Clearly, the mapping $h : L \to \times_{F \in P(\mathbf{L})} L/F$ given by $h(a) = \langle \dots, [a]_{\theta_F}, \dots \rangle$ is a morphism of $\mathbf{L}$ into $\times_{F \in P(\mathbf{L})} \mathbf{L}/F$ and thus $h(\mathbf{L})$ is a subalgebra of $\times_{F \in P(\mathbf{L})} \mathbf{L}/F$. Moreover, h is injective: if $a \neq b$ then $a \leftrightarrow b \neq 1$ and thus, Lemma 2.64 (iii) implies the existence of $F \in P(\mathbf{L})$ with $a \leftrightarrow b \notin F$, i.e. $[a]_{\theta_F} \neq [b]_{\theta_F}$ and so $h(a) \neq h(b)$. Since $[a]_{\theta_F} = \mathrm{pr}_F(h(a))$, $h(\mathbf{L})$ is a subdirect product of $\mathbf{L}/F$'s ($F \in P(\mathbf{L})$) establishing the assertion. □

Example 2.70 We show a simple application of Theorem 2.69. Instead of our definition of biresiduum in residuated lattices, one sometimes uses $a \sim b = (a \to b) \otimes (b \to a)$. This is, for example, how Hájek defined equivalence connective in his Basic Logic [105]. We show that in a residuated lattice $\mathbf{L}$ which satisfies prelinearity, $\leftrightarrow$ (biresiduum) and $\sim$ (alternative biresiduum) coincide. By Theorem 2.69, $\mathbf{L}$ is a subdirect product of linear residuated lattices $\mathbf{L}_i$'s. Obviously, if an identity is true in all $\mathbf{L}_i$'s, it is true also in $\mathbf{L}$. It is thus sufficient to show that

$a \leftrightarrow b = a \sim b$ holds in each linear residuated lattice. However, the linearity assumption implies that either $a \leq b$ or $b \leq a$. However, if $a \leq b$ then $a \leftrightarrow b = b \rightarrow a = a \sim b$; similarly for $b \leq a$, completing the proof. Thus, one could equivalently define equivalence in Basic Logic by connecting implications by $\wedge$ instead of $\otimes$.

Special subalgebras

In the following, we concentrate on selected problems of the following kind. We take a residuated lattice (satisfying additional conditions) and want to know its largest subalgebra satisfying some further conditions. Denote

$$D(\mathbf{L}) = \{a \in L \mid a = \neg\neg a\},$$

the set of all elements satisfying the law of double negation, and

$$H(\mathbf{L}) = \{a \in L \mid a = a \otimes a\},$$

the set of all elements idempotent w.r.t. conjunction.

Theorem 2.71 *If $\mathbf{L}$ is a Heyting algebra then $D(\mathbf{L})$ is (w.r.t. the order inherited from $\mathbf{L}$) a Boolean lattice, where the meet $\cap$ coincides with the meet $\wedge$ of $\mathbf{L}$ and for the join $\cup$ we have*

$$a \cup b = \neg\neg(a \vee b) = \neg(\neg a \wedge \neg b).$$

Proof. Clearly, $D(\mathbf{L}) = \{\neg a \mid a \in L\}$. Since $\neg a \wedge \neg b = \neg(a \vee b)$, we have $\neg a \wedge \neg b \in D(\mathbf{L})$, hence $\cap = \wedge$. Let $a, b, c \in D(\mathbf{L})$. If $a, b \leq c$ then $a \vee b \leq c$, whence $\neg\neg(a \vee b) \leq \neg\neg c = c$, and so $a \cup b = \neg\neg(a \vee b)$. By (2.76), $\neg\neg(a \vee b) = \neg(\neg a \wedge \neg b)$.

Now, for any $a \in D(\mathbf{L})$ we have $a \cap \neg a = a \wedge \neg a = 0$ and $a \cup \neg a = \neg(\neg a \wedge \neg\neg a) = \neg 0 = 1$. Since $D(\mathbf{L})$ is distributive (Theorem 2.39 and the fact that a Heyting algebra is divisible), the assertion follows by definition of a Boolean lattice. □

Remark 2.72 In the proof of Theorem 2.71 we showed that for $a, b \in D(\mathbf{L})$ ($\mathbf{L}$ a Heyting algebra), we have $a \cup b = \neg(\neg a \wedge \neg b)$. In fact, a stronger assertion holds: for any $a, b \in L$, we have $\neg a \cup \neg b = \neg(a \wedge b)$. Indeed, since $\neg a \cup \neg b = \neg\neg(\neg a \vee \neg b) = \neg(\neg\neg a \wedge \neg\neg b)$, we have to check $\neg(a \wedge b) = \neg(\neg\neg a \wedge \neg\neg b)$. As $a \leq \neg\neg a$, $b \leq \neg\neg b$, antitony of $\neg$ $\neg(a \wedge b) \geq \neg(\neg\neg a \wedge \neg\neg b)$. We verify the converse inequality: $\neg(a \wedge b) \leq \neg(\neg\neg a \wedge \neg\neg b)$ holds iff $\neg(a \wedge b) \wedge \neg\neg a \wedge \neg\neg b \leq 0$. Since $\neg(a \wedge b) \wedge a \wedge b \wedge \neg\neg a \wedge \neg\neg b \leq 0$, we have $\neg(a \wedge b) \wedge a \wedge \neg\neg a \wedge \neg\neg b \leq \neg b$ which together with $\neg(a \wedge b) \wedge a \wedge \neg\neg a \wedge \neg\neg b \leq \neg\neg b$ yields $\neg(a \wedge b) \wedge a \wedge \neg\neg a \wedge \neg\neg b \leq$

$\neg b \wedge \neg\neg b = 0$. Thus, by adjointness, $\neg(a \wedge b) \wedge \neg\neg a \wedge \neg\neg b \leq \neg a$. Putting together with $\neg(a \wedge b) \wedge \neg\neg a \wedge \neg\neg b \leq \neg\neg a$ yields $\neg(a \wedge b) \wedge \neg\neg a \wedge \neg\neg b \leq \neg a \wedge \neg\neg a = 0$, the required inequality.

Theorem 2.73 *If* $\mathbf{L}$ *is a BL-algebra then* $H(\mathbf{L})$ *is the largest subalgebra of* $\mathbf{L}$ *that is a G-algebra.*

Proof. First, $0, 1 \in H(\mathbf{L})$. Now, observe that if $a \in H(\mathbf{L})$ then $a \otimes b = a \wedge b$ for any $b \in L$. Indeed, $a \wedge b = a \otimes (a \to b) = a \otimes a \otimes (a \to b) = a \otimes (a \wedge b) \leq a \otimes b$; $a \otimes b \leq a \wedge b$ follows from the isotony of $\otimes$. We prove that $H(\mathbf{L})$ is a subalgebra. Take any $a, b \in H(\mathbf{L})$. Since $\otimes$ is distributive over $\wedge$, we have $(a \wedge b) \otimes (a \wedge b) = (a \otimes a) \wedge (a \otimes b) \wedge (b \otimes b) = a \wedge b$, i.e. $H(\mathbf{L})$ is closed under $\wedge$. Furthermore, $(a \vee b) \otimes (a \vee b) = (a \otimes a) \vee (a \otimes b) \vee (b \otimes b) = a \vee (a \wedge b) \vee b = a \vee b$, i.e. $H(\mathbf{L})$ is closed under $\vee$. Finally, $(a \otimes b) \otimes (a \otimes b) = (a \otimes a) \otimes (b \otimes b) = a \otimes b$, proving closedness under $\otimes$. We prove that $H(\mathbf{L})$ is closed under $\to$: By Theorem 2.69, each BL-algebra is a subdirect product of linearly ordered BL-algebras. We may therefore safely assume that $\mathbf{L}$ is linearly ordered. If $a \leq b$ then $a \to b = 1 \in H(\mathbf{L})$. Let $a > b$. We show that $a \to b = b$. Since $b \leq a \to b$ is always true, it suffices to show that $b < a \to b$ is impossible. Let then $b < a \to b$. Since $a \in H(\mathbf{L})$, we have $a \wedge (a \to b) = a \otimes (a \to b) \leq b$. By linearity of $\mathbf{L}$, $a \wedge (a \to b) = \min(a, a \to b) > b$, a contradiction.

If $H' \supseteq H(\mathbf{L})$ is another subalgebra of $\mathbf{L}$ that is a G-algebra then for any $a \in H'$, $a \otimes a = a$, i.e. $a \in H(\mathbf{L})$, thus $H' = H(\mathbf{L})$. This proves that $H(\mathbf{L})$ is the largest subalgebra that is a G-algebra. □

Theorem 2.74 *If* $\mathbf{L}$ *is a BL-algebra then* $D(\mathbf{L})$ *is the largest subalgebra of* $\mathbf{L}$ *that is an MV-algebra.*

Proof. First, we show that $D(\mathbf{L})$ is a subalgebra of $\mathbf{L}$. Since $\neg x = \neg\neg\neg x$ is valid in $\mathbf{L}$, $D(\mathbf{L}) = \{\neg a \mid a \in L\}$. Clearly, $0, 1 \in D(\mathbf{L})$. Since $(a \to 0) \wedge (b \to 0) = (a \vee b) \to 0$, $D(\mathbf{L})$ is closed w.r.t. $\wedge$.

To see that $D(\mathbf{L})$ is closed w.r.t. $\vee$, we verify $(a \to 0) \vee (b \to 0) = (a \wedge b) \to 0$: The "$\leq$" part follows by antitony of negation. Conversely, $(a \wedge b) \to 0 = ((a \wedge b) \to 0) \otimes ((a \to b) \vee (b \to a)) = ((a \to b) \otimes ((a \wedge b) \to 0)) \vee ((b \to a) \otimes ((a \wedge b) \to 0)) \leq (a \to 0) \vee (b \to 0)$.

$x \otimes (x \to y) \leq y$ yields $\neg a \to \neg b = \neg(\neg a \otimes b)$ (indeed, applying adjointness to $b \otimes (\neg a \otimes (\neg a \to \neg b)) \leq 0$ and to $(\neg a \otimes b) \otimes ((\neg a \otimes b) \to 0) \leq 0$ gives the "$\leq$" and "$\geq$" inequalities).

Now, introduce a binary operation $\odot$ on $D(\mathbf{L})$ by $a \odot b = \neg\neg(a \otimes b)$. We show that $\langle D(\mathbf{L}), \odot, 1 \rangle$ is a commutative monoid: Clearly, $a \odot b \in D(\mathbf{L})$. Furthermore, $\odot$ is obviously commutative and since $\neg\neg(\neg a \otimes 1) =$

$\neg a$, 1 is its neutral element. To verify associativity, we reason as follows: $\neg\neg(\neg\neg(a\otimes b)\otimes c) \leq \neg\neg(a\otimes\neg\neg(b\otimes c))$ iff $\neg(a\otimes\neg\neg(b\otimes c)) \leq \neg(\neg\neg(a\otimes b)\otimes c)$ iff $\neg\neg(a\otimes b)\otimes c\otimes\neg(a\otimes\neg\neg(b\otimes c)) \leq 0$ iff $c\otimes\neg(a\otimes \neg\neg(b\otimes c)) \leq \neg\neg\neg(a\otimes b) = \neg(a\otimes b)$ iff $a\otimes b\otimes c\otimes\neg(a\otimes\neg\neg(b\otimes c)) \leq 0$ which follows from $b\otimes c \leq \neg\neg(b\otimes c)$. We proved $(a\odot b)\odot c \leq a\odot(b\odot c)$, the converse inequality is symmetric. Therefore, $\langle D(\mathbf{L}),\odot,1\rangle$ is a commutative monoid.

Furthermore, as $\neg a\to\neg b = \neg(\neg a\otimes b)$, $D(\mathbf{L})$ is closed under $\to$. We now verify that $\odot$ and $\to$ satisfy adjointness: Since $a\otimes b \leq \neg\neg(a\otimes b)$, $a\odot b\leq c$ implies $a\leq b\to c$ by adjointness of $\otimes$ and $\to$. If $a\leq b\to c$ then $a\otimes b\leq c$, and so $a\odot b = \neg\neg(a\otimes b)\leq\neg\neg c = c$. Both $\langle\otimes,\to\rangle$ and $\langle\odot,\to\rangle$ satisfy adjointness; Remark 2.2 (2) thus implies $\otimes = \odot$. Therefore, $D(\mathbf{L})$ is a subalgebra of $\mathbf{L}$. Obviously, $D(\mathbf{L})$ satisfies $x = \neg\neg x$ and so $D(\mathbf{L})$ is an MV-algebra. It is the largest MV-algebra contained in $\mathbf{L}$ as a subalgebra since otherwise there is an $a\in L - D(\mathbf{L})$ such that $a = \neg\neg a$, a contradiction to the definition of $D(\mathbf{L})$. □

Remark 2.75 Note that divisibility of $\mathbf{L}$ was not used in the proof of Theorem 2.74 and thus $D(\mathbf{L})$ is the largest subalgebra satisfying the law of double negation for any residuated lattice satisfying prelinearity.

Theorem 2.76 (1) *If $\mathbf{L}$ is an MV-algebra then $D(\mathbf{L}) = L$ and $H(\mathbf{L})$ is the largest subalgebra of $\mathbf{L}$ that is a Boolean algebra.*

(2) *If $\mathbf{L}$ is a Gödel algebra then $H(\mathbf{L}) = L$ and $D(\mathbf{L})$ is the largest subalgebra of $\mathbf{L}$ that is a Boolean algebra.*

(3) *If $\mathbf{L}$ is a product algebra then $D(\mathbf{L}) = H(\mathbf{L})$ is the largest subalgebra of $\mathbf{L}$ that is a Boolean algebra.*

Proof. (1): If $\mathbf{L}$ is an MV-algebra then obviously $D(\mathbf{L}) = L$. The second part follows directly from Theorem 2.73.

(2): Analogously, $H(\mathbf{L}) = L$ for a Gödel algebra $\mathbf{L}$; the assertion follows from Theorem 2.74.

(3): By Theorem 2.69, every product algebra $\mathbf{L}$ is a subdirect product of linearly ordered product algebras $\mathbf{L}_i$'s. We identify each $a\in L$ with the corresponding element $\langle\ldots,a_i,\ldots\rangle$ of the direct product of $\mathbf{L}_i$'s.

Let $\mathbf{L}\in\mathcal{P}$. First, we show that $a = \langle\ldots,a_i,\ldots\rangle\in H(\mathbf{L})$ iff $a_i = 0$ or $a_i = 1$ for all i. The right-to-left part is evident. Conversely, let $a\in H(\mathbf{L})$ and $0 < a_i$. Since $\mathbf{L}_i$ is linearly ordered, $\neg a_i = 0$ by Theorem 2.46, thus $\neg\neg a_i = 1$. Therefore, putting $x = 1$, $y = a_i$, and $z = a_i$, $\neg\neg z \leq ((x\otimes z)\to(y\otimes z))\to(x\to y)$ yields $1\leq(a_i\to a_i)\to(1\to a_i)$, thus $a_i = 1$. Therefore, for each i, either $a_i = 0$ or $a_i = 1$.

Second, we verify that $a = \langle \dots, a_i, \dots \rangle \in D(\mathbf{L})$ iff $a_i = 0$ or $a_i = 1$ for all i. Again, the right-to-left part is evident. Conversely, since $\mathbf{L}_i$ is linearly ordered and $a_i \wedge \neg a_i = 0$, $0 < a_i$ implies $\neg a_i = 0$. It follows that $0 < a_i$ and $a_i \in D(\mathbf{L}_i)$ imply $a_i = \neg\neg a_i = 1$. Therefore, $H(\mathbf{L}) = D(\mathbf{L})$, and the claim directly follows by Theorems 2.73 and 2.74. □

Remark 2.77 (1) Note that (1) of Theorem 2.76 can also be proved by the subdirect representation method: $a = \langle \dots, a_i, \dots \rangle \in H(\mathbf{L})$ implies $a_i \in H(\mathbf{L}_i)$, i.e. $a_i \otimes a_i = a_i$. We claim that $a_i = 0$ or $a_i = 1$. By contradiction, let $0 < a_i < 1$. Since $\mathbf{L}_i$ is linearly ordered, $0 < a_i \otimes a_i$ yields $\neg a_i < a_i$ ($a_i \leq \neg a_i$ gives $a_i \otimes \neg a_i = 0$). As $x \vee y = (x \to y) \to y$ and $x \to \neg y = \neg(x \otimes y)$, we conclude $a = a \vee \neg a = (a \to \neg a) \to \neg a = \neg(a \otimes a) \to \neg a = \neg a \to \neg a = 1$, a contradiction to $a < 1$. The rest is clear. In a similar way, one can prove (2) of Theorem 2.76.

(2) A direct consequence of (2) of Theorem 2.76 is that if a Heyting algebra $\mathbf{L}$ satisfies $(x \to y) \vee (y \to x) = 1$ then the join in the Boolean algebra $D(\mathbf{L})$ coincides with the join in $\mathbf{L}$.

We therefore have the following corollary.

Corollary 2.78 *If $\mathbf{L}$ is a BL-algebra then $D(\mathbf{L}) \cap H(\mathbf{L})$ is the largest subalgebra of $\mathbf{L}$ which is a Boolean algebra.*

2.3.5 ALTERNATIVE DEFINITIONS OF SOME CLASSES OF RESIDUATED LATTICES

We turn our attention to alternative definitions of some classes of residuated lattices, namely, MV-algebras and Boolean algebras.

A commonly used definition of MV-algebras is the following one [60]:

Definition 2.79 A *(Chang) MV-algebra* is an algebra $\mathbf{A} = \langle A, \oplus, \neg, 0 \rangle$ such that $\langle A, \oplus, 0 \rangle$ is a commutative monoid and $\neg\neg x = x$, $x \oplus \neg 0 = x$, $\neg(\neg x \oplus y) \oplus y = \neg(\neg y \oplus x) \oplus x$.

Theorem 2.80 *If $\langle A, \oplus, \neg, 0 \rangle$ is a (Chang) MV-algebra then putting $x \otimes y := \neg(\neg x \oplus \neg y)$, $x \to y := \neg x \oplus y$, $1 := \neg 0$, $x \wedge y := \neg(\neg x \oplus \neg(\neg x \oplus y))$, and $x \vee y := \neg(\neg x \oplus y) \oplus y$, $\langle A, \wedge, \vee, \otimes, \to, 0, 1 \rangle$ is an MV-algebra. Conversely, given an MV-algebra $\langle A, \wedge, \vee, \otimes, \to, 0, 1 \rangle$, then letting $\neg x := x \to 0$ and $x \oplus y := \neg(\neg x \otimes \neg y)$, $\langle A, \oplus, 0 \rangle$ is a (Chang) MV-algebra.*

Proof. See [105]. □

Another way to MV-algebras is the following [78].

Definition 2.81 A *Wajsberg algebra* is an algebra $\mathbf{A} = \langle A, \to, 0 \rangle$ satisfying under the stipulation $-x := x \to 0$, $1 := 0 \to 0$ the following identities: $1 \to x = x$, $(x \to y) \to ((y \to z) \to (x \to z)) = 1$, $(-x \to -y) \to (y \to x) = 1$, and $(x \to y) \to y = (y \to x) \to x$.

Theorem 2.82 *If $\langle A, \to, 0 \rangle$ is a Wajsberg algebra then putting $x \otimes y :=$, $x \to y :=$, $1 :=$, $x \wedge y :=$, and $x \vee y :=$, $\langle A, \wedge, \vee, \otimes, \to, 0, 1 \rangle$ is an MV-algebra. Conversely, for an MV-algebra $\langle A, \wedge, \vee, \otimes, \to, 0, 1 \rangle$, $\langle A, \to, 0 \rangle$ is a Wajsberg algebra.*

Proof. See [105]. □

Definition 2.83 An algebra $\mathbf{L} = \langle L, \wedge, \vee, ', 0, 1 \rangle$ is called a Boolean lattice if $\langle L, \wedge, \vee, 0, 1 \rangle$ is a bounded distributive lattice and $'$ is a unary operation satisfying $x \wedge x' = 0$ and $x \vee x' = 1$.

Theorem 2.84 *If $\mathbf{L} = \langle L, \wedge, \vee, ', 0, 1 \rangle$ is a Boolean lattice then letting $x \to y := x' \vee y$, $\mathbf{L}' = \langle L, \wedge, \vee, \wedge, \to, 0, 1 \rangle$ is a Boolean algebra. If $\mathbf{L} = \langle L, \wedge, \vee, \otimes, \to, 0, 1 \rangle$ is a Boolean algebra then $\mathbf{L}' = \langle L, \wedge, \vee, \neg, 0, 1 \rangle$ is a Boolean lattice.*

Proof. Let $\mathbf{L}$ be a Boolean lattice. First we show that $\wedge$ and $\to$ for an adjoint pair: if $x \wedge y \leq z$ then $x \vee y' = (x \wedge y) \vee y' \leq z \vee y'$, and thus $x \leq y' \vee z = y \to z$; conversely $x \leq y \to z = y' \vee z$ implies $x \wedge y \leq (y' \vee z) \wedge y = y \wedge z$, and so $x \wedge y \leq z$. It remains to show that $x \vee y = (x \to y) \to y$: we have $(x \to y) \to y = (x \to y)' \vee y = (x' \vee y)' \vee y = (x \wedge y') \vee y = (x \vee y) \wedge (y' \vee y) = x \vee y$ (we used a well-known identity $(x \vee y)' = x' \wedge y'$ of Boolean lattices [41]).

Conversely, let $\mathbf{L}$ be a Boolean algebra. As $\mathbf{L}$ is an MV-algebra, it is divisible (Theorem 2.42) and thus distributive (Theorem 2.39). (2.73) of Theorem 2.30 implies $x \wedge \neg x = 0$. To show $x \vee \neg x = 1$ it suffices to check $x \vee \neg x = \neg(x \wedge \neg x)$. We have $\neg(x \wedge \neg x) = (x \wedge \neg x) \to 0 = (x \to 0) \wedge (\neg x \to 0) = \neg x \wedge \neg\neg x = \neg x \wedge x$, completing the proof. □

2.4 RESIDUATED LATTICES AND T-NORMS

2.4.1 RESIDUATED LATTICES ON $[0, 1]$

We now turn our attention to residuated lattices defined on [0,1]. It turns out that such residuated lattices are in one-to-one correspondence to left-continuous t-norms. Call a function $f(x, y) : [0, 1] \times [0, 1] \to [0, 1]$

nondecreasing (nonincreasing) in x if for any $a_1, a_2, b \in [0,1]$, $a_1 \leq a_2$ implies $f(a_1, b) \leq f(a_2, b)$ ($f(a_1, b) \geq f(a_2, b)$). The notions nondecreasing (nonincreasing) in y are defined analogously.

Lemma 2.85 *Let $f(x,y) : [0,1] \times [0,1] \to [0,1]$ be a binary operation on $[0,1]$. If f is nondecreasing in x then f is left-continuous (right-continuous) in x iff for any $\{a_j \mid j \in J\} \subseteq [0,1]$ and any $b \in [0,1]$, we have (2.119) ((2.120)) where*

$$f(\bigvee_{j\in J} a_j, b) = \bigvee_{j\in J} f(a_j, b), \tag{2.119}$$

$$f(\bigwedge_{j\in J} a_j, b) = \bigwedge_{j\in J} f(a_j, b). \tag{2.120}$$

Therefore, f is continuous in x iff both (2.119) and (2.120) hold true. We obtain analogous statements if one replaces "in x" by "in y"; a dual statement holds true for f nonincreasing.

Proof. We prove the assertion concerning the left-continuity of a nondecreasing function. The case of right-continuity is symmetric, the case of continuity is clear, and for nonincreasing f, the proof is symmetric. Let f be both nondecreasing and left-continuous in x. Put $a = \bigvee_{j\in J} a_j$ and consider the following two possibilities. First, for each $n \in \mathbf{N}$ there exists some $a_{j(n)} \in \{a_j \mid j \in J\} - \{a\}$ such that $a - a_{j(n)} < 1/n$. Clearly, we may safely assume $a_{j(n)} \leq a_{j(n+1)}$. Then we have $\lim_{n\to\infty} a_{j(n)} = a$ and $a_{j(n)} < a$. A moment's reflection shows that

$$\begin{aligned} f(\bigvee_{j\in J} a_j, b) &= f(\lim_{n\to\infty} a_{j(n)}, b) = \lim_{n\to\infty} f(a_{j(n)}, b) = \\ &= \bigvee_{n\in\mathbf{N}} f(a_{j(n)}, b) = \bigvee_{j\in J} f(a_j, b) \end{aligned}$$

by the definition of left-continuity in x. Second, there is some $n \in \mathbf{N}$ such that $a - a_k > 1/n$ for each $a_k \in \{a_j \mid j \in J\} - \{a\}$. Clearly, $a = a_l$ for some $l \in J$, hence $f(\bigvee_{j\in J} a_j, b) = f(a_l, b) \leq \bigvee_{j\in J} f(a_j, b)$. The converse inequality, i.e. $f(\bigvee_{j\in J} a_j, b) \geq \bigvee_{j\in J} f(a_j, b)$, holds since f is nondecreasing in x.

Conversely, let (2.119) be true. To show that f is left-continuous in x it suffices to show that for any $a \in [0,1]$ and any sequence $\{a_n \mid n \in \mathbf{N},\, a_n \leq a,\, a_n \leq a_{n+1}\}$ such that $\lim_{n\to\infty} a_n = a$ we have $\lim_{n\to\infty} f(a_n, b) = f(\lim_{n\to\infty} a_n, b)$. Using (2.119), this readily follows from $\lim_{n\to\infty} f(a_n, b) = \bigvee_{n\in\mathbf{N}} f(a_n, b)$ and $\lim_{n\to\infty} a_n = \bigvee_{n\in\mathbf{N}} a_n$. □

Lemma 2.86 *Let $f : [0,1]^2 \to [0,1]$ be a function such that for each $a \in [0,1]$, $f(x,a) : [0,1] \to [0,1]$ and $f(a,x) : [0,1] \to [0,1]$ are continuous. Then f is continuous.*

Proof. We have to show that for every $a, b \in [0,1]$, for each $\varepsilon > 0$ there is some $\delta > 0$ such that $|f(x,y) - f(a,b)| < \varepsilon$ for every $x \in (a-\delta, a+\delta)$, $y \in (b-\delta, b+\delta)$ (if, e.g., $x = 1$ then "$x \in (a-\delta, a+\delta)$" is to be replaced by "$x \in (a-\delta, a]$"; similarly for the other boundary cases). Since[1] for any $A, B \in [0,1]$, both $f(A,x)$ and $f(x,B)$ are continuous functions on a compact set (closed interval), they are uniformly continuous. Therefore, there are $\delta_1, \delta_2 > 0$ such that $|f(x_1,B) - f(x_2,B)| < \varepsilon/2$ and $|f(A,y_1) - f(A,y_2)| < \varepsilon/2$ for any $x_1, x_2, y_1, y_2 \in [0,1]$ such that $|x_1 - x_2| < \delta_1$, $|y_1 - y_2| < \delta_2$. Put $\delta = \min(\delta_1, \delta_2)$ and take any $x \in (a-\delta, a+\delta)$, $y \in (b-\delta, b+\delta)$. We have

$$\begin{aligned} & |f(x,y) - f(a,b)| = |f(x,y) - f(a,y) + f(a,y) - f(a,b)| \leq \\ \leq \; & |f(x,y) - f(a,y)| + |f(a,y) - f(a,b)| < \varepsilon/2 + \varepsilon/2 = \varepsilon \end{aligned}$$

completing the proof. □

Remark 2.87 Due to Lemma 2.86, asserting that a t-norm is continuous (as a real function of two arguments) is tantamount to asserting that it is continuous in both the first and the second variable.

Theorem 2.88 *If $\otimes$ is a left-continuous t-norm then letting*

$$a \to b = \bigvee \{c \mid a \otimes c \leq b\}, \tag{2.121}$$

$\langle [0,1], \min, \max, \otimes, \to, 0, 1\rangle$ is a complete residuated lattice. Conversely, if $\langle [0,1], \min, \max, \otimes, \to, 0, 1\rangle$ is a residuated lattice then $\otimes$ is a left-continuous t-norm.

Proof. The assertion is a direct consequence of Lemma 2.85 and Theorem 2.28. □

Remark 2.89 (1) An example of a left-continuous t-norm which is not continuous is the "nilpotent minimum" [77] T_0 defined by

$$T_0(x,y) = \begin{cases} \min(x,y) & \text{for } x+y \geq 1 \\ 0 & \text{for } x+y < 1. \end{cases}$$

[1]Thanks are due to Dr. B. Krajc who helped me to see this easy argument.

(2) By Theorem 2.88, residuated lattices on $[0,1]$ correspond to left-continuous t-norms. Note, however, that since $[0,1]$ is a chain, residuated lattices on $[0,1]$ satisfy prelinearity.

(3) An example of a t-norm that is even not left-continuous is the "drastic product" defined by

$$T_d(x,y) = \begin{cases} x \cdot y & \text{for } \max(x,y) = 1 \\ 0 & \text{otherwise.} \end{cases}$$

A moment's reflection shows that for each t-norm T we have

$$T_d(x,y) \leq T(x,y) \leq \min(x,y)$$

for every $x, y \in [0,1]$.

2.4.2 CONTINUOUS T-NORMS AND BL-ALGEBRAS

We now turn our attention to continuous t-norms and the corresponding residuated lattices.

Lemma 2.90 *A t-norm $\otimes$ is continuous iff $\langle [0,1], \wedge, \vee, \otimes, \rightarrow, 0, 1\rangle$ is a divisible residuated lattice.*

Proof. Let $\otimes$ be continuous. We verify (2.93) which is equivalent to divisibility. Let $a, b \in [0,1], a \leq b$ and consider a function $f(z) : [0,1] \rightarrow [0,b]$ defined by $f(z) = b \otimes z$. Since f is a nondecreasing function and since $0 \leq a \leq b$, continuity yields some $c \in [0,1]$ such that $a = f(c)$, i.e. $a = b \otimes c$.

Conversely, let (2.93) hold. For an arbitrary $b \in [0,1]$, let $f(z) = b \otimes z$. By (2.93), f is a surjection from $[0,1]$ onto $[0,b]$ and since it is nondecreasing, it is continuous. □

Theorem 2.91 *A residuated lattice defined on $[0,1]$ by a t-norm $\otimes$ is a BL-algebra iff $\otimes$ is continuous.*

Proof. Prelinearity is always satisfied in a residuated lattice on $[0,1]$ (Remark 2.89 (2)). The assertion therefore follows from Lemma 2.90. □

Gödel, Łukasiewicz, and product are basic continuous t-norms. One can easily verify that Gödel t-norm, Łukasiewicz t-norm, and product t-norm are continuous; their graphs are depicted in Fig. 2.2.

We are going show that these three t-norms are basic ones among continuous t-norms: Each continuous t-norm can be obtained from Gödel, Łukasiewicz, and product by a simple construction.

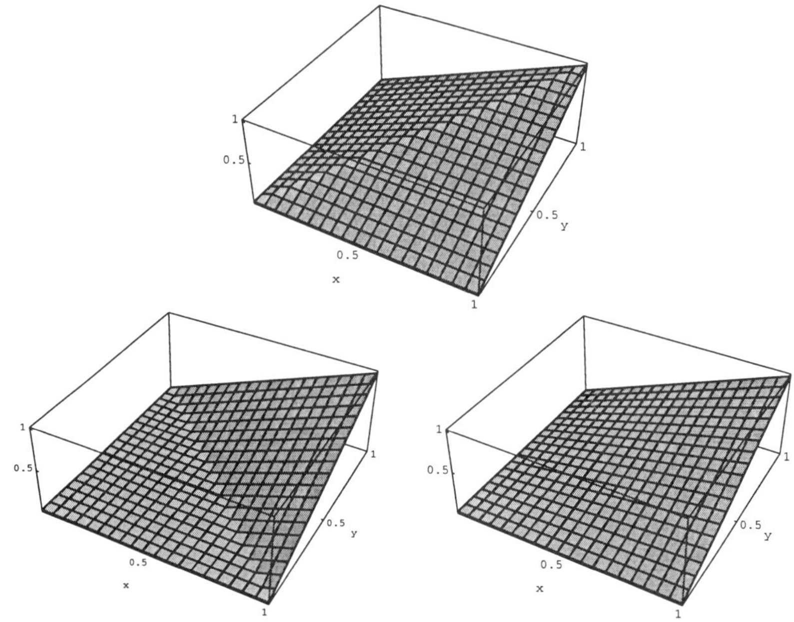

Figure 2.2. Gödel t-norm, Łukasiewicz t-norm (bottom left), and product t-norm (bottom right).

For a t-norm $\otimes$, denote by $E(\otimes)$ (or simply by E) the set of all idempotents of $\otimes$, i.e. $E(\otimes) = \{a \in [0,1] \mid a \otimes a = a\}$; in the following we assume that $\otimes$ is continuous.

Lemma 2.92 *For a continuous t-norm $\otimes$, $E(\otimes)$ is a closed set w.r.t. the natural topology on $[0,1]$ (i.e. topology induced by the standard metric $d(a,b) = |a - b|$).*

Proof. By the well-known property, E is closed iff the limit of any convergent sequence of elements of E belongs to E. Let thus a_n ($n \in \mathbf{N}$, $a_n \in E$) be convergent, let $b = \lim_n a_n$. Then

$$b \otimes b = (\lim_n a_n) \otimes (\lim_n a_n) = \lim_n (a_n \otimes a_n) = \lim_n a_n = b,$$

thus b is idempotent and hence $b \in E$ proving that E is closed. $\square$

Remark 2.93 Note that continuity of $\otimes$ is essential in Lemma 2.92: For nilpotent minimum T_0 (Remark 2.89) which is left-continuous but not continuous we have $E(T_0) = \{0\} \cup (\frac{1}{2}, 1]$ which is not a closed set.

Since E is closed, its complement $[0,1] - E$ is open.

Lemma 2.94 *Each open set $O \subseteq [0,1]$ is a union of countably many nonoverlapping open intervals.*

Proof. Each $a \in O$ is contained in some open interval O_a that is maximal among open intervals O' such that $a \in O'$ and $O' \subseteq O$. This is easy to see: consider $\mathcal{O}_a = \{(b_1, b_2) \mid a \in (b_1, b_2) \subseteq O\}$; since O is open, $\mathcal{O}_a$ is nonempty; then clearly, $O_a = \bigcup \mathcal{O}_a$. We thus have $O = \bigcup_{a \in O} O_a$. If $O_a \neq O_b$ then $O_a \cap O_b = \emptyset$, i.e. the open intervals O_a are non-overlapping. Finally, there is only a countable number of intervals O_a (indeed, we can pick one rational number from each O_a and there is only a countable number of rational numbers). □

Therefore, $[0,1] - E$ is the union of a system $\mathcal{I}_o(E)$ of open intervals. Denote by $\mathcal{I}(E)$ the corresponding system of closed intervals, i.e. $\mathcal{I}(E) = \{[a,b] \mid (a,b) \in \mathcal{I}_o(E)\}$. For $I \in \mathcal{I}(E)$ we denote by $\otimes|I$ the restriction of $\otimes$ to I. We have the following representation theorem.

Theorem 2.95 (Mostert-Shields representation) *For each continuous t-norm $\otimes$,*

(i) for each $I \in \mathcal{I}(E)$, $\otimes|I$ is isomorphic either to Łukasiewicz t-norm or to product t-norm;

(ii) if there is no $I \in \mathcal{I}(E)$ with $a, b \in I$, then $a \otimes b = \min(a,b)$.

Before we go to the proof we introduce some notions. A t-norm is called *Archimedean* if 0 and 1 are its only idempotents (i.e. $a \otimes a < a$ for any $0 \neq a \neq 1$). An Archimedean t-norm is called *strict* if 0 is its only nilpotent element (i.e. $a \otimes \cdots \otimes a > 0$ for any $a > 0$); an Archimedean t-norm is called *nilpotent* if it is not strict. We are now going to prove Theorem 2.95.

Proof. For each $I \in \mathcal{I}(E)$, $\otimes|I$ is isomorphic to an Archimedean t-norm (Lemma 2.97). Each Archimedean t-norm is either strict or nilpotent (by definition). Each strict Archimedean t-norm is isomorphic to the product t-norm; each nilpotent Archimedean t-norm is isomorphic to the Łukasiewicz t-norm (Lemma 2.102). If for $b < c$ there is no $I \in \mathcal{I}(E)$ with $b, c \in I$, then there is some idempotent a with $b \leq a \leq c$. By Lemma 2.96, $b = b \otimes a$, $a = a \otimes c$, whence $b \otimes c = b \otimes a \otimes c = b \otimes a = b = \min(b,c)$. On proving the lemmas we referred to, the proof is complete. □

Lemma 2.96 *If a is an idempotent of a continuous t-norm $\otimes$ then for each b we have $a \otimes b = \min(a,b)$.*

Proof. By Theorem 2.91, $\otimes$ makes $[0,1]$ into a BL-algebra; this BL-algebra is by definition divisible and the assertion follows from Theorem 2.39 (2.102) noticing that in $[0,1]$ we have $x \wedge y = \min(x,y)$. □

Lemma 2.97 *For a continuous t-norm $\otimes$ and $I \in \mathcal{I}(E)$, $\otimes|I$ is isomorphic to an Archimedean t-norm.*

Proof. Let $I = [a,b]$. Since b is an idempotent, Lemma 2.96 implies that for $c \in [a,b]$ we have $b \otimes c = c$. Consider the increasing linear mapping h sending $[a,b]$ bijectively onto $[0,1]$. h induces a binary operation $\otimes_h$ on $[0,1]$ by $x \otimes_h y = h(h^{-1}(x) \otimes h^{-1}(y))$. As $\otimes|I$ is associative, commutative, and monotone w.r.t. $\leq$, $\otimes_h$ is as well. Moreover, since for $c \in [a,b]$ we have $b \otimes c = c$ and since $h(1) = b$, we have $x \otimes_h 1 = x$. Furthermore, 0 and 1 are the only idempotents of $\otimes_h$. Therefore, $\otimes_h$ is an Archimedean t-norm isomorphic to $\otimes|I$. □

Earlier, we denoted the n-th power of a w.r.t. $\otimes$ by a^n. To prevent confusion with the usual power of elements from $[0,1]$, we denote the n-th power of $a \in [0,1]$ w.r.t. $\otimes$ by $a^{\otimes n}$, i.e. $a^{\otimes n} = a \otimes \cdots \otimes a$ (n-times).

Lemma 2.98 *For a continuous Archimedean t-norm $\otimes$ we have:*

(i) $a < 1$ *implies* $\lim_n a^{\otimes n} = 0$;

(ii) *if $\otimes$ is nilpotent then each $a < 1$ is nilpotent;*

(iii) *if $0 < a^{\otimes n} < 1$ and $n < m$ then $a^{\otimes m} < a^{\otimes n}$;*

(iv) *for each positive $a < 1$ and each positive n there is a unique b such that $b^{\otimes n} = a$.*

Proof. $\lim_n a^{\otimes n}$ exists since $(a^{\otimes n})_{n \in \mathbf{N}}$ is a nonincreasing sequence bounded from below by 0. Clearly, $\lim_n a^{\otimes n} < 1$; furthermore, $\lim_n a^{\otimes n} \otimes \lim_n a^{\otimes n} = \lim_n a^{\otimes n}$, i.e. $\lim_n a^{\otimes n}$ is an idempotent. Since 0 and 1 are the only idempotents, we have $\lim_n a^{\otimes n} = 0$, establishing (i).

If $\otimes$ is nilpotent then, by definition, some $b > 0$ is nilpotent (i.e. $b^{\otimes n} = 0$ for some n). If $a < b$ then a is clearly nilpotent since $a^{\otimes n} \leq b^{\otimes n} = 0$; $b < a < 1$ then, by (i), $a^{\otimes m} < b$ for some m and thus $a^{\otimes mn} \leq b^{\otimes n} = 0$, proving (ii).

(iii): If $a^{\otimes m} = a^{\otimes n}$ were the case, then for $b = a^{\otimes n}$, $c = a^{\otimes(m-n)}$ we have $b = b \otimes c = \cdots = b \otimes c^{\otimes k}$ for any k. As $\lim_k c^{\otimes k} = 0$ we have $b = \lim_k b = \lim_k (b \otimes c^{\otimes k}) = b \otimes \lim_k c^{\otimes k} = b \otimes 0 = 0$, a contradiction. Thus, $a^{\otimes m} < a^{\otimes n}$ is the case.

(iv): The function $f(x) = x^{\otimes n}$ is continuous and satisfies $f(0) = 0$, $f(1) = 1$; the existence of some b therefore follows. Clearly, $a < b$ ($\otimes$ is Archimedean). It suffices to show that for $a < c < b$ we cannot have $a = c^{\otimes n}$ (if $b < c$, just exchange roles of b and c). Suppose then $0 < a < c < b$ and $a = c^{\otimes n}$. Since $\otimes$ makes $[0,1]$ into a BL-algebra, Theorem 2.36 (2.93) implies that there is some d with $c = b \otimes d$. Therefore, $c^{\otimes n} = b^{\otimes n} \otimes d^{\otimes n} = \cdots = b^{\otimes n} \otimes (d^{\otimes n})^{\otimes k}$ for each k. Using (i) we get $\lim_k (d^{\otimes n})^{\otimes k} = 0$ from which we get $a = c^{\otimes n} = 0$, a contradiction.
□

Lemma 2.98 (iv) justifies the following definition.

Definition 2.99 For $a > 0$, $a^{\otimes \frac{1}{n}}$ is the unique b such that $b^{\otimes n} = a$; $0^{\otimes \frac{1}{n}} = 0$. For positive integers m, n and rational $r = \frac{m}{n}$ we define

$$a^{\otimes r} = (a^{\otimes \frac{1}{n}})^{\otimes m}.$$

Lemma 2.100 *For $a \in [0,1]$, m, m', n, n' positive integers, and r, s positive rationals we have*

(i) if $\frac{m}{n} = \frac{m'}{n'}$ then $a^{\otimes \frac{m}{n}} = a^{\otimes \frac{m'}{n'}}$;

(ii) $a^{\otimes r} \otimes a^{\otimes s} = a^{\otimes (r+s)}$;

(iii) if $a > 0$ then $\lim_n a^{\otimes \frac{1}{n}} = 1$.

Proof. (i): Assume first that $m' = km$, $n' = kn$; then

$$a^{\otimes \frac{m'}{n'}} = ((a^{\otimes \frac{1}{kn}})^{\otimes k})^{\otimes m} = (a^{\otimes \frac{1}{n}})^{\otimes m} = a^{\otimes \frac{m}{n}}.$$

This assertion can be applied for general m, n, m', n', to $\frac{m}{n} = \frac{nm'}{nn'}$ and to $\frac{m'}{n'} = \frac{n'm}{n'n}$; using $mn' = m'n$, (i) follows.

(ii): By (i), we may safely assume $r = \frac{m}{n}$, $s = \frac{m'}{n}$. Then

$$a^{\otimes r} \otimes a^{\otimes s} = (a^{\otimes \frac{1}{n}})^{\otimes m} \otimes (a^{\otimes \frac{1}{n}})^{\otimes m'} =$$
$$= (a^{\otimes \frac{1}{n}})^{\otimes (m+m')} = a^{\otimes \frac{m+m'}{n}} = a^{\otimes (r+s)}.$$

(iii): The sequence $(a^{\otimes \frac{1}{n}})_{n \in \mathbf{N}}$ is clearly non-decreasing and bounded from above by 1; moreover, it limit is an idempotent (indeed, $(\lim_n a^{\otimes \frac{1}{n}}) \otimes (\lim_n a^{\otimes \frac{1}{n}}) = \lim_n (a^{\otimes \frac{1}{n}} \otimes a^{\otimes \frac{1}{n}}) = \lim_n a^{\otimes \frac{2}{n}} = \lim_n a^{\otimes \frac{2}{2n}} = \lim_n a^{\otimes \frac{1}{n}}$); therefore, $\lim_n a^{\otimes \frac{1}{n}} = 1$. □

In the following, we denote by $*$ the binary operation on $[\frac{1}{4}, 1]$ defined by

$$a * b = \max(\frac{1}{4}, a \cdot b),$$

i.e. $*$ is the product "cut" at $\frac{1}{4}$.

Lemma 2.101 *$\langle[\frac{1}{4}, 1], *\rangle$ is isomorphic to $\langle[0,1], \otimes_Ł\rangle$ where $\otimes_Ł$ is the Łukasiewicz t-norm.*

Proof. The mapping $f(x) = 2^{2(x-1)}$ is an increasing bijection of $[0,1]$ to $[\frac{1}{4}, 1]$; for the thus induced operation $\odot$ on $[0,1]$ isomorphic to $*$, i.e. $a \odot b = f^{-1}(f(a) * f(b))$, we have: if $a + b - 1 \geq 0$ then

$$a \odot b = f^{-1}(f(a) * f(b)) = \frac{1}{2}\log_2(2^{2(a-1)} \cdot 2^{2(b-1)}) + 1 = a + b - 1;$$

if $a + b - 1 < 0$ then $a \odot b = 0$. Thus, $\odot$ is the Łukasiewicz t-norm. □

Lemma 2.102 *Each strict Archimedean t-norm is isomorphic to product t-norm; each nilpotent Archimedean t-norm is isomorphic to Łukasiewicz t-norm.*

Proof. Due to Lemma 2.101, it suffices to show that for each strict t-norm $\otimes$, $\langle[0,1], \otimes\rangle$ is isomorphic to $\langle[0,1], \cdot\rangle$, and for each nilpotent t-norm $\otimes$, $\langle[0,1], \otimes\rangle$ is isomorphic to $\langle[\frac{1}{4}, 1], *\rangle$. Since the operations in question are continuous, it suffices to show an isomorphism between (some) dense subsets (which will be subalgebras) of the respective algebras. The set $c = \{c_r = \frac{1}{2^r} \mid r \text{ a nonnegative rational }\}$ is clearly a dense subset of $[0,1]$; we have $c_r \cdot c_s = c_{r+s}$ and $c_r < c_s$ iff $r > s$. If $\otimes$ is strict, put $d = \frac{1}{2}$; if $\otimes$ is nilpotent, let d be the largest x satisfying $x \otimes x = 0$ (in either case we have $0 < d < 1$). Putting $d_r = d^{\otimes r}$ for nonnegative rational r, we have $d_r \otimes d_s = d_{r+s}$.

We claim that if $\otimes$ is strict then $s < r$ implies $d_r < d_s$; if $\otimes$ is nilpotent then $s < r$ implies $d_r < d_s$ for $s < r \leq 2$ and for $r \geq 2$ we have $d_r = 0$. Proof of the claim: For $\otimes$ strict, let $r = \frac{m_1}{n}$, $s = \frac{m_2}{n}$, put $a = d^{\otimes \frac{1}{n}}$; then $d_r = a^{\otimes m_1}$, $d_s = a^{\otimes m_2}$, $0 < a < 1$, $m_2 < m_1$, and from $a^{m_2} > 0$ ($\otimes$ is strict) we get by Lemma 2.98 (iii) that $a^{m_1} < a^{m_2}$. For $\otimes$ nilpotent, $d_2 = 0$ by definition of d and thus if $r \geq 2$ then $d_r = 0$; if $s < 2$ then $d_s > 0$ by the choice of d. Thus, if $s < r \leq 2$, the above argument can be repeated to show that $d_r < d_s$.

Therefore, we see that for strict $\otimes$, $c_r \leftrightarrow d_r$ (r a positive rational) is a bijective morphism between $[0,1] \cap c$ and $\{d_r \mid r \text{ a positive rational}\}$; for nilpotent $\otimes$, $c_r \leftrightarrow d_r$ (r a rational ≤ 2) is a bijective morphism between $[\frac{1}{4}, 1] \cap c$ and $\{d_r \mid r \text{ a rational} \leq 2\}$.

It remains to show that $D = \{d_r \mid r \text{ a rational}\}$ is a dense subset of $[0, 1]$; we will do this by showing that each $0 < a < 1$ can be approximated from above by d_r where r is of the form $r = \frac{m}{2^n}$. Take n_0 such that $d_{\frac{1}{2^{n_0}}} \geq a$ (this is possible due to Lemma 2.100 (iii)); for $n \geq n_0$, let r_n be $\frac{m}{2^n}$ for the largest m such that $d_{\frac{m}{2^n}} \geq a$. We have $d_{\frac{m+1}{2^n}} = d_{\frac{m}{2^n}} \otimes d_{\frac{1}{2^n}} = d_{r_n} \otimes d_{\frac{1}{2^n}} < a \leq d_{r_n}$. Finally, using $\lim_n d_{\frac{1}{2^n}} = 1$, we get $\lim_n (d_{\frac{1}{2^n}} \otimes d_{r_n}) = \lim_n d_{r_n} = a$. □

2.5 BIBLIOGRAPHICAL REMARKS

Section 2.1: Fuzzy approach to modeling and analysis of systems where human judgment plays a substantial role (so-called humanistic systems) is due to Lotfi A. Zadeh [128, 211]. The seminal paper on fuzzy sets is [216]. Although there are earlier papers (see Bibliographical remarks to Chapter 3) on many-valued sets, it was Zadeh's papers that caused a broad interest in fuzzy approach and development of fuzzy sets, fuzzy logic, and their applications. Klir and Yuan [127] consider the invention of fuzzy sets a paradigm shift in the sense of Kuhn [131]. Information about the development of fuzzy approach can be found in parts of [126, 129, 140] (history, applications) and [99, 105, 127] (fuzzy sets and fuzzy logic).

Note that the capability of fuzzy logic to solve paradoxes and sometimes also the whole idea of fuzzy logic is questioned by some philosophers (particularly, in the case of *sorites* paradoxes, questioned is the nonidempotency of conjunction which is crucial for resolving the paradox).

[106] puts forth important questions (with some answers).

Recently, fuzzy logic has been considered one of the main parts of so-called soft computing. Basically, *soft computing* is understood as a synthesis of disciplines that aim at utilizing "soft methods" of information processing and reasoning used by humans. Practically, this means a synthesis of fuzzy logic and fuzzy sets (a tool for dealing with vagueness, i.e. a particular type of indeterminacy typical for the use of a natural language), probabilistic reasoning (a tool for dealing with (probabilistic) uncertainty, i.e. a particular type of indeterminacy typical for predictions of events), artificial neural networks (a tool for learning of systems), and genetic algorithms and strategies (a tool for optimization by evolution). Information on related issues can be found e.g. in [68, 222, 221].

Section 2.2: The set of truth values proposed by Zadeh is the real unit interval $[0,1]$ equipped with min as conjunction (Zadeh himself mentions that one can use a partially ordered set; nevertheless, he does not develop this direction). The idea to use complete residuated lattices as structures of truth values is due to Goguen [90, 91]. In these papers, Goguen thoroughly discussed fuzzy approach and outlined a way to develop logic using fuzzy approach. Particularly, our justification of adjointness property is basically due to Goguen (cf. also beginning of Chapter 2 in [105]). Goguen's proposal was formally developed by Pavelka [158] (see Bibliographical remarks of Chapter 3).

Section 2.3: Residuated lattices have been introduced by Ward and Dilworth [202]. Several other names are used for residuated lattices: integral commutative residuated l-monoid by Birkhoff [41, pp. 324–325], residuated Abelian semigroup with a unit by Blyth and Janowitz [43, pp. 211–214], or commutative complete lattice ordered semigroup with infinity by Goguen [90, 91]. Residuated lattices have been studied from the point of view of fuzzy logic by Höhle [116]. The notion of a BL-algebra was introduced in [105] ("BL" stands for "basic logic") as a common generalization of Lukasiewicz, product, and Gödel logics—the logical systems that have the three basic t-norms as conjunction. [105] also contains chapters on MV-algebras, product algebras, and Gödel algebras. MV-algebras were introduced by C. C. Chang [56, 57] in his investigation of Łukasiewicz logic [135]; "MV" stands for "many-valued". Our definition of (Chang) MV-algebras is due to Mundici; it is a simplified version of the original definition of Chang, see [60]. [60] is a deep investigation of MV-algebras (with numerous references). Product algebras were introduced in [108]. Gödel algebras were used by Gödel in his investigation into intuitionistic logic [89]. Heyting algebras appear in the study of intuitionistic logic [109]. Girard monoids appear in so-called linear logic [87] which is of interest for computer science. Boolean algebras is a classical topic; they are the algebraic counterparts of classical logic and are treated in detail e.g. in [189].

The results of Section 2.3 can be found in the above-cited literature and in [25, 88].

Section 2.4: The notion of a t-norm goes back to the study of probabilistic metric spaces (until 1964 called statistical metric spaces); "t-norm" stands for "triangular norm". Probabilistic metric spaces were introduced by Menger in [141] where also some postulates close to those of t-norms appeared. Probabilistic metric spaces were significantly elaborated by Schweizer and Sklar (see their monograph [187]); it was Schweizer and Sklar who introduced the notion of a t-norm [186]. Note that in the original treatment, the values from $[0,1]$ on which t-norms operate are interpreted as probabilities, not degrees of truth. T-norms are well-covered in [125]. The standard algebras on $[0,1]$ have the following important property: An identity is true in the standard Łukasiewicz (product, Gödel) algebra over $[0,1]$ iff it is true in the class of all MV-algebras (product algebras, Gödel algebras), see e.g. [105]. [61] proves an analogous result for BL-algebras ([104]): an identity is true in all continuous t-norm algebras iff it is true in all BL-algebras; a continuous t-norm algebra is a residuated lattice on $[0,1]$ induced by a continuous t-norm $\otimes$. The fact that each continuous t-norm can be composed of Łukasiewicz, product, and Gödel is due to [147](see also [76]); our presentation follows [105]; for generalization to BL-algebras see [104, 61]. While the use of t-norms as many-valued conjunctions is more or less standard, there are also other approaches to many-valued implications than those via residua of t-norms (see e.g. Chapter 2 of [69]).

Chapter 3

FUZZY STRUCTURES

... the importance of fuzzy relations is almost self-evident. Science is, in a sense, the discovery of relations between observables ... Difficulties arise in so-called "soft" sciences because the relations involved do not appear to be "hard" as they are, say, in classical physics ...

— J. A. Goguen (1967)

... what might be called the principle of incompatibility. *Stated informally, the essence of this principle is that as the complexity of a system increases, our ability to make precise and yet significant statements about its behavior diminishes until a threshold is reached beyond which precision and significance (or relevance) become almost mutually exclusive characteristics.*

— L. A. Zadeh (1973)

This chapter is devoted to fuzzy structures — systems of fuzzy relations and functions on a set. We introduce and investigate basic structural notions connected with fuzzy structures, and basic relationships between fuzzy structures and a language of fuzzy predicate logic. As a matter of fact, what is developed here belongs traditionally to so-called model theory, in our case to model theory for fuzzy logic. However, this chapter is not to be understood to be a treatise on model theory for fuzzy logic. Rather, we present only notions and results that are more or less directly connected to fuzzy relational modeling. Notions and results of this chapter will be used in the subsequent chapters where several special fuzzy structures will be investigated.

3.1 FUZZY SETS AND FUZZY RELATIONS

We now develop basic notions of (naive) fuzzy set theory (with truth values for membership degrees taken from complete residuated lattices) needed in the sequel.

3.1.1 DEFINITIONS AND BASIC PROPERTIES

The central concept of this section is that of a fuzzy set. Fuzzy sets serve to model collections of objects such that the best answer to the question of whether an object belongs to this collection is a certain truth degree (selected) from an appropriate ordered scale of truth degrees. The scale of truth degrees may have more than two truth degrees 0 and 1 (representing the answers "no" and "yes"); there can be intermediate truth degrees. A fuzzy set A is given by specifying a universe X of all objects in question, a scale L of truth degrees, and a rule that associates to each object x from X a truth degree a from L that is thought of as the truth degree to which x belongs to A. For instance, fuzzy sets may be used to model collections like the collection of all big balls. One naturally considers some balls bigger than others: being a big ball is a graded property. A fuzzy set that represents the collection of big balls may be specified as follows: we let X denote the set of all balls in question, select an appropriate scale L of truth degrees (for instance, the real unit interval [0,1] or some finite scale), and specify a function that assigns a truth degree from L to any ball from X so that the bigger the ball, the bigger the truth degree. Since we want fuzzy sets to enter further reasoning, the scale of truth degrees has to be equipped with an appropriate structure. For reasons discussed in detail in Chapter 2, we assume that the structure of truth degrees forms a complete residuated lattice.

Definition 3.1 Let $\mathbf{L}$ be a complete residuated lattice, X be a nonempty set. An **L**-*set* in X is a mapping $A : X \to L$.

The set X is called a *universe*. $A(x)$ is called the *degree of membership of x in A*. **L**-sets are also called *fuzzy sets*. In this book we use the term **L**-set rather than fuzzy set (and in general, **L**-... rather than fuzzy ...) since it makes the structure of truth values explicit. The set of all **L**-sets in X will be denoted by L^X.

An **L**-set A in X is also denoted $\{ {}^{A(x)}/x \mid x \in X\}$. If for all elements y distinct from $x_1, x_2, \ldots,$ and x_n we have $A(y) = 0$, we also write $A = \{ {}^{A(x_1)}/x_1, \cdots, {}^{A(x_n)}/x_n\}$. An **L**-set A such that there is exactly one $x \in X$ with $A(x) > 0$ (i.e. $A = \{ {}^{a}/x\}$ for some $a > 0$) is called a *singleton*.

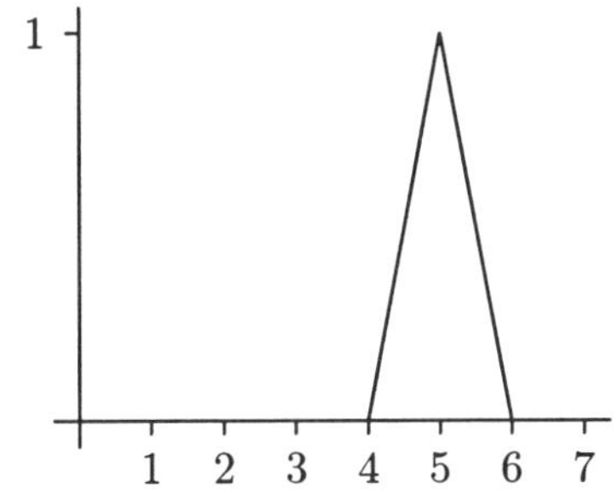

Figure 3.1. Fuzzy set representing "approximately 5".

For $K \subseteq L$, an **L**-set $A \in L^X$ is called *K-valent* if $A(x) \in K$ for any $x \in X$. For $K = \{0, 1\}$, K-valent **L**-sets are called *bivalent*. Bivalent **L**-sets are also called crisp. A *height* of an **L**-set A is defined by $\text{hgt}(A) = \bigvee_{x \in X} A(x)$.

Example 3.2 (1) Let $X = \{\text{circle, square, hexagon}\}$, **L** be the standard Gödel algebra on $[0, 1]$. Let $A : X \to [0, 1]$ be given by $A(\text{circle}) = 1$, $A(\text{square}) = 0$, $A(\text{hexagon}) = 0.5$. A is an **L**-set in X. A may represent the concept of "is circle-shaped" in the universe X.

(2) Let $X = \{\ldots, \text{Bonn}, \ldots, \text{London}, \ldots, \text{New York}, \ldots, \text{Prague}, \ldots\}$, **L** be a finite Gödel structure with $L = \{0, \frac{1}{5}, \ldots, \frac{4}{5}, 1\}$. The concept "large city" can be represented by a fuzzy set A with $A(\text{Bonn}) = \frac{1}{5}$, $A(\text{London}) = 1$, $A(\text{New York}) = 1$, $A(\text{Prague}) = \frac{3}{5}$, ...

(3) Let X be the set of all real numbers, **L** be the standard Łukasiewicz algebra on $[0, 1]$. Define an **L**-set $A \in L^X$ by

$$A(x) = \begin{cases} x - 4 & \text{for } 4 \leq x \leq 5 \\ 6 - x & \text{for } 5 < x \leq 6 \\ 0 & \text{otherwise.} \end{cases}$$

Then A is a fuzzy set that represents the concept "approximately 5"; A is depicted in Fig. 3.1.

(4) Let $X = [100, 250]$ be the range of people's heights in centimeters, let L be $[0, 1]$. Figure 3.2(left) shows fuzzy sets S, M, and T that intuitively represent the meaning of language expressions "small", "medium", and "tall", respectively. Moreover, fuzzy sets RT and VT representing the meaning of "roughly tall" and "very tall" are depicted in Fig. 3.2(right). We leave the detailed description of membership degrees unspecified (the reader may supply his own definition).

Remark 3.3 (1) Example 3.2 demonstrates that fuzzy sets are a natural formalism for modeling of the meaning of linguistic expressions.

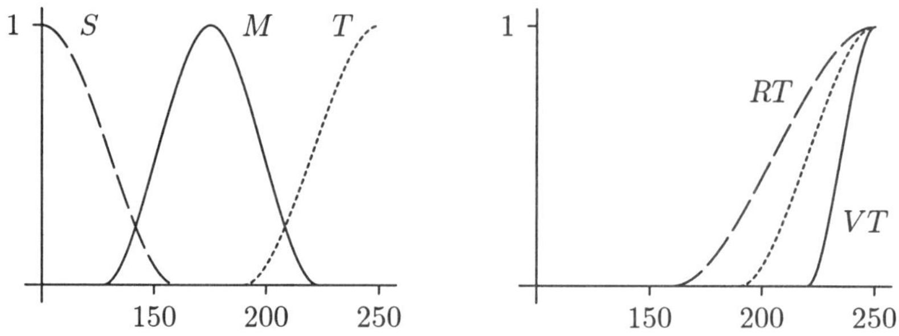

Figure 3.2. Left: fuzzy sets S ("small"), M ("medium"), and T ("tall"). Right: Fuzzy sets T ("tall"), RT ("roughly tall"), and VT ("very tall").

(2) At this point, the reader will probably ask two questions: First, where are the membership degrees coming from? Second, to what extent does the exact degree of membership matter?

The first question has, in fact, two parts: what are membership degrees, what is their nature and meaning? and how to determine a membership degree? The nature of membership degrees is a kind of philosophical and epistemological question that we do not attempt to answer in this text (we assume that the intuitive explanation from Chapter 2 of the notion of a degree of truth is clear and well acceptable). How to determine membership degrees is basically a question of expert knowledge (expert tells us the membership degrees) and/or experimental design (membership degrees are extracted from data). Since the process of determination of membership degrees is not the topic of our interest, we only refer to Chapter 10 of [127] (where one can find basic approaches and an extensive list of references). In the following we take membership degrees we are dealing with as given (supplied from outside).

The second question is particularly important when fuzzy sets enter further processing. A fuzzy set A may be, for instance, (a part of) an input to a reasoning process the output of which is, say, $r(\ldots, A, \ldots)$. In general, different fuzzy sets lead to different outputs. Does it matter (and to what extent) if we slightly change the membership degrees of an input fuzzy set A? That is to say, given two fuzzy sets A_1 and A_2 whose membership degrees are very close, what can be said about how close are the corresponding outputs $r(\ldots, A_1, \ldots)$ and $r(\ldots, A_2, \ldots)$? The issue in question is obviously the sensitivity to membership degrees. Later on, we provide an answer to this kind of questions for the case where reasoning is described using logical formulas. Note that the above-mentioned problem is the very immediate one encountered by an expert who is about to tell the membership degrees and wonders about whether

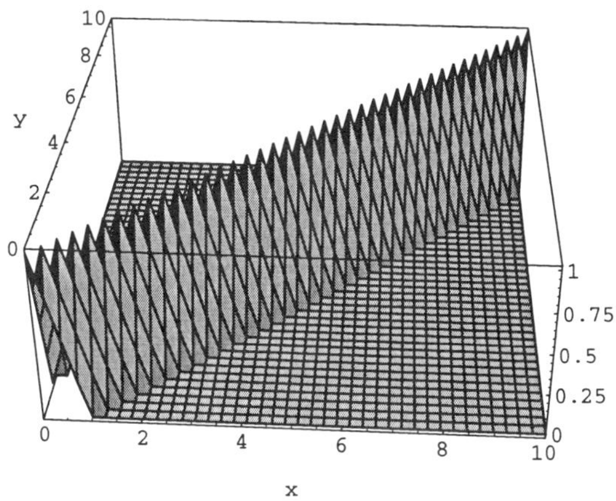

Figure 3.3. Fuzzy relation representing relationship "being close".

the membership degree should be 0.71 or 0.72 and whether it indeed matters.

If $X = X_1 \times \cdots \times X_n$, then an **L**-set in X is called an (n-ary) **L**-*relation* (or fuzzy relation) between X_1, X_2, ..., and X_n. If $X_1 = \cdots = X_n$, we speak of n-ary **L**-relation on a set X.

Example 3.4 Let X be the set of all reals, take any **L** with $L = [0, 1]$. Define $E : X \times X \to [0, 1]$ by putting $E(x, y) = \max(0, 1 - |x - y|)$. Then E is a binary **L**-relation (fuzzy relation) on X. Intuitively, E represents the meaning of a relationship between real numbers described verbally as "being close". The graph of E restricted to the interval $[0, 10]$ is depicted in Fig. 3.3.

Remark 3.5 (1) If **L** is the two-element Boolean algebra then an **L**-set is precisely what is called a characteristic function of a (classical) set. This is the way the concept of an **L**-set generalizes the concept of a (classical) set. Note that what we call a fuzzy set (i.e. a function mapping a universe X into a scale L of truth degrees) is sometimes called a *membership function* (or characteristic function) of a fuzzy set. Proceeding this way, one considers the notion of a fuzzy set an extramathematical notion; i.e. the notion of a fuzzy set is not defined, defined is only the notion of a characteristic function of a fuzzy set.

(2) If there is no danger of confusion, we do not distinguish between subsets of X and $\{0, 1\}$-valent **L**-sets in X. This means that if A is an **L**-set in X for which $A(x) = 0$ or $A(x) = 1$ for each $x \in X$ then we also write $x \in A$ for $A(x) = 1$ and $x \notin A$ for $A(x) = 0$, and conversely, if A is a subset of X then we also write $A(x) = 1$ for $x \in A$ and $A(x) \neq 0$ for

$x \notin A$. Particularly, the **L**-set A in X such that $A(x) = 1$ for each $x \in X$ will also be denoted by X, and the **L**-set A in X such that $A(x) = 0$ for each $x \in X$ will also be denoted by $\emptyset$.

Since L^X is the direct product of ($|X|$ copies of) L, it can be equipped with operations of **L**. Recall that an n-ary operation f defined on L extends to an n-ary operation $\overline{f} : (L^X)^n \to L^X$ by

$$(\overline{f}(A_1, \ldots, A_n))(x) = f(A_1(x), \ldots, A_n(x)), \tag{3.1}$$

i.e. $\overline{f}$ assigns to any **L**-sets $A_1, \ldots, A_n \in L^X$ an **L**-set $\overline{f}(A_1, \ldots, A_n) \in L^X$ in such a way that the membership degree of any $x \in X$ in $\overline{f}(A_1, \ldots, A_n)$ is given by (3.1), i.e. componentwise. The same way (i.e. componentwise) are defined arbitrary infima and suprema in L^X. Particularly, operations $\wedge, \vee, \otimes, \to, 0, 1$, and $\bigwedge$ and $\bigvee$ defined on L extend componentwise to operations $\wedge^{\mathbf{L}^X}, \vee^{\mathbf{L}^X}, \otimes^{\mathbf{L}^X}, \to^{\mathbf{L}^X}, 0^{\mathbf{L}^X}, 1^{\mathbf{L}^X}$, and $\bigwedge^{\mathbf{L}^X}$ and $\bigvee^{\mathbf{L}^X}$ on L^X. To comply with the usual notation we denote the operations on L^X that correspond to $\wedge$, $\vee$, $\otimes$, $\to$, 0, 1, $\bigwedge$, and $\bigvee$ by $\cap$, $\cup$, $\otimes$, $\to$, $\emptyset$, X, $\bigcap$, and $\bigcup$, respectively; the lattice order on L^X will be denoted by $\subseteq$. We therefore have

$$\begin{aligned}
(A \cap B)(x) &= A(x) \wedge B(x), \\
(A \cup B)(x) &= A(x) \vee B(x), \\
(A \otimes B)(x) &= A(x) \otimes B(x), \\
(A \to B)(x) &= A(x) \to B(x), \\
\emptyset(x) &= 0, \\
X(x) &= 1, \\
(\bigcap_{i \in I} A_i)(x) &= \bigwedge_{i \in I} A_i(x), \\
(\bigcup_{i \in I} A_i)(x) &= \bigvee_{i \in I} A_i(x), \\
A \subseteq B \quad & \textit{iff} \quad A(x) \leq B(x) \quad \textit{for any } x \in X.
\end{aligned}$$

for any $A, B, A_i \in L^X$ ($i \in I$) and $x \in X$. The thus resulting structure $\langle L^X, \cap, \cup, \otimes, \to, \emptyset, X \rangle$ will be denoted by $\mathbf{L}^X$. Since the direct product preserves the fulfillment of generalized identities (i.e. possibly involving arbitrary infima and suprema), we have

Theorem 3.6 *For any complete residuated lattice* **L** *and any set* $X \neq \emptyset$, $\mathbf{L}^X = \langle L^X, \cap, \cup, \otimes, \to, \emptyset, X \rangle$ *is a complete residuated lattice where infima and suprema are given by* $\bigcap$ *and* $\bigcup$*, respectively, and the corresponding lattice order is* $\subseteq$*. Moreover, all generalized identities satisfied by* **L** *are valid also in* $\mathbf{L}^X$.

Remark 3.7 (1) The preceding considerations generalize the classical set operations to fuzzy set operations. Theorem 3.6 generalizes the well-known fact that the set of all subsets of a given set is a complete Boolean algebra **2** (the last statement is a special case of Theorem 3.6; just take the two-element Boolean algebra for **L**).

(2) Since $\mathbf{L}^X$ is a compete residuated lattice, all results of Section 2.3 apply. Furthermore, if **L** satisfies any of the additional conditions for residuated lattices (like the law of double negation, being a Heyting algebra, BL-algebra, MV-algebra, etc.) then so does $\mathbf{L}^X$.

(3) The way we defined the basic operations with **L**-sets (i.e. $\cap, \cup$, etc.) is a special case of a more general approach. Each generalized term (i.e. a term containing possibly general infima and suprema) $t(x_1, \ldots, x_n)$ in the language of residuated lattices defines a mapping assigning to every $A_1, \ldots, A_n \in L^X$ an **L**-set $t^{\mathbf{L}^X}(A_1, \ldots, A_n)$ as follows: for $t = x$ we have $t^{\mathbf{L}^X}(A) = A$; for the two nullary terms 0 and 1 we have $0^{\mathbf{L}^X} = \emptyset$ and $1^{\mathbf{L}^X} = X$; if $t_1(x_1, \ldots, x_n)$ and $t_2(x_1, \ldots, x_n)$ are terms then $t^{\mathbf{L}^X}(A_1, \ldots, A_n) = t_1^{\mathbf{L}^X}(A_1, \ldots, A_n) \wedge^{\mathbf{L}^X} t_2^{\mathbf{L}^X}(A_1, \ldots, A_n)$ for $t = t_1 \wedge t_2$, ..., $t^{\mathbf{L}^X}(A_1, \ldots, A_n) = t_1^{\mathbf{L}^X}(A_1, \ldots, A_n) \to^{\mathbf{L}^X} t_2^{\mathbf{L}^X}(A_1, \ldots, A_n)$ for $t = t_1 \to t_2$; if t_i $(i \in I)$ are terms then $t^{\mathbf{L}^X}(A_1, \ldots, A_n) = \bigwedge_{i \in I}^{\mathbf{L}^X} t_i^{\mathbf{L}^X}(A_1, \ldots, A_n)$ for $t = \bigwedge_{i \in I} t_i$ and $t^{\mathbf{L}^X}(A_1, \ldots, A_n) = \bigvee_{i \in I}^{\mathbf{L}^X} t_i^{\mathbf{L}^X}(A_1, \ldots, A_n)$ for $t = \bigvee_{i \in I} t_i$. It is straightforward to verify that for each term t in the language of residuated lattices we have $(t^{\mathbf{L}^X}(A_1, \ldots, A_n))(x) = t^{\mathbf{L}^X}(A_1(x), \ldots, A_n(x))$ for any $x \in X$.

(4) Theorem 3.6 can be generalized; not only equations but also so-called Horn formulas valid in **L** are valid in $\mathbf{L}^X$ (this follows from classical model theory [59]). We will not need it.

Furthermore, for any **L**-set A we put

$$\neg A(x) = \neg(A(x)) \tag{3.2}$$

and call the **L**-set $\neg A$ a *complement* of A.

Example 3.8 Let **L** be the standard Łukasiewicz algebra on $[0, 1]$. Figure 3.4 shows $A \cap B$, $A \cup B$, $A \otimes B$, and $\neg A$ for **L**-sets A and B.

If $\mathcal{M}$ is an **L**-set in Y and each $y \in Y$ is an **L**-set in X, we define the **L**-sets $\bigcap \mathcal{M}$ and $\bigcup \mathcal{M}$ in X by

$$(\bigcap \mathcal{M})(x) = \bigwedge_{A \in Y} \mathcal{M}(A) \to A(x),$$
$$(\bigcup \mathcal{M})(x) = \bigvee_{A \in Y} \mathcal{M}(A) \otimes A(x).$$

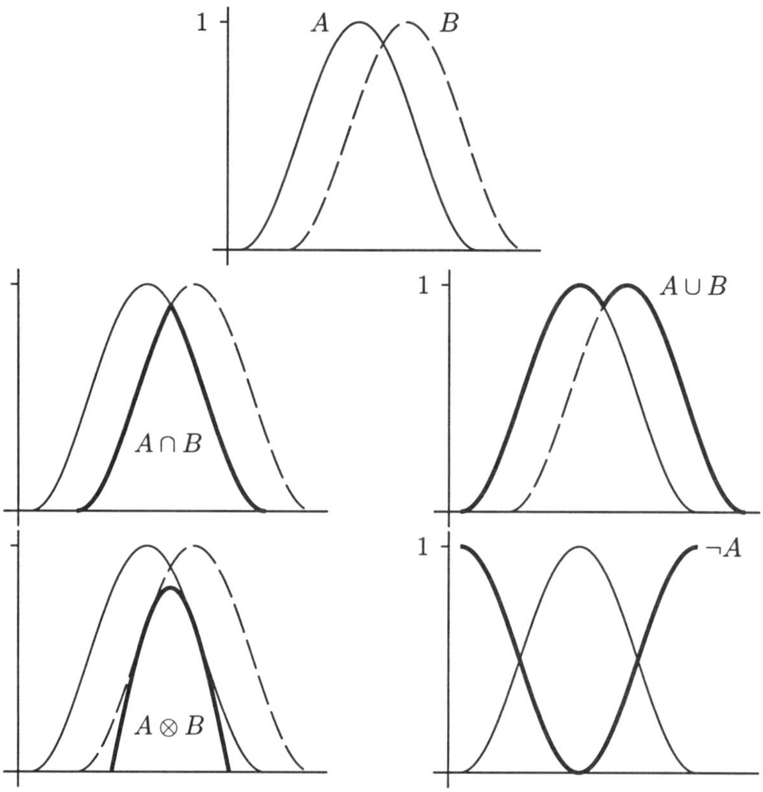

Figure 3.4. Operations with fuzzy sets.

Clearly, $\bigcap \mathcal{M}$ and $\bigcup \mathcal{M}$ are generalizations of an intersection and a union of a system of sets, respectively.

The relation $\subseteq$ on L^X is a bivalent one, i.e. for any two **L**-sets, either $A \subseteq B$ or not. From the point of view of graded approach it is natural to consider also a kind of a *graded subsethood* between **L**-sets. An immediate generalization yields the following one: For **L**-sets $A, B \in L^X$, put

$$S(A, B) = \bigwedge_{x \in X} (A(x) \to B(x)). \tag{3.3}$$

Hence, S is a binary **L**-relation on L^X; $S(A, B)$ is called a *subsethood degree of* A *in* B. Intuitively, $S(A, B)$ expresses the truth value of "each element of A is an element of B", cf. Example 3.70. As we will see, $\subseteq$ can be defined by S since $S(A, B) = 1$ iff $A \subseteq B$ (Theorem 3.12).

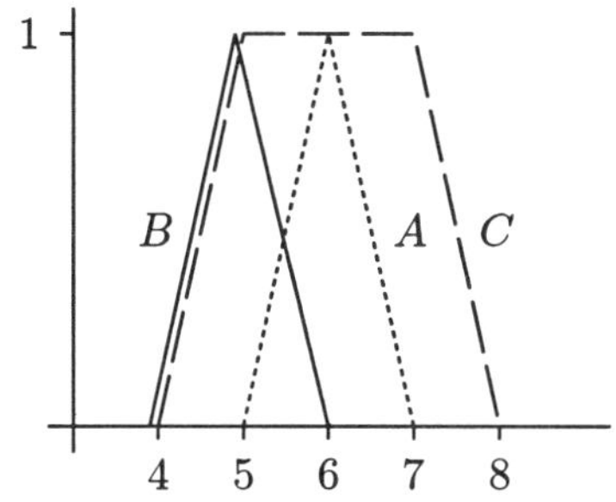

Figure 3.5. Graded subsethood.

Example 3.9 For Łukasiewicz, Gödel, and product structures on $[0,1]$, we have

$$\begin{aligned} S(A,B) &= \inf\{1 - A(x) + B(x) \mid x \in X, A(x) > B(x)\} \text{ (Łukasiewicz)}, \\ S(A,B) &= \inf\{B(x) \mid x \in X, A(x) > B(x)\} \quad \text{(Gödel)}, \\ S(A,B) &= \inf\{B(x)/A(x) \mid x \in X, A(x) > B(x)\} \quad \text{(product)}. \end{aligned}$$

Remark 3.10 Another way to look at $S(A,B)$ is the following one: Take the **L**-set $A \to B$ in X. $A \to B$ can be thought of as a truth value (note that L^X is a complete residuated lattice and can thus be considered as a structure of truth values), an aggregate of truth values $(A \to B)(x)$. The x-th coordinate $(A \to B)(x)$ bears the information about the subsethood w.r.t. x. The apparent disadvantage of taking $A \to B$ as the truth value expressing the subsethood of A in B is the increase of dimensionality in the structure of truth values. Taking $S(A,B)$, i.e. the infimum of all components of $A \to B$, is a natural "aggregation step".

Example 3.11 Consider the standard Łukasiewicz algebra **L** on $[0,1]$ and **L**-sets A, B, C in $X = [3,9]$ as depicted in Fig. 3.5. A and B are so-called *triangular* fuzzy sets, C a so-called *trapezoidal* fuzzy set. Generally, a triangular fuzzy set Tri in $X \subseteq \mathbf{R}$ is given by

$$\mathrm{Tri}(x) = \begin{cases} \frac{x-a}{b-a} & \text{for } a \le x < b \\ \frac{c-x}{c-b} & \text{for } b \le x \le c \\ 0 & \text{for } x < a \text{ or } c < x \end{cases}$$

where the real numbers $a \le b \le c$ determine Tri; a trapezoidal fuzzy set Tra in $X \subseteq \mathbf{R}$ is given by

$$\mathrm{Tra}(x) = \begin{cases} \frac{x-a}{b_1-a} & \text{for } a \le x < b_1 \\ 1 & \text{for } b_1 \le x < b_2 \\ \frac{c-x}{c-b_2} & \text{for } b_2 \le x \le c \\ 0 & \text{for } x < a \text{ or } c < x \end{cases}$$

with the real numbers $a \leq b_1 \leq b_2 \leq c$ determining Tra. Obviously, triangular fuzzy sets are special cases of trapezoidal fuzzy sets (for $b_1 = b_2$). In our case, A and B are determined by $a = 5, b = 6, c = 7$, and $a = 3.9, b = 4.9, c = 5.9$, respectively; C is given by $a = 4, b_1 = 5, b_2 = 7, c = 8$. Since $A(x) \leq C(x)$ for each $x \in X$, $S(A, C) = 1$, i.e. $A \subseteq C$, A is fully contained in C. On the other hand, $B \subseteq C$ is not the case. However, as one can easily see, $S(B, C) = 0.9$, i.e. the subsethood degree to which B is contained in C is 0.9; expressing the intuition that B is almost contained in C.

The following proposition summarizes the basic properties of S.

Theorem 3.12 (graded subsethood) *For* **L***-sets A, B, and C in X, we have*

$$S(A, A) = 1, \tag{3.4}$$

$$S(A, B) \otimes S(B, C) \leq S(A, C), \tag{3.5}$$

$$S(A, B) = 1 \quad \textit{iff} \quad A \subseteq B, \tag{3.6}$$

$$S(\emptyset, A) = 1, \quad S(A, X) = 1, \tag{3.7}$$

$$S(A, \bigcap_{i \in I} B_i) = \bigwedge_{i \in I} S(A, B_i), \quad S(\bigcup_{i \in I} A_i, B) = \bigwedge_{i \in I} S(A_i, B), \tag{3.8}$$

$$S(A, A \cap B) = S(A, B) = S(A \cup B, B), \tag{3.9}$$

$$S(A, B) \leq S(\neg B, \neg A), \tag{3.10}$$

if **L** *satisfies the law of double negation then* (3.11)

$$S(A, B) = S(\neg B, \neg A).$$

Proof. We will use Theorem 3.6, (3.3), and the fact that if $a_x \leq b_x$ ($a_x, b_x \in L$, $x \in X$) then $\bigwedge_{x \in X} a_x \leq \bigwedge_{x \in X} b_x$.

(3.4): $S(A, A) = \bigwedge_{x \in X} (A \rightarrow A)(x) = \bigwedge_{x \in X} 1 = 1$.

(3.5): $S(A, B) \otimes S(B, C) = (\bigwedge_{x \in X} A(x) \rightarrow B(x)) \otimes (\bigwedge_{y \in X} B(y) \rightarrow C(y)) \leq \bigwedge_{x \in X} \bigwedge_{y \in X} (A(x) \rightarrow B(x)) \otimes (B(y) \rightarrow C(y)) \leq \bigwedge_{x \in X} (A(x) \rightarrow B(x)) \otimes (B(x) \rightarrow C(x)) \leq \bigwedge_{x \in X} A(x) \rightarrow C(x) = S(A, C)$.

(3.6): If $S(A, B) = 1$ iff for each $x \in X$ we have $A(x) \rightarrow B(x) = 1$ iff for each $x \in X$ we have $A(x) \leq B(x)$ iff $A \subseteq B$.

(3.7): $S(\emptyset, A) = \bigwedge_{x \in X} 0 \rightarrow A(x) = 1$, and $S(A, X) = \bigwedge_{x \in X} A(x) \rightarrow 1 = 1$.

(3.8): By Theorem 3.6, $S(A, \bigcap_{i \in I} B_i) = \bigwedge_{x \in X} A \rightarrow \bigwedge_{i \in I} B_i = \bigwedge_{x \in X} \bigwedge_{i \in I} A \rightarrow B_i = \bigwedge_{i \in I} \bigwedge_{x \in X} A \rightarrow B_i = \bigwedge_{i \in I} S(A, B_i)$. Similarly, $S(\bigcup_{i \in I} A_i, B) = \bigwedge_{x \in X} (\bigvee_{i \in I} A_i \rightarrow B) = \bigwedge_{x \in X} \bigwedge_{i \in I} A_i \rightarrow B = \bigwedge_{i \in I} \bigwedge_{x \in X} A_i \rightarrow B = \bigwedge_{i \in I} S(A_i, B)$

(3.9): By (3.4) and (3.8), $S(A, A \cap B) = S(A, A) \wedge S(A, B) = 1 \wedge S(A, B) = S(A, B)$, and $S(A \cup B, B) = S(A, B) \wedge S(B, B) = S(A, B) \wedge 1 = S(A, B)$.

(3.10): By Theorem 3.6, $A \to B \subseteq \neg B \to \neg A$, therefore $S(A, B) = \bigwedge_{x \in X}(A \to B)(x) \leq \bigwedge_{x \in X}(\neg B \to \neg A) = S(\neg B, \neg A)$.

(3.11): Like (3.10) using $A \to B = \neg B \to \neg A$. □

The same reasons as those for considering graded subsethood lead us to consider graded equality of **L**-sets: for **L**-sets A and B in X put

$$\approx (A, B) \;=\; \bigwedge_{x \in X} (A(x) \leftrightarrow B(x)). \tag{3.12}$$

$\approx$ is a binary **L**-relation on L^X. $\approx (A, B)$ is called the *degree of equality* of A and B; usually we use infix notation and write $(A \approx B)$ instead of $\approx (A, B)$. Again, $(A \approx B)$ intuitively expresses the truth value of "A and B contain the same elements" (or, more precisely, "for all x: x belongs to A iff x belongs to B"). The equality $=$ can be defined by $\approx$ since $\approx (A, B) = 1$ iff $A = B$ (i.e. $A(x) = B(x)$ for all $x \in X$).

Example 3.13 In case of Łukasiewicz, Gödel, and product structures on $[0, 1]$, we have

$$\begin{aligned} A \approx B &= \inf_{x \in X} 1 - |A(x) - B(x)| \quad \text{(Łukasiewicz)}, \\ A \approx B &= \inf_{x \in X, A(x) \neq B(x)} \min(A(x), B(x)) \quad \text{(Gödel)}, \\ A \approx B &= \inf_{x \in X} \min(A(x)/B(x), B(x)/A(x)) \quad \text{(product)} \end{aligned}$$

where we put $0/0 = 1$ and $a/0 = \infty$ for $a \neq 0$.

Theorem 3.14 (graded equality) *Let A, B, C, D be **L**-sets in X. Then we have*

$$A \approx A \;=\; 1, \tag{3.13}$$

$$A \approx B \;=\; B \approx A, \tag{3.14}$$

$$(A \approx B) \otimes (B \approx C) \;\leq\; A \approx C, \tag{3.15}$$

$$A \approx B \;=\; 1 \quad \textit{iff} \quad A = B, \tag{3.16}$$

$$A \approx B \;=\; S(A, B) \wedge S(B, A), \tag{3.17}$$

$$A \approx \emptyset \;=\; S(A, \emptyset), \quad A \approx X \;=\; S(X, A), \tag{3.18}$$

$$A \approx A \cap B \;=\; A \approx B \;=\; A \approx A \cup B, \tag{3.19}$$

$$(A \approx B) \otimes S(B, C) \otimes (C \approx D) \;\leq\; S(A, D), \tag{3.20}$$

$$A \approx B \;\leq\; \neg B \approx \neg A, \tag{3.21}$$

*if **L** satisfies the law of double negation then*

$$A \approx B \;=\; \neg B \approx \neg A. \tag{3.22}$$

Proof. We use Theorem 3.6 and the properties of residuated lattices.

(3.13): Follows from the definition of $\approx$ by $A(x) \leftrightarrow A(x) = 1$.

(3.14): Follows from the definition of $\approx$ by $A(x) \approx B(x) = B(x) \leftrightarrow A(x)$.

(3.15): $(A \approx B) \otimes (B \approx C) = (\bigwedge_{x \in X} A(x) \leftrightarrow B(x)) \otimes (\bigwedge_{y \in X} B(y) \leftrightarrow C(y)) \leq \bigwedge_{x \in X} \bigwedge_{y \in X} (A(x) \leftrightarrow B(x)) \otimes (B(y) \leftrightarrow C(y)) \leq \bigwedge_{x \in X} (A(x) \leftrightarrow B(x)) \otimes (B(x) \leftrightarrow C(x)) \leq \bigwedge_{x \in X} A(x) \leftrightarrow C(x) = (A \approx C)$.

(3.16): $A \approx B = 1$ iff $A(x) \leftrightarrow B(x) = 1$ for each $x \in X$ iff $A(x) = B(x)$ for each $x \in X$ iff $A = B$.

(3.17): $A \approx B = \bigwedge_{x \in X} A(x) \leftrightarrow B(x) = \bigwedge((A(x) \rightarrow B(x)) \wedge (B(x) \rightarrow A(x))) = (\bigwedge_{x \in X} A(x) \rightarrow B(x)) \wedge (\bigwedge_{x \in X} B(x) \rightarrow A(x)) = S(A, B) \wedge S(B, A)$.

(3.18): Follows from (3.17) directly by (3.7).

(3.19): Follows from (3.17) by (3.9) using $S(A, A \cup B) = 1$ and $S(A \cap B, A) = 1$.

(3.20): By (3.17) and (3.5), we have $(A \approx B) \otimes S(B, C) \otimes (C \approx D) \leq S(A, B) \otimes S(B, C) \otimes S(C, D) \leq S(A, D)$.

(3.21) and (3.22) follow directly from (3.17) by (3.10) and (3.11), respectively. □

Remark 3.15 Notice how the assertions of Theorems 3.12 and 3.14 can be described verbally. For instance, take (3.5). If $\mathbf{L} = \mathbf{2}$ (bivalent case), (3.5) expresses exactly that if a set A is a subset of B, and B is a subset of C, then A is a subset of C. For a general $\mathbf{L}$, the meaning of (3.5) is the same, only interpreted in many-valued setting. The same holds for other assertions. So, for instance, (3.20) says that if A is equal to B, and B is a subset of C, and C is equal to D, then A is a subset of D.

Clearly, further properties of $\mathbf{L}$ can yield further properties of S and $\approx$. Generally, let $t_i(x_1, \ldots, x_n)$ $(i = 1, \ldots, 4)$ be generalized terms in the language of residuated lattices, and let, for example, $t_1 \rightarrow t_2 \leq t_3 \leftrightarrow t_4$ hold true in $\mathbf{L}$. Then, by Theorem 3.6, for any $A_1, \ldots, A_n \in L^X$, $(t_1(A_1, \ldots, A_n) \rightarrow t_2(A_1, \ldots, A_n)) \subseteq (t_3(A_1, \ldots, A_n) \leftrightarrow t_4(A_1, \ldots, A_n))$ holds, i.e. for each $x \in X$ we have $(t_1(A_1, \ldots, A_n) \rightarrow t_2(A_1, \ldots, A_n))(x) \leq (t_3(A_1, \ldots, A_n) \leftrightarrow t_4(A_1, \ldots, A_n))(x)$, and thus $S(t_1(A_1, \ldots, A_n), t_2(A_1, \ldots, A_n)) \leq (t_3(A_1, \ldots, A_n) \approx t_4(A_1, \ldots, A_n))$, etc. We therefore have the following theorem.

Theorem 3.16 *Let $t_1, \ldots, t_4$ be generalized terms in the language of residuated lattices, $\square$ and $\diamond$ be $\rightarrow$ or $\leftrightarrow$. If $t_1 \square t_2 \leq t_3 \diamond t_4$ is true in $\mathbf{L}$ then for every $A_1, \ldots, A_n \in L^X$ we have*

$$R_\square(t_1(A_1, \ldots, A_n), t_2(A_1, \ldots, A_n)) \leq R_\diamond(t_3(A_1, \ldots, A_n), t_4(A_1, \ldots, A_n))$$

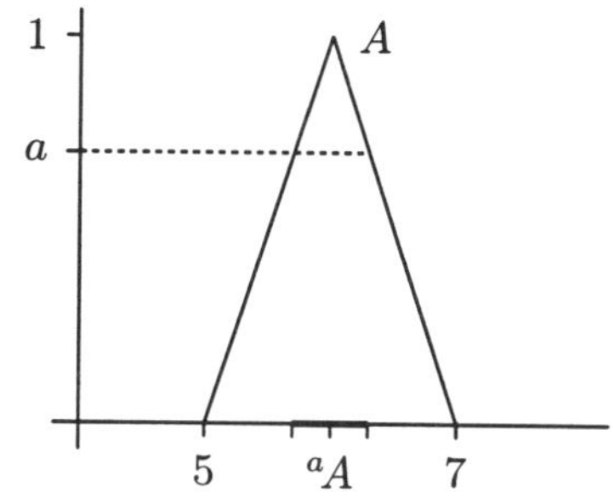

Figure 3.6. 0.7-cut of A.

where $R_{\rightarrow}$ *is* S *and* $R_{\leftrightarrow}$ *is* $\approx$.

Proof. By induction on terms; as indicated above. □

3.1.2 REPRESENTATION BY BIVALENT SETS

We are going to present two ways of representation of **L**-sets by **2**-sets (i.e. classical sets).

First way: cutlike representation . Fuzzy sets in X can be represented by special systems of subsets of X.

Definition 3.17 Let A be an **L**-set in X, $a \in L$. The *a-cut* of A is the subset ${}^aA$ of X defined by

$${}^aA=\{x \in X \mid A(x) \geq a\}.$$

Therefore, ${}^aA$ is the set of all $x \in X$ the membership degree of which is at least a.

Example 3.18 (1) Let $X = \{x_1, x_2, x_3, x_4\}$, $L = \{0, \frac{1}{3}, \frac{2}{3}, 1\}$, and consider an **L**-set A in X (with any structure of **L**) given by $A = \{{}^0/x_1, {}^1/x_2, {}^{\frac{2}{3}}/x_3, {}^{\frac{2}{3}}/x_4,\}$. Then ${}^0A = X$, ${}^{\frac{1}{3}}A = {}^{\frac{2}{3}}A = \{x_2, x_3, x_4\}$, ${}^1A = \{x_2\}$.

(2) Consider a triangular fuzzy set A determined by parameters $a = 5, b = 6, c = 7$ as depicted in Fig. 3.6, take $a = 0.7$. Then ${}^aA = [5.7, 6.3]$.

Definition 3.19 An L-indexed system $\mathcal{S} = \{A_a \subseteq X \mid a \in L\}$ is called **L**-*nested* if

(1) $a \leq b$ implies $A_b \subseteq A_a$,

(2) for each $x \in X$ the set $\{a \mid x \in A_a\}$ has a greatest element.

Remark 3.20 (1) Two L-indexed systems $\{A_a \mid a \in L\}$ and $\{B_a \mid a \in L\}$ are equal iff $A_a = B_a$ for each $a \in L$.

(2) Note that if $\mathcal{S}$ is $\mathbf{L}$-nested then $A_0 = X$. Indeed, by (2), for each $x \in X$ there is some a such that $x \in A_a$. By (1), $A_a \subseteq A_0$, thus $x \in A_0$.

Lemma 3.21 *For an $\mathbf{L}$-set A in X, $\{^aA \mid a \in L\}$ is an $\mathbf{L}$-nested system of subsets of X.*

Proof. The proof is almost evident. Let $a \leq b$. If $x \in {}^bA$ then $b \leq A(x)$, and thus also $a \leq A(x)$ whence $x \in {}^aA$. The greatest element of $\{a \mid x \in {}^aA\}$ is evidently $A(x)$. □

For an $\mathbf{L}$-set A in X, and a system $\mathcal{S} = \{A_a \subseteq X \mid a \in L\}$ of subsets of X we define a system $\mathcal{S}_A$ of subsets of X, and an $\mathbf{L}$-set $A_\mathcal{S}$ in X by

$$\begin{aligned} \mathcal{S}_A &= \{^aA \mid a \in L\}, \\ A_\mathcal{S}(x) &= \bigvee_{x \in A_a} a. \end{aligned}$$

Therefore, $\mathcal{S}_A$ is the system of all a-cuts.

Theorem 3.22 *Let A be an $\mathbf{L}$-set in X, $\mathcal{S}$ be an $\mathbf{L}$-nested system of subsets of X. Then $\mathcal{S}_A$ is $\mathbf{L}$-nested, and we have $A = A_{\mathcal{S}_A}$ and $\mathcal{S}_{A_\mathcal{S}}$.*

Proof. The fact that $\mathcal{S}_A$ is $\mathbf{L}$-nested follows directly from Lemma 3.21.

We show $A = A_{\mathcal{S}_A}$: $A(x) \leq A_{\mathcal{S}_A}(x)$ follows from $x \in {}^{A(x)}A \in \mathcal{S}_A$ by definition of $A_{\mathcal{S}_A}$. Conversely, $A_{\mathcal{S}_A}(x) \leq A(x)$ holds iff $\bigvee_{x \in {}^aA} a \leq A(x)$ which holds iff for each $a \in L$ such that $x \in {}^aA$ we have $a \in A(x)$. However, this is true since $x \in {}^aA$ means that $a \leq A(x)$.

It remains to show $\mathcal{S} = \mathcal{S}_{A_\mathcal{S}}$. That is, we have to verify that for each $a \in L$, $A_a = {}^aA_\mathcal{S}$. If $x \in A_a$ then $A_\mathcal{S}(x) = \bigvee_{x \in A_b} b \geq a$, and thus $x \in {}^aA_\mathcal{S}$. If, conversely, $x \in {}^aA_\mathcal{S}$, then $a \leq \bigvee_{x \in A_b} b$. By definition of an $\mathbf{L}$-nested system, the set $\{b \in L \mid x \in A_b\}$ has the greatest element, say b^*. Therefore, $\bigvee_{x \in A_b} b = b^*$ and thus $a \leq b^*$. Since $\mathcal{S}$ is $\mathbf{L}$-nested, we have $A_{b^*} \subseteq A_a$, and from $x \in A_{b^*}$ we thus get $x \in A_a$. □

Remark 3.23 Among others, Theorem 3.22 says that the mappings $A \mapsto \mathcal{S}_A$ and $\mathcal{S} \mapsto A_\mathcal{S}$ are mutually inverse mappings between the set of all $\mathbf{L}$-sets and the set of all $\mathbf{L}$-nested systems of subsets of X. Therefore, we have established the first way, namely representation of $\mathbf{L}$-sets in X by $\mathbf{L}$-nested systems of subsets of X.

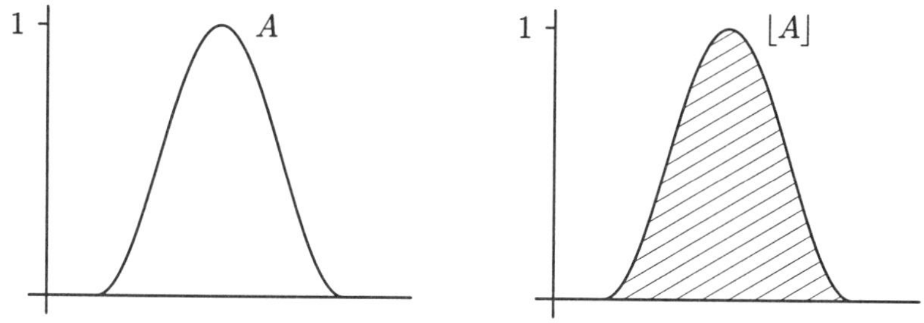

Figure 3.7. A and the corresponding $\lfloor A \rfloor$.

Second way: L-sets in X as subsets of $X \times L$.

Definition 3.24 A subset $A \subseteq X \times L$ is called ($\mathbf{L}$*-set*)*-representative* if

(1) for each $x \in X$, $\langle x, a\rangle \in A$ and $b \leq a$ imply $\langle x, b\rangle \in A$,

(2) for each $x \in X$ the set $\{a \in L \mid \langle x, a\rangle \in A\}$ has the greatest element.

For an $\mathbf{L}$-set $A \in L^X$ and a subset $A' \subseteq X \times L$ we put

$$\begin{aligned} \lfloor A \rfloor &= \{\langle x, a\rangle \in X \times L \mid a \leq A(x)\}, \\ \lceil A' \rceil &= \{\langle x, a\rangle \in X \times L \mid a = \bigvee_{\langle x,b\rangle \in A'} b\}. \end{aligned}$$

Figure 3.7 demonstrates operator $\lfloor\ \rfloor$. We have the following result.

Theorem 3.25 *Let $A \in L^X$ be an $\mathbf{L}$-set, $A' \subseteq X \times L$ be a representative set. Then (1) $\lfloor A \rfloor \subseteq X \times L$ is a representative set, (2) $\lceil A' \rceil$ is an $\mathbf{L}$-set such that (3) $A = \lceil \lfloor A \rfloor \rceil$, $A' = \lfloor \lceil A' \rceil \rfloor$.*

Proof. (1): If $\langle x, a\rangle \in \lfloor A \rfloor$ and $b \leq a$, then $b \leq a \leq A(x)$, and thus $\langle x, b\rangle \in \lfloor A \rfloor$, by definition. Clearly, $A(x)$ is the greatest a such that $\langle x, a\rangle \in \lfloor A \rfloor$.

(2): Immediately from the definition of $\lceil A' \rceil$.

(3): Since $A(x)$ the greatest a such that $\langle x, A(x)\rangle \in \lfloor A \rfloor$, we have $A(x) = \bigvee_{\langle x,a\rangle \in \lfloor A \rfloor} a = \lceil \lfloor A \rfloor \rceil(x)$. If $\langle x, a\rangle \in A'$ then $\lceil A' \rceil(x) \geq a$, and so $\langle x, a\rangle \in \lfloor \lceil A' \rceil \rfloor$. Conversely, let $\langle x, a\rangle \in \lfloor \lceil A' \rceil \rfloor$. Then $a \leq \lceil A' \rceil(x)$, i.e. $a \leq \bigvee_{\langle x,b\rangle \in A'} b$. Since A' is representative, $\{b \mid \langle x, b\rangle \in A'\}$ has the greatest element, say b^*. Therefore, $a \leq b^*$ and $\langle x, b^*\rangle \in A'$. Since A' is representative, we conclude $\langle x, a\rangle \in A'$. □

Remark 3.26 We established the second way, namely representation of fuzzy sets by sets that can be considered as areas under membership functions.

3.1.3 L-EQUIVALENCE AND L-EQUALITY RELATIONS

The purpose of this section is to present basic facts about fuzzy equivalence relations (we will see more on fuzzy equivalence relations in Sections 4.2 and 7.1). Basically, fuzzy equivalence relations are encountered when fuzzifying the notion of a (crisp) equivalence relation.

Definition 3.27 A binary **L**-relation E on X is called an **L**-*equivalence* if for any $x, y, z \in X$ we have

$$E(x,x) = 1, \tag{3.23}$$

$$E(x,y) = E(y,x), \tag{3.24}$$

$$E(x,y) \otimes E(y,z) \leq E(x,z). \tag{3.25}$$

An **L**-equivalence is called an **L**-*equality* if

$$E(x,y) = 1 \quad \text{implies} \quad x = y. \tag{3.26}$$

Remark 3.28 (1) There are various other terms used for **L**-equivalences: the most common are similarity (or fuzzy similarity) and indistinguishability. These terms, however, suggest a kind of an epistemic interpretation that may be disputable.

(2) Conditions (3.23), (3.24), and (3.25) are called reflexivity, symmetry, and transitivity, respectively. These conditions generalize their bivalent counterparts. Indeed, consider $\mathbf{L} = \mathbf{2}$: (3.23) and (3.24) obviously express reflexivity and symmetry (in bivalent sense); (3.25) implies that if $\langle x, y\rangle$ and $\langle y, z\rangle$ are in E (i.e. $E(x,y) = 1$ and $E(y,z) = 1$) then $1 = 1 \otimes 1 = E(x,y) \otimes E(y,z) \leq E(x,z)$, therefore $E(x,z) = 1$, i.e. $\langle x, z\rangle$ is in E, i.e. E is transitive (in bivalent sense). Conversely, let E be transitive (in bivalent sense). There are only two possibilities: Either $E(x,y) = 0$ or $E(y,z) = 0$ and then (3.25) is true; or both $E(x,y) = 1$ and $E(y,z) = 1$ in which case $\langle x, y\rangle$ and $\langle y, z\rangle$ are in E, and so also $\langle x, z\rangle$ is in E, i.e. $E(x,z) = 1$, i.e. (3.25) holds true.

(3) (3.23)–(3.25) are equivalent to saying that logical formulas expressing reflexivity, symmetry, and transitivity, are true in degree 1; see Example 3.70.

Example 3.29 (1) Let $X = \{x, y, z, u\}$, **L** be the standard product algebra on $[0,1]$. Let a binary **L**-relation E be given by Table 3.1. That is, $E(x,x) = 1, E(x,y) = 0.9, E(x,z) = 0.8, E(x,u) = 1$, etc. One easily verifies that E is an **L**-equivalence relation which is not an **L**-equality. Indeed, $E(x,u) = 1$ but $x \neq u$. On the other hand, restriction of E to $X - \{u\}$ is an **L**-equality on $\{x, y, z\}$.

E	x	y	z	u
x	1	0.9	0.8	1
y	0.9	1	0.7	0.9
z	0.8	0.7	1	0.8
u	1	0.9	0.8	1

Table 3.1. **L**-equivalence E.

(2) If we equip $[0,1]$ with min instead of product and consider the standard Gödel algebra $\mathbf{L}$ on $[0,1]$ then E given by Table 3.1 is not an $\mathbf{L}$-equivalence relation. Indeed, $0.8 = E(y,x) \otimes E(x,z) = \min(0.9, 0.8) \not\leq E(y,z) = 0.7$, and so E is not transitive.

We will need the following lemma (more general results will be presented in Chapter 4).

Lemma 3.30 *For any system $\mathcal{S} \subseteq L^X$ of $\mathbf{L}$-sets in X, the $\mathbf{L}$-relation $\approx_{\mathcal{S}}$ on X defined by*

$$(x \approx_{\mathcal{S}} y) = \bigwedge_{A \in \mathcal{S}} (A(x) \leftrightarrow A(y)) \tag{3.27}$$

is an $\mathbf{L}$-equivalence relation on X. Moreover, $\approx_{\mathcal{S}}$ is an $\mathbf{L}$-equality iff for every distinct x and y there is some $A \in \mathcal{S}$ such that $A(x) \neq A(y)$.

Proof. We show that $\approx_{\mathcal{S}}$ is an $\mathbf{L}$-equivalence: Reflexivity: Since $A(x) \leftrightarrow A(x) = 1$, we have $(x \approx_{\mathcal{S}} x) = \bigwedge_{A \in \mathcal{S}}(A(x) \leftrightarrow A(x)) = 1$. Symmetry follows from $A(x) \leftrightarrow A(y) = A(y) \leftrightarrow A(x)$. Transitivity:

$$\begin{aligned}
& (x \approx_{\mathcal{S}} y) \otimes (y \approx_{\mathcal{S}} z) = \\
= \; & \bigwedge_{A \in \mathcal{S}} (A(x) \leftrightarrow A(y)) \otimes \bigwedge_{B \in \mathcal{S}} (B(y) \leftrightarrow B(z)) \leq \\
\leq \; & \bigwedge_{A \in \mathcal{S}} \bigwedge_{B \in \mathcal{S}} (A(x) \leftrightarrow A(y)) \otimes (B(y) \leftrightarrow B(z)) \leq \\
\leq \; & \bigwedge_{A \in \mathcal{S}} (A(x) \leftrightarrow A(y)) \otimes (A(y) \leftrightarrow A(z)) \leq \\
\leq \; & \bigwedge_{A \in \mathcal{S}} (A(x) \leftrightarrow A(z)) = (x \approx_{\mathcal{S}} z).
\end{aligned}$$

The second statement: Let x and y be distinct (i.e. $x \neq y$). Then we have $(x \approx_{\mathcal{S}} y) \neq 1$ iff there is some $A \in \mathcal{S}$ such that $A(x) \leftrightarrow A(y) \neq 1$, i.e. $A(x) \neq A(y)$. □

Remark 3.31 The set $\mathcal{S}$ from Lemma 3.30 can be considered to be a set of attributes. Putting in more detail, each $A \in \mathcal{S}$ can be thought of as a (fuzzy) attribute, $A(x)$ being the degree to which the element $x \in X$ has the attribute A. Then $x \approx_{\mathcal{S}} y$ is the truth degree of "for each attribute A: x has A iff y has A" or, more succinctly, the truth degree of "x and y have the same attributes (of $\mathcal{S}$)". Having the same attributes has been considered by Leibniz [4] a criterion for being the same. Therefore, (3.27) can be considered a fuzzy counterpart of Leibniz definition. We will call $\approx_{\mathcal{S}}$ (and **L**-relations defined in a similar spirit) a *Leibniz* **L***-equivalence* (induced by $\mathcal{S}$). For more information on Leibniz equivalence see Section 4.2.

In Section 3.1 we defined the equality degree $A \approx B$ of **L**-sets $A, B \in L^X$. (3.4)–(3.7) show that $\approx$ is an **L**-equality on L^X. This fact is a special case of the following corollary of Lemma 3.30:

Corollary 3.32 *For $Z \subseteq Y$, the* **L***-relation $\approx_Z$ on L^Y defined by*

$$(A \approx_Z B) = \bigwedge_{y \in Z} (A(y) \leftrightarrow B(y))$$

is an **L***-equivalence. Moreover, $\approx_Z$ is an* **L***-equality iff for every distinct A and B there is some $y \in Z$ such that $A(y) \neq B(y)$. Particularly, $\approx_Y$ is an* **L***-equality.*

Proof. Put $X = L^Y$, $\mathcal{S} = Z$, and for $y \in Y$ and $A \in L^Y$ put $y(A) = A(y)$ and apply Lemma 3.30. □

Remark 3.33 Note that the proof of Corollary 3.32 may be turned the other way around. Starting from Corollary 3.32, one gets Lemma 3.30 as follows: put $Y = L^X$, $Z = \mathcal{S}$. Then $\approx_Z$ is an **L**-equivalence on L^Y (i.e. on L^{L^X}). For each $x \in X$ take one $A_x \in L^Y$ such that $A_x(z) = z(x)$ (note that $z \in L^X$). Then the restriction $\approx$ of $\approx_Z$ to $U = \{A_x \mid x \in X\}$ is an **L**-equivalence and one has

$$A_{x_1} \approx A_{x_2} = \bigwedge_{A \in Z} A_{x_1}(A) \approx A_{x_2}(A) =$$

$$= \bigwedge_{A \in \mathcal{S}} A(x_1) \approx A(x_2) = x_1 \approx_{\mathcal{S}} x_2,$$

i.e. $\approx_{\mathcal{S}}$ is an **L**-equivalence on X. The rest is immediate.

Definition 3.34 If $\approx$ is an **L**-equivalence relation on X, an **L**-set A in X is said to be *compatible* with $\approx$ if

$$A(x) \otimes (x \approx y) \leq A(y)$$

for all $x, y \in X$. If $\approx_i$ is an $\mathbf{L}$-equivalence relation on X_i $(i = 1, \ldots, n)$, an $\mathbf{L}$-relation R between $X_1, \ldots, X_n$ is said to be compatible with $\approx_1, \ldots, \approx_n$ if

$$(x_1 \approx_1 y_1) \otimes \cdots \otimes (x_n \approx_n y_n) \otimes R(x_1, \ldots, x_n) \leq R(y_1, \ldots, y_n)$$

for all $x_i, y_i \in X_i$ $(i = 1, \ldots, n)$.

Definition 3.35 An $\mathbf{L}$-set A in X is called an $\approx$-singleton ($\approx$ is an $\mathbf{L}$-equivalence) if there is $x_0 \in X$ such that $A(x) = (x_0 \approx x)$ for each $x \in X$.

Remark 3.36 (1) Obviously, each $\approx$-singleton is compatible with $\approx$.

(2) The notion of an $\approx$-singleton generalizes the above-introduced notion of a singleton (denoted $\{ {}^1/x \}$). Namely, $\{ {}^1/x \}$ is an $\approx$-singleton for $\approx$ being the crisp identity (i.e. $(x \approx y) = 1$ for $x = y$ and $(x \approx y) = 0$ for $x \neq y$).

(3) If A is an $\approx$-singleton with $A(x_0) = 1$, then A is the least compatible $\mathbf{L}$-set with $A(x_0) = 1$.

For $A \in L^X$ and an $\mathbf{L}$-equivalence $\approx$ on X we define the $\mathbf{L}$-set $\mathrm{C}_\approx(A)$ in X (denoted also simply $\mathrm{C}(A)$) by

$$(\mathrm{C}_\approx(A))(x) = \bigvee_{y \in X} A(y) \otimes (y \approx x). \tag{3.28}$$

It is easy to show that $\mathrm{C}_\approx(A)$ is the least $\mathbf{L}$-set in X that contains A and is compatible with $\approx$ (we will prove this in Section 7.1).

Instead of saying that A is compatible with $\approx$ we also say that $\approx$ is compatible with A; similarly for R. The set of all $\mathbf{L}$-relations R between $X_1, \ldots, X_n$ that are compatible with $\approx_1, \ldots, \approx_n$ will be denoted by $L^{\langle X_1, \approx_1 \rangle \times \cdots \times \langle X_n, \approx_n \rangle}$. Clearly, $L^{X_1 \times \cdots \times X_n}$ is exactly $L^{\langle X_1, \approx_1 \rangle \times \cdots \times \langle X_n, \approx_n \rangle}$ for $\approx_i = \mathrm{id}_{X_i}$ (i.e. $\approx_i (x_i, y_i) = 1$ if $x_i = y_i$ and $\approx_i (x_i, y_i) = 0$ if $x_i \neq y_i$).

Lemma 3.37 *An $\mathbf{L}$-relation $R \in L^{X_1 \times \cdots \times X_n}$ is compatible with $\approx_1, \ldots, \approx_n$ iff for every $x_i, y_i \in X_i$ $(i = 1, \ldots, n)$ we have*

$$(x_1 \approx_1 y_1) \otimes \cdots \otimes (x_n \approx_n y_n) \leq R(x_1, \ldots, x_n) \leftrightarrow R(y_1, \ldots, y_n).$$

Proof. If R is compatible then adjointness yields $(x_1 \approx_1 y_1) \otimes \cdots \otimes (x_n \approx_n y_n) \leq R(x_1, \ldots, x_n) \to R(y_1, \ldots, y_n)$ and thus also $(y_1 \approx_1 x_1) \otimes \cdots \otimes (y_n \approx_n x_n) \leq R(y_1, \ldots, y_n) \to R(x_1, \ldots, x_n)$. Since $(x_i \approx_i y_i) = (y_i \approx_i x_i)$, we get $(x_1 \approx_1 y_1) \otimes \cdots \otimes (x_n \approx_n y_n) \leq (R(x_1, \ldots, x_n) \to R(y_1, \ldots, y_n)) \wedge (R(y_1, \ldots, y_n) \to R(x_1, \ldots, x_n)) = R(x_1, \ldots, x_n) \leftrightarrow R(y_1, \ldots, y_n)$.

Conversely, if R satisfies the condition of lemma, it also satisfies $(x_1 \approx_1 y_1) \otimes \cdots \otimes (x_n \approx_n y_n) \leq R(x_1, \ldots, x_n) \to R(y_1, \ldots, y_n)$. Now, compatibility of R follows immediately by adjointness. □

Compatibility of a fuzzy relation with fuzzy equivalence relations expresses the intuitively acceptable condition "if there is a relationship between $x_1, \ldots, x_n$ and if $y_1, \ldots, y_n$ are pairwise similar to $x_1, \ldots, x_n$, then there is also the relationship between $y_1, \ldots, y_n$"; see Example 3.70. The condition of compatibility may also be formulated for functions.

Definition 3.38 Let $\approx$ and $\approx'$ be equivalence **L**-relations on X and Y, respectively. A function $f : X \to Y$ is said to be *compatible* with $\approx$ and $\approx'$ if

$$(x \approx y) \leq (f(x) \approx' f(y))$$

for any $x, y \in X$. More generally, if $\approx_i$ and $\approx$ are **L**-equivalence **L**-relations on X_i $(i = 1, \ldots, n)$ and Y, respectively, a function $f : X_1 \times \cdots \times X_n \to Y$ is said to be compatible with $\approx_1, \ldots, \approx_n, \approx$ if

$$(x_1 \approx_1 y_1) \otimes \cdots \otimes (x_n \approx_n y_n) \leq (f(x_1, \ldots, x_n) \approx f(y_1, \ldots, y_n))$$

for all $x_1, y_1 \in X_1, \ldots, x_n, y_n \in X_n$.

We also write $f : \langle X_1, \approx_1 \rangle \times \cdots \times \langle X_n, \approx_n \rangle \to \langle Y, \approx \rangle$ to denote the fact that the function $f : X_1 \times \cdots \times X_n \to Y$ is compatible with **L**-equivalences $\approx_1, \ldots, \approx_n, \approx$ on $X_1, \ldots, X_n, Y$, respectively.

For a binary **L**-relation $\approx$ on X and a positive integer k, we introduce the **L**-relation $\approx^k$ on X defined by $(x \approx^k y) = (x \approx y)^k$. For binary **L**-relations $\approx_1, \ldots, \approx_n$ on $X_1, \ldots, X_n$, respectively, we introduce the **L**-relation $\bigotimes_{i=1}^n \approx_i$ on $X_1 \times \cdots \times X_n$ by $(\langle x_1, \ldots, x_n \rangle \bigotimes_{i=1}^n \approx_i \langle x'_1, \ldots, x'_n \rangle) = \bigotimes_{i=1}^n (x_i \approx_i x'_i)$. Basic properties of these constructions for the case of **L**-equivalence relations are summarized in the following lemma.

Lemma 3.39 *Let $\approx_i$ be an **L**-equivalence relation on X_i $(i = 1, \ldots, n)$. Then for every nonnegative integer $k_1, \ldots, k_n$, the **L**-relation $\bigotimes_{i=1}^n \approx_i^{k_i}$ on $X_1 \times \cdots \times X_n$ defined for $x = \langle x_1, \ldots, x_n \rangle$ and $x' = \langle x'_1, \ldots, x'_n \rangle$ by*

$$(x \bigotimes_{i=1}^n \approx_i^{k_i} x') = \bigotimes_{i=1}^n (x_i \approx_i x'_i)^{k_i}$$

*is an **L**-equivalence on $X_1 \times \cdots \times X_n$. Moreover, $\bigotimes_{i=1}^n \approx_i^{k_i}$ is an **L**-equality iff each $\approx_i$ is an **L**-equality.*

Proof. We prove that $\bigotimes_{i=1}^{n} \approx_i^{k_i}$ is an **L**-equivalence. Reflexivity follows easily from reflexivity of $\approx_i$'s by $1^n = 1$ (for any $n \geq 0$). Symmetry follows from symmetry of $\approx_i$'s. Transitivity: For $x_j = \langle x_{j1}, \ldots, x_{jn} \rangle$ ($j = 1, 2, 3$) we have

$$
\begin{aligned}
& (x_1 \bigotimes_{i=1}^{n} \approx_i^{k_i} x_2) \otimes (x_2 \bigotimes_{i=1}^{n} \approx_i^{k_i} x_3) = \\
= \; & \bigotimes_{i=1}^{n} (x_{1i} \approx_i x_{2i})^{k_i} \otimes \bigotimes_{i=1}^{n} (x_{2i} \approx_i x_{3i})^{k_i} = \\
= \; & \bigotimes_{i=1}^{n} ((x_{1i} \approx_i x_{2i}) \otimes (x_{2i} \approx_i x_{3i}))^{k_i} \leq \\
\leq \; & \bigotimes_{i=1}^{n} (x_{1i} \approx_i x_{3i})^{k_i} = (x_1 \bigotimes_{i=1}^{n} \approx_i^{k_i} x_3).
\end{aligned}
$$

Take now any $x_j = \langle x_{j1}, \ldots, x_{jn} \rangle$ ($j = 1, 2$) such that $(x_1 \bigotimes_{i=1}^{n} \approx_i^{k_i} x_3) = 1$. If all $\approx_i$ are **L**-equalities then since (as is easily seen) $(x_{1i} \approx_i x_{2i}) = 1$ we conclude $x_{1i} = x_{2i}$ for each $i = 1, \ldots, n$, whence $x_1 = x_2$. If some $\approx_i$ (say $\approx_1$, for simplicity) is not an **L**-equality then there are $x_1, x_1' \in X_1$, $x_1 \neq x_1'$ with $(x_1 \approx_1 x_1') = 1$. Taking any $x_2 \in X_2, \ldots, x_n \in X_n$, we can easily see that for $x = \langle x_1, \ldots, x_n \rangle$ and $x' = \langle x_1', x_2, \ldots, x_n \rangle$ we have $x \neq x'$ but $(x \bigotimes_{i=1}^{n} \approx_i^{k_i} x') = 1$. Thus, $\bigotimes_{i=1}^{n} \approx_i^{k_i}$ is not an **L**-equality. □

Remark 3.40 Lemma 3.39 has two immediate consequences. First, if $\approx$ is an **L**-equivalence relation on X then so is $\approx^k$ for any positive integer k (and $\approx^k$ is an **L**-equality iff $\approx$ is an **L**-equality). Second, if $\approx_i$ are **L**-equivalences on X_i ($i = 1, \ldots, n$) then $\bigotimes_{i=1}^{n} \approx_i$ is an **L**-equivalence on $X_1 \times \cdots \times X_n$ which is an **L**-equality iff each $\approx_i$ ($i = 1 \ldots, n$) is an **L**-equality.

We may now generalize the notion of compatibility as follows:

Definition 3.41 Let $\approx$ be an **L**-equivalence relation on X, $k_1, \ldots, k_n, k$ be positive integers. An n-ary **L**-relation R on X is said to be $(k_1, \ldots, k_n)$-*compatible* with $\approx$ if it is compatible with $\approx^{k_1}, \ldots, \approx^{k_n}$; an n-ary function $f : X^n \to X$ is called $(k_1, \ldots, k_n, k)$-compatible with $\approx$ if it is compatible with $\approx^{k_1}, \ldots, \approx^{k_n}$, and $\approx^k$.

Therefore, $(1, \ldots, 1)$-compatibility coincides with the above notion of compatibility with an **L**-equivalence on X.

Theorem 3.22 can be easily extended to obtain a bijection between fuzzy sets compatible with a fuzzy equivalence and special nested systems of ordinary sets.

Definition 3.42 An **L**-nested system $\mathcal{S} = \{A_a \mid a \in L\}$ of subsets of X is said to be compatible with an **L**-equivalence relation $\approx$ on X if

$$x \in A_a \quad \text{implies} \quad y \in A_{a \otimes (x \approx y)}$$

for every $a \in L$ and $x, y \in X$.

Theorem 3.43 *An* **L***-set* A *in* X *is compatible with an* **L***-equivalence* $\approx$ *iff the corresponding* **L***-nested system* $\mathcal{S}_A$ *is compatible with* $\approx$.

Proof. Let A be compatible with $\approx$. $x \in A_a = {}^aA$ means that $a \leq A(x)$ and thus $a \otimes (x \approx y) \leq A(x) \otimes (x \approx y)$. Compatibility of A yields $A(x) \otimes (x \approx y) \leq A(y)$ whence $a \otimes (x \approx y) \leq A(y)$ and so $y \in {}^{a \otimes (\approx y)}A = A_{a \otimes (x \approx y)}$.

Conversely, let $\mathcal{S}_A$ be compatible with $\approx$. We have $A(x) \otimes (x \approx y) = (\bigvee_{x \in A_a} a) \otimes (x \approx y) = \bigvee_{x \in A_a} (a \otimes (x \approx y)) \leq \bigvee_{y \in A_b} b = A(y)$, the last inequality being true by compatibility of $\mathcal{S}_A$. □

Remark 3.44 A simple consequence is this: There is a bijection between **L**-relations between $X_1, \ldots, X_n$ which are compatible with **L**-equivalences $\approx_1, \ldots, \approx_n$, respectively, and **L**-nested systems of relations between $X_1, \ldots, X_n$ which are compatible with $\bigotimes_{i=1}^n \approx_i$.

3.2 FUZZY LOGIC: SYNTAX AND SEMANTICS

3.2.1 SYNTAX

When communicating knowledge, people use language. The particular languages may have various forms. For example, the language can be a certain natural language (like English, Russian, etc.), some artificial language like a programming language, etc. A characteristic feature of a language is that it has its *syntax* and *semantics*. Basically, syntax concerns the rules for manipulating symbols of the language (it determines how correct sentences are formed from words in the case of natural languages etc.) whereas semantics deals with the meaning of symbols and composites of symbols (for instance, sentence "3+3 is 6 and Paris is in France" is true if and only if both "3+3 is 6" and "Paris is in France" are true).

The study of formalized languages, i.e. the study of syntax, semantics, and related aspects is the subject of (mathematical) logic. Logic studies consequence (i.e. how statements—called formulas—follow from other statements) and other relationships between statements; it studies the notion of proof and provability (i.e. how statements can be derived from other statements using simple symbolic manipulation—inference rules); it studies structures in which statements are interpreted (i.e. in which statements may be true); it studies the (formalized) notion of truth, conditions of truth, and related topics. From this point of view, we are interested in a special part of the above-sketched topics traditionally studied by logic: we are interested in developing the syntax of a first-order fuzzy logic (i.e. logic with truth values taken from a scale of truth degrees which is equipped with a structure of a complete residuated lattice); in fuzzy relational and functional systems (i.e. structures in which statements—formulas of our language—can be true or have some degree of truth); in the definition of truth degree of a statement and in some of the related aspects. We concentrate mainly on issues that are connected to fuzzy relational modeling and on issues that are hidden in the classical case (bivalent one). The question of syntactical consequence (provability) is not addressed here; the interested reader is referred to [99, 105, 154].

We are going to introduce a first-order fuzzy logic. In this section we develop its syntax. A language $\mathcal{J}$ of our first-order fuzzy logic consists of (and is given by) a nonempty set R of *relation symbols*, each $r \in R$ with a nonnegative integer $\sigma(r)$ called the *arity* of r; a (possibly empty) set F of *function symbols*, each $f \in F$ with a nonnegative integer $\sigma(f)$ called the *arity* of f, such that R and F have no symbols in common (i.e. $R \cap F = \emptyset$); a denumerable number of *object variables* (we denote object variables by $x_1, x_2, \ldots, x, y, z, \ldots, \mathbb{x}_1, \mathbb{x}_2, \ldots, \mathbb{x}, \mathbb{y}, \mathbb{z}, \ldots$, etc.); *logical connectives* $\mathbin{\&\!\!\wedge}$, $\mathbin{\vee\!\!\!\vee}$, $\otimes$, $\Rightarrow$; a set $\mathbb{L}$ of *truth constants* containing $\mathbb{0}$ and $\mathbb{1}$ (and possibly other truth constants); *quantifiers* $\forall$ and $\exists$ (universal and existential); and auxiliary symbols (,), and , (left and right parenthesis, and a comma). If R contains a binary relation symbol $\approx$ then $\mathcal{J}$ is called a *language with equality*. Unless otherwise specified, we assume that the language in question is a language with equality and that $\mathbb{0}$ and $\mathbb{1}$ are the only truth constants in our language, i.e. $\mathbb{L} = \{\mathbb{0}, \mathbb{1}\}$.

Remark 3.45 (1) If $r \in R$ and $\sigma(r) = n$, then r is said to be n-ary; similarly for $f \in F$.

(2) 0-ary function symbols are also called symbols of constants.

(3) Relation symbols are also called predicate symbols, function symbols are also called operation symbols.

Clearly, two languages may differ only in their relation or function symbols. Each language is thus determined by R, F, and σ. The triple

$\langle R, F, \sigma \rangle$ will be called a *type*. A language $\mathcal{J}$ with R, F, and σ will also be called to be of type $\langle R, F, \sigma \rangle$.

The basic syntactic units built out of symbols of a given language are terms and formulas. Terms represent applications of functions to their elements. Formulas represent statements that concern elements of a universe.

Definition 3.46 A *term* of a language of type $\langle R, F, \sigma \rangle$ is inductively defined as follows:

(i) each variable x is a term;

(ii) if $f \in F$ is n-ary and $t_1, \ldots, t_n$ are terms then $f(t_1, \ldots, t_n)$ is a term.

Remark 3.47 If $f \in F$ is binary, we use also infix notation and write xfy or (xfy) instead of $f(x, y)$ (e.g. $2 + 3$ instead of $+(2, 3)$); in nested terms, we use parentheses (e.g. $(2 + 3) \cdot 5$).

Definition 3.48 A *formula* of a language of type $\langle R, F, \sigma \rangle$ is inductively defined as follows:

(i) if $r \in R$ is n-ary and $t_1, \ldots, t_n$ are terms then $r(t_1, \ldots, t_n)$ is a formula, in particular, $\approx (t_1, t_2)$ is a formula; truth constants ($\mathbb{0}$ and $\mathbb{1}$) are formulas;

(ii) if φ and ψ are formulas then $(\varphi \wedge \psi)$, $(\varphi \vee \psi)$, $(\varphi \otimes \psi)$, $(\varphi \Rightarrow \psi)$ are as well;

(iii) if φ is a formula and x a variable then both $(\forall x)\varphi$ and $(\exists x)\varphi$ are formulas.

Remark 3.49 (1) Formulas formed by (i) are called *atomic*. If $r \in R$ is binary then we write also $t_1 r t_2$ or $(t_1 r t_2)$ instead of $r(t_1, t_2)$ (thus, e.g. $x \leq y$ instead of $\leq (x, y)$). Particularly, we write also $t_1 \approx t_2$ instead of $\approx (t_1, t_2)$.

(2) We adopt usual conventions about writing formulas in a simple way to help readability (in addition to (and) we use also other types of parentheses like [or {; we write $\ldots (\forall x, y) \ldots$ instead of $\ldots (\forall x)(\forall y) \ldots$, etc).

(3) $\neg\, \varphi$ is an abbreviation for $(\varphi \Rightarrow \mathbb{0})$; $(\varphi \Leftrightarrow \psi)$ is an abbreviation for $((\varphi \Rightarrow \psi) \wedge (\psi \Rightarrow \varphi))$. Clearly, the intended meaning of thus defined connectives $\neg$ and $\Leftrightarrow$ is negation and biresiduum in the structure of truth values.

(4) We could, equivalently, have only $\mathbb{0}$ as a truth constant in our language and consider $\mathbb{1}$ as abbreviation for the formula $\mathbb{0} \Rightarrow \mathbb{0}$. This is

the way we proceed in Section 7.3 where we present an axiomatic system for our logic.

Example 3.50 Let $R = \{\text{big}, \leq, \approx\}$, $F = \{c, +\}$, $\sigma(\text{big}) = 1$, $\sigma(\leq) = 2$, $\sigma(c) = 0$, $\sigma(+) = 2$. A language $\mathcal{J}$ of type $\langle R, F, \sigma \rangle$ may be used to formulate statements about numbers, their addition, order, and a property "being a big number". Notice, however, that this interpretation of symbols from R and F is but an intended interpretation. The actual interpretation of $\mathcal{J}$ may be arbitrary. Expressions x, $x+y$ (which is our way of writing $+(x, y)$), or $c+(x+y)$ are terms, $((x, (x+y)+$, or $x++y$ are not; expressions $x \leq (x+y)$, $(\forall x)(\text{big}(x) \Rightarrow \text{big}(x+c))$, $(\forall x)(\text{big}(x+c) \Rightarrow \text{big}(x))$, $(\text{big}(x) \otimes (x \approx y) \Rightarrow \text{big}(y))$ are formulas, $\text{big} \Rightarrow (x \leq y)$, $x+y$, $\text{big}(\forall x(x \leq x))$ are not. Keeping the intended interpretation of symbols of $\mathcal{J}$ in mind, formula $(\forall x)(\text{big}(x) \Rightarrow \text{big}(x+c))$ represents the statement "for all x: if x is big then $x + c$ is also big".

Terms and formulas are built inductively. Every term used in construction of a term t is called a *subterm* of t; every formula used in construction of a formula φ is called a *subformula* of φ. More precisely: The set $\text{sub}(t)$ of all subterms of a term t is defined as follows: for a variable x, $\text{sub}(x) = \{x\}$; $\text{sub}(f(t_1, \ldots, t_n)) = \{f(t_1, \ldots, t_n)\} \cup \text{sub}(t_1) \cup \cdots \cup \text{sub}(t_n)$. The set $\text{sub}(\varphi)$ of all subformulas of a formula φ is defined as follows: if φ is atomic then $\text{sub}(\varphi) = \{\varphi\}$; $\text{sub}(\varphi \wedge \psi) = \{\varphi \wedge \psi\} \cup \text{sub}(\varphi) \cup \text{sub}(\psi)$ and similarly for $\vee$, $\otimes$, and $\Rightarrow$; $\text{sub}((\forall x)\varphi) = \{(\forall x)\varphi\} \cup \text{sub}(\varphi)$ and similarly for $(\exists x)\varphi$.

Definition 3.51 For a term t and a variable x we define $|t|_x$ as follows:

(i) if t is a variable then

$$|t|_x = \begin{cases} 1 & \text{if } t = x \\ 0 & \text{if } t \neq x; \end{cases}$$

(ii) if $f \in F$ is n-ary and $t_1, \ldots, t_n$ are terms then $|f(t_1, \ldots, t_n)|_x = |t_1|_x + \cdots + |t_n|_x$.

$|t|_x$ is called the *degree of x in t*. It is easy to see that $|t|_x$ is the number of occurrences of a variable x in a term t.

Definition 3.52 For a term t and a function symbol f we define $|t|_f$ as follows:

(i) if t is a variable then $|t|_f = 0$;

(ii) if $t = g(t_1, \ldots, t_n)$ then

$$|t|_f = \begin{cases} 1 + |t_1|_f + \cdots + |t_n|_f & \text{if } f = g \\ |t_1|_f + \cdots + |t_n|_f & \text{if } f \neq g; \end{cases}$$

$|t|_f$ is called the *degree of f in t*.

Definition 3.53 For a formula φ and a variable x we define $|\varphi|_x$ as follows:

(i) for atomic formulas: for a truth constant $\mathbb{t}$, $|\mathbb{t}|_x = 0$ (in particular, $|\mathbb{0}|_x = 0$, $|\mathbb{1}|_x = 0$); if $\varphi = s(t_1, \ldots, t_n)$ then

$$|\varphi|_x = \begin{cases} |t_1|_x + \cdots + |t_n|_x & \text{if } s = r \\ 0 & \text{if } s \neq r; \end{cases}$$

(ii) if φ and ψ are formulas then

$$\begin{aligned} |\varphi \wedge \psi|_x &= \max(|\varphi|_x, |\psi|_x), \\ |\varphi \vee \psi|_x &= \max(|\varphi|_x, |\psi|_x), \\ |\varphi \otimes \psi|_x &= |\varphi|_x + |\psi|_x, \\ |\varphi \Rightarrow \psi|_x &= |\varphi|_x + |\psi|_x; \end{aligned}$$

(iii) if φ is a formula and y is a variable then

$$\begin{aligned} |(\forall y)\varphi|_x &= |\varphi|_x, \\ |(\exists y)\varphi|_x &= |\varphi|_x. \end{aligned}$$

$|\varphi|_x$ is called the *degree of x in φ*.

Definition 3.54 For a formula φ and a relation symbol r we define $|\varphi|_r$ as follows:

(i) for atomic formulas: for a truth constant t, $|t|_r = 0$ (in particular, $|\mathbb{0}|_r = 0$, $|\mathbb{1}|_r = 0$); if $\varphi = s(t_1, \ldots, t_n)$ then

$$|\varphi|_r = \begin{cases} 1 & \text{if } s = r \\ 0 & \text{if } s \neq r; \end{cases}$$

(ii) if φ and ψ are formulas then

$$\begin{aligned} |\varphi \wedge \psi|_r &= \max(|\varphi|_r, |\psi|_r), \\ |\varphi \vee \psi|_r &= \max(|\varphi|_r, |\psi|_r), \\ |\varphi \otimes \psi|_r &= |\varphi|_r + |\psi|_r, \\ |\varphi \Rightarrow \psi|_r &= |\varphi|_r + |\psi|_r; \end{aligned}$$

(iii) if φ is a formula and x is a variable then

$$\begin{aligned} |(\forall x)\varphi|_r &= |\varphi|_r, \\ |(\exists x)\varphi|_r &= |\varphi|_r. \end{aligned}$$

$|\varphi|_r$ is called the *degree of r in φ* .

Definition 3.55 For a formula φ and a function symbol f we define $|\varphi|_f$ as follows:

(i) for atomic formulas: for a truth constant t, $|t|_f = 0$ (in particular, $|\mathbb{0}|_f = 0$, $|\mathbb{1}|_f = 0$); if $\varphi = r(t_1, \ldots, t_n)$ then $|\varphi|_f = |t_1|_f + \cdots + |t_n|_f$.

(ii) if φ and ψ are formulas then

$$\begin{aligned} |\varphi \wedge \psi|_f &= \max(|\varphi|_f, |\psi|_f), \\ |\varphi \vee \psi|_f &= \max(|\varphi|_f, |\psi|_f), \\ |\varphi \otimes \psi|_f &= |\varphi|_f + |\psi|_f, \\ |\varphi \Rightarrow \psi|_f &= |\varphi|_f + |\psi|_f; \end{aligned}$$

(iii) if φ is a formula and x is a variable then

$$\begin{aligned} |(\forall x)\varphi|_f &= |\varphi|_f, \\ |(\exists x)\varphi|_f &= |\varphi|_f. \end{aligned}$$

$|\varphi|_f$ is called the *degree of f in φ* .

We say that a variable x occurs in a term t if $|t|_x \geq 1$ (and the same for a function symbol f). Similarly, a relation symbol r occurs in a formula φ if $|\varphi|_r \geq 1$ (and the same for a function symbol f or a variable x). Clearly, the term "occurs in" as just defined exactly expresses the intuitive notion of an occurrence of a symbol in a string of symbols (except for $\ldots(\forall x)\ldots$ and $\ldots(\exists x)\ldots$ which is not considered as an occurrence of x).

We denote by $\mathrm{var}(t)$ the set of all variables which occur in the term t, i.e. $\mathrm{var}(t) = \{x \mid |t|_x \geq 1\}$; $\mathrm{var}(\varphi)$ denotes the set of all variables that occur in a formula φ, i.e. $\mathrm{var}(\varphi) = \{x \mid |\varphi|_x \geq 1\}$. Following the common usage we write $t(x_1, \ldots, x_n)$ to stress that each variable that occurs in t is among $x_1, \ldots, x_n$.

Variables from $\mathrm{var}(\varphi)$ may occur in φ free or bound. The set $\mathrm{free}(\varphi)$ of *free variables* of a formula φ is defined as follows: if φ is atomic then $\mathrm{free}(\varphi) = \mathrm{var}(\varphi)$; $\mathrm{free}(\varphi \wedge \psi) = \mathrm{free}(\varphi) \cup \mathrm{free}(\psi)$, $\mathrm{free}(\varphi \vee \psi) = \mathrm{free}(\varphi) \cup \mathrm{free}(\psi)$, $\mathrm{free}(\varphi \otimes \psi) = \mathrm{free}(\varphi) \cup \mathrm{free}(\psi)$, $\mathrm{free}(\varphi \Rightarrow \psi) = \mathrm{free}(\varphi) \cup \mathrm{free}(\psi)$; $\mathrm{free}((\forall x)\varphi) = \mathrm{free}(\varphi) - \{x\}$, $\mathrm{free}((\exists x)\varphi) = \mathrm{free}(\varphi) - \{x\}$. The set $\mathrm{bound}(\varphi)$ of *bound variables* of a formula φ is defined as follows: if φ is atomic then $\mathrm{bound}(\varphi) = \emptyset$; $\mathrm{bound}(\varphi \wedge \psi) = \mathrm{bound}(\varphi) \cup \mathrm{bound}(\psi)$, $\mathrm{bound}(\varphi \vee \psi) = \mathrm{bound}(\varphi) \cup \mathrm{bound}(\psi)$, $\mathrm{bound}(\varphi \otimes \psi) = \mathrm{bound}(\varphi) \cup \mathrm{bound}(\psi)$, $\mathrm{bound}(\varphi \Rightarrow \psi) = \mathrm{bound}(\varphi) \cup \mathrm{bound}(\psi)$; $\mathrm{bound}((\forall x)\varphi) = \mathrm{bound}(\varphi) \cup \{x\}$, $\mathrm{bound}((\exists x)\varphi) = \mathrm{bound}(\varphi) \cup \{x\}$.

A variable may have several occurrences in a formula. Each variable of $\mathrm{var}(\varphi)$ may occur in φ as free or bound (or both). We write $\varphi(x_1,\ldots,x_n)$ to emphasize that all variables that have a free occurrence in a formula φ are among $x_1,\ldots,x_n$, i.e. $\mathrm{free}(\varphi)\subseteq\{x_1,\ldots,x_n\}$.

Example 3.56 Consider the language $\mathcal{J}$ from Example 3.50, consider terms $t_1 = x_1 + ((c + x_1) + x_2)$, $t_2 = (c + c) + c$, and formulas $\varphi_1 = (\forall x)(\forall y)((\mathrm{big}(x) \otimes (x \leq y)) \Rightarrow \mathrm{big}(y))$, $\varphi_2 = (c \leq x) \otimes (\forall x)(\exists y)(x + x \leq x + y)$. Then $|t_1|_x = 0$, $|t_1|_{x_1} = 2$, $|t_1|_{x_2} = 1$, $|t_1|_+ = 3$, $|t_2|_+ = 2$, $\mathrm{sub}(t_1) = \{t_1, (c+x_1)+x_2, c+x_1, x_1, x_2, c\}$, $\mathrm{var}(t_1) = \{x_1, x_2\}$, $\mathrm{var}(t_2) = \emptyset$; $|\varphi_1|_{\mathrm{big}} = 2$, $|\varphi_1|_{\leq} = 1$, $|\varphi_1|_{\approx} = 0$, $|\varphi_1|_+ = 0$, $|\varphi_2|_+ = 2$, $|\varphi_1|_x = 2$, $|\varphi_1|_y = 2$, $|\varphi_2|_x = 4$, $\mathrm{var}(\varphi_1) = \{x, y\}$, $\mathrm{var}(\varphi_2) = \{x, y\}$, $\mathrm{free}(\varphi_1) = \emptyset$, $\mathrm{bound}(\varphi_1) = \{x, y\}$, $\mathrm{free}(\varphi_2) = \{x\}$, $\mathrm{bound}(\varphi_2) = \{x, y\}$ (variable x has both a free and a bound occurrence in φ_2).

Definition 3.57 Let t and s be terms, x be a variable. The result of a *substitution of s for x in t* is a term $t(x/s)$ defined as follows:

(i) if t is a variable then

$$t(x/s) = \begin{cases} s & \text{if } t = x \\ t & \text{if } t \neq x; \end{cases}$$

(ii) if $t = f(t_1,\ldots,t_n)$ where $f \in F$ is n-ary and $t_1,\ldots,t_n$ are terms then $t(x/s) = f(t_1(x/s),\ldots,t_n(x/s))$.

Definition 3.58 For a formula φ, a term s, and a variable x, the result of a *substitution of s for x in φ* is a formula $\varphi(x/s)$ defined as follows:

(i) if $\varphi = r(t_1,\ldots,t_n)$ then $\varphi(x/s) = r(t_1(x/s),\ldots,t_n(x/s))$, $\mathbb{0}(x/s) = \mathbb{0}$, $\mathbb{1}(x/s) = \mathbb{1}$;

(ii) $(\varphi \wedge \psi)(x/s) = \varphi(x/s) \wedge \psi(x/s)$, $(\varphi \vee \psi)(x/s) = \varphi(x/s) \vee \psi(x/s)$, $(\varphi \otimes \psi)(x/s) = \varphi(x/s) \otimes \psi(x/s)$, $(\varphi \Rightarrow \psi)(x/s) = \varphi(x/s) \Rightarrow \psi(x/s)$;

(iii) $((\forall y)\varphi)(x/s) = (\forall y)\varphi$ if $y = x$, $((\forall y)\varphi)(x/s) = (\forall y)(\varphi(x/s))$ if $y \neq x$, and similarly for $(\exists y)\varphi$.

Obviously, the preceding definitions can be extended to the definition of substitution of a term for a term.

Substitution of a term for a variable may lead to undesirable situations. Consider the formula $\varphi = (\forall x)(x \leq y)$ which expresses that y is greater or equal to any x. The substitution of $x + x$ for y leads to $\varphi(y/x + x) = (\forall x)(x \leq x + x)$. To prevent cases like this, one can allow only correct substitutions in accord with the following definition.

Definition 3.59 A substitution of a term t for a variable x in a formula φ is *correct* if for any $y \in \mathrm{var}(t)$, no subformula of φ which is of the form $(\forall y)\psi$ or $(\exists y)\psi$ contains an occurrence of x which is a free occurrence of x in φ.

Example 3.60 For the language $\mathcal{J}$ from Example 3.50, let $t = x + (c + y)$, $\varphi_1 = (\forall x)(c \leq x)$, $\varphi_2 = (\exists y)(y \leq x)$, $\varphi_3 = (\exists z)(z \leq x)$. We have $\varphi_1(x/t) = (\forall x)(c \leq x)$, $\varphi_2(x/t) = (\exists y)(y \leq x + (c + y))$, $\varphi_3(x/t) = (\exists z)(z \leq x + (c + y))$. Substitution of t for x is correct in φ_1 and φ_3, it is not correct in φ_2.

3.2.2 SEMANTICS

A language of fuzzy logic contains relation and function symbols, i.e. symbols of fuzzy relations and functions. Assigning particular fuzzy relations and functions is a matter of *interpretation* of language. A language may be interpreted in many ways; the particular fuzzy relation (or function) assigned to a relation (or function) symbol r (or f) is not determined by r (or f). That is, given a universe M (i.e. a set of elements to which object variables refer) of interpretation, an interpretation may assign any fuzzy relation (or function) on M to a relation (or function) symbol. The universe M of interpretation and the fuzzy relations and functions assigned to relation and function symbols of a language $\mathcal{J}$ of a given type form a so-called structure for $\mathcal{J}$. The fuzzy relation assigned to $\approx$ is required to be a fuzzy equivalence relation. Moreover, we require that the fuzzy relations and functions are in a natural way compatible with this fuzzy equivalence relation.

Definition 3.61 Let $\mathbf{L}$ be a complete residuated lattice, let $\mathcal{J}$ be a language of type $\langle R, F, \sigma\rangle$. An $\mathbf{L}$-*structure* for $\mathcal{J}$ is a triple $\mathbf{M} = \langle M, R^{\mathbf{M}}, F^{\mathbf{M}}\rangle$ which consists of a nonempty set M, a set $R^{\mathbf{M}} = \{r^{\mathbf{M}} \in L^{M^n} \mid r \in R, \sigma(r) = n\}$ of $\mathbf{L}$-relations, and a set $F^{\mathbf{M}} = \{f^{\mathbf{M}} : M^n \to M \mid f \in F, \sigma(r) = n\}$ of functions such that $\approx^{\mathbf{M}}$ is an $\mathbf{L}$-equivalence relation on M and each $r^{\mathbf{M}} \in R^{\mathbf{M}}$ and $f^{\mathbf{M}} \in F^{\mathbf{M}}$ is compatible w.r.t. $\approx^{\mathbf{M}}$.

Remark 3.62 (1) In other words, an $\mathbf{L}$-structure $\mathbf{M}$ for a language $\mathcal{J}$ of type $\langle R, F, \sigma\rangle$ is a system of $\mathbf{L}$-relations and (ordinary) functions on a certain set M; to each n-ary relation symbol $r \in R$ there is an n-ary $\mathbf{L}$-relation $r^{\mathbf{M}}$ in M, to each n-ary function symbol $f \in F$ there is an n-ary function $f^{\mathbf{M}}$ in M; $\approx^{\mathbf{M}}$ is an $\mathbf{L}$-equivalence compatible with all $\mathbf{L}$-relations and function of $\mathbf{M}$.

(2) For the sake of brevity we omit the superscripts in $r^{\mathbf{M}}$, $f^{\mathbf{M}}$, and $\approx^{\mathbf{M}}$, and write only r, f, and $\approx$; there will be no danger of confusion.

(3) Instead of $\mathbf{L}$-structure we say also fuzzy structure.

Recall that compatibility (i.e. $\langle 1,\dots,1\rangle$-compatibility) of $r^{\mathbf{M}} \in R^{\mathbf{M}}$ and $f^{\mathbf{M}} \in F^{\mathbf{M}}$ w.r.t. $\approx^{\mathbf{M}}$ means that

$$(a_1 \approx^{\mathbf{M}} b_1) \otimes \cdots \otimes (a_n \approx^{\mathbf{M}} b_n) \leq r^{\mathbf{M}}(a_1,\dots,a_n) \leftrightarrow r^{\mathbf{M}}(b_1,\dots,b_n) \tag{3.29}$$

for any n-ary $r \in R$ and any $a_i, b_i \in M$ $(i = 1,\dots,n)$ and

$$(a_1 \approx^{\mathbf{M}} b_1) \otimes \cdots \otimes (a_n \approx^{\mathbf{M}} b_n) \leq f^{\mathbf{M}}(a_1,\dots,a_n) \approx^{\mathbf{M}} f^{\mathbf{M}}(b_1,\dots,b_n) \tag{3.30}$$

for any n-ary $f \in F$ and any $a_i, b_i \in M$ $(i = 1,\dots,n)$. Note also that compatibility of $\approx^{\mathbf{M}}$ with $\approx^{\mathbf{M}}$ is equivalent to transitivity of $\approx^{\mathbf{M}}$ (easy to verify). One could also allow for a more general condition of $\langle k_1,\dots,k_n\rangle$- and $\langle k_1,\dots,k_n,k\rangle$-compatibility of $r^{\mathbf{M}}$ and $f^{\mathbf{M}}$. This would result in a possibility to consider fuzzy relations and functions with "different sensitivity to **L**-equivalence $\approx^{\mathbf{M}}$ in different arguments". We will comment on this possibility later. Note also that if $r^{\mathbf{M}}$ is $\langle k_1,\dots,k_n\rangle$-compatible with $\approx^{\mathbf{M}}$ then it is $\langle 1,\dots,1\rangle$-compatible with $(\approx^{\mathbf{M}})^{\max(k_1,\dots,k_n)}$; if $f^{\mathbf{M}}$ is $\langle k_1,\dots,k_n,1\rangle$-compatible with $\approx^{\mathbf{M}}$ then it is $\langle 1,\dots,1\rangle$-compatible with $(\approx^{\mathbf{M}})^{\max(k_1,\dots,k_n)}$.

Example 3.63 (1) Let $\mathcal{J}$ be the language from Example 3.50. Let **L** be the standard Łukasiewicz algebra on $[0,1]$, M be the set of all positive integers. Let a unary **L**-relation $\mathrm{big}^{\mathbf{M}}$ be given by

$$\mathrm{big}^{\mathbf{M}}(m) = \begin{cases} 0 & \text{for } m \in [1,100] \\ \frac{m-100}{100} & \text{for } m \in [101,200] \\ 1 & \text{for } m \in [201,\infty), \end{cases}$$

$\leq^{\mathbf{M}}$ be the usual order (i.e. $(m_1 \leq^{\mathbf{M}} m_2) = 1$ if m_1 is less than or equal to m_2, otherwise $(m_1 \leq^{\mathbf{M}} m_2) = 0$), $\approx^{\mathbf{M}}$ be the crisp identity on M (i.e. $(m_1 \approx^{\mathbf{M}} m_2) = 1$ if $m_1 = m_2$, $(m_1 \approx^{\mathbf{M}} m_2) = 0$ otherwise), $+^{\mathbf{M}}$ be the addition of positive integers (i.e. $3 +^{\mathbf{M}} 11$ is 14 etc.), $c^{\mathbf{M}}$ be 1. Then $\mathbf{M} = \langle M, \{\leq^{\mathbf{M}}, \mathrm{big}^{\mathbf{M}}, \approx^{\mathbf{M}}\}, \{+^{\mathbf{M}}, c^{\mathbf{M}}\}\rangle$ is an **L**-structure for $\mathcal{J}$.

(2) Another **L**-structure for the same language: Let $M = \{A,B,C,D\}$; $\mathrm{big}^{\mathbf{M}}(A) = 0$, $\mathrm{big}^{\mathbf{M}}(B) = 0.5$, $\mathrm{big}^{\mathbf{M}}(C) = 0.6$, $\mathrm{big}^{\mathbf{M}}(D) = 0$; let $(A \leq^{\mathbf{M}} D) = 1$, $(D \leq^{\mathbf{M}} A) = 0$, $(m \leq^{\mathbf{M}} n) = 0.5$ otherwise; $\approx^{\mathbf{M}}$ be the crisp identity on M; let $(m +^{\mathbf{M}} n) = A$ for all $m,n \in M$; put $c^{\mathbf{M}} = D$. $\mathbf{M} = \langle M, \{\leq^{\mathbf{M}}, \mathrm{big}^{\mathbf{M}}, \approx^{\mathbf{M}}\}, \{+^{\mathbf{M}}, c^{\mathbf{M}}\}\rangle$ is another **L**-structure for $\mathcal{J}$.

We see that there are several (in fact, infinitely many) **L**-structures for a given language.

An **L**-structure for a language $\mathcal{J}$ of a given type represents a particular interpretation of relation and function symbols. Language $\mathcal{J}$ allows for formulation of statements about the elements of the particular universe of discourse (for instance, statement "if x is big and if x is less than or equal to y then y is also big" is represented by a formula $[\mathrm{big}(x) \otimes (x \leq y)] \Rightarrow \mathrm{big}(y))$ and for denoting of elements that result by application of functions (for instance, the result of addition of x and y is represented by a term $x+y$). Thus, syntactic expressions that represent statements are formulas, expressions that represent applications of functions are terms. We are now going to present rules that specify how terms and formulas are evaluated in a given **L**-structure, i.e. what is the truth degree of a given formula and what element is denoted by a term under an interpretation given by the **L**-structure.

Let **M** be an **L**-structure for a language $\mathcal{J}$. An *$\mathbf{M}$-valuation of object variables* is a mapping v assigning an element $v(x) \in M$ to any variable x. If v and v' are valuations, and x a variable, we write $v =_x v'$ if for each variable $y \neq x$ we have $v(y) = v'(y)$, i.e. v and v' differ at most in what value they assign to x (in general, $v =_{x,\ldots,y} v'$ means that v and v' differ at most in $x, \ldots, y$).

Definition 3.64 Let v be an **M**-valuation. A value $\|t\|_{\mathbf{M},v}$ of a term t under the **M**-valuation v is defined as follows:

(i) for a variable x, $\|x\|_{\mathbf{M},v} = v(x)$;

(ii) for $t = f(t_1, \ldots, t_n)$, $\|t\|_{\mathbf{M},v} = f^{\mathbf{M}}(\|t_1\|_{\mathbf{M},v}, \ldots, \|t_n\|_{\mathbf{M},v})$.

Lemma 3.65 *The value $\|t\|_{\mathbf{M},v}$ does not depend on the values assigned by v to variables that do not occur in t, i.e. for any **M**-valuations v_1, v_2 such that $v_1(x) = v_2(x)$ for each $x \in \mathrm{var}(t)$ we have $\|t\|_{\mathbf{M},v_1} = \|t\|_{\mathbf{M},v_2}$.*

Proof. The assertion is obvious and can be proved by induction over the complexity of t. □

By Definition 3.64 and Lemma 3.65, each term $t(x_1, \ldots, x_n)$ induces a so-called *term function* $t^{\mathbf{M}}$: for $m_1, \ldots m_n \in M$ we put

$$t^{\mathbf{M}}(m_1, \ldots, m_n) = \|t\|_{\mathbf{M},v}$$

where v is an **M**-valuation such that $v(x_1) = m_1, \ldots, v(x_n) = m_n$.

Remark 3.66 The notion of a term function induced by a term $t(x_1, \ldots, x_n)$ can be defined directly as follows: (i) if $t = x_i$ (t is a variable) then $t^{\mathbf{M}}(m_1, \ldots, m_n) = m_i$; (ii) if $t = f(t_1, \ldots, t_p)$ then $t^{\mathbf{M}}(m_1, \ldots, m_n) = f^{\mathbf{M}}(t_1^{\mathbf{M}}(m_1, \ldots, m_n), \ldots, t_p^{\mathbf{M}}(m_1, \ldots, m_n))$.

Definition 3.67 A *truth value* $\|\varphi\|_{\mathbf{M},v}$ *of a formula* φ *under an* **M**-*valuation* v is defined as follows:

(i) for atomic formulas:

$$\begin{aligned}\|r(t_1,\ldots,t_n)\|_{\mathbf{M},v} &= r^{\mathbf{M}}(\|t_1\|_{\mathbf{M},v},\ldots,\|t_n\|_{\mathbf{M},v}),\\ \|\mathbb{0}\|_{\mathbf{M},v} &= 0, \quad \|\mathbb{1}\|_{\mathbf{M},v} = 1\end{aligned}$$

(and similarly for other truth constants);

(ii) if φ and ψ are formulas then

$$\begin{aligned}\|\varphi \mathbin{\&\!\!\wedge} \psi\|_{\mathbf{M},v} &= \|\varphi\|_{\mathbf{M},v} \wedge \|\psi\|_{\mathbf{M},v},\\ \|\varphi \mathbin{\vee\!\!\vee} \psi\|_{\mathbf{M},v} &= \|\varphi\|_{\mathbf{M},v} \vee \|\psi\|_{\mathbf{M},v},\\ \|\varphi \mathbin{\otimes} \psi\|_{\mathbf{M},v} &= \|\varphi\|_{\mathbf{M},v} \otimes \|\psi\|_{\mathbf{M},v},\\ \|\varphi \Rightarrow \psi\|_{\mathbf{M},v} &= \|\varphi\|_{\mathbf{M},v} \to \|\psi\|_{\mathbf{M},v};\end{aligned}$$

(iii) if φ is a formula and x a variable then

$$\begin{aligned}\|(\forall x)\varphi\|_{\mathbf{M},v} &= \bigwedge\{\|\varphi\|_{\mathbf{M},v'} \mid v' =_x v\},\\ \|(\exists x)\varphi\|_{\mathbf{M},v} &= \bigvee\{\|\varphi\|_{\mathbf{M},v'} \mid v' =_x v\}.\end{aligned}$$

Lemma 3.68 *The value* $\|\varphi\|_{\mathbf{M},v}$ *does not depend on the values assigned by* v *to variables that do not occur in* φ *as free, i.e. for any* **M**-*valuations* v_1, v_2 *such that* $v_1(x) = v_2(x)$ *for each* $x \in \text{free}(\varphi)$ *we have* $\|\varphi\|_{\mathbf{M},v_1} = \|\varphi\|_{\mathbf{M},v_2}$.

Proof. The assertion is immediate and can be proved by induction over the complexity of φ. □

If $\|\varphi\|_{\mathbf{M},v_1} = \|\varphi\|_{\mathbf{M},v_2}$ (or $\|t\|_{\mathbf{M},v_1} = \|t\|_{\mathbf{M},v_2}$) for all v_1, v_2, we may simply write $\|\varphi\|_{\mathbf{M}}$ (or $\|t\|_{\mathbf{M}}$).

Example 3.69 Consider the **L**-structure **M** from Example 3.63 (1). Let v be an **M**-valuation such that $v(x) = 2$, $v(y) = 10$, $v(z) = 110$. Then for terms $t_1 = (x+y)+x$ and $t_2 = c+x$ we have $\|t_1\|_{\mathbf{M},v} = 14$, $\|t_2\|_{\mathbf{M},v} = 3$; for formulas $\varphi_1 = (x \leq y)$, $\varphi_2 = (y \leq x)$, $\varphi_3 = \text{big}(x)$, $\varphi_4 = \text{big}(z)$, $\varphi_5 = (\forall x)(\forall y)((\text{big}(x) \otimes (x \leq y)) \Rightarrow \text{big}(y))$, $\varphi_6 = (\forall x)(\text{big}(x+c) \Rightarrow \text{big}(x))$ we have $\|\varphi_1\|_{\mathbf{M},v} = 1$, $\|\varphi_2\|_{\mathbf{M},v} = 0$, $\|\varphi_3\|_{\mathbf{M},v} = 0$, $\|\varphi_4\|_{\mathbf{M},v} = 0.1$, $\|\varphi_5\|_{\mathbf{M},v} = 1$, $\|\varphi_6\|_{\mathbf{M},v} = 0.99$. Note that the truth values of φ_5 and φ_6 do not depend on v (Lemma 3.68).

Example 3.70 (1) Let A, B be **L**-sets in M. Consider a language $\mathcal{J}$ with R containing unary relation symbols r_A, r_B, and r_C; let **M** be an **L**-structure for $\mathcal{J}$. Let $A = r_A^{\mathbf{M}}$, $B = r_B^{\mathbf{M}}$, $C = r_C^{\mathbf{M}}$. The subsethood degree $S(A, B)$ of A in B is exactly the truth degree of a formula $(\forall x)(r_A(x) \Rightarrow r_B(x))$ in **M** under any **M**-valuation v, i.e.

$$S(A, B) = \|(\forall x)(r_A(x) \Rightarrow r_B(x))\|_{\mathbf{M}}.$$

Furthermore $S(A, B) \otimes S(B, C) \leq S(A, C)$ (see Theorem 3.12) is equivalent to (and may be reformulated by) saying that formula

$$((\forall x)(r_A(x) \Rightarrow r_B(x)) \otimes (\forall x)(r_B(x) \Rightarrow r_C(x))) \Rightarrow (\forall x)(r_A(x) \Rightarrow r_C(x))$$

has truth degree 1 in **M** (independent of v).

(2) Let E be a binary **L**-relation in M. Consider a language $\mathcal{J}$ with R containing a binary relation symbol r_E; let **M** be an **L**-structure for $\mathcal{J}$ with $E = r_E^{\mathbf{M}}$. Then E is an **L**-equivalence in **M** iff each of the formulas

$$(\forall x) r_E(x, x),$$
$$(\forall x, y)(r_E(x, y) \Rightarrow r_E(y, x)),$$
$$(\forall x, y, z)((r_E(x, y) \otimes r_E(y, z)) \Rightarrow r_E(x, z))$$

has truth degree 1 in **M** (the truth degrees of these formulas do not depend on valuations).

(3) Similarly, compatibility of an n-ary **L**-relation $r^{\mathbf{M}}$ w.r.t. **L**-equivalence $\approx^{\mathbf{M}}$ is equivalent to require that formula

$$(\forall x_1, y_1, \ldots, x_n, y_n)[r(x_1, \ldots, x_n) \otimes (x_1 \approx y_1) \otimes \cdots \otimes (x_n \approx y_n) \Rightarrow r(y_1, \ldots, y_n)]$$

has truth degree 1.

Now, we are going to make explicit a simple but still very useful notion. By Definition 3.67 and Lemma 3.68, given an **L**-structure **M**, each formula $\varphi(x_1, \ldots, x_n)$ induces an n-ary **L**-relation $\varphi^{\mathbf{M}}$ in M: for $m_1, \ldots m_n \in M$ we put

$$\varphi^{\mathbf{M}}(m_1, \ldots, m_n) = \|\varphi\|_{\mathbf{M},v} \tag{3.31}$$

where v is an **M**-valuation such that $v(x_1) = m_1, \ldots, v(x_n) = m_n$. Particularly, if φ does not contain free variables (such a formula is called a sentence), $\varphi^{\mathbf{M}}$ is the truth value of φ in **M**.

More generally, let φ be a formula with all free variables of φ among $x_1, \ldots, x_n$, all relation symbols of φ among $r_1, \ldots, r_p$, and all function symbols of φ among $f_1, \ldots, f_q$; denote this fact by writing $\varphi(r_1, \ldots, r_p, f_1, \ldots, f_q, x_1, \ldots, x_n)$. Given a nonempty set M, $\varphi(r_1, \ldots, r_p, f_1, \ldots, f_q, x_1, \ldots, x_n)$ induces a mapping that assigns to any **L**-relations R_i in $M^{\sigma(r_i)}$ $(i = 1, \ldots, p)$, and functions $F_i : M^{\sigma(f_i)} \to M$ $(i = 1, \ldots, q)$ an **L**-relation

$$\varphi^{(\mathbf{L},M)}(R_1, \ldots, R_p, F_1, \ldots, F_q)$$

in M^n defined for $m_1, \ldots, m_n \in M$ by

$$(\varphi^{(\mathbf{L},M)}(R_1, \ldots, R_p, F_1, \ldots, F_q))(m_1, \ldots, m_n) = \|\varphi\|_{\mathbf{M},v} \tag{3.32}$$

where $\mathbf{M} = \langle M, R^{\mathbf{M}}, F^{\mathbf{M}} \rangle$ is any **L**-structure such that $r_i^{\mathbf{M}} = R_i$ $(i = 1, \ldots, p)$, $f_i^{\mathbf{M}} = F_i$ $(i = 1, \ldots, q)$, and the **M**-valuation v is such that $v(x_1) = m_1, \ldots, v(x_n) = m_n$ (this is correct, cf. Lemma 3.68). We assume that if some r_i is $\approx$ then R_i is an **L**-equivalence relation on M and $\approx^{\mathbf{M}}$ is just R_i; otherwise we let $\approx^{\mathbf{M}}$ be the crisp identity on M. Obviously, fixing some R_i's, F_i's (and possibly also some m_i's), we get a function of less arguments (and possibly of less arity).

Similarly, given a set M, a term $t(f_1, \ldots, f_p, x_1, \ldots, x_n)$ (i.e. with function symbols among $f_1, \ldots, f_p$ and variables among $x_1, \ldots, x_n$) induces a function t^M assigning to functions $F_i : M^{\sigma(f_i)} \to M$ $(i = 1, \ldots, p)$ a function $t^M(F_1, \ldots, F_p)$ given by

$$(t^M(F_1, \ldots, F_p))(m_1, \ldots, m_n) = \|t\|_{\mathbf{M},v}$$

where $\mathbf{M} = \langle M, R^{\mathbf{M}}, F^{\mathbf{M}} \rangle$ is any **L**-structure such that $f_i^{\mathbf{M}} = F_i$ $(i = 1, \ldots, p)$, v is such that $v(x_1) = m_1, \ldots, v(x_n) = m_n$ (again, this is correct).

Example 3.71 Let $g : M \to M$ be a function, **L** be the standard Gödel structure, i.e. $L = [0, 1]$ with its natural order, $a \otimes b = \min(a, b)$ (and the rest thus uniquely determined). The so-called *extension principle* tells how to obtain a function $\overline{g} : L^M \to L^M$ which is a natural extension of g ($\overline{g}$ extends g in that both coincide on elements of M, i.e. for each $m \in M$ we have $\overline{g}(\{{}^1/m\}) = \{{}^1/g(m)\}$). Given a fuzzy set A in M, $\overline{g}(A)$ is a fuzzy set in M given by

$$(\overline{g}(A))(m) = \sup\{A(m') \mid g(m') = m\}.$$

Consider now a language $\mathcal{J}$ with a unary relation symbol r, a binary relation symbol $\approx$ (equality), and a unary function symbol f. Let φ be $(\exists x)(r(x) \otimes (f(x) \approx y))$. $\varphi(r, \approx, f, y)$ induces a mapping assigning to any fuzzy set A in X, a fuzzy equality E in X, and a function $g : M \to M$ a fuzzy set $\varphi^{(\mathbf{L},M)}$ in M given as follows: Let **M** be an **L**-structure with

$r^{\mathbf{M}} = A$, $\approx^{\mathbf{M}} = E$, $f^{\mathbf{M}} = g$, let furthermore v be an $\mathbf{M}$-valuation such that $v(y) = m$. Then

$$\begin{aligned}(\varphi^{(\mathbf{L},M)}(A, E, g))(m) &= \|\varphi\|_{\mathbf{M},v} = \\ &= \bigvee\{\|r(x) \otimes (f(x) \approx y)\|_{\mathbf{M},v'} \mid v' =_y v\} = \\ &= \bigvee\{A(m') \otimes E(g(m'), m) \mid m' \in M\}.\end{aligned}$$

Particularly, if we take E to be id_M (crisp identity on M) and keep both g and E fixed, we may consider mapping $h : L^M \to L^M$ defined by $h(A) = \varphi^{(\mathbf{L},M)}(A, E, g)$. It is easy to see that

$$(h(A))(m) = \bigvee\{A(m') \mid g(m') = m\} = (\overline{g}(A))(m),$$

i.e. h is the function obtained from g by the extension principle. The important point is to see that, put in loose terms, extension principle is defined by logical formula (as a consequence, one may apply several general results and thus get automatically properties of the function $\overline{g}$ obtained by extension principle, see Section 7.2).

Definition 3.72 For a formula φ, an $\mathbf{L}$-structure $\mathbf{M}$, and a class $\mathcal{K}$ of $\mathbf{L}$-structures we put

$$\|\varphi\|_{\mathbf{M}} = \bigwedge_{v:X \to M} \|\varphi\|_{\mathbf{M},v}$$

and

$$\|\varphi\|_{\mathcal{K}} = \bigwedge_{\mathbf{M} \in \mathcal{K}} \|\varphi\|_{\mathbf{M}}.$$

$\|\varphi\|_{\mathbf{M}}$ is called the truth degree to which φ is true in $\mathbf{M}$; $\|\varphi\|_{\mathcal{K}}$ is called the truth degree to which φ is true in $\mathcal{K}$. Intuitively, $\|\varphi\|_{\mathbf{M}}$ is the truth degree of "for each valuation v: φ is true in $\mathbf{M}$ under v"; $\|\varphi\|_{\mathcal{K}}$ is the truth degree of "φ is true in each structure $\mathbf{M}$ from $\mathcal{K}$". In general, we can consider $\mathcal{K}$ an $\mathbf{L}$-class of $\mathbf{L}$-structures and define

$$\|\varphi\|_{\mathcal{K}} = \bigwedge_{\mathbf{M}} \mathcal{K}(\mathbf{M}) \to \|\varphi\|_{\mathbf{M}}.$$

For an $\mathbf{L}$-set Σ of formulas and a class (or $\mathbf{L}$-class) $\mathcal{K}$ of $\mathbf{L}$-structures we put

$$\|\Sigma\|_{\mathcal{K}} = \bigwedge_{\varphi} \Sigma(\varphi) \to \|\varphi\|_{\mathcal{K}}.$$

$\|\Sigma\|_{\mathcal{K}}$ is the truth degree of "each formula of Σ is true in each structure from $\mathcal{K}$". Particularly, we denote $\|\Sigma\|_{\{\mathbf{M}\}}$ by $\|\Sigma\|_{\mathbf{M}}$ etc.

Thus, an **L**-set Σ of formulas induces an **L**-set $\mathrm{Mod}(\Sigma)$ of **L**-structures given by $(\mathrm{Mod}(\Sigma))(\mathbf{M}) = \|\Sigma\|_{\mathbf{M}}$; $(\mathrm{Mod}(\Sigma))(\mathbf{M})$ is the degree to which $\mathbf{M}$ is a *model* of Σ. Particularly, $\mathbf{M} \in {}^1\mathrm{Mod}(\Sigma)$ will be called models of Σ. An **L**-class $\mathcal{K}$ of **L**-structures induces an **L**-set $\mathrm{Th}(\mathcal{K})$ of formulas by $(\mathrm{Th}(\mathcal{K}))(\varphi) = \|\varphi\|_{\mathcal{K}}$.

Remark 3.73 As we will see in Section 7.1, Mod and Th satisfy $S(\Sigma_1, \Sigma_2) \leq S(\mathrm{Mod}(\Sigma_2), \mathrm{Mod}(\Sigma_1))$, $S(\mathcal{K}_1, \mathcal{K}_2) \leq S(\mathrm{Th}(\mathcal{K}_2), \mathrm{Th}(\mathcal{K}_1))$, $\Sigma \subseteq \mathrm{Th}(\mathrm{Mod}(\Sigma))$, and $\mathcal{K} \subseteq \mathrm{Mod}(\mathrm{Th}(\mathcal{K}))$.

So far, we kept **L**-fixed. However, we can consider a class $\mathcal{K}$ of $\mathbf{L}_i$-structures $\mathbf{M}_i$ with possibly different $\mathbf{L}_i$ and consider the set $\mathrm{Th}_{\mathbf{2}}(\mathcal{K})$ of all formulas φ such that $\|\varphi\|_{\mathbf{M}} = 1$. Similarly, for a set Σ of formulas, we can consider the class $\mathrm{Mod}_{\mathbf{2}}(\Sigma)$ of all **L**-structures $\mathbf{M}$ such that $\|\Sigma\|_{\mathbf{M}} = 1$.

Remark 3.74 Again, results of Section 7.1 imply that $\mathrm{Mod}_{\mathbf{2}}$ and $\mathrm{Th}_{\mathbf{2}}$ satisfy: $\Sigma_1 \subseteq \Sigma_2$ implies $\mathrm{Mod}_{\mathbf{2}}(\Sigma_2) \subseteq \mathrm{Mod}_{\mathbf{2}}(\Sigma_1)$, $\mathcal{K}_1 \subseteq \mathcal{K}_2$ implies $\mathrm{Th}_{\mathbf{2}}(\mathcal{K}_2) \subseteq \mathrm{Th}_{\mathbf{2}}(\mathcal{K}_1)$, $\Sigma \subseteq \mathrm{Th}_{\mathbf{2}}(\mathrm{Mod}_{\mathbf{2}}(\Sigma))$, and $\mathcal{K} \subseteq \mathrm{Mod}_{\mathbf{2}}(\mathrm{Th}_{\mathbf{2}}(\mathcal{K}))$.

3.3 SIMILARITY ISSUES

One of the issues that is to great extent degenerate in bivalent approach is the issue of similarity. A fuzzy structure may be thought of as representing an (initial) knowledge about some domain (relevant relational and functional correspondences between the elements of domain M). Specifying a fuzzy structure means assigning truth degrees to tuples of elements (membership degrees to which tuples of elements belong to respective fuzzy relations) and assigning elements of the domain to tuples of elements (function values assigned to tuples of elements by respective functions). It is natural to consider two fuzzy structures similar if these assignments do not differ much in the sense that membership degrees assigned by corresponding fuzzy relations are close (in the lattice of truth degrees) and elements assigned by corresponding functions are similar (in some way). Put in more detail, it is natural to consider fuzzy relations R_1 and R_2 (of the same arity, defined on the same set) similar if the truth degree to which elements $m_1, \ldots, m_n$ belong to R_1 is about the same as the truth degree to which $m_1, \ldots, m_n$ belong to R_2 (for every $m_1, \ldots, m_n$); analogously, functions F_1 and F_2 are considered similar if $F_1(m_1, \ldots, m_n)$ and $F_2(m_1, \ldots, m_n)$ are about the same (for every $m_1, \ldots, m_n$). We may thus consider two fuzzy structures (of the same type) with a common universe similar if the corresponding fuzzy relations and functions they contain are pairwise similar. Since fuzzy

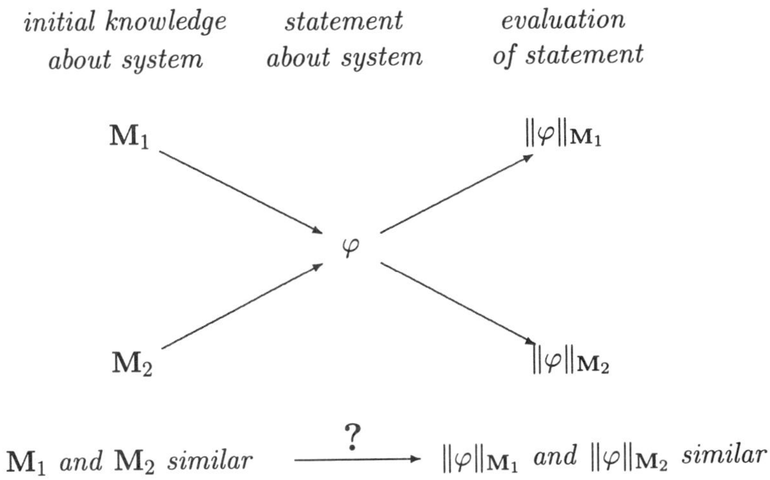

Figure 3.8. Do similar fuzzy structures lead to similar evaluations of a statement?

structures may enter further processing and/or analysis, natural questions arise as to what extent it is true that further processing and/or analysis of similar fuzzy structures leads to similar results. Needless to say, this issue may be thought of as a kind of *sensitivity analysis* and is of crucial importance. For example, one of the immediate questions we are asking when specifying a truth degree High(50) (say a truth degree to which a speed of 50 mph is high) is what difference it makes (w.r.t. further processing and/or analysis) saying "High(50) is 0.75" or "High(50) is 0.76". Further processing and/or analysis of the initial information provided by a fuzzy structure is very often performed by evaluation of truth degrees of formulas that represent statements about the domain (formulas are also being used when describing input/output relationships, for the definition of functions, etc.; cf. Example 3.71). In this way, the above question reduces to what is the relationship between the truth degrees to which a given formula evaluates in two fuzzy structures that are similar (and also, what is the relationship between elements to which a given term evaluates in these fuzzy structures). Put informally, the basic question is depicted in Fig.3.8. In the following we attempt to provide answers to the above-sketched questions connected to similarity issues. Results we obtain will be applied to several particular fuzzy relational systems later on.

Lemma 3.75 *Let $\approx_i$ $(i \in I)$ be* **L***-equivalences on M. The* **L***-relation $\bigwedge_{i\in I} \approx_i$ on M defined by*

$$(m\ (\bigwedge_{i\in I} \approx_i)\ n) \quad = \quad \bigwedge_{i\in I}(m \approx_i n)$$

is an **L***-equivalence. Moreover, if some $\approx_i$ is an* **L***-equality then so is $\bigwedge_{i\in I} \approx_i$.*

Proof. Reflexivity and symmetry of $\bigwedge_{i\in I} \approx_i$ follow immediately from reflexivity and symmetry of $\approx_i$. Transitivity:

$$\begin{aligned} & (m_1(\bigwedge_{i\in I} \approx_i)m_2) \otimes (m_2(\bigwedge_{i\in I} \approx_i)m_3) = \\ = \quad & (\bigwedge_{i\in I}(m_1 \approx_i m_2)) \otimes (\bigwedge_{i\in I}(m_2 \approx_i m_3)) \leq \\ \leq \quad & \bigwedge_{i\in I}((m_1 \approx_i m_2) \otimes (m_2 \approx_i m_3)) \leq \\ \leq \quad & \bigwedge_{i\in I}(m_1 \approx_i m_3) = (m_1(\bigwedge_{i\in I} \approx_i)m_3). \end{aligned}$$

Let now $\approx_i$ be an **L**-equality. If $m_1(\bigwedge_{i\in I} \approx_i)m_2 = 1$ then also $(m_1 \approx_i m_2) = 1$, whence $m_1 = m_2$ proving that $\bigwedge_{i\in I} \approx_i$ is an **L**-equality. □

Given n-ary **L**-relations r, s in M and n-ary functions $f, g : M^n \to M$ on a set with **L**-equivalence $\approx^M$, put

$$(r \approx s) = \bigwedge_{m_1,\ldots,m_n \in M} r(m_1, \ldots, m_n) \leftrightarrow s(m_1, \ldots, m_n) \tag{3.33}$$

and

$$(f \approx g) = \bigwedge_{m_1,\ldots,m_n \in M} f(m_1, \ldots, m_n) \approx^M g(m_1, \ldots, m_n). \tag{3.34}$$

If there is an **L**-equivalence $\approx_f^M$ related to f and an **L**-equivalence $\approx_g^M$ related to g, we will take $\approx^M = \approx_f^M \wedge \approx_g^M$, but this will be explicitly mentioned whenever used.

Lemma 3.76 *The* **L***-relations defined by (3.33) and (3.34) are* **L***-equivalences on the set of n-ary* **L***-relations and n-ary functions on M, respectively. Moreover, (3.33) is an* **L***-equality; (3.34) is an* **L***-equality whenever $\approx^M$ is an* **L***-equality.*

Proof. The fact that (3.33) is an **L**-equality follows directly from Corollary 3.32. Consider (3.34): Reflexivity and symmetry are obvious. Transitivity:

$$
\begin{aligned}
&(f \approx g) \otimes (g \approx h) = \\
= &\Big(\bigwedge_{m_1,\ldots,m_n \in M} f(m_1,\ldots,m_n) \approx^M g(m_1,\ldots,m_n)\Big) \otimes \\
&\otimes \Big(\bigwedge_{m'_1,\ldots,m'_n \in M} g(m'_1,\ldots,m'_n) \approx^M h(m'_1,\ldots,m'_n)\Big) \leq \\
\leq &\Big(\bigwedge_{m_1,\ldots,m_n \in M} (f(m_1,\ldots,m_n) \approx^M g(m_1,\ldots,m_n)) \otimes \\
&\otimes (g(m_1,\ldots,m_n) \approx^M h(m_1,\ldots,m_n))\Big) \leq \\
\leq &\bigwedge_{m_1,\ldots,m_n \in M} f(m_1,\ldots,m_n) \approx^M h(m_1,\ldots,m_n) = (f \approx h).
\end{aligned}
$$

Let now $\approx^M$ be an **L**-equality. If $(f \approx g) = 1$ then for every $m_1,\ldots,m_n \in M$ we have $(f(m_1,\ldots,m_n) \approx^M g(m_1,\ldots,m_n)) = 1$. As $\approx^M$ is an **L**-equality, we conclude $f(m_1,\ldots,m_n) = g(m_1,\ldots,m_n)$, and thus $f = g$. □

We are now going to define a similarity degree of two **L**-structures (with the same support set). Different **L**-structures $\mathbf{M}_1$ and $\mathbf{M}_2$ (even with the same support set) may have different **L**-equivalence relations. In order to be able to apply (3.34), one can use Lemma 3.75 and measure the degree of similarity between two functions $f^{\mathbf{M}_1}$ and $f^{\mathbf{M}_2}$ w.r.t. the intersection of $\approx^{\mathbf{M}_1}$ and $\approx^{\mathbf{M}_2}$. In fact, we have the following.

Lemma 3.77 *Let $\approx_1$ and $\approx_2$ be* **L***-equivalences on a set M such that $(m \approx_1 n) \leq (m \approx_2 n)$ for any $m, n \in M$. Then any* **L***-relation on M which is compatible w.r.t. $\approx_2$ is compatible w.r.t. $\approx_1$ as well. The same holds true for functions on M, i.e. any function on M compatible w.r.t. $\approx_2$ is also compatible w.r.t. $\approx_1$.*

Proof. The assertions follow immediately from the definition of compatibility of **L**-relations and functions. □

Therefore, any $r^{\mathbf{M}_i}$ and $f^{\mathbf{M}_i}$ $(i = 1, 2)$ is compatible w.r.t. $\approx^{\mathbf{M}_1} \wedge \approx^{\mathbf{M}_2}$.

Definition 3.78 Let $\mathbf{M}_1$ and $\mathbf{M}_2$ be **L**-structures of type $\langle R, F, \sigma\rangle$ such that $M_1 = M_2$, let $S \subseteq R \cup F$. The *degree of similarity* of $\mathbf{M}_1$ and $\mathbf{M}_2$ w.r.t. S is defined by

$$
(\mathbf{M}_1 \approx_S \mathbf{M}_2) = \bigwedge_{r \in S \cap R} (r^{\mathbf{M}_1} \approx r^{\mathbf{M}_2}) \wedge \bigwedge_{f \in S \cap F} (f^{\mathbf{M}_1} \approx f^{\mathbf{M}_2}), \qquad (3.35)
$$

where $r^{\mathbf{M}_1} \approx r^{\mathbf{M}_2}$ and $f^{\mathbf{M}_1} \approx f^{\mathbf{M}_2}$ are defined by (3.33) and (3.34) with $\approx^M$ being $\approx^{\mathbf{M}_1} \wedge \approx^{\mathbf{M}_2}$.

If $S = \{s_1, \ldots, s_n\}$, we write also $\approx_{s_1,\ldots,s_n}$ instead of $\approx_{\{s_1,\ldots,s_n\}}$.

Theorem 3.79 *Let $\langle R, F, \sigma \rangle$ be a type, M be a nonempty set, $S \subseteq R \cup F$. Then $\approx_S$ defined by (3.35) is an $\mathbf{L}$-equivalence relation on the set of all $\mathbf{L}$-structures of type $\langle R, F, \sigma \rangle$ with support set M. Moreover, if $S = R \cup F$, then the restriction of $\approx_S$ to $\mathbf{L}$-structures with $\mathbf{L}$-equality is an $\mathbf{L}$-equality.*

Proof. From Lemma 3.76 we easily see that for each $s \in S$, $\approx_{\{s\}}$ is an $\mathbf{L}$-equivalence. Since $\approx_S$ is $\bigwedge_{s\in S} \approx_{\{s\}}$, Lemma 3.75 yields that $\approx_S$ is an $\mathbf{L}$-equivalence. Let now $S = R \cup F$, $\approx^{\mathbf{M}_1}$ and $\approx^{\mathbf{M}_2}$ be $\mathbf{L}$-equalities, let $(\mathbf{M}_1 \approx_S \mathbf{M}_2) = 1$. By Lemma 3.75, $\approx^{\mathbf{M}_1} \wedge \approx^{\mathbf{M}_2}$ is an $\mathbf{L}$-equality and, by Lemma 3.76, $r^{\mathbf{M}_1} = r^{\mathbf{M}_2}$ for each $r \in R$, $f^{\mathbf{M}_1} = f^{\mathbf{M}_2}$ for each $f \in F$, thus $\mathbf{M}_1 = \mathbf{M}_2$. □

In the following, we present results that, roughly speaking, show how similarity is preserved by terms and formulas. The results present answers to the questions stated in the beginning of this section. We write $\bigotimes_{i=1}^n a_i$ to denote $a_1 \otimes \cdots \otimes a_n$ (which is correct due to associativity of $\otimes$). More generally, if only a finite number of a_i's ($i \in I$) are different from 1, we write $\bigotimes_{i\in I} a_i$ to denote $a_{i_1} \otimes \cdots \otimes a_{i_k}$ where $i_1, \ldots, i_k$ are pairwise different, $\{i_1, \ldots, i_k\} = \{i \in I \mid a_i \neq 1\}$ (and define $\bigotimes_{i\in I} a_i = 1$ if $a_i = 1$ for all $i \in I$).

Remark 3.80 Observe that for the biresidua corresponding to the Łukasiewicz, Gödel, and product structures on $[0,1]$ we have

$$(a \leftrightarrow b)^n = 1 - n|a-b| \quad \text{(Łukasiewicz)}, \tag{3.36}$$

$$(a \leftrightarrow b)^n = \begin{cases} 1 & \text{for } a = b \\ \min(a,b) & \text{otherwise} \end{cases} \quad \text{(Gödel)}, \tag{3.37}$$

$$(a \leftrightarrow b)^n = \left(\frac{\min(a,b)}{\max(a,b)}\right)^n \quad \text{(product)} \tag{3.38}$$

(we put $0/0 = 1$).

Lemma 3.81 *For an $\mathbf{L}$-structure $\mathbf{M}$ of type $\langle R, F, \sigma \rangle$, a term t, and $\mathbf{M}$-valuations v_1 and v_2 we have*

$$\bigotimes_{x\in \mathrm{var}(t)} (v_1(x) \approx^{\mathbf{M}} v_2(x))^{|t|_x} \leq \|t\|_{\mathbf{M},v_1} \approx^{\mathbf{M}} \|t\|_{\mathbf{M},v_2}.$$

Proof. The proof goes by induction on complexity of t:

Induction base: If t is a variable x_i then $|t|_{x_i} = 1$ and the assertion follows from the compatibility of $f^{\mathbf{M}}$.

Induction step: Let $t = f(t_1, \ldots, t_m)$ and let the assertion be valid for t_j, i.e. $\bigotimes_{x \in \mathrm{var}(t_j)} (v_1(x) \approx^{\mathbf{M}} v_2(x))^{|t_j|_x} \leq \|t_j\|_{\mathbf{M},v_1} \approx^{\mathbf{M}} \|t_j\|_{\mathbf{M},v_2}$ $(j = 1, \ldots, m)$. Then we have

$$\begin{aligned}
&\bigotimes_{x\in\mathrm{var}(t)} (v_1(x) \approx^{\mathbf{M}} v_2(x))^{|t|_x} = \bigotimes_{x\in\mathrm{var}(t)} (v_1(x) \approx^{\mathbf{M}} v_2(x))^{\sum_{j=1}^{m} |t_j|_x} = \\
= &\bigotimes_{x\in\mathrm{var}(t)} (v_1(x) \approx^{\mathbf{M}} v_2(x))^{|t_1|_x} \otimes \cdots \otimes (v_1(x) \approx^{\mathbf{M}} v_2(x))^{|t_m|_x} = \\
= &\bigotimes_{j=1}^{m} \bigotimes_{x\in\mathrm{var}(t)} (v_1(x) \approx^{\mathbf{M}} v_2(x))^{|t_j|_x} = \\
= &\bigotimes_{j=1}^{m} \bigotimes_{x\in\mathrm{var}(t_j)} (v_1(x) \approx^{\mathbf{M}} v_2(x))^{|t_j|_x} \leq \bigotimes_{j=1}^{m} (\|t_j\|_{\mathbf{M},v_1} \approx^{\mathbf{M}} \|t_j\|_{\mathbf{M},v_2}) \leq \\
\leq\ & f^{\mathbf{M}}(\|t_1\|_{\mathbf{M},v_1}, \ldots, \|t_1\|_{\mathbf{M},v_1}) \approx^{\mathbf{M}} f^{\mathbf{M}}(\|t_1\|_{\mathbf{M},v_2}, \ldots, \|t_1\|_{\mathbf{M},v_2}) = \\
= &\|t\|_{\mathbf{M},v_1} \approx^{\mathbf{M}} \|t\|_{\mathbf{M},v_2},
\end{aligned}$$

by compatibility of $f^{\mathbf{M}}$. □

Remark 3.82 Stated another way, Lemma 3.81 says that for any term $t(x_1, \ldots, x_n)$ and $m_1, m_1', \ldots, m_n, m_n' \in M$ we have

$$\bigotimes_{i=1}^{n} (m_i \approx^{\mathbf{M}} m_i')^{|t|_{x_i}} \leq (t^{\mathbf{M}}(m_1, \ldots, m_n) \approx^{\mathbf{M}} t^{\mathbf{M}}(m_1', \ldots, m_n')).$$

Lemma 3.83 *For an* $\mathbf{L}$*-structure* $\mathbf{M}$ *of type* $\langle R, F, \sigma\rangle$*, a formula* φ*, and* $\mathbf{M}$*-valuations* v_1 *and* v_2 *we have*

$$\bigotimes_{x\in\mathrm{free}(\varphi)} (v_1(x) \approx^{\mathbf{M}} v_2(x))^{|\varphi|_x} \quad \leq \quad \|\varphi\|_{\mathbf{M},v_1} \leftrightarrow \|\varphi\|_{\mathbf{M},v_2}.$$

Proof. First, observe that the assertion is equivalent to

$$\bigotimes_{x\in\mathrm{free}(\varphi)} (v_1(x) \approx^{\mathbf{M}} v_2(x))^{|\varphi|_x} \otimes \|\varphi\|_{\mathbf{M},v_1} \leq \|\varphi\|_{\mathbf{M},v_2}.$$

We prove the assertion by induction over the complexity of φ.

The assertion is trivial if φ is a truth degree. If $\varphi = r(t_1, \ldots, t_m)$ then, by Lemma 3.81, we have

$$\begin{aligned}
& \bigotimes_{x\in\mathrm{free}(\varphi)} (v_1(x) \approx^{\mathbf{M}} v_2(x))^{|\varphi|_x} = \\
= \;& \bigotimes_{x\in\mathrm{free}(\varphi)} (v_1(x) \approx^{\mathbf{M}} v_2(x))^{|t_1|_x+\cdots+|t_m|_x} = \\
= \;& \bigotimes_{x\in\mathrm{var}(t_1)} (v_1(x) \approx^{\mathbf{M}} v_2(x))^{|t_1|_x} \otimes\cdots\otimes \\
& \otimes \bigotimes_{x\in\mathrm{var}(t_m)} (v_1(x) \approx^{\mathbf{M}} v_2(x))^{|t_m|_x} \leq \\
\leq \;& (\|t_1\|_{\mathbf{M},v_1} \approx^{\mathbf{M}} \|t_1\|_{\mathbf{M},v_2}) \otimes\cdots\otimes (\|t_m\|_{\mathbf{M},v_1} \approx^{\mathbf{M}} \|t_m\|_{\mathbf{M},v_2}) \leq \\
\leq \;& r^{\mathbf{M}}(\|t_1\|_{\mathbf{M},v_1}, \ldots, \|t_m\|_{\mathbf{M},v_1}) \leftrightarrow r^{\mathbf{M}}(\|t_1\|_{\mathbf{M},v_2}, \ldots, \|t_m\|_{\mathbf{M},v_2}) = \\
= \;& \|\varphi\|_{\mathbf{M},v_1} \leftrightarrow \|\varphi\|_{\mathbf{M},v_2}.
\end{aligned}$$

Let the assertion be valid for φ and ψ. We have

$$\begin{aligned}
& \bigotimes_{x\in\mathrm{free}(\varphi\mathbin{\&}\psi)} (v_1(x) \approx^{\mathbf{M}} v_2(x))^{|\varphi\mathbin{\&}\psi|_x} = \\
= \;& \bigotimes_{x\in\mathrm{free}(\varphi\mathbin{\&}\psi)} (v_1(x) \approx^{\mathbf{M}} v_2(x))^{\max(|\varphi|_x,|\psi|_x)} \leq \\
\leq \;& \bigotimes_{x\in\mathrm{free}(\varphi)} (v_1(x) \approx^{\mathbf{M}} v_2(x))^{|\varphi|_x} \leq \\
\leq \;& \|\varphi\|_{\mathbf{M},v_1} \leftrightarrow \|\varphi\|_{\mathbf{M},v_2},
\end{aligned}$$

and similarly,

$$\bigotimes_{x\in\mathrm{free}(\varphi\mathbin{\&}\psi)} (v_1(x) \approx^{\mathbf{M}} v_2(x))^{|\varphi\mathbin{\&}\psi|_x} \leq \|\psi\|_{\mathbf{M},v_1} \leftrightarrow \|\psi\|_{\mathbf{M},v_2}.$$

Since

$$\begin{aligned}
& (\|\varphi\|_{\mathbf{M},v_1} \leftrightarrow \|\varphi\|_{\mathbf{M},v_2}) \wedge (\|\psi\|_{\mathbf{M},v_1} \leftrightarrow \|\psi\|_{\mathbf{M},v_2}) \leq \\
\leq \;& (\|\varphi\|_{\mathbf{M},v_1} \wedge \|\psi\|_{\mathbf{M},v_1}) \leftrightarrow (\|\varphi\|_{\mathbf{M},v_2} \wedge \|\psi\|_{\mathbf{M},v_2}) = \\
= \;& \|\varphi \mathbin{\&} \psi\|_{\mathbf{M},v_1} \leftrightarrow \|\varphi \mathbin{\&} \psi\|_{\mathbf{M},v_2},
\end{aligned}$$

we have

$$\bigotimes_{x\in\mathrm{free}(\varphi\mathbin{\&}\psi)} (v_1(x) \approx^{\mathbf{M}} v_2(x))^{|\varphi\mathbin{\&}\psi|_x} \leq \|\varphi \mathbin{\&} \psi\|_{\mathbf{M},v_1} \leftrightarrow \|\varphi \mathbin{\&} \psi\|_{\mathbf{M},v_2},$$

i.e. the assertion is valid for $\varphi \mathbin{\&} \phi$.

In order to show that the assertion is valid for $\varphi \vee \phi$, we have to show

$$\bigotimes_{x\in\mathrm{free}(\varphi\vee\psi)} (v_1(x) \approx^{\mathbf{M}} v_2(x))^{|\varphi\vee\psi|_x} \otimes \|\varphi \vee \psi\|_{\mathbf{M},v_1} \leq \|\varphi \vee \psi\|_{\mathbf{M},v_2}.$$

We have

$$\begin{aligned}
&\bigotimes_{x\in\mathrm{free}(\varphi\vee\psi)} (v_1(x) \approx^{\mathbf{M}} v_2(x))^{|\varphi\vee\psi|_x} \otimes \|\varphi \vee \psi\|_{\mathbf{M},v_1} = \\
&= \Big(\bigotimes_{x\in\mathrm{free}(\varphi\vee\psi)} (v_1(x) \approx^{\mathbf{M}} v_2(x))^{\max(|\varphi|_x,|\psi|_x)} \otimes \|\varphi\|_{\mathbf{M},v_1}\Big) \vee \\
&\quad \vee\Big(\bigotimes_{x\in\mathrm{free}(\varphi\vee\psi)} (v_1(x) \approx^{\mathbf{M}} v_2(x))^{\max(|\varphi|_x,|\psi|_x)} \otimes \|\psi\|_{\mathbf{M},v_1}\Big) \leq \\
&\leq \Big(\bigotimes_{x\in\mathrm{free}(\varphi)} (v_1(x) \approx^{\mathbf{M}} v_2(x))^{|\varphi|_x} \otimes \|\varphi\|_{\mathbf{M},v_1}\Big) \vee \\
&\quad \vee\Big(\bigotimes_{x\in\mathrm{free}(\psi)} (v_1(x) \approx^{\mathbf{M}} v_2(x))^{|\psi|_x} \otimes \|\psi\|_{\mathbf{M},v_1}\Big) \leq \\
&\leq \|\varphi\|_{\mathbf{M},v_2} \vee \|\psi\|_{\mathbf{M},v_2} = \|\varphi \vee \psi\|_{\mathbf{M},v_2}
\end{aligned}$$

which was to be proved.

Consider $\varphi \otimes \psi$: We have

$$\begin{aligned}
&\Big(\bigotimes_{x\in\mathrm{free}(\varphi\otimes\psi)} (v_1(x) \approx^{\mathbf{M}} v_2(x))^{|\varphi\otimes\psi|_x}\Big) \otimes \|\varphi \otimes \psi\|_{\mathbf{M},v_1} = \\
&= \Big(\bigotimes_{x\in\mathrm{free}(\varphi\otimes\psi)} (v_1(x) \approx^{\mathbf{M}} v_2(x))^{|\varphi|_x+|\psi|_x}\Big) \otimes \|\varphi\|_{\mathbf{M},v_1} \otimes \|\varphi\|_{\mathbf{M},v_1} \leq \\
&\leq \Big(\bigotimes_{x\in\mathrm{free}(\varphi)} (v_1(x) \approx^{\mathbf{M}} v_2(x))^{|\varphi|_x} \otimes \|\varphi\|_{\mathbf{M},v_1}\Big) \otimes \\
&\quad \otimes\Big(\bigotimes_{x\in\mathrm{free}(\psi)} (v_1(x) \approx^{\mathbf{M}} v_2(x))^{|\psi|_x} \otimes \|\psi\|_{\mathbf{M},v_1}\Big) \leq \\
&\leq \|\varphi\|_{\mathbf{M},v_2} \otimes \|\psi\|_{\mathbf{M},v_2} = \|\varphi \otimes \psi\|_{\mathbf{M},v_2},
\end{aligned}$$

i.e. the assertion holds for $\varphi \otimes \psi$.

We show that the assertion is true for $\varphi \Rightarrow \psi$. We have to show

$$\Big(\bigotimes_{x\in\mathrm{free}(\varphi\Rightarrow\psi)} (v_1(x) \approx^{\mathbf{M}} v_2(x))^{|\varphi\Rightarrow\psi|_x}\Big) \otimes \|\varphi \Rightarrow \psi\|_{\mathbf{M},v_1} \leq \|\varphi \Rightarrow \psi\|_{\mathbf{M},v_2}.$$

Since $\|\varphi \Rightarrow \psi\|_{\mathbf{M},v_2} = \|\varphi\|_{\mathbf{M},v_2} \to \|\varphi\|_{\mathbf{M},v_2}$, adjointness yields that the required inequality is equivalent to

$$\begin{aligned}
&\Big(\bigotimes_{x\in\mathrm{free}(\varphi\Rightarrow\psi)} (v_1(x) \approx^{\mathbf{M}} v_2(x))^{|\varphi\Rightarrow\psi|_x}\Big) \otimes \|\varphi \Rightarrow \psi\|_{\mathbf{M},v_1} \otimes \|\varphi\|_{\mathbf{M},v_2} \leq \\
&\leq \|\psi\|_{\mathbf{M},v_2},
\end{aligned}$$

which is true. Indeed,

$$
\begin{aligned}
&(\bigotimes_{x\in \mathrm{free}(\varphi\Rightarrow\psi)} (v_1(x)\approx^{\mathbf{M}} v_2(x))^{|\varphi\Rightarrow\psi|_x})\otimes\|\varphi\Rightarrow\psi\|_{\mathbf{M},v_1}\otimes\|\varphi\|_{\mathbf{M},v_2} =\\
=\ &(\bigotimes_{x\in \mathrm{free}(\psi)} (v_1(x)\approx^{\mathbf{M}} v_2(x))^{|\psi|_x})\otimes\\
&\otimes(\bigotimes_{x\in \mathrm{free}(\varphi)} (v_1(x)\approx^{\mathbf{M}} v_2(x))^{|\varphi|_x})\otimes\|\varphi\|_{\mathbf{M},v_2}\otimes(\|\varphi\|_{\mathbf{M},v_1}\to\|\psi\|_{\mathbf{M},v_1})\\
\leq\ &(\bigotimes_{x\in \mathrm{free}(\psi)} (v_1(x)\approx^{\mathbf{M}} v_2(x))^{|\psi|_x})\otimes\|\varphi\|_{\mathbf{M},v_1}\otimes(\|\varphi\|_{\mathbf{M},v_1}\to\|\psi\|_{\mathbf{M},v_1})\\
\leq\ &(\bigotimes_{x\in \mathrm{free}(\psi)} (v_1(x)\approx^{\mathbf{M}} v_2(x))^{|\psi|_x})\otimes\|\psi\|_{\mathbf{M},v_1}\leq\|\psi\|_{\mathbf{M},v_2}.
\end{aligned}
$$

Consider now quantifiers.

$$
\begin{aligned}
&(\bigotimes_{x\in \mathrm{free}((\forall y)\varphi)} (v_1(x)\approx^{\mathbf{M}} v_2(x))^{|(\forall y)\varphi|_x})\otimes\|(\forall y)\varphi\|_{\mathbf{M},v_1} =\\
=\ &(\bigotimes_{x\in \mathrm{free}((\forall y)\varphi)} (v_1(x)\approx^{\mathbf{M}} v_2(x))^{|\varphi|_x})\otimes\bigwedge_{v_1'=_y v_1}\|\varphi\|_{\mathbf{M},v_1'}\leq\\
\leq\ &\bigwedge_{v_1'=_y v_1}((\bigotimes_{x\in \mathrm{free}((\forall y)\varphi)} (v_1(x)\approx^{\mathbf{M}} v_2(x))^{|\varphi|_x})\otimes\|\varphi\|_{\mathbf{M},v_1'}) =\\
=\ &\bigwedge_{v_1'=_y v_1}((\bigotimes_{x\in \mathrm{free}((\forall y)\varphi)} (v_1'(x)\approx^{\mathbf{M}} v_2'(x))^{|\varphi|_x})\otimes\|\varphi\|_{\mathbf{M},v_1'})\leq\\
\leq\ &\|\varphi\|_{\mathbf{M},v_2'}
\end{aligned}
$$

for any valuation v_2' such that $v_2' =_y v_2$. Therefore, we also have

$$
\begin{aligned}
&(\bigotimes_{x\in \mathrm{free}((\forall y)\varphi)} (v_1(x)\approx^{\mathbf{M}} v_2(x))^{|(\forall y)\varphi|_x})\otimes\|(\forall y)\varphi\|_{\mathbf{M},v_1}\leq\\
\leq\ &\bigwedge_{v_2'=_y v_2}\|\varphi\|_{\mathbf{M},v_2'} = \|(\forall y)\varphi\|_{\mathbf{M},v_2}.
\end{aligned}
$$

We used the obvious fact that if $v_1 =_y v_1'$ and $v_2 =_y v_2'$, then $(v_1(x)\approx^{\mathbf{M}} v_2(x)) = (v_1'(x)\approx^{\mathbf{M}} v_2'(x))$ for any $x\in\mathrm{free}((\forall y)\varphi)$. Thus, the assertion is valid for $(\forall y)\varphi$.

Finally, for $(\exists y)\varphi$ we similarly have

$$
\begin{aligned}
&(\bigotimes_{x\in \mathrm{free}((\exists y)\varphi)} (v_1(x)\approx^{\mathbf{M}} v_2(x))^{|(\exists y)\varphi|_x})\otimes\|(\exists y)\varphi\|_{\mathbf{M},v_1} =\\
=\ &(\bigotimes_{x\in \mathrm{free}((\exists y)\varphi)} (v_1(x)\approx^{\mathbf{M}} v_2(x))^{|\varphi|_x})\otimes\bigvee_{v_1'=_y v_1}\|\varphi\|_{\mathbf{M},v_1'} =
\end{aligned}
$$

$$
\begin{aligned}
&= \bigvee_{v_1' =_y v_1} \Big(\Big(\bigotimes_{x \in \text{free}((\exists y)\varphi)} (v_1(x) \approx^{\mathbf{M}} v_2(x))^{|\varphi|_x} \Big) \otimes \|\varphi\|_{\mathbf{M},v_1'} \Big) = \\
&= \bigvee_{v_1' =_y v_1} \Big(\Big(\bigotimes_{x \in \text{free}((\exists y)\varphi)} (v_1'(x) \approx^{\mathbf{M}} (v_2'(v_1'))(x))^{|\varphi|_x} \Big) \otimes \|\varphi\|_{\mathbf{M},v_1'} \Big) \leq \\
&\leq \bigvee_{v_1' =_y v_1} \|\varphi\|_{\mathbf{M},v_2'(v_1')} = \bigvee_{v_2' =_y v_2} \|\varphi\|_{\mathbf{M},v_2'} = \|(\exists y)\varphi\|_{\mathbf{M},v_2},
\end{aligned}
$$

where $v_2'(v_1')$ is a valuation such that $v_2'(v_1') =_y v_2$ and $(v_2'(v_1'))(y) = v_1'(y)$. □

For **L**-structures $\mathbf{M}_1$ and $\mathbf{M}_2$ with $M_1 = M_2$ we define $(f^{\mathbf{M}_1} \approx f^{\mathbf{M}_2})$ by (3.34) with $\approx^M$ being $\approx^{\mathbf{M}_1} \wedge \approx^{\mathbf{M}_2}$.

Lemma 3.84 *Let $\mathbf{M}_1$, $\mathbf{M}_2$ be **L**-structures of type $\langle R, F, \sigma\rangle$ that have the same support set M (i.e. $M_1 = M_2$), t be a term, v be a valuation. Then, for $i = 1, 2$, we have*

$$
\bigotimes_{f \in F} (f^{\mathbf{M}_1} \approx f^{\mathbf{M}_2})^{|t|_f} \leq \|t\|_{\mathbf{M}_1,v} \approx^{\mathbf{M}_i} \|t\|_{\mathbf{M}_2,v}.
$$

Proof. We prove the assertion by induction over complexity of t. If t is a variable, then the assertion is trivial. For the induction step, let $t = g(t_1, \ldots, t_n)$ and assume the assertion is valid for $t_1, \ldots, t_n$. For $i = 1, 2$, we have

$$
\begin{aligned}
&\bigotimes_{f \in F} (f^{\mathbf{M}_1} \approx f^{\mathbf{M}_2})^{|t|_f} = \\
&= (g^{\mathbf{M}_1} \approx g^{\mathbf{M}_2})^{|t|_g} \otimes \bigotimes_{f \neq g} (f^{\mathbf{M}_1} \approx f^{\mathbf{M}_2})^{|t|_f} = \\
&= (g^{\mathbf{M}_1} \approx g^{\mathbf{M}_2})^{1+\sum_{i=1}^n |t_i|_g} \otimes \bigotimes_{f \neq g, |t|_f} (f^{\mathbf{M}_1} \approx f^{\mathbf{M}_2})^{\sum_{i=1}^n |t_i|_f} = \\
&= (g^{\mathbf{M}_1} \approx g^{\mathbf{M}_2}) \otimes \bigotimes_{i=1}^n \bigotimes_{f \in F, |t_i|_f > 0} (f^{\mathbf{M}_1} \approx f^{\mathbf{M}_2})^{|t_i|_f} \leq \\
&\leq (g^{\mathbf{M}_1} \approx g^{\mathbf{M}_2}) \otimes \bigotimes_{i=1}^n (\|t_i\|_{\mathbf{M}_1,v} \approx^{\mathbf{M}_i} \|t_i\|_{\mathbf{M}_2,v}) \leq \\
&\leq (g^{\mathbf{M}_1} \approx g^{\mathbf{M}_2}) \otimes (g^{\mathbf{M}_2}(\|t_1\|_{\mathbf{M}_1,v}, \ldots, \|t_n\|_{\mathbf{M}_1,v}) \approx^{\mathbf{M}_i} \\
&\quad \approx^{\mathbf{M}_i} (g^{\mathbf{M}_2}(\|t_1\|_{\mathbf{M}_2,v}, \ldots, \|t_n\|_{\mathbf{M}_2,v})) = \\
&= \Big(\bigwedge_{m_1,\ldots,m_n} g^{\mathbf{M}_1}(m_1, \ldots, m_n)(\approx^{\mathbf{M}_1} \wedge \approx^{\mathbf{M}_2}) g^{\mathbf{M}_2}(m_1, \ldots, m_n)\Big) \otimes
\end{aligned}
$$

$$\begin{aligned}
& (g^{\mathbf{M}_2}(\|t_1\|_{\mathbf{M}_1,v},\dots,\|t_n\|_{\mathbf{M}_1,v}) \approx^{\mathbf{M}_i} g^{\mathbf{M}_2}(\|t_1\|_{\mathbf{M}_2,v},\dots,\|t_n\|_{\mathbf{M}_2,v}))\\
\leq\ & (g^{\mathbf{M}_1}(\|t_1\|_{\mathbf{M}_1,v},\dots,\|t_n\|_{\mathbf{M}_1,v}) \approx^{\mathbf{M}_i} g^{\mathbf{M}_2}(\|t_1\|_{\mathbf{M}_1,v},\dots,\|t_n\|_{\mathbf{M}_1,v})) \otimes\\
& (g^{\mathbf{M}_2}(\|t_1\|_{\mathbf{M}_1,v},\dots,\|t_n\|_{\mathbf{M}_1,v}) \approx^{\mathbf{M}_i} g^{\mathbf{M}_2}(\|t_1\|_{\mathbf{M}_2,v},\dots,\|t_n\|_{\mathbf{M}_2,v}))\\
\leq\ & g^{\mathbf{M}_1}(\|t_1\|_{\mathbf{M}_1,v},\dots,\|t_n\|_{\mathbf{M}_1,v}) \approx^{\mathbf{M}_i} g^{\mathbf{M}_2}(\|t_1\|_{\mathbf{M}_2,v},\dots,\|t_n\|_{\mathbf{M}_2,v}) =\\
=\ & \|t\|_{\mathbf{M}_1,v} \approx^{\mathbf{M}_i} \|t\|_{\mathbf{M}_2,v},
\end{aligned}$$

completing the proof. □

Lemma 3.85 *Let $\mathbf{M}_1$, $\mathbf{M}_2$ be $\mathbf{L}$-structures of type $\langle R, F, \sigma\rangle$ that have the same support set M (i.e. $M_1 = M_2$), φ be a formula, v be a valuation. Then we have*

$$\bigotimes_{r\in R}(r^{\mathbf{M}_1} \approx r^{\mathbf{M}_2})^{|\varphi|_r} \otimes \bigotimes_{f\in F}(f^{\mathbf{M}_1} \approx f^{\mathbf{M}_2})^{|\varphi|_f} \leq \|\varphi\|_{\mathbf{M}_1,v} \leftrightarrow \|\varphi\|_{\mathbf{M}_2,v}.$$

Proof. It is easy to see that we have to prove

$$\bigotimes_{r\in R}(r^{\mathbf{M}_1} \approx r^{\mathbf{M}_2})^{|\varphi|_r} \otimes \bigotimes_{f\in F}(f^{\mathbf{M}_1} \approx f^{\mathbf{M}_2})^{|\varphi|_f} \otimes \|\varphi\|_{\mathbf{M}_1,v} \leq \|\varphi\|_{\mathbf{M}_2,v}.$$

The proof goes by induction over the complexity of φ. If $\varphi = s(t_1,\dots,t_m)$ then, by Lemma 3.84,

$$\begin{aligned}
& \bigotimes_{r\in R}(r^{\mathbf{M}_1} \approx r^{\mathbf{M}_2})^{|\varphi|_r} \otimes \bigotimes_{f\in F}(f^{\mathbf{M}_1} \approx f^{\mathbf{M}_2})^{|\varphi|_f} =\\
=\ & (s^{\mathbf{M}_1} \approx s^{\mathbf{M}_2}) \otimes \bigotimes_{f\in F}(f^{\mathbf{M}_1} \approx f^{\mathbf{M}_2})^{|t_1|_f+\dots+|t_m|_f} =\\
=\ & (s^{\mathbf{M}_1} \approx s^{\mathbf{M}_2}) \otimes \bigotimes_{i=1}^{m}\bigotimes_{f\in F}(f^{\mathbf{M}_1} \approx f^{\mathbf{M}_2})^{|t_i|_f} \leq\\
\leq\ & (s^{\mathbf{M}_1} \approx s^{\mathbf{M}_2}) \otimes \bigotimes_{i=1}^{m}(\|t_i\|_{\mathbf{M}_1,v} \approx^{\mathbf{M}_2} \|t_i\|_{\mathbf{M}_2,v}) \leq\\
\leq\ & (s^{\mathbf{M}_1} \approx s^{\mathbf{M}_2}) \otimes\\
& (s^{\mathbf{M}_2}(\|t_1\|_{\mathbf{M}_1,v},\dots,\|t_m\|_{\mathbf{M}_1,v}) \leftrightarrow s^{\mathbf{M}_2}(\|t_1\|_{\mathbf{M}_2,v},\dots,\|t_m\|_{\mathbf{M}_2,v})) \leq\\
\leq\ & (s^{\mathbf{M}_1}(\|t_1\|_{\mathbf{M}_1,v},\dots,\|t_m\|_{\mathbf{M}_1,v}) \leftrightarrow s^{\mathbf{M}_2}(\|t_1\|_{\mathbf{M}_1,v},\dots,\|t_m\|_{\mathbf{M}_1,v})) \otimes\\
& \otimes(s^{\mathbf{M}_2}(\|t_1\|_{\mathbf{M}_1,v},\dots,\|t_m\|_{\mathbf{M}_1,v}) \leftrightarrow s^{\mathbf{M}_2}(\|t_1\|_{\mathbf{M}_2,v},\dots,\|t_m\|_{\mathbf{M}_2,v}))\\
\leq\ & (s^{\mathbf{M}_1}(\|t_1\|_{\mathbf{M}_1,v},\dots,\|t_1\|_{\mathbf{M}_1,v}) \leftrightarrow s^{\mathbf{M}_2}(\|t_1\|_{\mathbf{M}_2,v},\dots,\|t_1\|_{\mathbf{M}_2,v})) =\\
=\ & \|\varphi\|_{\mathbf{M}_1,v} \leftrightarrow \|\varphi\|_{\mathbf{M}_2,v}.
\end{aligned}$$

Let the assertion be valid for φ and ψ. We have

$$\begin{aligned}
& \bigotimes_{r\in R}(r^{\mathbf{M}_1} \approx r^{\mathbf{M}_2})^{|\varphi \wedge \psi|_r} \otimes \bigotimes_{f\in F}(f^{\mathbf{M}_1} \approx f^{\mathbf{M}_2})^{|\varphi \wedge \psi|_f} \otimes \|\varphi \wedge \psi\|_{\mathbf{M}_1,v} = \\
= \;& \bigotimes_{r\in R}(r^{\mathbf{M}_1} \approx r^{\mathbf{M}_2})^{|\varphi \wedge \psi|_r} \otimes \bigotimes_{f\in F}(f^{\mathbf{M}_1} \approx f^{\mathbf{M}_2})^{|\varphi \wedge \psi|_f} \otimes \\
& \otimes(\|\varphi\|_{\mathbf{M}_1,v} \wedge \|\psi\|_{\mathbf{M}_1,v}) \leq \\
\leq \;& (\bigotimes_{r\in R}(r^{\mathbf{M}_1} \approx r^{\mathbf{M}_2})^{|\varphi|_r} \otimes \bigotimes_{f\in F}(f^{\mathbf{M}_1} \approx f^{\mathbf{M}_2})^{|\varphi|_f} \otimes \|\varphi\|_{\mathbf{M}_1,v}) \wedge \\
& \wedge(\bigotimes_{r\in R}(r^{\mathbf{M}_1} \approx r^{\mathbf{M}_2})^{|\psi|_r} \otimes \bigotimes_{f\in F}(f^{\mathbf{M}_1} \approx f^{\mathbf{M}_2})^{|\psi|_f} \otimes \|\psi\|_{\mathbf{M}_1,v}) \leq \\
\leq \;& \|\varphi\|_{\mathbf{M}_2,v} \wedge \|\psi\|_{\mathbf{M}_2,v} = \|\varphi \wedge \psi\|_{\mathbf{M}_2,v}.
\end{aligned}$$

The proof for $\varphi \vee \psi$ can be done in the same way as for $\varphi \wedge \psi$ (just replace $\wedge$ by $\vee$).

Consider $\varphi \otimes \psi$:

$$\begin{aligned}
& \bigotimes_{r\in R}(r^{\mathbf{M}_1} \approx r^{\mathbf{M}_2})^{|\varphi \otimes \psi|_r} \otimes \bigotimes_{f\in F}(f^{\mathbf{M}_1} \approx f^{\mathbf{M}_2})^{|\varphi \otimes \psi|_f} \otimes \|\varphi \otimes \psi\|_{\mathbf{M}_1,v} = \\
= \;& \bigotimes_{r\in R}(r^{\mathbf{M}_1} \approx r^{\mathbf{M}_2})^{|\varphi|_r} \otimes \bigotimes_{f\in F}(f^{\mathbf{M}_1} \approx f^{\mathbf{M}_2})^{|\varphi|_f} \otimes \|\varphi\|_{\mathbf{M}_1,v} \otimes \\
& \otimes \bigotimes_{r\in R}(r^{\mathbf{M}_1} \approx r^{\mathbf{M}_2})^{|\varphi|_r} \otimes \bigotimes_{f\in F}(f^{\mathbf{M}_1} \approx f^{\mathbf{M}_2})^{|\varphi|_f} \otimes \|\psi\|_{\mathbf{M}_1,v} \leq \\
\leq \;& \|\varphi\|_{\mathbf{M}_2,v} \otimes \|\psi\|_{\mathbf{M}_2,v} = \|\varphi \otimes \psi\|_{\mathbf{M}_2,v}.
\end{aligned}$$

For $\varphi \Rightarrow \psi$, adjointness implies that we have to show

$$\begin{aligned}
& \bigotimes_{r\in R}(r^{\mathbf{M}_1} \approx r^{\mathbf{M}_2})^{|\varphi \Rightarrow \psi|_r} \otimes \bigotimes_{f\in F}(f^{\mathbf{M}_1} \approx f^{\mathbf{M}_2})^{|\varphi \Rightarrow \psi|_f} \otimes \\
& \otimes \|\varphi \Rightarrow \psi\|_{\mathbf{M}_1,v} \otimes \|\varphi\|_{\mathbf{M}_2,v} \leq \|\psi\|_{\mathbf{M}_2,v}.
\end{aligned}$$

This inequality is true:

$$\begin{aligned}
& \bigotimes_{r\in R}(r^{\mathbf{M}_1} \approx r^{\mathbf{M}_2})^{|\varphi \Rightarrow \psi|_r} \otimes \bigotimes_{f\in F}(f^{\mathbf{M}_1} \approx f^{\mathbf{M}_2})^{|\varphi \Rightarrow \psi|_f} \otimes \\
& \otimes \|\varphi \Rightarrow \psi\|_{\mathbf{M}_1,v} \otimes \|\varphi\|_{\mathbf{M}_2,v} = \\
= \;& \bigotimes_{r\in R}(r^{\mathbf{M}_1} \approx r^{\mathbf{M}_2})^{|\varphi|_r + |\psi|_r} \otimes \bigotimes_{f\in F}(f^{\mathbf{M}_1} \approx f^{\mathbf{M}_2})^{|\varphi|_f + |\psi|_f} \otimes \\
& \otimes \|\varphi\|_{\mathbf{M}_2,v} \otimes(\|\varphi\|_{\mathbf{M}_1,v} \to \|\psi\|_{\mathbf{M}_1,v}) \leq \\
\leq \;& \bigotimes_{r\in R}(r^{\mathbf{M}_1} \approx r^{\mathbf{M}_2})^{|\psi|_r} \otimes \bigotimes_{f\in F}(f^{\mathbf{M}_1} \approx f^{\mathbf{M}_2})^{|\psi|_f} \otimes
\end{aligned}$$

$$
\begin{aligned}
&\otimes \|\varphi\|_{\mathbf{M}_1,v} \otimes (\|\varphi\|_{\mathbf{M}_1,v} \to \|\psi\|_{\mathbf{M}_1,v}) \leq \\
\leq\ & \bigotimes_{r\in R}(r^{\mathbf{M}_1} \approx r^{\mathbf{M}_2})^{|\psi|_r} \otimes \bigotimes_{f\in F}(f^{\mathbf{M}_1} \approx f^{\mathbf{M}_2})^{|\psi|_f} \otimes \|\psi\|_{\mathbf{M}_1,v} \leq \\
\leq\ & \|\psi\|_{\mathbf{M}_1,v}.
\end{aligned}
$$

For $(\forall y)\varphi$ we have

$$
\begin{aligned}
& \bigotimes_{r\in R}(r^{\mathbf{M}_1} \approx r^{\mathbf{M}_2})^{|(\forall y)\varphi|_r} \otimes \bigotimes_{f\in F}(f^{\mathbf{M}_1} \approx f^{\mathbf{M}_2})^{|(\forall y)\varphi|_f} \otimes \|(\forall y)\varphi\|_{\mathbf{M}_1,v} = \\
=\ & \bigotimes_{r\in R}(r^{\mathbf{M}_1} \approx r^{\mathbf{M}_2})^{|\varphi|_r} \otimes \bigotimes_{f\in F}(f^{\mathbf{M}_1} \approx f^{\mathbf{M}_2})^{|\varphi|_f} \otimes \bigwedge_{v'=_y v} \|\varphi\|_{\mathbf{M}_1,v'} \leq \\
\leq\ & \bigwedge_{v'=_y v} (\bigotimes_{r\in R}(r^{\mathbf{M}_1} \approx r^{\mathbf{M}_2})^{|\varphi|_r} \otimes \bigotimes_{f\in F}(f^{\mathbf{M}_1} \approx f^{\mathbf{M}_2})^{|\varphi|_f} \otimes \|\varphi\|_{\mathbf{M}_1,v'}) \leq \\
\leq\ & \bigwedge_{v'=_y v} \|\varphi\|_{\mathbf{M}_2,v'} = \|(\forall y)\varphi\|_{\mathbf{M}_2,v}.
\end{aligned}
$$

For $(\exists y)\varphi$, the above proof for $(\forall y)\varphi$ works as well (just replace $\forall$ by $\exists$). □

To sum up, we obtain the following theorems.

Theorem 3.86 (preservation of similarity by terms) *For* **L**-*structures* $\mathbf{M}_1, \mathbf{M}_2$ *of type* $\langle R, F, \sigma\rangle$ *with* $M_1 = M_2$, *a term* t, *and valuations* v_1 *and* v_2, *we have*

$$
\begin{aligned}
\bigotimes_{x\in \mathrm{var}(t)} (v_1(x) \approx^{\mathbf{M}_i} v_2(x))^{|t|_x} \otimes \bigotimes_{f\in F}(f^{\mathbf{M}_1} \approx f^{\mathbf{M}_2})^{|t|_f} \quad \leq \\
\leq \|t\|_{\mathbf{M}_1,v_1} \approx^{\mathbf{M}_i} \|t\|_{\mathbf{M}_2,v_2}
\end{aligned}
$$

for $i = 1, 2$.

Proof. By Lemmas 3.81 and 3.84 we have

$$
\begin{aligned}
& \bigotimes_{x\in \mathrm{var}(t)} (v_1(x) \approx^{\mathbf{M}_1} v_2(x))^{|t|_x} \otimes \bigotimes_{f\in F}(f^{\mathbf{M}_1} \approx f^{\mathbf{M}_2})^{|t|_f} \quad \leq \\
\leq\ & (\|t\|_{\mathbf{M}_1,v_1} \approx^{\mathbf{M}_1} \|t\|_{\mathbf{M}_1,v_2}) \otimes (\|t\|_{\mathbf{M}_1,v_2} \approx^{\mathbf{M}_i} \|t\|_{\mathbf{M}_2,v_2}) \leq \\
\leq\ & \|t\|_{\mathbf{M}_1,v_1} \approx^{\mathbf{M}_i} \|t\|_{\mathbf{M}_2,v_2},
\end{aligned}
$$

proving the assertion for $i = 1$; for $i = 2$ one proceeds analogously. □

Theorem 3.87 (preservation of similarity by formulas) *For* **L**-*structures* $\mathbf{M}_1, \mathbf{M}_2$ *of type* $\langle R, F, \sigma\rangle$ *with* $M_1 = M_2$, *a formula* φ, *and valuations* v_1 *and* v_2, *we have*

$$\begin{aligned} &\bigotimes_{x\in \mathrm{free}(\varphi)} (v_1(x) \approx^{\mathbf{M}_i} v_2(x))^{|\varphi|_x} \otimes \\ &\otimes \bigotimes_{r\in R} (r^{\mathbf{M}_1} \approx r^{\mathbf{M}_2})^{|\varphi|_r} \otimes \bigotimes_{f\in F} (f^{\mathbf{M}_1} \approx f^{\mathbf{M}_2})^{|\varphi|_f} \quad \leq \\ &\qquad \leq \|\varphi\|_{\mathbf{M}_1, v_1} \leftrightarrow \|\varphi\|_{\mathbf{M}_2, v_2} \end{aligned}$$

for $i = 1, 2$.

Proof. By Lemmas 3.83 and 3.85 we get for $i = 1$:

$$\begin{aligned} &\bigotimes_{x\in \mathrm{free}(\varphi)} (v_1(x) \approx^{\mathbf{M}_1} v_2(x))^{|\varphi|_x} \otimes \\ &\otimes \bigotimes_{r\in R} (r^{\mathbf{M}_1} \approx r^{\mathbf{M}_2})^{|\varphi|_r} \otimes \bigotimes_{f\in F} (f^{\mathbf{M}_1} \approx f^{\mathbf{M}_2})^{|\varphi|_f} \quad \leq \\ \leq\ & (\|\varphi\|_{\mathbf{M}_1, v_1} \leftrightarrow \|\varphi\|_{\mathbf{M}_1, v_2}) \otimes (\|\varphi\|_{\mathbf{M}_1, v_2} \leftrightarrow \|\varphi\|_{\mathbf{M}_2, v_2}) \leq \\ \leq\ & \|\varphi\|_{\mathbf{M}_1, v_1} \leftrightarrow \|\varphi\|_{\mathbf{M}_2, v_2}. \end{aligned}$$

The proof is analogous for $i = 2$. □

Remark 3.88 (1) Note that Lemmas 3.81–3.85 are consequences of Theorems 3.86 and 3.87.

(2) Consider Lemma 3.83 and what it says: Given a formula φ with free variables $x_1, \ldots, x_n$, i.e. $\varphi(x_1, \ldots, x_n)$, and an **L**-structure **M**, the **L**-relation $\varphi^{\mathbf{M}}$ given by (3.31) is $(|\varphi|_{x_1}, \ldots, |\varphi|_{x_n})$-compatible with $\approx^{\mathbf{M}}$. Similarly, Lemma 3.81 says that for a term $t(x_1, \ldots, x_n)$ and an **L**-structure **M**, $t^{\mathbf{M}}$ is a function on M that is $(|t|_{x_1}, \ldots, |t|_{x_n})$-compatible with $\approx^{\mathbf{M}}$.

(3) Consider now function $\varphi^{(\mathbf{L},M)}$ defined by a formula φ by (3.32). The definition of compatibility of functions can be extended in a natural way to functions like $\varphi^{(\mathbf{L},M)}$ (since the arguments, i.e. **L**-relations and functions defined on a set M and compatible with an **L**-equivalence relation $\approx^M$ on M, are taken from sets with **L**-equivalences (namely those defined by (3.33) and (3.34))). Theorem 3.87 then says that $\varphi^{(\mathbf{L},M)}$ is $(|\varphi|_{r_1}, \ldots, |\varphi|_{r_p}, |\varphi|_{f_1}, \ldots, |\varphi|_{f_q})$-compatible and, moreover, the **L**-relation $\varphi^{(\mathbf{L},M)}(R_1, \ldots, R_p, F_1, \ldots, F_q)$ is $(|\varphi|_{x_1}, \ldots, |\varphi|_{x_n})$-compatible with $\approx^M$.

Note that Theorem 3.86 implies a similar property for t^M.

Formulas describing how similarities are preserved can be thought of as estimations; the greater the left-hand side of the inequality (estimation formula) the better the estimation. Consider, e.g., formula given by Theorem 3.87. Clearly, the left-hand side increases if, instead of $\approx^{\mathbf{M}}$, one uses a greater **L**-equivalence relation. If, in addition, φ (formula appearing on the right-hand side) does not contain $\approx$ then the right-hand side remains the same no matter what **L**-equivalence $\approx^{\mathbf{M}}$ we have. Therefore, if we find another **L**-equivalence on M which is greater than $\approx^{\mathbf{M}}$ and still compatible with all $r^{\mathbf{M}}$ and $f^{\mathbf{M}}$ then the estimating formula is still valid and provides a better estimation. We show that for each **L**-structure **M** there always is the greatest **L**-equivalence on M compatible with all $r^{\mathbf{M}}$ ($r \in R$) and all $f^{\mathbf{M}}$ ($f \in F$) and that this **L**-equivalence can be explicitly described.

Let **M** be an **L**-structure of type $\langle R, F, \sigma\rangle$. For each $r^{\mathbf{M}}$ and each i, $1 \leq i \leq n = \sigma(r)$, we define an **L**-relation $\approx_{r,i}$ by

$$\begin{aligned}(m \approx_{r,i} n) = &\\ \bigwedge_{m_1,\dots,m_{i-1},m_{i+1},\dots,m_n \in M} & r^{\mathbf{M}}(m_1,\dots,m_{i-1},m,m_{i+1},\dots,m_n) \leftrightarrow \\ &\leftrightarrow r^{\mathbf{M}}(m_1,\dots,m_{i-1},n,m_{i+1},\dots,m_n).\end{aligned}$$

For each $f^{\mathbf{M}}$ and each i, $1 \leq i \leq n = \sigma(f)$, we define an **L**-relation $\approx_{f,i}$ by

$$\begin{aligned}(m \approx_{r,i} n) = &\\ \bigwedge_{m_1,\dots,m_{i-1},m_{i+1},\dots,m_n \in M} & f^{\mathbf{M}}(m_1,\dots,m_{i-1},m,m_{i+1},\dots,m_n) \approx^{\mathbf{M}} \\ &\approx^{\mathbf{M}} f^{\mathbf{M}}(m_1,\dots,m_{i-1},n,m_{i+1},\dots,m_n).\end{aligned}$$

Lemma 3.89 *$\approx_{r,i}$ and $\approx_{f,i}$ (as defined above) are* **L**-*equivalence relations.*

Proof. Reflexivity and symmetry of $\approx_{r,i}$ and $\approx_{f,i}$ are immediate from definitions. Transitivity of $\approx_{r,i}$:

$$\begin{aligned}&(m \approx_{r,i} m') \otimes (m' \approx_{r,i} n) \leq \\ \leq & \bigwedge_{m_1,\dots,m_{i-1},m_{i+1},\dots,m_n \in M} (r^{\mathbf{M}}(m_1,\dots,m_{i-1},m,m_{i+1},\dots,m_n) \leftrightarrow \\ &\leftrightarrow r^{\mathbf{M}}(m_1,\dots,m_{i-1},m',m_{i+1},\dots,m_n)) \otimes \\ &\otimes (r^{\mathbf{M}}(m_1,\dots,m_{i-1},m',m_{i+1},\dots,m_n) \leftrightarrow \\ &\leftrightarrow r^{\mathbf{M}}(m_1,\dots,m_{i-1},n,m_{i+1},\dots,m_n)) \leq\end{aligned}$$

$$\leq \bigwedge_{m_1,\ldots,m_{i-1},m_{i+1},\ldots,m_n \in M} (r^{\mathbf{M}}(m_1,\ldots,m_{i-1},m,m_{i+1},\ldots,m_n) \leftrightarrow$$
$$\leftrightarrow r^{\mathbf{M}}(m_1,\ldots,m_{i-1},n,m_{i+1},\ldots,m_n)) = (m \approx_{r,i} n).$$

Transitivity of $\approx_{f,i}$ can be obtained in a similar way using transitivity of $\approx^{\mathbf{M}}$. □

Theorem 3.90 *For an* $\mathbf{L}$*-structure* $\mathbf{M}$*, the* $\mathbf{L}$*-relation* $\approx_{\text{ind}}$ *defined by*

$$(m \approx_{\text{ind}} n) = \bigwedge_{r\in R, 1\leq i\leq \sigma(r)} (m \approx_{r,i} n) \wedge \bigwedge_{f\in F, 1\leq i\leq \sigma(f)} (m \approx_{f,i} n)$$

is an $\mathbf{L}$*-equivalence;* $\approx_{\text{ind}}$ *is the largest* $\mathbf{L}$*-equivalence on* M *that is compatible with each* $r^{\mathbf{M}} \in R^{\mathbf{M}}$ *and* $f^{\mathbf{M}} \in F^{\mathbf{M}}$*.*

Proof. Since $\approx_{\text{ind}}$ is an intersection of $\mathbf{L}$-equivalence relations on M, it is itself an $\mathbf{L}$-equivalence by Lemma 3.75. Let $r \in R$, $\sigma(r) = p$, take $m_i, n_i \in M$ $(i = 1,\ldots,p)$; we have

$$\bigotimes_{i=1}^{p}(m_i \approx_{\text{ind}} n_i) \leq \bigotimes_{i=1}^{p}(m_i \approx_{r,i} n_i) \leq$$
$$\leq \bigotimes_{i=1}^{p} r^{\mathbf{M}}(m_1,\ldots,m_{i-1},m_i,m_{i+1},\ldots,m_p) \leftrightarrow$$
$$\leftrightarrow r^{\mathbf{M}}(m_1,\ldots,m_{i-1},n_i,n_{i+1},\ldots,n_p) \leq$$
$$\leq r^{\mathbf{M}}(m_1,\ldots,m_p) \leftrightarrow r^{\mathbf{M}}(n_1,\ldots,n_p)$$

proving compatibility of $\approx_{\text{ind}}$ with $r^{\mathbf{M}}$. Compatibility with functions $f^{\mathbf{M}}$ can be proved in a similar way. If $\approx'$ is another $\mathbf{L}$-equivalence compatible with all $r^{\mathbf{M}}$ and $f^{\mathbf{M}}$ then it is less than or equal to each $\approx_{r,i}$ and $\approx_{f,i}$ (immediately follows from compatibility of $\approx'$ and the definition of $\approx_{r,i}$ and $\approx_{f,i}$) and hence it is less than or equal to the infimum of all $\approx_{r,i}$ and $\approx_{f,i}$, i.e. $\approx'$ is less than or equal to $\approx_{\text{ind}}$ proving that $\approx_{\text{ind}}$ is the largest one. □

Remark 3.91 Relation $\approx_{\text{ind}}$ represents "indistinguishability" on M that is induced by the $\mathbf{L}$-relations and functions of $\mathbf{M}$.

3.4 MORPHISMS, PRODUCTS, SUBSTRUCTURES

3.4.1 MORPHISMS (AND CONGRUENCES)

Let us consider an $\mathbf{L}$-structure $\mathbf{M}$ of type $\langle R, F, \sigma\rangle$. That is, we have a set M, a collection of functions on M, and a collection of fuzzy relations in M with truth values from a structure $\mathbf{L}$. For various reasons, the $\mathbf{L}$-structure $\mathbf{M}$ may be too complex; one may wish to have a simplified version of $\mathbf{M}$. Basically, there are two ways to simplify $\mathbf{M}$.

The first one is based on the fact that it is not always necessary to distinguish all elements of M; that is to say, it may be desirable to identify some elements of M (those which are equivalent or indistinguishable from a certain point of view): instead of distinguishing some different elements $m_1, m_2 \in M$, one substitutes both m_1 and m_2 by a single element m'; instead of m_1 and m_2 (and possibly other elements from M), one considers a single element m' (a common abstraction of m_1 and m_2 (and possibly others) which represents only those features that are relevant in a particular case). Such a substitution may be represented by a mapping h_M assigning to any $m \in M$ an element $h_M(m)$ (in the above example, $h_M(m_1) = h_M(m_2) = m'$); elements $h_M(m)$ are then the abstractions of elements from M. Collection $h_M(M) = \{h_M(m) \mid m \in M\}$ of all these elements may be a part of another set M' on which an $\mathbf{L}$-structure of the same type as that of $\mathbf{M}$ is defined. Therefore, the idea of substituting elements by their abstractions leads to the situation where one has two $\mathbf{L}$-structures $\mathbf{M}$ and $\mathbf{M}'$ and a mapping $h_M : M \to M'$ (the substitution). Note that the process of substituting elements by their abstractions is common in everyday life: A typical example is the substitution of all even numbers by the concept of an even number (element e) and all odd numbers by the concept of an odd number (element o). In this case, $\mathbf{M}$ is the structure of all integers, and h_M assigns e to all even numbers and o to all odd numbers. This makes it possible to perform reasoning that concerns properties of being even and being odd in a two-element system whose elements are e and o; e is the common abstraction of all even numbers and similarly for o. We make use of this kind of simplification, consciously or unconsciously, whenever we say "the sum of two even numbers is an even number" etc. In order to simplify $\mathbf{M}$ in a systematic manner, h_M has to preserve the structure defined on M by means of functions and fuzzy relations. Let us just verbally illustrate (we will present exact definitions shortly): If the type contains a binary relation symbol r then it is natural to require that if m_1 and m_2 are in the relation $r^{\mathbf{M}}$ then $h_M(m_1)$ and $h_M(m_2)$ are in the relation $r^{\mathbf{M}'}$; h_M is thus in a sense compatible with $r^{\mathbf{M}}$ and $r^{\mathbf{M}'}$ and $r^{\mathbf{M}'}$ can be considered an abstract version of $r^{\mathbf{M}}$. The thus-described simplification is known as *simplification by abstraction.*

The second way of simplification of an $\mathbf{L}$-structure $\mathbf{M}$ is based on the fact that it is not always desirable to distinguish all the truth values of $\mathbf{L}$, i.e. one may wish to substitute two (or more) truth values a_1 and a_2 (that are in some sense close in $\mathbf{L}$, e.g. 0.8 and 0.81 in $[0,1]$)

by their abstraction a'. Similarly as above, what takes place here is a mapping $h_L : L \to L'$ between support sets of two structures $\mathbf{L}$ and $\mathbf{L}'$ of truth values. That is, while considering the same elements (those of M), instead of using the set L of truth values, one may wish to use another set L' of truth values (both L and L' with appropriate structures); instead of an $\mathbf{L}$-structure with support set M we have an $\mathbf{L}'$-structure with the same support set. Now, $\mathbf{L}$-relations on M induce $\mathbf{L}'$-relations on M in a very natural way: if $a \in L$ is the truth degree to which m_1 and m_2 belong to a binary $\mathbf{L}$-relation $r^{\mathbf{M}}$ then $h_L(a)$ may be taken for the truth degree to which m_1 and m_2 belong to the $\mathbf{L}'$-relation $r^{h_L(\mathbf{M})}$. In order for the substitution h_L to be systematic, it should be compatible with the structure of the respective residuated lattices, i.e. h_L should be a kind of a morphism. Given M, the structure $\mathbf{L}$ of truth values gives a kind of framework for logical precision we can allow when asserting statements about elements from M: the more elements $\mathbf{L}$ has, the finer (and thus in a sense more precise) our judgment can be. Going from $\mathbf{L}$ to $\mathbf{L}'$ therefore means a *change of logical precision* we have at our disposal. If $h_L(L) = L'$ (and if h_L is not a bijection), the change is, in fact, a decrease of logical precision; a simplification due to the shift from L to L' may therefore be thought of as a *simplification by a decrease of logical precision.*

Both of the above ways of simplification can be performed in parallel. The formal treatment follows. As we have seen, the problem of replacing a given structure by a simpler structure is closely connected to the study of structure-preserving mappings. Structure-preserving mappings are called morphisms. Morphisms between fuzzy relational and functional structures are the content of the present section.

We will need the following notation. We will deal with mappings h_M of M_1 to M_2 (M_1 and M_2 are supports of fuzzy structures) and mappings h_L of L_1 to L_2 (L_1 and L_2 will be supports of complete residuated lattices $\mathbf{L}_1$ and $\mathbf{L}_2$, respectively) and since we need to consider a pair of such like mappings as a unit we denote it by $h = \langle h_M, h_L\rangle$. Composition of these pairs is defined componentwise, i.e. we put

$$\langle h_M, h_L\rangle \circ \langle h'_M, h'_L\rangle = \langle h_M \circ h'_M, h_L \circ h'_L\rangle.$$

h_M and h_L will represent substitutions realizing abstraction and change of logical precision, respectively. If $M_1 = M_2$ and h_M is the identity id_{M_1} (i.e. $h_M(m) = m$), we also write just h_L instead of $\langle h_M, h_L\rangle$; similarly, we write only h_M instead of $\langle h_M, h_L\rangle$ if $L_1 = L_2$ and h_L is the identity. Pairs of the form $\langle \mathrm{id}, h_L\rangle$ and $\langle h_M, \mathrm{id}\rangle$ are canonical:

Lemma 3.92 *For any* $\langle h_M, h_L\rangle$ *(where* $h_M : M_1 \to M_2$, $h_L : L_1 \to L_2$*) we have* $\langle h_M, h_L\rangle = \langle \mathrm{id}_{M_1}, h_L\rangle \circ \langle h_M, \mathrm{id}_{L_2}\rangle$ *and* $\langle h_M, h_L\rangle = \langle h_M, \mathrm{id}_{L_1}\rangle \circ \langle \mathrm{id}_{M_2}, h_L\rangle$.

The foregoing lemma is trivial; it shows that abstraction and change of logical precision can be done equivalently in parallel or one after the other.

Consider now $\mathbf{L}_1$ and $\mathbf{L}_2$, $h_L : L_1 \to L_2$, and a set M. Each (n-ary) $\mathbf{L}_1$-relation R on M induces an (n-ary) $\mathbf{L}_2$-relation $h_L(R)$ in M by

$$(h_L(R))(m_1, \ldots, m_n) = h_L(R(m_1, \ldots, m_n)).$$

That is, $h_L(R)$ is the composition of $R : M^n \to L_1$ and $h_L : L_1 \to L_2$, i.e. $h_L(R) = R \circ h_L : M^n \to L_2$.

Given $\mathbf{L}_1, \mathbf{L}_2$, an $\mathbf{L}_1$-structure $\mathbf{M}$, each complete morphism $h_L : \mathbf{L}_1 \to \mathbf{L}_2$ induces an $\mathbf{L}_2$-structure $h_L(\mathbf{M})$ with the universe M, $\mathbf{L}_2$-relations $r^{h_L(\mathbf{M})} = h_L(r^{\mathbf{M}})$, and functions $f^{h_L(\mathbf{M})} = f^{\mathbf{M}}$. $h_L(\mathbf{M})$ thus arises from $\mathbf{M}$ solely by a decrease of logical precision. The definition of $h_L(\mathbf{M})$ is correct on account of the following assertion.

Lemma 3.93 *Let* $h_L : \mathbf{L}_1 \to \mathbf{L}_2$ *be a complete morphism,* F *and* R *be an* n*-ary function and an* n*-ary* $\mathbf{L}_1$*-relation on* M *that are compatible with an* $\mathbf{L}_1$*-equivalence* $\approx$*. Then (1)* $h_L(\approx)$ *is an* $\mathbf{L}_2$*-equivalence; (2)* $h_L(R)$ *and* F *are compatible with* $h_L(\approx)$.

Proof. (1) Reflexivity of $h_L(\approx)$ follows from $h_L(1) = 1$, symmetry is obvious. Transitivity: $(m_1 h_L(\approx) m_2) \otimes (m_2 h_L(\approx) m_3) = h_L(m_1 \approx m_2) \otimes h_L(_2\!\approx m_3) = h_L((m_1 \approx m_2) \otimes (m_2 \approx m_3)) \leq h_L(m_1 \approx m_3) = (m_1 h_L(\approx) m_3)$.

(2) Compatibility of $h_L(R)$: $\bigotimes_{i=1}^n (m_i h_L(\approx) m_i') \otimes (h_L(R))(m_1, \ldots, m_n) = h_L(\bigotimes_{i=1}^n (m_i \approx m_i') \otimes R(m_1, \ldots, m_n)) \leq h_L(R(m_1', \ldots, m_n')) = (h_L(R))(m_1', \ldots, m_n')$; compatibility of F can be verified analogously. □

Evaluation of truth degrees of formulas in $h(\mathbf{M})$ is as follows.

Theorem 3.94 *If* $h_L : \mathbf{L}_1 \to \mathbf{L}_2$ *is a complete morphism then, under the above notation, we have* $\|\varphi\|_{h_L(\mathbf{M}),v} = h_L(\|\varphi\|_{\mathbf{M},v})$ *for each formula* φ *and each valuation* v.

Proof. The proof is by induction over φ. If φ is atomic, i.e. $\varphi = r(t_1, \ldots, t_n)$ for some n-ary $r \in R$ then we have $\|\varphi\|_{h_L(\mathbf{M}),v} = r^{h_L(\mathbf{M})}(\|t_1\|_{h_L(\mathbf{M}),v}, \ldots) = r^{h_L(\mathbf{M})}(\|t_1\|_{\mathbf{M},v}, \ldots) = h_L(r^{\mathbf{M}}(\|t_1\|_{\mathbf{M},v}, \ldots))$

by definition of $r^{h_L(\mathbf{M})}$. If φ is not atomic, one proves directly using the assumption that h_L is a complete morphism. For illustration, suppose the assertion is valid for φ and ψ and consider $\varphi \mathbin{\&} \psi$: $\|\varphi \mathbin{\&} \psi\|_{h_L(\mathbf{M}),v} = \|\varphi\|_{h_L(\mathbf{M}),v} \otimes \|\psi\|_{h_L(\mathbf{M}),v} = h_L(\|\varphi\|_{\mathbf{M},v}) \otimes h_L(\|\psi\|_{h_L(\mathbf{M}),v}) = h_L(\|\varphi\|_{\mathbf{M},v} \otimes \|\psi\|_{h_L(\mathbf{M}),v}) = h_L(\|\varphi \mathbin{\&} \psi\|_{h_L(\mathbf{M}),v})$. The proof is similar for other connectives and for quantifiers. □

Given $\mathbf{L}_1$ and a mapping $h_M : M_1 \to M_2$, each (n-ary) $\mathbf{L}_1$-relation R in M_1 induces an (n-ary) $\mathbf{L}_1$-relation $h_M(R)$ in M_2 by

$$(h_M(R))(m_1, \ldots, m_n) = \bigvee_{m'_i \in M_1, h(m'_i) = m_i} R(m'_1, \ldots, m'_n).$$

Intuitively, the degree to which $\langle m_1, \ldots, m_n\rangle$ belongs to $h_M(R)$ equals the degree to which there exist $m'_1, \ldots, m'_n$, mapped by h_M to $m_1, \ldots, m_n$, respectively, such that $\langle m'_1, \ldots, m'_n\rangle$ belongs to R.

Remark 3.95 Notice that the mapping sending R to $h_M(R)$ is exactly that one obtained from h_M by the extension principle.

Now, given $\mathbf{L}_1$, $\mathbf{L}_2$, M_1, M_2, $h_L : L_1 \to L_2$, $h_M : M_1 \to M_2$, an (n-ary) $\mathbf{L}_1$-relation R in M_1 induces an (n-ary) $\mathbf{L}_2$-relation R' in M_2 in two possible ways. Either we can take $R' = h_M(h_L(R))$ or $R' = h_L(h_M(R))$.

Lemma 3.96 *Under the above notation, if h_L is a $\bigvee$-morphism of $\mathbf{L}_1$ to $\mathbf{L}_2$ then $h_L(h_M(R)) = h_M(h_L(R))$, i.e. the two ways of inducing R' are equivalent.*

Proof. Since h_L preserves suprema, we have

$$\begin{aligned} & (h_L(h_M(R)))(m_1, \ldots, m_n) = h_L((h_M(R))(m_1, \ldots, m_n)) = \\ = \; & h_L(\bigvee_{m'_i \in M_1, h(m'_i) = m_i} R(m'_1, \ldots, m'_n)) = \\ = \; & \bigvee_{m'_i \in M_1, h(m'_i) = m_i} h_L(R(m'_1, \ldots, m'_n)) = \\ = \; & (h_M(h_L(R)))(m'_1, \ldots, m'_n). \end{aligned}$$

□

Due to Lemma 3.96, the $\mathbf{L}_2$-relation in M_2 induced by R and $\langle h_M, h_L\rangle$ will also be denoted by $\langle h_M, h_L\rangle(R)$.

Our goal is now the following: given an $\mathbf{L}_1$-structure $\mathbf{M}_1$ and an $\mathbf{L}_2$-structure $\mathbf{M}_2$ of the same type $\langle R, F, \sigma\rangle$, and mappings $h_M : M_1 \to M_2$ and $h_L : L_1 \to L_2$, find a natural condition expressing that $\langle h_M, h_L\rangle$ is structure-preserving. We assume that h_L is a complete morphism of $\mathbf{L}_1$ into $\mathbf{L}_2$ (one could proceed more generally and assume that h_L is an (appropriately defined) almost morphism preserving the structure of complete residuated lattices only to a certain degree; however, this is cumbersome; the reader may verify). Moreover, we require that $\langle h_M, h_L\rangle$ preserve fuzzy equivalences $\approx^{\mathbf{M}_1}$ and $\approx^{\mathbf{M}_2}$ in that we have

$$h_L(m_1 \approx^{\mathbf{M}_1} m_2) \leq (h_M(m_1) \approx^{\mathbf{M}_2} h_M(m_2)).$$

Thus, let $\mathbf{M}_i$ be $\mathbf{L}_i$-structures ($i = 1, 2$) of type $\langle R, F, \sigma\rangle$. Denote by $\mathrm{Adm}_{\mathrm{hom}}(\mathbf{M}_1, \mathbf{M}_2)$ the set of all pairs $\langle h_M, h_L\rangle$ such that $h_M : M_1 \to M_2$ preserves $\approx^{\mathbf{M}_1}$ and $\approx^{\mathbf{M}_2}$, and $h_L : \mathbf{L}_1 \to \mathbf{L}_2$ is a complete morphism ($\mathrm{Adm}_{\mathrm{hom}}$ stands for "admissible w.r.t. the property of being a morphism"). For $\langle h_M, h_L\rangle \in \mathrm{Adm}_{\mathrm{hom}}(\mathbf{M}_1, \mathbf{M}_2)$, $r \in R$ ($\sigma(r) = n$) and $f \in F$ ($\sigma(f) = p$) we put

$$\begin{aligned}(\mathrm{Hom}_r(\mathbf{M}_1, \mathbf{M}_2))(h_M, h_L) &= \bigwedge_{m_1,\ldots,m_n \in M_1} h_L(r^{\mathbf{M}_1}(m_1, \ldots, m_n)) \to \\ &\quad \to r^{\mathbf{M}_2}(h_M(m_1), \ldots, h_M(m_n)), \\ (\mathrm{Hom}_f(\mathbf{M}_1, \mathbf{M}_2))(h_M, h_L) &= \bigwedge_{m_1,\ldots,m_p \in M_1} (h_M(f^{\mathbf{M}_1}(m_1, \ldots, m_p)) \approx^{\mathbf{M}_2} \\ &\quad \approx^{\mathbf{M}_2} f^{\mathbf{M}_2}(h_M(m_1), \ldots, h_M(m_p))).\end{aligned}$$

Moreover, for $S \subseteq R \cup F$ we put

$$(\mathrm{Hom}_S(\mathbf{M}_1, \mathbf{M}_2))(h_M, h_L) = \bigwedge_{s \in S} (\mathrm{Hom}_s(\mathbf{M}_1, \mathbf{M}_2))(h_M, h_L).$$

If $S = R \cup F$, we write also $\mathrm{Hom}(\mathbf{M}_1, \mathbf{M}_2)$ instead of $\mathrm{Hom}_{R\cup F}(\mathbf{M}_1, \mathbf{M}_2)$. Thus, for each $s \in R \cup F$, $\mathrm{Hom}_s(\mathbf{M}_1, \mathbf{M}_2)$ is an $\mathbf{L}_2$-set in the universe $\mathrm{Adm}_{\mathrm{hom}}(\mathbf{M}_1, \mathbf{M}_2)$; for $S \subseteq R \cup F$, $\mathrm{Hom}_S(\mathbf{M}_1, \mathbf{M}_2)$ is the intersection of all $\mathrm{Hom}_s(\mathbf{M}_1, \mathbf{M}_2)$ over $s \in S$. If $\mathbf{M}_1$ and $\mathbf{M}_2$ are obvious from context, we write only Hom_s instead of $\mathrm{Hom}_s(\mathbf{M}_1, \mathbf{M}_2)$ and thus only $\mathrm{Hom}_s(h_M, h_L)$ instead of $(\mathrm{Hom}_s(\mathbf{M}_1, \mathbf{M}_2))(h_M, h_L)$.

$(\mathrm{Hom}_S(\mathbf{M}_1, \mathbf{M}_2))(h_M, h_L)$ is called the *degree to which* $\langle h_M, h_L\rangle$ *is a morphism* of $\mathbf{M}_1$ to $\mathbf{M}_2$ w.r.t. S; $\langle h_M, h_L\rangle$ will be called an a-morphism w.r.t. S if $a \leq (\mathrm{Hom}_S(\mathbf{M}_1, \mathbf{M}_2))(h_M, h_L)$; 1-morphisms w.r.t. S will be called *morphisms* w.r.t. S. We omit the phrase "w.r.t. S" if $S = R \cup F$.

Remark 3.97 (1) The meaning of $\mathrm{Hom}_r(h_M, h_L)$ and $\mathrm{Hom}_f(h_M, h_L)$: First, suppose for simplicity that h_L is an identity (thus $\mathbf{L}_1 = \mathbf{L}_2$).

The degrees $\mathrm{Hom}_r(h_M, h_L)$ and $\mathrm{Hom}_f(h_M, h_L)$ are intuitively truth degrees of propositions "for all $m_1, \ldots, m_n$: if $\langle m_1, \ldots, m_n \rangle$ belongs to $r^{\mathbf{M}_1}$ then $\langle h_M(m_1), \ldots, h_M(m_n) \rangle$ belongs to $r^{\mathbf{M}_2}$" and "for all $m_1, \ldots, m_n$: $h_M(f^{\mathbf{M}_1}(m_1, \ldots, m_n))$ is equal to $f^{\mathbf{M}_2}(h_M(m_1), \ldots, h_M(m_n))$", respectively. Thus, if we take the last two propositions for criteria of compatibility with r and f, $\mathrm{Hom}_r(h_M, h_L)$ and $\mathrm{Hom}_f(h_M, h_L)$ can be naturally interpreted as the degrees to which (h_M, h_L) is compatible with fuzzy relations $r^{\mathbf{M}_1}, r^{\mathbf{M}_2}$, and functions $f^{\mathbf{M}_1}, f^{\mathbf{M}_2}$, respectively. This interpretation may be kept even in the general case, only the truth degrees a from L_1 are converted by h_L to truth degrees $h_L(a)$ from L_2.

(2) If $\langle h_M, h_L \rangle$ is a morphism (i.e. its morphism degree is 1) then since $a \leq b$ iff $a \to b = 1$, we have

$$h_L(r^{\mathbf{M}_1}(m_1, \ldots, m_n)) \leq r^{\mathbf{M}_2}(h_M(m_1), \ldots, h_M(m_n))$$

and

$$h_M(f^{\mathbf{M}_1}(m_1, \ldots, m_p)) \approx^{\mathbf{M}_2} f^{\mathbf{M}_2}(h_M(m_1), \ldots, h_M(m_p)) = 1.$$

If, moreover, $\approx^{\mathbf{M}_2}$ is an $\mathbf{L}_2$-equality, we have $h_M(f^{\mathbf{M}_1}(m_1, \ldots, m_p)) = f^{\mathbf{M}_2}(h_M(m_1), \ldots, h_M(m_p))$.

(3) For a morphism $h = \langle h_M, h_L \rangle$ of an $\mathbf{L}_1$-structure $\mathbf{M}_1$ to an $\mathbf{L}_2$-structure $\mathbf{M}_2$, term t, and an $\mathbf{M}_1$-valuation v we have $(h_M(\|t\|_{\mathbf{M}_1,v}) \approx^{\mathbf{M}_2} \|t\|_{\mathbf{M}_2, v \circ h_M}) = 1$ (easy to verify by induction over t).

We now turn to isomorphisms. To this end, denote by $\mathrm{Adm}_{\mathrm{iso}}(\mathbf{M}_1, \mathbf{M}_2)$ the set of all pairs $\langle h_M, h_L \rangle$ such that $h_M : M_1 \to M_2$ is a bijection satisfying

$$h_L(m_1 \approx^{\mathbf{M}_1} m_2) = (h_M(m_1) \approx^{\mathbf{M}_1} h_M(m_2))$$

and $h_L : \mathbf{L}_1 \to \mathbf{L}_2$ is an isomorphism. For $\langle h_M, h_L \rangle \in \mathrm{Adm}_{\mathrm{iso}}(\mathbf{M}_1, \mathbf{M}_2)$ put

$$\begin{aligned} & (\mathrm{Iso}_r(\mathbf{M}_1, \mathbf{M}_2))(h_M, h_L) = \\ = \; & (\mathrm{Hom}_r(\mathbf{M}_1, \mathbf{M}_2))(h_M, h_L) \wedge (\mathrm{Hom}_r(\mathbf{M}_2, \mathbf{M}_1))(h_M^{-1}, h_L^{-1}) \end{aligned}$$

for $r \in R$ and $(\mathrm{Iso}_f(\mathbf{M}_1, \mathbf{M}_2))(h_M, h_L) = (\mathrm{Hom}_f(\mathbf{M}_1, \mathbf{M}_2))(h_M, h_L)$ for $f \in F$. We use the same conventions as for Hom: we write only Iso_s instead of $\mathrm{Iso}_s(\mathbf{M}_1, \mathbf{M}_2)$ if $\mathbf{M}_i$ are obvious, Iso_S for the intersection of Iso_s over $s \in S$, etc.

Remark 3.98 (1) Observe that for $\langle h_M, h_L \rangle \in \mathrm{Adm}_{\mathrm{iso}}(\mathbf{M}_1, \mathbf{M}_2)$ we have

$$\begin{aligned} & \mathrm{Iso}_r(h_M, h_L) = \\ = \; & \bigwedge_{m_1, \ldots, m_n} h_L(r^{\mathbf{M}_1}(m_1, \ldots, m_n)) \leftrightarrow r^{\mathbf{M}_2}(h_M(m_1), \ldots, h_M(m_n)) \end{aligned}$$

for $r \in R$,

$$\mathrm{Iso}_S(h_M, h_L) = \mathrm{Hom}_S(h_M, h_L) \wedge \mathrm{Hom}_S(h_M^{-1}, h_L^{-1}),$$

and $h_L^{-1}(\mathrm{Iso}_S(h_M, h_L)) = \mathrm{Iso}_S(h_M^{-1}, h_L^{-1})$.

(2) $\mathrm{Iso}_r(h_M, h_L) = 1$ implies

$$h_L(r^{\mathbf{M}_1}(m_1, \ldots, m_n)) = r^{\mathbf{M}_2}(h_M(m_1), \ldots, h_M(m_n)),$$

and, as for Hom, $\mathrm{Iso}_f(h_M, h_L) = 1$ implies

$$(h_M(f^{\mathbf{M}_1}(m_1, \ldots, m_p)) \approx f^{\mathbf{M}_2}(h_M(m_1), \ldots, h_M(m_p))) = 1.$$

$(\mathrm{Iso}_S(\mathbf{M}_1, \mathbf{M}_2))(h_M, h_L)$ is called the degree to which $\langle h_M, h_L \rangle$ is an isomorphism of $\mathbf{M}_1$ to $\mathbf{M}_2$ w.r.t. S. Terms a-isomorphism, isomorphism (i.e. 1-isomorphism), etc. are used analogously as for morphisms. We say that $\mathbf{M}_1$ and $\mathbf{M}_2$ are a-isomorphic if there is an a-isomorphism of $\mathbf{M}_1$ to $\mathbf{M}_2$.

We are to define the concept of a degree of isomorphism between $\mathbf{M}_1$ and $\mathbf{M}_2$. Intuitively, this should be the best possible of $\mathrm{Iso}(h_M, h_L)$ over all $\langle h_M, h_L \rangle \in \mathrm{Adm}_{\mathrm{iso}}(\mathbf{M}_1, \mathbf{M}_2)$, i.e. one could take

$$\mathrm{I}(\mathbf{M}_1, \mathbf{M}_2) = \bigvee_{\langle h_M, h_L \rangle \in \mathrm{Adm}_{\mathrm{iso}}(\mathbf{M}_1, \mathbf{M}_2)} (\mathrm{Iso}(\mathbf{M}_1, \mathbf{M}_2))(h_M, h_L),$$

i.e. $\mathrm{I}(\mathbf{M}_1, \mathbf{M}_2) = \mathrm{hgt}(\mathrm{Iso}(\mathbf{M}_1, \mathbf{M}_2))$. However, taking supremum over all h_L may lead to undesirable situations (for instance: let $M_1 = M_2 = \{m\}$, $\mathbf{L}_1 = \mathbf{L}_2$ a four-element Boolean algebra with $L_1 = \{0, a, b, 1\}$, let the only relation symbol be a unary r and let $r^{\mathbf{M}_1}(m) = 1$, $r^{\mathbf{M}_2}(m) = a$; there is only one bijection h_M and two isomorphisms h_L and h'_L where $h_L(a) = a$ and $h'_L(a) = b$; thus $\mathrm{I}(\mathbf{M}_1, \mathbf{M}_2) = \mathrm{Iso}(h_M, h_L) \vee \mathrm{Iso}(h_M, h'_L) = a \vee b = 1$ which is counterintuitive). Therefore, in order to speak about a degree of isomorphism it seems appropriate to fix some isomorphism h_L between the structures of truth values and to take the suprema over all admissible h_M. Thus, we can take

$$\mathrm{I}_{h_L}(\mathbf{M}_1, \mathbf{M}_2) = \bigvee_{\langle h_M, h'_L \rangle \in \mathrm{Adm}_{\mathrm{iso}}(\mathbf{M}_1, \mathbf{M}_2)} (\mathrm{Iso}(\mathbf{M}_1, \mathbf{M}_2))(h_M, h_L)$$

and call it the degree of isomorphism of $\mathbf{M}_1$ to $\mathbf{M}_2$ w.r.t. h_L. One can also define the degree of isomorphism only w.r.t. some $S \subseteq R \cup F$ by

$$\mathrm{I}_{h_L, S}(\mathbf{M}_1, \mathbf{M}_2) = \bigvee_{\langle h_M, h'_L \rangle \in \mathrm{Adm}_{\mathrm{iso}}(\mathbf{M}_1, \mathbf{M}_2)} (\mathrm{Iso}_S(\mathbf{M}_1, \mathbf{M}_2))(h_M, h_L).$$

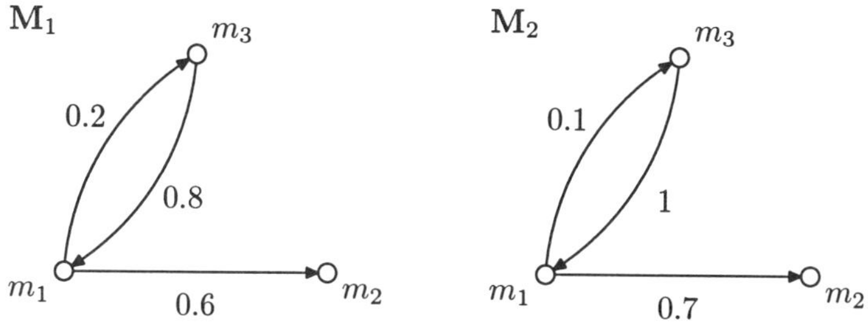

Figure 3.9. Two "fuzzy graphs" from Example 3.100. They are not isomorphic; however, they are "almost isomorphic" (isomorphic in degree 0.8).

Remark 3.99 It is immediate that for **L**-structure $\mathbf{M}_1$ and $\mathbf{M}_2$ with the same support set (i.e. $M_1 = M_2$), the degree of similarity $(\mathbf{M}_1 \approx_S \mathbf{M}_2)$ w.r.t. S equals the degree $\mathrm{Iso}_S(\mathbf{M}_1, \mathbf{M}_2)(\mathrm{id}_{M_1}, \mathrm{id}_L)$ to which $\langle \mathrm{id}_{M_1}, \mathrm{id}_L \rangle$ is an isomorphism of $\mathbf{M}_1$ to $\mathbf{M}_2$.

Example 3.100 (1) Let **L** be the standard Łukasiewicz algebra on $[0,1]$, let the type $\langle R, F, \sigma \rangle$ be given by $R = \{\approx, r\}$ with $\sigma(r) = 2$. Let $\mathbf{M}_1$ and $\mathbf{M}_2$ be **L**-structures with $M_1 = M_2 = \{m_1, m_2, m_3\}$ and $\approx^{\mathbf{M}_1} = \approx^{\mathbf{M}_2} = \mathrm{id}_M$. Let $r^{\mathbf{M}_1}(m_1, m_2) = 0.6$, $r^{\mathbf{M}_2}(m_1, m_2) = 0.7$, $r^{\mathbf{M}_1}(m_1, m_3) = 0.2$, $r^{\mathbf{M}_2}(m_1, m_3) = 0.1$, $r^{\mathbf{M}_1}(m_3, m_1) = 0.8$, $r^{\mathbf{M}_2}(m_3, m_1) = 1$, and $r^{\mathbf{M}_i}(m_1, m_1) = 0$, $r^{\mathbf{M}_i}(m_2, m_1) = 0$, $r^{\mathbf{M}_i}(m_2, m_2) = 0$, $r^{\mathbf{M}_i}(m_2, m_3) = 0$, $r^{\mathbf{M}_i}(m_3, m_2) = 0$, $r^{\mathbf{M}_i}(m_3, m_3) = 0$, for $i = 1, 2$. Thus, each of $\mathbf{M}_1$ and $\mathbf{M}_2$ represents a kind of a oriented "fuzzy graph". Let $h_M(m_1) = m_1$, $h_M(m_2) = m_2$, and $h_L(a) = a$ for $a \in L$. Since $h_L(r^{\mathbf{M}_1}(m_1, m_2)) \leftrightarrow r^{\mathbf{M}_2}(h_M(m_1), h_M(m_2)) = r^{\mathbf{M}_1}(m_1, m_2) \leftrightarrow r^{\mathbf{M}_2}(m_1, m_2) = 0.6 \leftrightarrow 0.7 = 0.9$. One easily verifies that (h_M, h_L) is an isomorphism in degree 0.8 (the "critical" pair is $\langle m_3, m_1 \rangle$: $r^{\mathbf{M}_1}(m_3, m_1) \leftrightarrow r^{\mathbf{M}_2}(m_3, m_1) = 0.8 \leftrightarrow 1 = 0.8$). Observe that according to the classical view, $\mathbf{M}_1$ and $\mathbf{M}_2$ are not isomorphic (they do not differ "only by labels"). Fuzzy approach allows us to precisely express the intuition saying that $\mathbf{M}_1$ and $\mathbf{M}_2$ are almost isomorphic. One also easily verifies that $\mathrm{I}_{\mathrm{id}_L}(\mathbf{M}_1, \mathbf{M}_2) = 0.8$, i.e. we cannot find better h_M. The situation is depicted in Fig. 3.9.

(2) In general, it may happen that $\mathrm{I}_{h_L}(\mathbf{M}_1, \mathbf{M}_2)$ is strictly greater than any Iso (either if the universe and the set of truth values are infinite or if the structure of truth values has noncomparable elements); the reader may construct his own examples.

We now consider the question of composition of morphisms. In general, a composition of two a-morphisms need not be an a-morphism. However, the following assertion gives an estimation of a degree of morphism of a composition of mappings in terms of degrees of morphism of these mappings.

Lemma 3.101 *For* $g = \langle g_M, g_L \rangle \in \mathrm{Adm}_{\mathrm{hom}}(\mathbf{M}_1, \mathbf{M}_2)$ *and* $h = \langle h_M, h_L \rangle \in \mathrm{Adm}_{\mathrm{hom}}(\mathbf{M}_2, \mathbf{M}_3)$ *we have*

$$h_L(\mathrm{Hom}_s(g)) \otimes \mathrm{Hom}_s(h) \leq \mathrm{Hom}_s(g \circ h)$$

for any $s \in R \cup F$.

Proof. For an n-ary $r \in R$, the inequality is equivalent (use definitions and adjointness) to

$$\begin{aligned} & g_L \circ h_L(r^{\mathbf{M}_1}(m_1, \ldots, m_n)) \otimes h_L(\mathrm{Hom}_s(g_M, g_L)) \otimes \mathrm{Hom}_s(h_M, h_L) \leq \\ \leq \; & r^{\mathbf{M}_3}(g_M \circ h_M(m_1), \ldots, g_M \circ h_M(m_n)) \end{aligned}$$

which is true by

$$\begin{aligned} & g_L \circ h_L(r^{\mathbf{M}_1}(m_1, \ldots, m_n)) \otimes h_L(\mathrm{Hom}_s(g_M, g_L)) \otimes \mathrm{Hom}_s(h_M, h_L) \leq \\ \leq \; & h_L(g_L(r^{\mathbf{M}_1}(m_1, \ldots, m_n))) \otimes \\ & (h_L(g_L(r^{\mathbf{M}_1}(m_1, \ldots, m_n))) \to h_L(r^{\mathbf{M}_2}(g_M(m_1), \ldots, g_M(m_n)))) \otimes \\ & \otimes (h_L((r^{\mathbf{M}_2}(g_M(m_1), \ldots, g_M(m_n)))) \to \\ & \quad \to r^{\mathbf{M}_3}(h_M(g_M(m_1)), \ldots, h_M(g_M(m_n)))) \leq \\ \leq \; & r^{\mathbf{M}_3}(g_M \circ h_M(m_1), \ldots, g_M \circ h_M(m_n)). \end{aligned}$$

For $s \in F$, one can proceed analogously (we omit the proof). □

A direct corollary is the following theorem.

Theorem 3.102 *For* $g = \langle g_M, g_L \rangle \in \mathrm{Adm}_{\mathrm{hom}}(\mathbf{M}_1, \mathbf{M}_2)$, $h = \langle h_M, h_L \rangle \in \mathrm{Adm}_{\mathrm{hom}}(\mathbf{M}_2, \mathbf{M}_3)$, *and* $S \subseteq R \cup F$ *we have*

$$h_L(\mathrm{Hom}_S(g)) \otimes \mathrm{Hom}_S(h) \leq \mathrm{Hom}_S(gh).$$

Proof. The assertion follows directly from Lemma 3.101 and $a \otimes \bigwedge b_i \leq \bigwedge (a \otimes b_i)$ by definition of Hom_S. □

Let us now have $\mathbf{M}_1, \mathbf{M}_2$, and $\langle h_M, h_L \rangle \in \mathrm{Adm}_{\mathrm{hom}}(\mathbf{M}_1, \mathbf{M}_2)$. Intuitively, if we take $\mathbf{M}'_1$ and $\mathbf{M}'_2$ that are pairwise similar (in the sense we understood earlier) to $\mathbf{M}_1$ and $\mathbf{M}_2$, respectively, and $\langle h'_M, h_L \rangle \in \mathrm{Adm}_{\mathrm{hom}}(\mathbf{M}'_1, \mathbf{M}'_2)$ such that h_M and h'_M are similar (in that $h_M(m)$ is similar to $h'_M(m)$ for any $m \in M$), then if $\langle h_M, h_L \rangle$ is a morphism, $\langle h'_M, h_L \rangle$ is a morphism as well. Formal treatment of this situation follows.

Recall that for two mappings $h, h' : M_1 \to M_2$ and an $\mathbf{L}_2$-equivalence relation $\approx^{M_2}$ on M_2 we put

$$(h \approx h') = \bigwedge_{m \in M_1} (h(m) \approx^{M_2} h'(m)).$$

The thus-defined $\mathbf{L}_2$-relation $\approx$ on the set of all mappings of M_1 to M_2 is an $\mathbf{L}_2$-equivalence relation.

Lemma 3.103 *Let $\mathbf{M}_i$ be $\mathbf{L}_i$-structures ($i = 1,2$). Then for $\langle h_M, h_L\rangle, \langle h'_M, h_L\rangle \in \mathrm{Adm}_{\mathrm{hom}}(\mathbf{M}_1, \mathbf{M}_2)$, and $s \in R \cup F$ we have*

$$\mathrm{Hom}_s(h_M, h_L) \otimes (h_M \approx h'_M)^k \leq \mathrm{Hom}_s(h'_M, h_L)$$

where $k = \sigma(s)$ if $s \in R$ and $k = \sigma(s) + 1$ if $s \in F$.

Proof. For an n-ary $r \in R$, the required inequality is true iff for each $m_1, \dots, m_n \in M_1$ we have

$$\begin{aligned} & h_L(r^{\mathbf{M}_1}(m_1, \dots, m_n)) \otimes \mathrm{Hom}_r(h_M, h_L) \otimes (h_M \approx h'_M)^n \leq \\ \leq \; & r^{\mathbf{M}_2}(h'_M(m_1), \dots, h'_M(m_n)). \end{aligned}$$

(use adjointness). This inequality is true since

$$\begin{aligned} & h_L(r^{\mathbf{M}_1}(m_1, \dots, m_n)) \otimes \mathrm{Hom}_r(h_M, h_L) \otimes (h_M \approx h'_M)^n \leq \\ \leq \; & h_L(r^{\mathbf{M}_1}(m_1, \dots, m_n)) \otimes (h_L(r^{\mathbf{M}_1}(m_1, \dots, m_n)) \to \\ & \to r^{\mathbf{M}_2}(h'_M(m_1), \dots, h'_M(m_n))) \otimes (h_M \approx h'_M)^n \leq \\ \leq \; & r^{\mathbf{M}_2}(h'_M(m_1), \dots, h'_M(m_n)) \otimes \\ & \otimes (h_M(m_1) \approx^{\mathbf{M}_2} h'_M(m_1)) \otimes \cdots \otimes (h_M(m_n) \approx^{\mathbf{M}_2} h'_M(m_n)) \leq \\ \leq \; & r^{\mathbf{M}_2}(h'_M(m_1), \dots, h'_M(m_n)). \end{aligned}$$

For an n-ary $f \in F$ and for each $m_1, \dots, m_n \in M_1$ we have

$$\begin{aligned} & \mathrm{Hom}_f(h_M, h_L) \otimes (h_M \approx h'_M)^{n+1} \leq \\ \leq \; & (h_M(f^{\mathbf{M}_1}(m_1, \dots, m_n)) \approx^{\mathbf{M}_2} f^{\mathbf{M}_2}(h_M(m_1), \dots, h_M(m_n))) \otimes \\ & \bigotimes_{i=1}^{n} (h_M(m_i) \approx^{\mathbf{M}_2} h'_M(m_i)) \otimes \\ & (h'_M(f^{\mathbf{M}_1}(m_1, \dots, m_n)) \approx^{\mathbf{M}_2} h_M(f^{\mathbf{M}_1}(m_1, \dots, m_n))) \leq \\ \leq \; & (h'_M(f^{\mathbf{M}_1}(m_1, \dots, m_n)) \approx^{\mathbf{M}_2} h_M(f^{\mathbf{M}_1}(m_1, \dots, m_n))) \otimes \\ & \otimes (h_M(f^{\mathbf{M}_1}(m_1, \dots, m_n)) \approx^{\mathbf{M}_2} f^{\mathbf{M}_2}(h_M(m_1), \dots, h_M(m_n))) \otimes \\ & \otimes (f^{\mathbf{M}_2}(h_M(m_1), \dots, h_M(m_n)) \approx^{\mathbf{M}_2} f^{\mathbf{M}_2}(h'_M(m_1), \dots, h'_M(m_n))) \\ \leq \; & (h'_M(f^{\mathbf{M}_1}(m_1, \dots, m_n)) \approx^{\mathbf{M}_2} f^{\mathbf{M}_2}(h'_M(m_1), \dots, h'_M(m_n))) \end{aligned}$$

from which the assertion readily follows. □

Lemma 3.104 *Let* $\mathbf{M}_i, \mathbf{M}'_i$ *be* $\mathbf{L}_i$*-structures with* $M_i = M'_i$ *and* $\approx^{\mathbf{M}_i} = \approx^{\mathbf{M}'_i}$ $(i = 1, 2)$*. Then for* $\langle h_M, h_L \rangle \in \mathrm{Adm}_{\mathrm{hom}}(\mathbf{M}_1, \mathbf{M}_2)$ *and* $s \in R \cup F$ *we have*

$$(\mathrm{Hom}_s(\mathbf{M}_1, \mathbf{M}_2))(h_M, h_L) \otimes h_L(\mathbf{M}_1 \approx_s \mathbf{M}'_1) \otimes (\mathbf{M}_2 \approx_s \mathbf{M}'_2) \leq$$
$$\leq (\mathrm{Hom}_s(\mathbf{M}'_1, \mathbf{M}'_2))(h_M, h_L).$$

Proof. For an n-ary $r \in R$, we have to prove (use adjointness)

$$h_L(r^{\mathbf{M}'_1}(m_1, \ldots, m_n)) \otimes (\mathrm{Hom}_r(\mathbf{M}_1, \mathbf{M}_2))(h_M, h_L) \otimes$$
$$\otimes\, h_L(\mathbf{M}_1 \approx_r \mathbf{M}'_1) \otimes (\mathbf{M}_2 \approx_r \mathbf{M}'_2) \leq r^{\mathbf{M}'_2}(h_M(m_1), \ldots, h_M(m_n)).$$

This inequality is true:

$$
\begin{aligned}
& h_L(r^{\mathbf{M}'_1}(m_1, \ldots, m_n)) \otimes (\mathrm{Hom}_r(\mathbf{M}_1, \mathbf{M}_2))(h_M, h_L) \otimes \\
& \otimes\, h_L(\mathbf{M}_1 \approx_r \mathbf{M}'_1) \otimes (\mathbf{M}_2 \approx_r \mathbf{M}'_2) \leq \\
\leq\; & h_L(r^{\mathbf{M}'_1}(m_1, \ldots, m_n)) \otimes h_L(r^{\mathbf{M}'_1}(m_1, \ldots, m_n) \to r^{\mathbf{M}_1}(m_1, \ldots, m_n)) \\
& \otimes (\mathrm{Hom}_r(\mathbf{M}_1, \mathbf{M}_2))(h_M, h_L) \otimes (\mathbf{M}_2 \approx_r \mathbf{M}'_2) \leq \\
\leq\; & h_L(r^{\mathbf{M}_1}(m_1, \ldots, m_n)) \otimes \\
& \otimes (h_L(r^{\mathbf{M}_1}(m_1, \ldots, m_n)) \to r^{\mathbf{M}_2}(h_M(m_1), \ldots, h_M(m_n))) \otimes \\
& \otimes (r^{\mathbf{M}_2}(h_M(m_1), \ldots, h_M(m_n)) \to r^{\mathbf{M}'_2}(h_M(m_1), \ldots, h_M(m_n))) \leq \\
\leq\; & r^{\mathbf{M}'_2}(h_M(m_1), \ldots, h_M(m_n)).
\end{aligned}
$$

For an n-ary $f \in F$ we have for any $m_1, \ldots, m_n \in M_1$

$$
\begin{aligned}
& (\mathrm{Hom}_f(\mathbf{M}_1, \mathbf{M}_2))(h_M, h_L) \otimes h_L(\mathbf{M}_1 \approx_f \mathbf{M}'_1) \otimes (\mathbf{M}_2 \approx_f \mathbf{M}'_2) \leq \\
\leq\; & (h_M(f^{\mathbf{M}_1}(m_1, \ldots, m_n)) \approx^{\mathbf{M}_2} f^{\mathbf{M}_2}(h_M(m_1), \ldots, h_M(m_n))) \otimes \\
& \otimes\, h_L(f^{\mathbf{M}_1}(m_1, \ldots, m_n) \approx^{\mathbf{M}_1} f^{\mathbf{M}'_1}(m_1, \ldots, m_n)) \otimes (\mathbf{M}_2 \approx_f \mathbf{M}'_2) \leq \\
\leq\; & (h_M(f^{\mathbf{M}_1}(m_1, \ldots, m_n)) \approx^{\mathbf{M}_2} f^{\mathbf{M}_2}(h_M(m_1), \ldots, h_M(m_n))) \otimes \\
& \otimes (h_M(f^{\mathbf{M}_1}(m_1, \ldots, m_n)) \approx^{\mathbf{M}_2} h_M(f^{\mathbf{M}'_1}(m_1, \ldots, m_n))) \otimes \\
& \otimes (f^{\mathbf{M}_2}(h_M(m_1), \ldots, h_M(m_n)) \approx^{\mathbf{M}_2} f^{\mathbf{M}'_2}(h_M(m_1), \ldots, h_M(m_n))) \\
\leq\; & h_M(f^{\mathbf{M}'_1}(m_1, \ldots, m_n)) \approx^{\mathbf{M}_2} f^{\mathbf{M}'_2}(h_M(m_1), \ldots, h_M(m_n)),
\end{aligned}
$$

whence the assertion follows. □

Lemma 3.105 *Let* $\mathbf{M}_i, \mathbf{M}'_i$ *be* $\mathbf{L}_i$*-structures with* $M_i = M'_i$ *and* $\approx^{\mathbf{M}_i} = \approx^{\mathbf{M}'_i}$ $(i = 1, 2)$*. Then for* $\langle h_M, h_L\rangle \in \mathrm{Adm}_{\mathrm{hom}}(\mathbf{M}_1, \mathbf{M}_2)$, $\langle h'_M, h_L\rangle \in \mathrm{Adm}_{\mathrm{hom}}(\mathbf{M}'_1, \mathbf{M}'_2)$*, and* $s \in R \cup F$ *we have*

$$(\mathrm{Hom}_s(\mathbf{M}_1, \mathbf{M}_2))(h_M, h_L) \otimes (h_M \approx h'_M)^k \otimes h_L(\mathbf{M}_1 \approx_s \mathbf{M}'_1) \otimes \\ \otimes(\mathbf{M}_2 \approx_s \mathbf{M}'_2) \leq (\mathrm{Hom}_s(\mathbf{M}'_1, \mathbf{M}'_2))(h'_M, h_L)$$

where $k = \sigma(s)$ *for* $s \in R$, $k = \sigma(s) + 1$ *for* $s \in F$*, and* $(h_M \approx h'_M)$ *is given by* $\approx^{\mathbf{M}_2}$.

Proof. The assertion is a direct consequence of Lemmas 3.103 and 3.104. □

Remark 3.106 Note that the condition $\approx^{\mathbf{M}_i} = \approx^{\mathbf{M}'_i}$ is not too restrictive. In general, one may take $\approx_i = \approx^{\mathbf{M}_i} \wedge \approx^{\mathbf{M}'_i}$; then each $r^{\mathbf{M}_i}$ and $r^{\mathbf{M}'_i}$ ($f^{\mathbf{M}_i}$ and $f^{\mathbf{M}'_i}$) is compatible with $\approx_i$ and we may apply the above assertions.

We have the following consequence.

Theorem 3.107 *Let* $\mathbf{M}_i, \mathbf{M}'_i$ *be* $\mathbf{L}_i$*-structures with* $M_i = M'_i$ *and* $\approx^{\mathbf{M}_i} = \approx^{\mathbf{M}'_i}$ $(i = 1, 2)$, $\langle h_M, h_L\rangle \in \mathrm{Adm}_{\mathrm{hom}}(\mathbf{M}_1, \mathbf{M}_2)$, $\langle h'_M, h_L\rangle \in \mathrm{Adm}_{\mathrm{hom}}(\mathbf{M}'_1, \mathbf{M}'_2)$*. If for* $S \subseteq R \cup F$ *there exists* $k = \max(\max_{r\in R} \sigma(r), \max_{f\in F} \sigma(f) + 1)$ *we have*

$$(\mathrm{Hom}_S(\mathbf{M}_1, \mathbf{M}_2))(h_M, h_L) \otimes (h_M \approx h'_M)^k \otimes h_L(\mathbf{M}_1 \approx_S \mathbf{M}'_1) \otimes \\ \otimes(\mathbf{M}_2 \approx_S \mathbf{M}'_2) \leq (\mathrm{Hom}_S(\mathbf{M}'_1, \mathbf{M}'_2))(h'_M, h_L)$$

where $(h_M \approx h'_M)$ *is given by* $\approx^{\mathbf{M}_2}$.

Proof. The assertion easily follows from Lemma 3.105 by $a \otimes \bigwedge b_i \leq \bigwedge(a \otimes b_i)$. □

Similar assertions hold also for degree of isomorphism; they are easy consequences of the above for degree of morphism since $\mathrm{Iso}_S(h_M, h_L) = \mathrm{Hom}_S(h_M, h_L) \wedge \mathrm{Hom}_S(h_M^{-1}, h_L^{-1})$ and $a \otimes (b_1 \wedge b_2) \leq (a \otimes b_1) \wedge (a \otimes b_2)$. We thus have

Theorem 3.108 *Let* $\mathbf{M}_i, \mathbf{M}'_i$ *be* $\mathbf{L}_i$*-structures with* $M_i = M'_i$ *and* $\approx^{\mathbf{M}_i} = \approx^{\mathbf{M}'_i}$ $(i = 1, 2)$, $\langle h_M, h_L\rangle \in \mathrm{Adm}_{\mathrm{iso}}(\mathbf{M}_1, \mathbf{M}_2)$, $\langle h'_M, h_L\rangle \in \mathrm{Adm}_{\mathrm{iso}}(\mathbf{M}'_1, \mathbf{M}'_2)$*. If for* $S \subseteq R \cup F$ *there exists* $k = \max(\max_{r\in R} \sigma(r), \max_{f\in F} \sigma(f) + 1)$ *we have*

$$(\mathrm{Iso}_S(\mathbf{M}_1, \mathbf{M}_2))(h_M, h_L) \otimes (h_M \approx h'_M)^k \otimes (\mathbf{M}_1 \approx_S \mathbf{M}'_1) \otimes \\ \otimes(\mathbf{M}_2 \approx_S \mathbf{M}'_2) \leq (\mathrm{Iso}_S(\mathbf{M}'_1, \mathbf{M}'_2))(h'_M, h_L)$$

where $(h_M \approx h'_M)$ *is given by* $\approx^{\mathbf{M}_2}$.

Having $\mathbf{L}_i$-structures $\mathbf{M}_i$ $(i = 1, 2)$ and an $\mathbf{M}_1$-valuation v, a mapping $h_M : M_1 \to M_2$ induces an $\mathbf{M}_2$-valuation $v \circ h_M$. Verbally, if a formula φ is true in $\mathbf{M}_1$ under v and if $\langle h_M, h_L\rangle$ is an isomorphism of $\mathbf{M}_1$ to $\mathbf{M}_2$, then φ is true in $\mathbf{M}_2$ under $v \circ h_M$. In degrees, we have

Lemma 3.109 *For* $\langle h_M, h_L\rangle \in \mathrm{Adm}_{\mathrm{iso}}(\mathbf{M}_1, \mathbf{M}_2)$, *a formula* φ, *and an* $\mathbf{M}_1$*-valuation* v *we have*

$$h_L(\|\varphi\|_{\mathbf{M}_1,v}) \otimes \bigotimes_{r\in R} \mathrm{Iso}_r(h_M, h_L)^{|\varphi|_r} \otimes \bigotimes_{f\in F} \mathrm{Iso}_f(h_M, h_L)^{|\varphi|_f} \le \|\varphi\|_{\mathbf{M}_2, v\circ h_M}.$$

Proof. The proof is completely analogous to that of Theorem 3.87 (thus to those of preceding lemmas); in fact, Theorem 3.87 is a special case of the assertion, cf. Remark 3.99. □

Congruences. Let us now consider the notion of a congruence for a fuzzy structure. For an $\mathbf{L}$-structure $\mathbf{M}$ denote by $\mathrm{Adm}_{\mathrm{con}}(\mathbf{M})$ the set of all pairs $\theta = \langle \theta_M, \theta_L\rangle$ such that θ_L is a complete congruence on $\mathbf{L}$ and θ_M is an $\mathbf{L}/\theta_L$-equivalence on M such that

$$\theta_M(m_1, m_2) \otimes [m_1 \approx^{\mathbf{M}} m'_1]_{\theta_L} \otimes [m_2 \approx^{\mathbf{M}} m'_2]_{\theta_L} \le \theta_M(m'_1, m'_2) \quad (3.39)$$

for all $m_1, m'_1, m_2, m'_2 \in M$.

Remark 3.110 (1) Condition (3.39) says that $\langle \theta_M, \theta_L\rangle$ is compatible w.r.t. the equivalence $\approx^{\mathbf{M}}$ on M.

(2) Putting $m_1 = m_2 = m$, $m'_1 = m'$, $m'_2 = m$, (3.39) yields $[m \approx^{\mathbf{M}} m']_{\theta_L} \le \theta(m, m')$.

For $\theta = \langle \theta_M, \theta_L\rangle \in \mathrm{Adm}_{\mathrm{con}}(\mathbf{M})$ and an n-ary function symbol $f \in F$ we put

$$(\mathrm{Con}_f(\mathbf{M}))(\theta) = \bigwedge_{m_i, m'_i \in M} (\theta_M(m_1, m'_1) \otimes \cdots \otimes \theta_M(m_n, m'_n)) \to$$
$$\to \theta_M(f^{\mathbf{M}}(m_1, \ldots, m_n), f^{\mathbf{M}}(m'_1, \ldots, m'_n));$$

for an n-ary $r \in R$ we put

$$(\mathrm{Con}_r(\mathbf{M}))(\theta) = \bigwedge_{m_i, m'_i \in M} (\bigvee_{\theta_M(m_i, m''_i)=1} [r^{\mathbf{M}}(m''_1, \ldots, m''_n)]_{\theta_L} \otimes$$
$$\otimes \bigotimes_{i=1}^{n} \theta_M(m_i, m'_i)) \to \bigvee_{\theta_M(m'_i, m'''_i)=1} [r^{\mathbf{M}}(m'''_1, \ldots, m'''_n)]_{\theta_L};$$

for $S \subseteq R \cup F$ we put

$$(\mathrm{Con}_S(\mathbf{M}))(\theta) = \bigwedge_{s \in S} (\mathrm{Con}_s(\mathbf{M}))(\theta);$$

$\mathrm{Con}_s(\mathbf{M})(\theta)$ and $\mathrm{Con}_S(\mathbf{M})(\theta)$ are called the *degree to which θ is a congruence* on $\mathbf{M}$ w.r.t. s and S, respectively. Thus, $\mathrm{Con}_s(\mathbf{M})$ and $\mathrm{Con}_S(\mathbf{M})$ are $\mathbf{L}/\theta_L$-sets in $\mathrm{Adm}_{\mathrm{con}}(\mathbf{M})$; $\mathrm{Con}_S(\mathbf{M})$ is the intersection of all $\mathrm{Con}_s(\mathbf{M})$, $s \in S$. Clearly, $\mathrm{Con}_S(\mathbf{M})(\theta) = 1$ ($1 \in \mathbf{L}/\theta_L$) iff for every $m_i, m'_i \in M$ we have

$$\begin{aligned} &\theta_M(m_1, m'_1) \otimes \cdots \otimes \theta_M(m_n, m'_n) \leq \\ \leq\ &\theta_M(f^{\mathbf{M}}(m_1, \ldots, m_n), f^{\mathbf{M}}(m_1, \ldots, m_n)) \end{aligned}$$

and

$$\begin{aligned} &\Big(\bigvee_{\theta_M(m_i, m''_i)=1} [r^{\mathbf{M}}(m''_1, \ldots, m''_n)]_{\theta_L} \otimes \bigotimes_{i=1}^{n} \theta_M(m_i, m'_i) \Big) \leq \\ \leq\ &\bigvee_{\theta_M(m'_i, m'''_i)=1} [r^{\mathbf{M}}(m'''_1, \ldots, m'''_n)]_{\theta_L} \end{aligned}$$

for each $f \in S \cap F$ and $r \in S \cap R$; in this case θ will be called a congruence on $\mathbf{M}$ w.r.t. S (we omit "w.r.t. S" if $S = R \cup F$). We write only Con_S instead of $\mathrm{Con}_S(\mathbf{M})$ if there is no danger of confusion.

Remark 3.111 (1) Notice that if θ_M is crisp (i.e. for each m, m' we have either $\theta_M(m, m') = 0$ or $\theta_M(m, m') = 1$) then $\mathrm{Con}_r(\theta) = 1$.

(2) Observe that saying that $f^{\mathbf{M}}$ ($f \in F$) is compatible with $\approx^{\mathbf{M}}$ is equivalent to saying that $\mathrm{Con}_f(\approx^{\mathbf{M}} \circ I, \mathrm{id}_L)$ is 1 (where $I(a) = \{a\}$); the same is true for $r \in R$.

(3) Similarly as Lemma 3.81, one can prove the following statement: For a term t, valuations v_1, v_2, and a congruence $\theta = \langle \theta_M, \theta_L \rangle$ on an $\mathbf{L}$-structure $\mathbf{M}$ we have

$$\bigotimes_{x \in \mathrm{var}(t)} \theta_M(v_1(x), v_2(x))^{|t|_x} \quad \leq \quad \theta_M(\|t\|_{\mathbf{M},v_1}, \|t\|_{\mathbf{M},v_2}).$$

Moreover, one can generalize for arbitrary $\theta \in \mathrm{Adm}_{\mathrm{con}}(\mathbf{M})$ (i.e. also for congruence in degree strictly less than 1). The same applies to Theorem 3.86.

Lemma 3.112 *For $\theta_i \in \mathrm{Adm}_{\mathrm{con}}(\mathbf{M})$ and $S \subseteq R \cup F$ we have*

$$\bigwedge_{i \in I} \mathrm{Con}_S(\theta_i) \leq \mathrm{Con}_S(\bigwedge_{i \in I} \theta_i).$$

Proof. Clearly, it is sufficient to verify for $S = \{f\}$ $(f \in F)$ and $S = \{r\}$ $(r \in R)$. $\bigwedge_{i\in I} \mathrm{Con}_f(\theta_i) \leq \mathrm{Con}_f(\bigwedge_{i\in I} \theta_i)$ is true iff for every $m_i, m'_i \in M$ we have $\bigotimes_{k=1}^n(\bigwedge_{i\in I} \theta_i(m_k, m'_k)) \otimes \bigwedge_{i\in I} \mathrm{Con}_f(\theta_i) \leq \theta_j(f^{\mathbf{M}}(m_1,\ldots,m_n), f^{\mathbf{M}}(m'_1,\ldots,m'_n))$ for each $j \in I$. This is indeed the case: $\bigotimes_{k=1}^n(\bigwedge_{i\in I} \theta_i(m_k, m'_k)) \otimes \bigwedge_{i\in I} \mathrm{Con}_f(\theta_i) \leq \bigotimes_{k=1}^n \theta_j(m_k, m'_k) \otimes \mathrm{Con}_f(\theta_j) \leq \theta_j(f^{\mathbf{M}}(m_1,\ldots,m_n), f^{\mathbf{M}}(m'_1,\ldots,m'_n))$. The proof for Con_r is equally easy. □

Theorem 3.113 *The intersection of an arbitrary set of congruences of an* $\mathbf{L}$*-structure* $\mathbf{M}$ *is again a congruence of* $\mathbf{M}$.

Proof. Directly from Lemma 3.112. □

We are now going to investigate the relationship between morphisms and congruences of fuzzy structures. First, let $h = \langle h_M, h_L\rangle \in \mathrm{Adm}_{\mathrm{hom}}(\mathbf{M}_1, \mathbf{M}_2)$. As h_L is a complete morphism, θ_{h_L} is a complete congruence on $\mathbf{L}_1$ and $\mathbf{L}_1/\theta_{h_L}$ is isomorphic to the complete subalgebra $h_L(\mathbf{L}_1)$ of $\mathbf{L}_2$; denote by $l : \mathbf{L}_1/\theta_{h_L} \to h_L(\mathbf{L}_1)$ the isomorphism in question (i.e. $l([a]_{\theta_{h_L}}) = h_L(a)$). Suppose that $\approx^{\mathbf{M}_2} (h_M(M_1), h_M(M_1)) \subseteq h_L(L_1)$, i.e. $(h_M(m_1) \approx^{\mathbf{M}_2} h_M(m_2)) = h_L(a)$ for some $a \in L_1$ (this is automatically satisfied if, for instance, h_L is an onto mapping). Furthermore, denote by θ_{h_M} the $\mathbf{L}_1/\theta_{h_L}$-relation on M_1 given by

$$\theta_{h_M}(m, m') = l^{-1}(h_M(m) \approx^{\mathbf{M}_2} h_M(m')).$$

We denote the pair $\langle\theta_{h_M}, \theta_{h_L}\rangle$ induced in the above way by $h = \langle h_M, h_L\rangle$ simply by θ_h.

For $\langle h_M, h_L\rangle \in \mathrm{Adm}_{\mathrm{hom}}(\mathbf{M}_1, \mathbf{M}_2)$ and $r \in R$ we we define the *degree* $(\mathrm{Str}_r(\mathbf{M}_1, \mathbf{M}_2))(h_M, h_L)$ (denoted also simply by $\mathrm{Str}_r(h_M, h_L)$) *to which* $\langle h_M, h_L\rangle$ *is strong* w.r.t. r by

$$(\mathrm{Str}_r(\mathbf{M}_1, \mathbf{M}_2))(h_M, h_L) = \bigwedge_{m_1,\ldots,m_n\in M} (r^{\mathbf{M}_2}(h_M(m_1),\ldots,h_M(m_n)) \to$$
$$\to \bigvee_{m'_i\in M_1,(h_M(m'_i)\approx^{\mathbf{M}_2} h_M(m_i))=1} h_L(r^{\mathbf{M}_1}(m'_1,\ldots,m'_n))).$$

Similarly as before, for $S \subseteq R \cup F$ we put

$$\mathrm{Str}_S(h_M, h_L) = \bigwedge_{r\in S\cap R} \mathrm{Str}_r(h_M, h_L)$$

and write only $\mathrm{Str}(h_M, h_L)$ if $R \subseteq S$. $\langle h_M, h_L\rangle$ is called strong w.r.t. S if $\mathrm{Str}_S(h_M, h_L) = 1$; this means that for each $r \in S \cap R$ and $m_1, \ldots, m_n \in$

M_1 we have

$$r^{\mathbf{M}_2}(h_M(m_1), \ldots, h_M(m_n)) \leq \bigvee_{m_i' \in M_1, h_M(m_i') = h_M(m_i)} h_L(r^{\mathbf{M}_1}(m_1', \ldots, m_n')).$$

We say that $\langle h_M, h_L \rangle$ is strong if it is strong w.r.t. R.

We have the following assertion.

Lemma 3.114 *Under the above notation we have for any* $\langle h_M, h_L \rangle \in \mathrm{Adm}_{\mathrm{hom}}(\mathbf{M}_1, \mathbf{M}_2)$ *that* $\langle \theta_{h_M}, \theta_{h_L} \rangle \in \mathrm{Adm}_{\mathrm{con}}(\mathbf{M})$*,*

$$(\mathrm{Hom}_f(h_M, h_L))^2 \leq l(\mathrm{Con}_f(\theta_{h_M}, \theta_{h_L}))$$

and

$$\mathrm{Hom}_r(h_M, h_L) \otimes \mathrm{Str}_r(h_M, h_L) \leq l(\mathrm{Con}_r(\theta_{h_M}, \theta_{h_L})).$$

Proof. The fact that θ_{h_M} is an $\mathbf{L}/\theta_{h_L}$-equivalence follows easily by definition; (3.39) follows from $\langle h_M, h_L \rangle \in \mathrm{Adm}_{\mathrm{hom}}(\mathbf{M}_1, \mathbf{M}_2)$. Indeed, $\theta_{h_M}(m_1, m_2) \otimes [m_1 \approx^{\mathbf{M}_1} m_1']_{\theta_{h_L}} \otimes [m_2 \approx^{\mathbf{M}_1} m_2']_{\theta_{h_L}} = l^{-1}(h_M(m_1) \approx^{\mathbf{M}_2} h_M(m_2)) \otimes l^{-1}(h_L(m_1 \approx^{\mathbf{M}_1} m_1')) \otimes l^{-1}(h_L(m_2 \approx^{\mathbf{M}_1} m_2')) \leq l^{-1}(h_M(m_1) \approx^{\mathbf{M}_2} h_M(m_2)) \otimes l^{-1}(h_M(m_1) \approx^{\mathbf{M}_1} h_M(m_1')) \otimes l^{-1}(h_M(m_2) \approx^{\mathbf{M}_1} h_M(m_2')) = l^{-1}((h_M(m_1) \approx^{\mathbf{M}_2} h_M(m_2)) \otimes (h_M(m_1) \approx^{\mathbf{M}_1} h_M(m_1')) \otimes (h_M(m_2) \approx^{\mathbf{M}_1} h_M(m_2'))) \leq l^{-1}((h_M(m_1') \approx^{\mathbf{M}_2} h_M(m_2'))) = \theta_M(m_1', m_2')$. Using adjointness and the fact that l is an isomorphism, the required inequality is true iff for any m_i, m_i' we have

$$\begin{aligned} & l(\theta_{h_M}(m_1, m_1') \otimes \cdots \otimes \theta_{h_M}(m_n, m_n')) \otimes (\mathrm{Hom}_f(h_M, h_L))^2 \leq \\ \leq \; & l(\theta_{h_M}(f^{\mathbf{M}_1}(m_1, \ldots, m_n), f^{\mathbf{M}_1}(m_1', \ldots, m_n'))) \end{aligned}$$

which is true. Indeed,

$$\begin{aligned} & l(\theta_{h_M}(m_1, m_1') \otimes \cdots \otimes \theta_{h_M}(m_n, m_n')) \otimes (\mathrm{Hom}_f(h_M, h_L))^2 \leq \\ \leq \; & \bigotimes_{i=1}^{n} (h_M(m_i) \approx^{\mathbf{M}_2} h_M(m_i')) \otimes \\ & \otimes (h_M(f^{\mathbf{M}_1}(m_1, \ldots, m_n)) \approx^{\mathbf{M}_2} f^{\mathbf{M}_2}(h_M(m_1), \ldots, h_M(m_n))) \otimes \\ & \otimes (h_M(f^{\mathbf{M}_1}(m_1', \ldots, m_n')) \approx^{\mathbf{M}_2} f^{\mathbf{M}_2}(h_M(m_1'), \ldots, h_M(m_n'))) \leq \\ \leq \; & (h_M(f^{\mathbf{M}_1}(m_1, \ldots, m_n)) \approx^{\mathbf{M}_2} h_M(f^{\mathbf{M}_1}(m_1', \ldots, m_n'))) = \\ = \; & l \circ l^{-1}(h_M(f^{\mathbf{M}_1}(m_1, \ldots, m_n)) \approx^{\mathbf{M}_2} h_M(f^{\mathbf{M}_1}(m_1', \ldots, m_n'))) = \\ = \; & l(\theta_{h_M}(f^{\mathbf{M}_1}(m_1, \ldots, m_n), f^{\mathbf{M}_1}(m_1', \ldots, m_n'))). \end{aligned}$$

$\mathrm{Hom}_r(h_M,h_L)\otimes\mathrm{Str}_r(h_M,h_L)\leq l(\mathrm{Con}_r(\theta_{h_M},\theta_{h_L}))$ is true iff for every m_i, m'_i we have

$$\begin{aligned} & l(\bigvee_{\theta_M(m_i,m''_i)=1}[r^{\mathbf{M}_1}(m''_1,\dots,m''_n)]_{\theta_L}\otimes\bigotimes_{i=1}^{n}\theta_M(m_i,m'_i))\otimes \\ & \otimes\mathrm{Hom}_r(h_M,h_L)\otimes\mathrm{Str}_r(h_M,h_L)\leq \\ \leq\ & l(\bigvee_{\theta_M(m'_i,m'''_i)=1}[r^{\mathbf{M}_1}(m'''_1,\dots,m'''_n)]_{\theta_L}) \end{aligned}$$

and this inequality is true. Indeed,

$$\begin{aligned} & l(\bigvee_{\theta_{h_M}(m_i,m''_i)=1}[r^{\mathbf{M}_1}(m''_1,\dots,m''_n)]_{\theta_{h_L}}\otimes\bigotimes_{i=1}^{n}\theta_{h_M}(m_i,m'_i))\otimes \\ & \otimes\mathrm{Hom}_r(h_M,h_L)\otimes\mathrm{Str}_r(h_M,h_L)= \\ =\ & \bigvee_{\theta_{h_M}(m_i,m''_i)=1}h_L(r^{\mathbf{M}_1}(m''_1,\dots,m''_n))\otimes\mathrm{Hom}_r(h_M,h_L)\otimes \\ & \bigotimes_{i=1}^{n}(h_M(m_i)\approx^{\mathbf{M}_2}h_M(m'_i)))\otimes\mathrm{Str}_r(h_M,h_L)\leq \\ \leq\ & \bigvee_{(h_M(m_i)\approx^{\mathbf{M}_2}m''_i)=1}r^{\mathbf{M}_2}(h_M(m''_1),\dots,h_M(m''_n))\otimes \\ & \bigotimes_{i=1}^{n}(h_M(m_i)\approx^{\mathbf{M}_2}h_M(m'_i)))\otimes\mathrm{Str}_r(h_M,h_L)= \\ =\ & r^{\mathbf{M}_2}(h_M(m_1),\dots,h_M(m_n))\otimes \\ & \bigotimes_{i=1}^{n}(h_M(m_i)\approx^{\mathbf{M}_2}h_M(m'_i))\otimes\mathrm{Str}_r(h_M,h_L)\leq \\ \leq\ & r^{\mathbf{M}_2}(h_M(m'_1),\dots,h_M(m'_n))\otimes\mathrm{Str}_r(h_M,h_L)\leq \\ \leq\ & \bigvee_{(h_M(m'_i)\approx^{\mathbf{M}_2}h_M(m'''_i))=1}h_L(r^{\mathbf{M}_1}(m'''_1,\dots,m'''_n))= \\ =\ & l(\bigvee_{\theta_{h_M}(m'_i,m'''_i)=1}[r^{\mathbf{M}_1}(m'''_1,\dots,m'''_n)]_{\theta_{h_L}}) \end{aligned}$$

□

Remark 3.115 (1) Lemma 3.114 shows to what extent it is true that $\langle\theta_{h_M},\theta_{h_L}\rangle$ induced in a natural way by a morphism $\langle h_M,h_L\rangle$ is a congruence. Note that since $1=1\otimes 1$, $\langle\theta_{h_M},\theta_{h_L}\rangle$ is a congruence (in degree 1) whenever $\langle h_M,h_L\rangle$ is a morphism (in degree 1).

(2) An immediate verification shows that $(\mathrm{Hom}_S(h_M, h_L))^2 \leq l(\mathrm{Con}_S(\theta_{h_M}, \theta_{h_L}))$ holds true for any $S \subseteq R \cup F$ (just use $a \otimes \bigwedge_i b_i \leq \bigwedge_i (a \otimes b_i)$).

In the following we are going to define factor fuzzy structures by congruences; a factor fuzzy structure can be thought of as resulting from the original fuzzy structure by abstraction. Among others, it will allow us to get morphisms from congruences. Thus, let $\mathbf{M}$ be an $\mathbf{L}$-structure, let $\theta = \langle \theta_M, \theta_L \rangle$ be a congruence on $\mathbf{M}$ w.r.t. $R \cup F$ (i.e. its congruence degree is 1). A moment reflection shows that the 1-cut ${}^1E$ of a fuzzy equivalence relation E on X is an equivalence relation on X. Therefore, ${}^1\theta_M$ is an equivalence on M with equivalence classes $[m]_{^1\theta_M} = \{m' \mid \theta_M(m, m') = 1 \in \mathbf{L}/\theta_L\}$ (denoted also simply by $[m]_\theta$). Now, we define the factor structure $\mathbf{M}/\theta$ of $\mathbf{M}$ by θ as follows: $\mathbf{M}/\theta$ is an $\mathbf{L}/\theta_L$-structure of the same type as $\mathbf{M}$ with support set $M/\theta = M/{}^1\theta_M$; $[m]_\theta \approx^{\mathbf{M}/\theta} [m']_\theta = \theta_M(m, m')$; operations: for an n-ary $f \in F$ and $[m_i]_\theta \in M/\theta$ $(i = 1, \ldots, n)$ we put

$$f^{\mathbf{M}/\theta}([m_1]_\theta, \ldots, [m_n]_\theta) = [f^{\mathbf{M}}(m_1, \ldots, m_n)]_\theta;$$

fuzzy relations: for an n-ary $r \in R$ and $[m_i]_\theta \in M/\theta$ $(i = 1, \ldots, n)$ we put

$$r^{\mathbf{M}/\theta}([m_1]_\theta, \ldots, [m_n]_\theta) = \bigvee_{m'_i \in [m_i]_\theta} [r^{\mathbf{M}}(m'_1, \ldots, m'_n)]_{\theta_L}.$$

We have to verify that the definition of $\mathbf{M}/\theta$ is correct. If $[m_1]_\theta = [m'_1]_\theta$ and $[m_2]_\theta = [m'_2]_\theta$, then $\theta_M(m_1, m'_1) = 1$ and $\theta_M(m_2, m'_2) = 1$, and thus $\theta_M(m_1, m_2) = \theta_M(m'_1, m_1) \otimes \theta_M(m_1, m_2) \otimes \theta_M(m_2, m'_2) \leq \theta_M(m'_1, m'_2)$, thus $\theta_M(m_1, m_2) \leq \theta_M(m'_1, m'_2)$. In a similar way we obtain $\theta_M(m'_1, m'_2) \leq \theta_M(m_1, m_2)$ whence $\theta_M(m_1, m_2) = \theta_M(m'_1, m'_2)$ proving that $\approx^{\mathbf{M}/\theta}$ is defined correctly. If $[m_i]_\theta = [m'_i]_\theta$ $(i = 1, \ldots, n)$ then, since θ is a congruence, we have $1 = \theta_M(m_1, m'_1) \otimes \cdots \otimes \theta_M(m_n, m'_n) \leq \theta_M(f^{\mathbf{M}}(m_1, \ldots, m_n), f^{\mathbf{M}}(m'_1, \ldots, m'_n))$, thus $[f^{\mathbf{M}}(m_1, \ldots, m_n)]_\theta = [f^{\mathbf{M}}(m'_1, \ldots, m'_n)]_\theta$ which means that $f^{\mathbf{M}/\theta}$ (for each n-ary $f \in F$) is defined correctly. The fact that $r^{\mathbf{M}/\theta}$ $(r \in R)$ is defined correctly is obvious. It remains to verify that each $f^{\mathbf{M}/\theta}$ and $r^{\mathbf{M}/\theta}$ $(f \in F, r \in R)$ is compatible with $\approx^{\mathbf{M}/\theta}$. As $([m]_\theta \approx^{\mathbf{M}/\theta} [m']_\theta) = \theta_M(m, m')$, compatibility of $f^{\mathbf{M}/\theta}$ follows directly from the fact that θ is a congruence in degree 1. Furthermore,

$$r^{\mathbf{M}/\theta}([m_1]_\theta, \ldots, [m_n]_\theta) \otimes \bigotimes_{i=1}^{n} ([m_i]_\theta \approx^{\mathbf{M}/\theta} [m'_i]_\theta) =$$

$$= (\bigvee_{m_i'' \in [m_i]_\theta} [r^{\mathbf{M}}(m_1'', \ldots, m_n'')]_{\theta_L}) \otimes \bigotimes_{i=1}^{n} ([m_i]_\theta \approx^{\mathbf{M}/\theta} [m_i']_\theta) =$$

$$= \bigvee_{m_i'' \in [m_i]_\theta} ([r^{\mathbf{M}}(m_1'', \ldots, m_n'')]_{\theta_L} \otimes \bigotimes_{i=1}^{n} \theta_M(m_i, m_i')) =$$

$$= \bigvee_{m_i'' \in [m_i]_\theta} ([r^{\mathbf{M}}(m_1'', \ldots, m_n'')]_{\theta_L} \otimes \bigotimes_{i=1}^{n} \theta_M(m_i'', m_i')) \leq$$

$$\leq \bigvee_{m_i' \in [m_i]} [r^{\mathbf{M}}(m_1', \ldots, m_n')]_{\theta_L} = r^{\mathbf{M}/\theta}([m_1']_\theta, \ldots, [m_n']_\theta)$$

proving compatibility of $r^{\mathbf{M}/\theta}$.

Definition 3.116 The structure $\mathbf{M}/\theta$ defined above is called the *factor structure of* $\mathbf{M}$ *by* θ.

Now, for an $\mathbf{L}$-structure $\mathbf{M}$ and $\theta = \langle \theta_M, \theta_L \rangle \in \mathrm{Adm}_{\mathrm{con}}(\mathbf{M})$, denote by $h_\theta = \langle h_{\theta_M}, h_{\theta_L} \rangle$ the pair of mappings $h_{\theta_M} : M \to M/^1\theta_M$ and $h_{\theta_L} : L \to L/\theta_L$ defined by $h_{\theta_M}(m) = [m]_{^1\theta_M}$ and $h_{\theta_L}(a) = [a]_{\theta_L}$. We have the following result.

Lemma 3.117 *If* $\theta = \langle \theta_M, \theta_L \rangle$ *is a congruence on an* $\mathbf{L}$*-structure* $\mathbf{M}$ *then* $h_\theta = \langle h_{\theta_M}, h_{\theta_L} \rangle$ *is a strong morphism of* $\mathbf{M}$ *to* $\mathbf{M}/\theta$.

Proof. For an n-ary $f \in F$ we have $\mathrm{Hom}_f(h_\theta) = 1$ iff for each $m_i \in M$ we have $(h_{\theta_M}(f^{\mathbf{M}}(m_1, \ldots, m_n)) \approx^{\mathbf{M}_2} f^{\mathbf{M}/\theta}(h_{\theta_M}(m_1), \ldots, h_{\theta_M}(m_n))) = 1$. Since $h_{\theta_M}(f^{\mathbf{M}}(m_1, \ldots, m_n)) = [f^{\mathbf{M}}(m_1, \ldots, m_n)]_\theta = f^{\mathbf{M}/\theta}(h_{\theta_M}(m_1), \ldots, h_{\theta_M}(m_n)) = [f^{\mathbf{M}}(m_1, \ldots, m_n)]_\theta$, the required equality follows. For an n-ary $r \in R$ we have $\mathrm{Hom}_r(h_\theta) = 1$ iff for every $m_i \in M$ we have $h_{\theta_L}(r^{\mathbf{M}_1}(m_1, \ldots, m_n)) \leq r^{\mathbf{M}/\theta}(h_{\theta_M}(m_1), \ldots, h_{\theta_M}(m_n))$. By definition, this is equivalent to $[r^{\mathbf{M}}(m_1, \ldots, m_n)]_{\theta_L} \leq \bigvee_{\theta_M(m_i, m_i')=1} [r^{\mathbf{M}}(m_1', \ldots, m_n')]_{\theta_L}$ which is obviously true. It remains to verify that h_θ is strong, i.e. to show that for every $m_i \in M$ we have $r^{\mathbf{M}/\theta}(h_{\theta_M}(m_1), \ldots, h_{\theta_M}(m_n)) \leq \bigvee_{\theta_M(m_i', m_i)=1} [r^{\mathbf{M}}(m_1', \ldots, m_n'))]_{\theta_L}$ which is true since $r^{\mathbf{M}/\theta}(h_{\theta_M}(m_1), \ldots, h_{\theta_M}(m_n)) = r^{\mathbf{M}/\theta}([m_1]_\theta, , \ldots, [m_n]_\theta) = \bigvee_{\theta_M(m_i', m_i)=1} [r^{\mathbf{M}}(m_1', \ldots, m_n'))]_{\theta_L}$. □

A morphism $\langle h_M, h_L \rangle$ is said to be surjective (onto) if both h_M and h_L are surjective.

Theorem 3.118 (first isomorphism theorem) *Let* θ *be a congruence on an* $\mathbf{L}$*-structure* $\mathbf{M}$, h *be a strong morphism of an* $\mathbf{L}_1$*-structure*

$\mathbf{M}_1$ *to an* $\mathbf{L}_2$*-structure* $\mathbf{M}_2$ *with* $\approx^{\mathbf{M}_2}$ *being an* $\mathbf{L}_2$*-equality. Then* h_θ *is a strong morphism of* $\mathbf{M}$ *onto* $\mathbf{M}/\theta$*;* θ_h *is a congruence on* $\mathbf{M}_1$*;* $\theta = \theta_{h_\theta}$ *and if* h *is surjective,* $\mathbf{M}_1/\theta_h$ *is isomorphic to* $\mathbf{M}_2$*.*

Proof. By Lemmas 3.114 and 3.117 it suffices to verify that $\theta = \theta_{h_\theta}$ and that there is an isomorphism between $\mathbf{M}_1/\theta_h$ and $\mathbf{M}_2$. We have $\langle a_1, a_2\rangle \in (\theta_{h_\theta})_L$ iff $h_\theta(a_1) = h_\theta(a_2)$ iff $\langle a_1, a_2\rangle \in \theta_L$, i.e. $(\theta_{h_\theta})_L = \theta_L$; furthermore, $(\theta_{h_\theta})_M(m, m') = ([m]_\theta \approx^{\mathbf{M}/\theta} [m']_\theta) = \theta_M(m, m')$, i.e. $(\theta_{h_\theta})_M = \theta_M$, verifying $\theta = \theta_{h_\theta}$. Now, we claim that the pair $\langle g_M, g_L\rangle$ where $g_M : M_1/\theta_h \to M_2$ and $g_L : L_1/\theta_{h_L} \to L_2$ defined by $g_M([m]_{\theta_h}) = h_M(m)$ and $g_L([a]_{\theta_{h_L}}) = h_L(a)$ is an isomorphism. First, as $\approx^{\mathbf{M}_2}$ is an $\mathbf{L}_2$-equality, g_M is defined correctly. Indeed, if $[m]_{\theta_h} = [m']_{\theta_h}$ then $(h_M(m) \approx^{\mathbf{M}_2} h_M(m')) = 1$ and thus $h_M(m) = h_M(m')$. Clearly, g_L is an isomorphism of $\mathbf{L}_1/\theta_{h_L}$ to $\mathbf{L}_2$. As h_M is surjective, g_M is as well. If $[m]_{\theta_h} \neq [m']_{\theta_h}$ then $(h_M(m) \approx^{\mathbf{M}_2} h_M(m')) \neq 1$ and thus $g_M([m]_{\theta_h}) = h_M(m) \neq h_M(m') = g_M([m']_{\theta_h})$, whence g_M is a bijection. Furthermore, $g_L([m]_{\theta_h} \approx^{\mathbf{M}_1/\theta_h} [m']_{\theta_h}) = g_L(\theta_{h_M}(m, m')) = g_L(g_L^{-1}(h_M(m) \approx^{\mathbf{M}_2} h_M(m'))) = (g_M([m]_{\theta_h}) \approx^{\mathbf{M}_2} g_M([m']_{\theta_h}))$. For an n-ary $r \in R$ we have $g_L(r^{\mathbf{M}_1/\theta_h}([m_1]_{\theta_h}, \ldots, [m_n]_{\theta_h})) = h_L(\bigvee_{m'_i \in [m_i]_{\theta_h}} r^{\mathbf{M}_1}(m'_1, \ldots, m'_n)) = \bigvee_{(h_M(m_i) \approx^{\mathbf{M}_2} h_M(m'_i))=1} h_L(r^{\mathbf{M}_1}(m'_1, \ldots, m'_n)) = r^{\mathbf{M}_2}(h_M(m_1), \ldots, h_M(m_n)) = r^{\mathbf{M}_2}(g_M([m_1]_{\theta_h}), \ldots, g_M([m_n]_{\theta_h}))$ (we used the fact that h is a strong morphism). Thus, $\mathrm{Iso}_r(\langle g_M, g_L\rangle) = 1$. For each n-ary $f \in F$, we have $g_M(f^{\mathbf{M}_1/\theta_h}([m_1]_{\theta_h}, \ldots, [m_n]_{\theta_h})) = g_M([f^{\mathbf{M}_1}(m_1, \ldots, m_n)]_{\theta_h}) = h_M(f^{\mathbf{M}}(m_1, \ldots, m_n))$ and $f^{\mathbf{M}_2}(g_M([m_1]_{\theta_h}), \ldots, g_M([m_n]_{\theta_h}) = f^{\mathbf{M}_2}(h_M(m_1), \ldots, h_M(m_n))$, $(f^{\mathbf{M}_2}(h_M(m_1), \ldots, h_M(m_n) \approx^{\mathbf{M}_2} h_M(f^{\mathbf{M}_1}(m_1, \ldots, m_n)) = 1$ whence $g_M(f^{\mathbf{M}_1/\theta_h}([m_1]_{\theta_h}, \ldots, [m_n]_{\theta_h})) \approx^{\mathbf{M}_2} f^{\mathbf{M}_2}(g_M([m_1]_{\theta_h}), \ldots, g_M([m_n]_{\theta_h})) = 1$. Therefore, $\mathrm{Iso}_f(g_M, g_L) = 1$ completing the proof. □

Let now $\theta = \langle \theta_M, \theta_L\rangle$ and $\phi = \langle \phi_M, \phi_L\rangle$ be congruences on an $\mathbf{L}$-structure $\mathbf{M}$. If $\theta_L \subseteq \phi_L$, we put $(\phi/\theta)_L = \phi_L/\theta_L$, i.e. $(\phi/\theta)_L$ is a relation on L/θ_L defined by $\langle [a]_{\theta_L}, [a']_{\theta_L}\rangle \in (\phi/\theta)_L$ iff $\langle a, a'\rangle \in \phi_L$. $(\phi/\theta)_L$ is a complete congruence on $\mathbf{L}/\theta_L$ and $(\mathbf{L}/\theta_L)/(\phi/\theta)_L$ is isomorphic to $\mathbf{L}/\phi_L$ (cf. Proposition 1.8); denote by l_θ the isomorphism of $(\mathbf{L}/\theta_L)/(\phi/\theta)_L$ to $\mathbf{L}/\phi_L$ sending $[[a]_{\theta_L}]_{(\phi/\theta)_L}$ to $[a]_{\phi_L}$. By $\theta_M \subseteq \phi_M$ we denote the fact that $l_\theta([\theta_M(m_1, m_2)]_{(\phi/\theta)_L}) \leq \phi_M(m_1, m_2)$. Now, if $\theta_M \subseteq \phi_M$ and $\theta_L \subseteq \phi_L$ (which will be denoted by $\theta \subseteq \phi$), we define $\phi/\theta = \langle (\phi/\theta)_M, (\phi/\theta)_L\rangle$ as follows. First, $(\phi/\theta)_L$ is defined as above; second, $(\phi/\theta)_M$ is an $(\mathbf{L}/\theta)/(\phi/\theta)$-relation on M/θ defined by

$$(\phi/\theta)_M([m]_\theta, [m']_\theta) = l_\theta^{-1}(\phi_M(m, m')).$$

This definition is correct: If $m'' \in [m]_\theta$ and $m''' \in [m']_\theta$, then $\theta_M(m, m'') = 1$ and $\theta_M(m', m''') = 1$ whence also $\phi_M(m, m'') = 1$ and $\phi_M(m', m''') = 1$ from which one easily gets $\phi_M(m, m') = \phi_M(m'', m''')$ (namely, $\phi_M(m, m') = \phi_M(m'', m) \otimes \phi_M(m, m') \otimes \phi_M(m', m''') \leq \phi_M(m'', m''')$ and similarly for the converse inequality).

Theorem 3.119 (second isomorphism theorem) *For congruences $\theta = \langle \theta_M, \theta_L \rangle$ and $\phi = \langle \phi_M, \phi_L \rangle$ such that $\theta \subseteq \phi$, ϕ/θ is a congruence on $\mathbf{M}/\theta$. Furthermore, taking $h_M : [[m]_\theta]_{\phi/\theta} \mapsto [m]_\phi$ and $h_L : [[a]_{\theta_L}]_{(\phi/\theta)_L} \mapsto [a]_{\phi_L}$, $h = \langle h_M, h_L \rangle$ is an isomorphism of $(\mathbf{M}/\theta)/(\phi/\theta)$ to $\mathbf{M}/\phi$.*

Proof. By definition, $\mathbf{M}/\theta$ is an $\mathbf{L}/\theta_L$-structure and $(\phi/\theta)_L$ is a congruence on $\mathbf{L}/\theta_L$. Furthermore,

$$
\begin{aligned}
& (\phi/\theta)_M([m_1]_\theta, [m_2]_\theta) \otimes [([m_1]_\theta \approx [m_1']_\theta)]_{(\phi/\theta)_L} \otimes \\
& \quad [([m_2]_\theta \approx [m_2']_\theta)]_{(\phi/\theta)_L} \leq \\
\leq\ & l_\theta^{-1}(\phi_M(m_1, m_2)) \otimes l_\theta^{-1} \circ l_\theta([\theta_M(m_1, m_1')]_{(\phi/\theta)_L}) \otimes \\
& l_\theta^{-1} \circ l_\theta([\theta_M(m_1, m_1')]_{(\phi/\theta)_L}) \leq \\
\leq\ & l_\theta^{-1}(\phi_M(m_1, m_2)) \otimes l_\theta^{-1}(\phi_M(m_1, m_1')) \otimes l_\theta^{-1}(\phi_M(m_2, m_2')) = \\
=\ & l_\theta^{-1}(\phi_M(m_1, m_2) \otimes \phi_M(m_1, m_1') \otimes \phi_M(m_2, m_2')) \leq \\
\leq\ & l_\theta^{-1}(\phi_M(m_1', m_2') = (\phi/\theta)_M([m_1']_\theta, [m_2']_\theta)
\end{aligned}
$$

follows by definition of $(\phi/\theta)_M$ using the fact that ϕ and θ are congruences. Therefore, $\phi/\theta \in \mathrm{Adm}_{\mathrm{con}}(\mathbf{M}/\theta)$.

For an n-ary $f \in F$, we have

$$
\begin{aligned}
& (\phi/\theta)_M([m_1]_\theta, [m_1']_\theta) \otimes \cdots \otimes (\phi/\theta)_M([m_n]_\theta, [m_n']_\theta) = \\
=\ & l_\theta^{-1}(\phi_M(m_1, m_1')) \otimes \cdots \otimes l_\theta^{-1}(\phi_M(m_n, m_n')) = \\
=\ & l_\theta^{-1}(\phi_M(m_1, m_1') \otimes \cdots \otimes \phi_M(m_n, m_n')) \leq \\
\leq\ & l_\theta^{-1}(\phi_M(f^{\mathbf{M}}(m_1, \ldots, m_n), f^{\mathbf{M}}(m_1', \ldots, m_n'))) = \\
=\ & (\phi/\theta)_M([f^{\mathbf{M}}(m_1, \ldots, m_n)]_\theta, [f^{\mathbf{M}}(m_1', \ldots, m_n')]_\theta) = \\
=\ & (\phi/\theta)_M(f^{\mathbf{M}/\theta}([m_1]_\theta, \ldots, [m_n]_\theta), f^{\mathbf{M}/\theta}([m_1']_\theta, \ldots, [m_n']_\theta)),
\end{aligned}
$$

i.e. $\mathrm{Con}_f(\phi/\theta) = 1$. Similarly, one can prove $\mathrm{Con}_r(\phi/\theta) = 1$ for $r \in R$. Therefore, ϕ/θ is a congruence on $\mathbf{M}/\theta$. We now show that $h = \langle h_M, h_L \rangle$ is an isomorphism.

First, it is easy to see that $h \in \mathrm{Adm}_{\mathrm{iso}}((M/\theta)/(\phi/\theta), M/\phi)$: By definition,

$$
\begin{aligned}
& ([[m_1]_\theta]_{(\phi/\theta)} \approx^{(M/\theta)/(\phi/\theta)} [[m_1]_\theta]_{(\phi/\theta)}) = (\phi/\theta)_M([m_1]_\theta, [m_2]_\theta) = \\
=\ & l_\theta^{-1}(\phi_M(m_1, m_2)) = h_L^{-1}([m_1]_\phi \approx^{M/\phi} [m_2]_\phi);
\end{aligned}
$$

furthermore, h_L is an isomorphism and h_M a bijection. Indeed, $[m]_\phi$ is an image of $[[m]_\theta]_{\phi/\theta}$ and if $[m_1]_\phi = [m_2]_\phi$ then $\phi_M(m_1, m_2) = 1$ and thus $(\phi/\theta)_M([m_1]_\theta, [m_2]_\theta) = l_\theta^{-1}(\phi_M(m_1, m_2)) = 1$, i.e. $[[m_1]_\theta]_{\phi/\theta} = [[m_2]_\theta]_{\phi/\theta}$ proving that h_M is a bijection. Second,

$$
\begin{aligned}
& h_L(r^{(M/\theta)/(\phi/\theta)}([[m_1]_\theta]_{\phi/\theta}, \ldots, [[m_n]_\theta]_{\phi/\theta})) = \\
= \; & h_L(\bigvee_{[m_i']_\theta \in [[m_i]_\theta]_{\phi/\theta}} [r^{M/\theta}([m_1']_\theta, \ldots, [m_n']_\theta)]_{(\phi/\theta)_L}) = \\
= \; & h_L(\bigvee_{[m_i']_\theta \in [[m_i]_\theta]_{\phi/\theta}} \bigvee_{m_i'' \in [m_i']_\theta} [[r^M(m_1'', \ldots, m_n'')]_\theta]_{(\phi/\theta)_L}) = \\
& \bigvee_{[m_i']_\theta \in [[m_i]_\theta]_{\phi/\theta}} \bigvee_{m_i'' \in [m_i']_\theta} [r^M(m_1'', \ldots, m_n'')]_{\phi_L} = \\
& \bigvee_{m_i'' \in [m_i]_\phi} [r^M(m_1'', \ldots, m_n'')]_{\phi_L} = r^{M/\phi}(m_1, \ldots, m_n).
\end{aligned}
$$

Finally, the fact $(h_M(f^{(M/\theta)/(\phi/\theta)}([[m_1]_\theta]_{\phi/\theta}, \ldots)) \approx^{\mathbf{M}/\phi} f^{\mathbf{M}/\phi}(h_M([[m_1]_\theta]_{\phi/\theta}, \ldots))) = 1$ follows easily by definitions. □

3.4.2 PRODUCTS

We are now to introduce direct products of fuzzy structures. Having $\mathbf{L}_i$-structures $\mathbf{M}_i$ $(i \in I)$, our aim is to define an appropriate $\mathbf{L}$-structure on the direct product $\times_{i\in I} M_i$. Particularly, we will be interested in the case $\mathbf{L}_i = \mathbf{L}_j$ for all $i, j \in I$.

Since a direct product of complete residuated lattices is a complete residuated lattice, in general one may take $\mathbf{L} = \times_{i\in I}\mathbf{L}_i$ (i.e. the direct product of $\mathbf{L}_i$'s). Operations in $\times_{i\in I} M_i$ will be defined as follows.

Definition 3.120 Let $f_i : M_i^n \to M_i$ be n-ary mappings $(i \in I)$. The direct product of f_i's is an n-ary mapping $\times_{i\in I} f_i$ on $\times_{i\in I} M_i$ defined for $m_1, \ldots, m_n \in \times_{i\in I} M_i$ by

$$(\times_{i\in I} f_i(m_1, \ldots, m_n))(i) = f_i(m_1(i), \ldots, m_n(i))$$

for each $i \in I$.

For n-ary $\mathbf{L}_i$-relations R_i on M_i we define an n-ary $(\times_{i\in I}\mathbf{L}_i)$-relation $\Pi_{i\in I} R_i$ by

$$((\Pi_{i\in I} R_i)(m_1, \ldots, m_n))(i) = R_i(m_1(i), \ldots, m_n(i)).$$

Definition 3.121 For $\mathbf{L}_i$-structures $\mathbf{M}_i$ we define an $\times_{i\in I}\mathbf{L}_i$-structure $\Pi_{i\in I}\mathbf{M}_i$ as follows: the universe of $\Pi_{i\in I}\mathbf{M}_i$ is $\times_{i\in I}M_i$, $f^{\Pi_{i\in I}\mathbf{M}_i}$ is $\times_{i\in I}f^{\mathbf{M}_i}$ for each $f\in F$, and $r^{\Pi_{i\in I}\mathbf{M}_i}$ is $\Pi_{i\in I}r^{\mathbf{M}_i}$ for each $r\in R$.

The definition of $\Pi_{i\in I}\mathbf{M}_i$ is correct: It is almost immediate that $\approx^{\Pi_{i\in I}\mathbf{M}_i}$ is an $\times_i\mathbf{L}_i$-equivalence relation which is compatible with all $r^{\Pi_{i\in I}\mathbf{M}_i}$ and $f^{\Pi_{i\in I}\mathbf{M}_i}$.

If all $\mathbf{M}_i$ are $\mathbf{L}$-structures (i.e. $\mathbf{L}_i=\mathbf{L}_j$ for all $i,j\in I$), it is natural to require that the product be again an $\mathbf{L}$-structure. The crucial point is how to define an appropriate (n-ary) $\mathbf{L}$-relation R on $\times_{i\in I}M_i$ using (n-ary) $\mathbf{L}$-relations R_i defined on M_i. What we need is an appropriate "aggregation" operator $g: L^I\to L$ and define $R(m_1,\ldots,m_n)$ to be $g(\ldots,R_i(m_1(i),\ldots,m_n(i)),\ldots)$. In general, an appropriate candidate for g is $\bigwedge$ (infimum); if I is finite, one may consider also $\bigotimes$.

Definition 3.122 Let $R_i\in L^{M_i^n}$ be n-ary $\mathbf{L}$-relations ($i\in I$). The $\times$-product ($\bigwedge$-product) of R_i's is an n-ary $\mathbf{L}$-relation $\times_{i\in I}R_i$ on $\times_{i\in I}M_i$ defined for $m_1,\ldots,m_n\in\times_{i\in I}M_i$ by

$$\times_{i\in I}R_i(m_1,\ldots,m_n)=\bigwedge_{i\in I}R_i(m_1(i),\ldots,m_n(i)).$$

If I is finite, the $\bigotimes$-product of R_i is an n-ary $\mathbf{L}$-relation on $\times_{i\in I}M_i$ defined for $m_1,\ldots,m_n\in\times_{i\in I}M_i$ by

$$\bigotimes_{i\in I}R_i(m_1,\ldots,m_n)=\bigotimes_{i\in I}R_i(m_1(i),\ldots,m_n(i)).$$

The corresponding definitions of direct products of $\mathbf{L}$-structures follow.

Definition 3.123 Let $\mathbf{M}_i$ be $\mathbf{L}$-structures ($i\in I$). The $\times$-product of $\mathbf{M}_i$'s, denoted by $\times_{i\in I}\mathbf{M}_i$, is defined as follows: the universe is $\times_{i\in I}M_i$, $r^{\times_{i\in I}\mathbf{M}_i}$ is $\times_{i\in I}r^{\mathbf{M}_i}$ ($r\in R$), and $f^{\times_{i\in I}\mathbf{M}_i}$ is $\times_{i\in I}f^{\mathbf{M}_i}$ ($f\in F$). If I is finite, then the $\bigotimes$-product of $\mathbf{M}_i$'s, denoted by $\bigotimes_{i\in I}\mathbf{M}_i$, is defined as follows: the universe is $\times_{i\in I}M_i$, $r^{\bigotimes_{i\in I}\mathbf{M}_i}$ is $\bigotimes_{i\in I}r^{\mathbf{M}_i}$ ($r\in R$), and $f^{\times_{i\in I}\mathbf{M}_i}$ is $\times_{i\in I}f^{\mathbf{M}_i}$ ($f\in F$).

By a *direct product* of $\mathbf{L}$-structures $\mathbf{M}_i$ we mean their $\times$-product. Before we show that the definitions of an $\bigwedge$-product and an $\bigotimes$-product are correct we give a few remarks.

Remark 3.124 (1) If I is obvious from the context, we also write only $\times_i\cdots$ or even $\times\cdots$ instead of $\times_{i\in I}\cdots$; similarly for $\bigotimes$.

(2) The foregoing definitions say: $\times f_i$ is defined componentwise, i.e. $\times f_i(m_1,\ldots,m_n)$ is an element of $\times M_i$ whose i-th component is

equal to f_i applied to the i-th components of $m_1, \ldots, m_n$, i.e. to $f_i(m_1(i), \ldots, m_n(i))$. $\times R_i$ and $\bigotimes R_i$ are defined by "aggregating" the components by $\bigwedge$ and $\bigotimes$, respectively. This gives: $\times R_i(m_1, \ldots, m_n)$ is $\geq a$ iff $R_i(m_1(i), \ldots, m_n(i)) \geq a$ for all $i \in I$.

(3) In a sense, $\Pi\mathbf{M}_i$ consists of all $\mathbf{M}_i$'s. Particularly, the truth degree $\Pi R_i(m_1, \ldots, m_n)$ is the "vector" consisting of all $R_i(m_1(i), \ldots, m_n(i))$. The set of all such possible vectors is just the support set of $\times\mathbf{L}_i$. If $\mathbf{M}_i$ represent knowledge supplied by experts, $\Pi\mathbf{M}_i$ can be thought of as representing the overall knowledge. It is immediate that for each formula φ and a valuation $v : X \to \times M_i$ (that may be thought of as consisting of valuations $v_i : x \mapsto (v(x))(i)$) we have $\|\varphi\|_{\Pi\mathbf{M}_i,v} = \langle \ldots, \|\varphi\|_{\mathbf{M}_i,v_i}, \ldots \rangle$; thus the truth value of any statement interpreted in $\Pi\mathbf{M}_i$ consists of the truth values of this statement interpreted in $\mathbf{M}_i$'s.

We are now going to show that $\times\mathbf{M}_i$ and $\bigotimes\mathbf{M}_i$ are defined correctly. We have to verify that $\times \approx^{\mathbf{M}_i}$ and $\bigotimes \approx^{\mathbf{M}_i}$ are $\mathbf{L}$-equivalence relations and that they are compatible with all the $\mathbf{L}$-relations and functions.

Lemma 3.125 *If $\approx_i$ is an $\mathbf{L}$-equivalence relation on M_i $(i \in I)$ then $\times_{i\in I} \approx_i$ is an $\mathbf{L}$-equivalence relation on $\times_{i\in I} M_i$. Moreover, $\times_{i\in I} \approx_i$ is the greatest $\mathbf{L}$-equivalence on $\times_i M_i$ such that each projection $\mathrm{pr}_i : \times M_i \to M_i$ is a morphism. If I is finite then $\bigotimes_{i\in I} \approx_i$ is an $\mathbf{L}$-equivalence relation on $\times_{i\in I} M_i$ as well.*

Proof. Reflexivity and symmetry of $\times \approx_i$ follow directly by definitions. Transitivity of $\times \approx_i$: For any $m_1, m_2, m_3 \in \times M_i$ we have

$$\begin{aligned}
& (m_1 \times \approx_i m_2) \otimes (m_2 \times \approx_i m_3) = \\
= \; & \Big(\bigwedge_{i\in I} (m_1(i) \approx_i m_2(i))\Big) \otimes \Big(\bigwedge_{i\in I} (m_2(i) \approx_i m_3(i))\Big) \leq \\
\leq \; & \bigwedge_{i\in I} \bigwedge_{j\in I} ((m_1(i) \approx_i m_2(i)) \otimes (m_2(j) \approx_j m_3(j))) \leq \\
\leq \; & \bigwedge_{i\in I} ((m_1(i) \approx_i m_2(i)) \otimes (m_2(i) \approx_i m_3(i))) \leq \\
\leq \; & \bigwedge_{i\in I} m_1(i) \approx_i m_3(i) = (m_1 \times \approx_i m_3).
\end{aligned}$$

If $\approx$ is another binary $\mathbf{L}$-relation on $\times M_i$ such that all pr_i are morphisms, then $(m_1 \approx m_2) \leq (m_1(i) \approx_i m_2(i))$ for each $i \in I$, by definition, and so $(m_1 \approx m_2) \leq \bigwedge_{i\in I} (m_1(i) \approx_i m_2(i)) = (m_1 \times \approx_i m_2)$.

That $\bigotimes \approx_i$ is an $\mathbf{L}$-equivalence relation follows from Lemma 3.39. □

Lemma 3.126 *Let R_i be an n-ary* **L***-relation in M_i, f_i be an n-ary mapping in M_i, $\approx_i$ be an* **L***-equivalence relation on M_i ($i \in I$). If each R_i and f_i are compatible with $\approx_i$ then $\times_i R_i$ and $\times_i f_i$ are compatible with $\times_i \approx_i$. If, moreover, I is finite then $\times_i f_i$ and $\bigotimes_i R_i$ are compatible with $\bigotimes_i \approx_i$.*

Proof. The first part: Take any $m_1, m_1', \ldots, m_n, m_n' \in \times M_i$. For simplicity, we write $\approx$, R, and f instead of $\times \approx_i$, $\times R_i$, and $\times f_i$, respectively. To verify compatibility of R w.r.t. $\approx$ we have to show $(m_1 \approx m_1') \otimes \cdots \otimes (m_n \approx m_n') \otimes R(m_1, \ldots, m_n) \leq R(m_1', \ldots, m_n')$. By definition of R, the last inequality holds iff for each $i \in I$ we have $(m_1 \approx m_1') \otimes \cdots \otimes (m_n \approx m_n') \otimes R(m_1, \ldots, m_n) \leq R_i(m_1'(i), \ldots, m_n'(i))$ which is true. Indeed,

$$\begin{aligned} & (m_1 \approx m_1') \otimes \cdots \otimes (m_n \approx m_n') \otimes R(m_1, \ldots, m_n) \leq \\ \leq\ & (m_1(i) \approx m_1'(i)) \otimes \cdots \otimes (m_n(i) \approx m_n'(i)) \otimes R_i(m_1(i), \ldots, m_n(i)) \leq \\ \leq\ & R_i(m_1'(i), \ldots, m_n'(i)), \end{aligned}$$

by compatibility of R_i. Compatibility of f: We have to show $(m_1 \approx m_1') \otimes \cdots \otimes (m_n \approx m_n') \leq f(m_1, \ldots, m_n) \approx f(m_1', \ldots, m_n')$. By definition, the inequality is true iff for each $i \in I$ we have $(m_1 \approx m_1') \otimes \cdots \otimes (m_n \approx m_n') \leq (f(m_1, \ldots, m_n))(i) \approx_i (f(m_1', \ldots, m_n'))(i)$ which is true. Indeed, by compatibility of f_i we get

$$\begin{aligned} & (m_1 \approx m_1') \otimes \cdots \otimes (m_n \approx m_n') \leq \\ \leq\ & (m_1(i) \approx m_1'(i)) \otimes \cdots \otimes (m_n(i) \approx m_n'(i)) \leq \\ \leq\ & f_i(m_1(i), \ldots, m_n(i)) \approx_i f_i(m_1'(i), \ldots, m_n'(i)) = \\ =\ & (f(m_1, \ldots, m_n))(i) \approx_i (f(m_1', \ldots, m_n'))(i). \end{aligned}$$

Consider now the compatibility w.r.t. $\bigotimes_i \approx_i$. Again, we write $\approx$, R, and f instead of $\bigotimes_i \approx_i$, $\bigotimes_i R_i$, and $\times_i f_i$, respectively. We have

$$\begin{aligned} & (m_1 \approx m_1') \otimes \cdots \otimes (m_n \approx m_n') \otimes R(m_1, \ldots, m_n) = \\ =\ & \bigotimes_i (m_1(i) \approx_i m_1'(i)) \otimes \cdots \otimes \bigotimes_i (m_n(i) \approx_i m_n'(i)) \otimes \\ & \otimes \bigotimes_i R_i(m_1(i), \ldots, m_n(i)) = \\ =\ & \bigotimes_i ((m_1(i) \approx_i m_1'(i)) \otimes \cdots \otimes (m_n(i) \approx_i m_n'(i))) \otimes \\ & \otimes R_i(m_1(i), \ldots, m_n(i)) \leq \\ \leq\ & \bigotimes_i R_i(m_1'(i), \ldots, m_n'(i)) = R(m_1', \ldots, m_n') \end{aligned}$$

and

$$\begin{aligned} & (m_1 \approx m_1') \otimes \cdots \otimes (m_n \approx m_n') = \\ = \; & \bigotimes_i (m_1(i) \approx_i m_1'(i)) \otimes \cdots \otimes \bigotimes_i (m_n(i) \approx_i m_n'(i)) = \\ = \; & \bigotimes_i ((m_1(i) \approx_i m_1'(i)) \otimes \cdots \otimes (m_n(i) \approx_i m_n'(i))) \leq \\ \leq \; & \bigotimes_i (f_i(m_1(i), \ldots, m_n(i)) \approx_i f_i(m_1'(i), \ldots, m_n'(i))) = \\ = \; & f(m_1, \ldots, m_n) \approx f(m_1', \ldots, m_n'). \end{aligned}$$

□

We verified that $\times_i \mathbf{M}_i$ and $\bigotimes_i \mathbf{M}_i$ are defined correctly. Not surprisingly, products of pairwise similar fuzzy structures are similar. In degrees, we have

Lemma 3.127 *For* $\mathbf{L}$*-structures* $\mathbf{M}_i, \mathbf{M}_i'$ *with* $M_i = M_i'$ *and* $\approx^{\mathbf{M}_i} = \approx^{\mathbf{M}_i'}$ *(*$i \in I$*) and* $S \subseteq R \cup F$ *we have*

$$\bigwedge_{i \in I} (\mathbf{M}_i \approx_S \mathbf{M}_i') \leq \times_i \mathbf{M}_i \approx_S \times_i \mathbf{M}_i'.$$

If I *is finite then we have*

$$\bigwedge_{s \in S} \bigotimes\nolimits_{i \in I} (\mathbf{M}_i \approx_s \mathbf{M}_i') \leq \bigotimes_i \mathbf{M}_i \approx_S \bigotimes_i \mathbf{M}_i'.$$

where $f^{\bigotimes \mathbf{M}_i} \approx f^{\bigotimes \mathbf{M}_i'}$ is defined by (3.34) with $\approx^M$ being $(\otimes_i \approx_i)$.

Proof. The first inequality follows easily from definitions. For the second one, it is enough to show that for $s \in S$ we have $\bigotimes_{i \in I} (\mathbf{M}_i \approx_s \mathbf{M}_i') \leq \bigotimes_i \mathbf{M}_i \approx_s \bigotimes_i \mathbf{M}_i'$ which, again, can be easily verified. □

3.4.3 SUBSTRUCTURES

For an $\mathbf{L}$-structure $\mathbf{M}$ and a subset $M' \subseteq M$ it may happen that the result of any function of $\mathbf{M}$ applied to elements from M' is an element of M'. If this is the case, one may equip M' with restrictions of $\mathbf{L}$-relations and functions of $\mathbf{M}$ and consider it a substructure of the original structure $\mathbf{M}$.

Definition 3.128 A *subuniverse* of an $\mathbf{L}$-structure $\mathbf{M}$ is a subset M' of M closed w.r.t. each function of $\mathbf{M}$ and satisfying for every $m_1, m_2 \in M$ that $m_1 \in M$ and $(m_1 \approx^{\mathbf{M}} m_2) = 1$ implies $m_2 \in M'$.

Note that closedness of M' means that for each n-ary $f \in F$ and $m_1, \ldots, m_n \in M'$ we have $f^{\mathbf{M}}(m_1, \ldots, m_n) \in M'$. The set of all subuniverses of an $\mathbf{L}$-structure $\mathbf{M}$ will be denoted by Sub($\mathbf{M}$). Trivially, $\emptyset \in \mathrm{Sub}(\mathbf{M})$ and $M \in \mathrm{Sub}(\mathbf{M})$.

A nonempty subuniverse M' of Sub($\mathbf{M}$) may be naturally turned into an $\mathbf{L}$-structure $\mathbf{M}'$: for n-ary $r \in R$ and $m_1, \ldots, m_n \in M'$ we put $r^{\mathbf{M}'}(m_1, \ldots, m_n) = r^{\mathbf{M}}(m_1, \ldots, m_n)$, and for n-ary $f \in F$ and $m_1, \ldots, m_n \in M'$ we put $f^{\mathbf{M}'}(m_1, \ldots, m_n) = f^{\mathbf{M}}(m_1, \ldots, m_n)$. Obviously, $\approx^{\mathbf{M}'}$ is an $\mathbf{L}$-equivalence relation on M' which is compatible with each $r^{\mathbf{M}'}$ and $f^{\mathbf{M}'}$. Therefore, $\mathbf{M}'$ is indeed an $\mathbf{L}$-structure.

Definition 3.129 An $\mathbf{L}$-structure $\mathbf{M}'$ is called a *substructure* of an $\mathbf{L}$-structure $\mathbf{M}$ if $M' \subseteq M$; $m_1 \in M'$ and $(m_1 \approx^{\mathbf{M}} m_2) = 1$ implies $m_2 \in M'$; for each $r \in R$, $r^{\mathbf{M}'}$ is a restriction of $r^{\mathbf{M}}$ to M', and for each $f \in F$, $f^{\mathbf{M}'}$ is a restriction of $f^{\mathbf{M}}$ to M'.

If M' is a nonempty subuniverse of an $\mathbf{L}$-structure $\mathbf{M}$ then, clearly, $\mathbf{M}'$ as defined above is a substructure of $\mathbf{M}$; for a substructure $\mathbf{M}'$ of $\mathbf{M}$, M' is a nonempty subuniverse of $\mathbf{M}$. We thus see that there is a bijection between the set of nonempty subuniverses of an $\mathbf{L}$-structure $\mathbf{M}$ and the set of all substructures of $\mathbf{M}$.

Remark 3.130 (1) A substructure $\mathbf{M}'$ of an $\mathbf{L}$-structure $\mathbf{M}$ is itself an $\mathbf{L}$-structure. However, it can happen that there is a proper complete subalgebra $\mathbf{L}'$ of $\mathbf{L}$ (i.e. $L' \subset L$) such that for each n-ary $r \in R$ and $m_1, \ldots, m_n \in M'$ we have $r^{\mathbf{M}'} \in L'$ (fuzzy relations take only truth values from L'). Therefore, one could consider $\mathbf{M}'$ an $\mathbf{L}'$-structure and say that an $\mathbf{L}'$-structure $\mathbf{M}'$ is a substructure of an $\mathbf{L}$-structure $\mathbf{M}$ if $\mathbf{L}'$ is a complete subuniverse of $\mathbf{L}$ and ...(continued obviously). We will not proceed this way. However, given an $\mathbf{L}$-structure $\mathbf{M}$ and letting $\mathbf{L}'$ denote the complete subalgebra of $\mathbf{L}$ generated by $R^{\mathbf{M}}(M) = \{r^{\mathbf{M}}(M^{\sigma(r)}) \mid r \in R\}$, $\mathbf{L}'$ is the least complete subalgebra of $\mathbf{L}$ such that $\mathbf{M}$ can be considered an $\mathbf{L}'$-structure.

(2) If $F = \emptyset$ then each subset M' of M satisfying $m_2 \in M'$ whenever $(m_1 \approx^{\mathbf{M}} m_2) = 1$ and $m_1 \in M'$, is a subuniverse of $\mathbf{M}$.

Theorem 3.131 *For each* $\mathbf{L}$*-structure* $\mathbf{M}$, Sub($\mathbf{M}$) *is closed w.r.t. arbitrary intersections, i.e.* $M_i \in \mathrm{Sub}(\mathbf{M})$ *implies* $\bigcap_{i \in I} M_i \in \mathrm{Sub}(\mathbf{M})$.

Proof. If $m_1 \in \bigcap_{i \in I} M_i$ and $(m_1 \approx^{\mathbf{M}} m_2) = 1$, then $m_1 \in M_j$ and $(m_1 \approx^{\mathbf{M}} m_2) = 1$ for each $j \in J$, thus $m_2 \in M_j$ for each $j \in J$ and so $m_2 \in \bigcap_{i l I} M_i$. Take n-ary $f \in F$. If $m_1, \ldots, m_n \in \bigcap_{i \in I} M_i$, then $m_1, \ldots, m_n \in M_j$ (for each $j \in I$) and thus, by closedness of M_j, $f^{\mathbf{M}}(m_1, \ldots, m_n) \in M_j$ whence $f^{\mathbf{M}}(m_1, \ldots, m_n) \in \bigcap_{i \in I} M_i$. □

Remark 3.132 (1) By Theorem 3.131, Sub$(\mathbf{M})$ is a complete lattice w.r.t. $\subseteq$. It is thus possible to speak of the least subuniverse $[M']$ of $\mathbf{M}$ containing a subset $M' \subseteq M$ (note that $[M'] = \bigcap\{M'' \in \mathrm{Sub}(\mathbf{M}) \mid M' \subseteq M''\}$).

(2) It is easy to see that for any $\mathbf{M}$ and $M' \subseteq M$, $[M']$ can be inductively defined as follows: For $N \subseteq M$, put $F^0(N) = N$ and $F^{i+1}(N) = \{f^{\mathbf{M}}(m_1, \dots, m_{\sigma(f)}) \mid f \in F,\ m_i \in F^i(N)\} \cup \{m \mid (m \approx^{\mathbf{M}} m') = 1$ for some $m' \in F^i(N)\}$. Then, $[M'] = \bigcup_{i=0}^{\infty} F^i(M')$.

(3) A substructure $\mathbf{M}''$ of $\mathbf{M}$ is said to be generated by $M' \subseteq M$ if $M'' = [M']$; in this case, M' is called a set of generators of $\mathbf{M}''$. One can easily verify that if $h_1 = \langle h_{1M}, h_{1L}\rangle$, $h_2 = \langle h_{2M}, h_{2L}\rangle$ are morphisms of $\mathbf{M}_1$ to $\mathbf{M}_2$, M' is a set of generators of $\mathbf{M}_1$, and if for the restrictions h'_{iM} of h_{iM} to M' $(i = 1, 2)$ we have $(h'_{1M} \approx h'_{2M}) = 1$, then $(h_{1M} \approx h_{2M}) = 1$.

Theorem 3.133 *Let $h = \langle h_M, h_L\rangle$ be a morphism of an $\mathbf{L}_1$-structure $\mathbf{M}_1$ to an $\mathbf{L}_2$-structure $\mathbf{M}_2$.*

(1) If $\approx^{\mathbf{M}_2}$ is an $\mathbf{L}_2$-equality then $h_M(M')$ is a subuniverse of $\mathbf{M}_2$ whenever M' is a subuniverse of $\mathbf{M}$.

(2) If M' is a subuniverse of $\mathbf{M}_2$ then $h_M^{-1}(M')$ is a subuniverse of $\mathbf{M}_1$.

Proof. (1) We verify that for n-ary $f \in F$ and $m_1, \dots, m_n \in M'$, $f^{\mathbf{M}_2}(h_M(m_1), \dots, h_M(m_n)) \in h(M')$. As M' is a subuniverse of $\mathbf{M}_1$, we have $f^{\mathbf{M}_1}(m_1, \dots, m_n) \in M'$. Since h is a morphism (i.e. $\mathrm{Hom}(h) = 1$) we have $(h_M(f^{\mathbf{M}_1}(m_1, \dots, m_n)) \approx^{\mathbf{M}_2} f^{\mathbf{M}_2}(h_M(m_1), \dots, h_M(m_n)) = 1)$ and since $\approx^{\mathbf{M}_2}$ is an $\mathbf{L}_2$-equality, $h_M(f^{\mathbf{M}_1}(m_1, \dots, m_n)) = f^{\mathbf{M}_2}(h_M(m_1), \dots, h_M(m_n))$ proving $f^{\mathbf{M}_2}(h_M(m_1), \dots, h_M(m_n)) \in h(M')$. Furthermore, if $m_1 \in h_M(M')$ and $(m_1 \approx^{\mathbf{M}_2} m_2) = 1$, then since $\approx^{\mathbf{M}_2}$ is an $\mathbf{L}_2$-equality, we have $m_1 = m_2$ and thus $m_2 \in h_M(M')$.

(2) We verify that for n-ary $f \in F$ and $m_1, \dots, m_n \in M_1$ such that $h_M(m_i) \in M'$ we have $h_M(f^{\mathbf{M}_1}(m_1, \dots, m_n)) \in M'$. As h is a morphism and M' a subuniverse, we have $(h_M(f^{\mathbf{M}_1}(m_1, \dots, m_n)) \approx^{\mathbf{M}_2} f^{\mathbf{M}_2}(h_M(m_1), \dots, h_M(m_n)) = 1)$ and $f^{\mathbf{M}_2}(h_M(m_1), \dots, h_M(m_n) \in M'$ and thus $h_M(f^{\mathbf{M}_1}(m_1, \dots, m_n)) \in M'$. Furthermore, if $(m \approx^{\mathbf{M}_1} m') = 1$ for some $m' \in h_M^{-1}(M')$ then as $1 = h_L(m \approx^{\mathbf{M}_1} m') \leq (h_M(m) \approx^{\mathbf{M}_2} h_M(m'))$, we have $h_M(m) \in M'$ and thus $m \in h_M^{-1}(M)$. □

For a subset M' of M and an $\mathbf{L}$-equivalence θ_M on M, denote $M'^{\theta_M} = \{m \in M \mid \theta_M(m, m') = 1$ for some $m' \in M'\}$. Intuitively, M'^{θ_M} is the subset of all elements of M indistinguishable from (some element of) M' by θ_M.

Theorem 3.134 (third isomorphism theorem) *For an* $\mathbf{L}$*-structure* $\mathbf{M}$*, a congruence* $\theta = \langle\theta_M, \theta_L\rangle$ *on* $\mathbf{M}$*, and* $M' \in \mathrm{Sub}(\mathbf{M})$ *we have* $M'^{\theta_M} \in \mathrm{Sub}(\mathbf{M})$*. For the corresponding substructure* $\mathbf{M}'^{\theta_M}$*,* $\mathbf{M}'/\langle\theta_M \cap M' \times M', \theta_L\rangle$ *is isomorphic to* $\mathbf{M}'^{\theta_M}/\langle\theta_M \cap M'^{\theta_M} \times M'^{\theta_M}, \theta_L\rangle$*.*

Proof. First, we show that $M'^{\theta_M} \in \mathrm{Sub}(\mathbf{M})$. Let $m_1 \in M'^{\theta_M}$ and $(m_1 \approx^{\mathbf{M}} m_2) = 1$. Then there is an $m' \in M$ such that $\theta_M(m', m_1) = 1$. As $(m_1 \approx^{\mathbf{M}} m_2) \leq \theta_M(m_1, m_2)$, we have $\theta_M(m_1, m_2) = 1$ and thus $1 = \theta_M(m', m_1) \otimes \theta_M(m_1, m_2) \leq \theta_M(m', m_2)$, i.e. $\theta_M(m', m_2) = 1$. Therefore, $m_2 \in M'^{\theta_M}$. Furthermore, if $m_1, \ldots, m_n \in M'^{\theta_M}$ then there are $m'_1, \ldots, m'_n \in M'$ with $\theta_M(m_i, m'_i) = 1$ ($i \in I$) and so for an n-ary $f \in F$ we have $1 = \bigotimes_i \theta_M(m_i, m'_i) \leq \theta_M(f^{\mathbf{M}}(m_1, \ldots, m_n), f^{\mathbf{M}}(m'_1, \ldots, m'_n))$. Therefore, since $f^{\mathbf{M}}(m'_1, \ldots, m'_n) \in M'$ ($M' \in \mathrm{Sub}(\mathbf{M})$), we have $f^{\mathbf{M}}(m_1, \ldots, m_n) \in M'^{\theta_M}$.

Furthermore, we claim that for a mapping h_M sending $[m]_{\theta_M \cap M' \times M'}$ to $[m]_{\theta_M \cap M'^{\theta_M} \times M'^{\theta_M}}$ and h_L being the identity mapping on $\mathbf{L}/\theta_L$, $h = \langle h_M, h_L\rangle$ is an isomorphism. We leave this as an exercise. □

Remark 3.135 A substructure of a fuzzy structure $\mathbf{M}$ is given by a crisp subset of M. We thus do not consider fuzzy subsets of M which is the core of so-called fuzzy subalgebras. Fuzzy subalgebras started with [174]; see [65] for a general approach. In our context, the main notion can be presented as follows: A fuzzy subalgebra ($\mathbf{L}$-subalgebra) of an ordinary algebra $\mathbf{M} = \langle M, F\rangle$ is an $\mathbf{L}$-subset A of M which is closed under all operations from F in that for each n-ary $f \in F$ and $m_1, \ldots, m_n \in M$ we have $A(m_1) \otimes \cdots \otimes A(m_n) \leq A(f(m_1, \ldots, m_n))$ (this reads: if m_i are from A then $f(m_1, \ldots, m_n)$ is from A). There are natural examples of fuzzy subalgebras. For example, for an $\mathbf{L}$-structure $\mathbf{M}$, $\mathrm{Adm}_{\mathrm{hom}}(\mathbf{M}, \mathbf{M})$ equipped with composition (see above) is a monoid (semigroup with the neutral element being $\langle \mathrm{id}_M, \mathrm{id}_L\rangle$). Lemma 3.101 says that Hom_S is an $\mathbf{L}$-subalgebra of $\mathrm{Adm}_{\mathrm{hom}}(\mathbf{M}, \mathbf{M})$. We will, however, not pursue fuzzy subalgebras any further.

3.5 REPRESENTATION BY CRISP STRUCTURES: CUTLIKE SEMANTICS

Theorem 3.22 says that fuzzy sets in M can be thought of as (special types of) nested systems of (ordinary) sets in M; Theorem 3.43 extends this to fuzzy sets which are compatible with fuzzy equivalences. Therefore, each $\mathbf{L}$-structure $\mathbf{M}$ of type $\langle R, F, \sigma\rangle$ can be represented by a system $\mathcal{C}_{\mathbf{M}} = \{{}^a\mathbf{M} \mid a \in L\}$ of $\mathbf{2}$-structures ${}^a\mathbf{M}$ defined by ${}^aM = M$; $f^{{}^a\mathbf{M}} = f^{\mathbf{M}}$ for $f \in F$; $r^{{}^a\mathbf{M}} = {}^a r^{\mathbf{M}}$ for $r \in R$. Thus, ${}^a\mathbf{M}$ has the same universe and functions as $\mathbf{M}$; relation in ${}^a\mathbf{M}$ corresponding to r is the a-cut of the $\mathbf{L}$-relation that corresponds to r in $\mathbf{M}$. Conversely, having an $\mathbf{L}$-indexed system $\mathcal{C} = \{\mathbf{M}_a \mid a \in L\}$ of $\mathbf{2}$-structures with a common universe M (i.e. $M_a = M$ for each $a \in L$) and common functions (i.e. $f^{\mathbf{M}_a} = f^{\mathbf{M}_b}$ for each $a, b \in L$) such that for each $r \in R$, $\{r^{\mathbf{M}_a} \mid a \in L\}$ is an $\mathbf{L}$-nested system compatible with $(\approx^{\mathbf{M}})^{\sigma(r)} = \bigotimes_{i=1}^{\sigma(r)} \approx^{\mathbf{M}}$ (cf. Definitions 3.19, 3.42, and Remark 3.44), $\mathcal{C}$ induces an $\mathbf{L}$-structure $\mathbf{M}_{\mathcal{C}}$ given by $M_{\mathcal{C}} = M$; $f^{\mathbf{M}_{\mathcal{C}}} = f^{\mathbf{M}_a}$ (for any $a \in L$);

$$r^{\mathbf{M}_{\mathcal{C}}}(m_1, \ldots, m_n) = \bigvee\{a \mid \langle m_1, \ldots, m_n\rangle \in r^{\mathbf{M}_a}\}.$$

Definition 3.136 An $\mathbf{L}$-indexed system $\{\mathbf{M}_a \mid a \in L\}$ of $\mathbf{2}$-structures (i.e. ordinary structures) of the same type is called $\mathbf{L}$-*nested* if for each $a, b \in L$ we have $M_a = M_b$ (common universe), $f^{\mathbf{M}_a} = f^{\mathbf{M}_b}$ for each $f \in F$ (common functions), and for each $r \in R$, $\{r^{\mathbf{M}_a} \mid a \in L\}$ is an $\mathbf{L}$-nested system.

Therefore, we have

Lemma 3.137 *The mappings sending an* $\mathbf{L}$*-structure* $\mathbf{M}$ *to an* $\mathbf{L}$*-nested system* $\mathcal{C}_{\mathbf{M}}$*, and an* $\mathbf{L}$*-nested system* $\mathcal{C}$ *of structures into an* $\mathbf{L}$*-structure* $\mathbf{M}_{\mathcal{C}}$ *are mutually inverse.*

Proof. Directly by the above considerations, Theorem 3.22, and Theorem 3.43. □

Let us have an $\mathbf{L}$-structure $\mathbf{M}$. Given a valuation $v : X \to M$, a formula φ evaluates to $||\varphi||_{\mathbf{M},v}$. As $\mathbf{M}_a$ ($a \in L$) is a $\mathbf{2}$-structure, one can consider the truth degree $||\varphi||_{\mathbf{M}_a,v}$ to which φ evaluates under v. By definition, $||\varphi||_{\mathbf{M}_a,v}$ is either 0 or 1. There is a natural question of whether the truth degree $||\varphi||_{\mathbf{M},v}$ of φ under v can be in some way composed of truth degrees $||\varphi_i||_{\mathbf{M}_a,v}$ that are assigned to φ or its subformulas in $\mathbf{M}_a$ under v. One way to answer this question is elaborated in the following. Denote

$${}^a||\varphi||_{\mathbf{M},v} = \begin{cases} 1 & \text{if } a \leq ||\varphi||_{\mathbf{M},v} \\ 0 & \text{otherwise} . \end{cases}$$

Thus, ${}^a||\varphi||_{\mathbf{M},v}$ is the a-cut of $||\varphi||_{\mathbf{M},v}$. Note that $||\varphi||_{\mathbf{M},v}$ can be seen as a truth degree assigned to v given $\mathbf{M}$ and φ; this way a formula and an $\mathbf{L}$-structure induce an $\mathbf{L}$-set in the set of all valuations.

Example 3.138 Theorem 3.22 presented in another way serves as a demonstration: Consider a language with a unary relation symbol r. In an **L**-structure **M**, $r^{\mathbf{M}}$ (**L**-relation corresponding to r) is an **L**-set in M. Clearly, for a valuation v with $v(x) = m$ and $\varphi = r(x)$, we have $r^{\mathbf{M}}(m) = ||\varphi||_{\mathbf{M},v}$. Therefore, Theorem 3.22 then says

$$||\varphi||_{\mathbf{M},v} = \bigvee\{a \mid {}^{a}||\varphi||_{\mathbf{M},v} = 1\},$$

which is a way to obtain $||\varphi||_{\mathbf{M},v}$ from ${}^{a}||\varphi||_{\mathbf{M},v}$ for the simple case of atomic formula.

In the following we denote the operations in **2** (the two-element Boolean algebra) by $\wedge^{\mathbf{2}}$, $\vee^{\mathbf{2}}$, $\otimes^{\mathbf{2}}$, $\rightarrow^{\mathbf{2}}$, $\bigwedge^{\mathbf{2}}$, and $\bigvee^{\mathbf{2}}$ (note that $\otimes^{\mathbf{2}} = \wedge^{\mathbf{2}}$).

Let $\mathcal{C} = \{\mathbf{M}_a \mid a \in L\}$ be an **L**-nested system with a universe M. For a term t we put

$$||t||_{\mathcal{C},v} = ||t||_{\mathbf{M}_a,v}$$

for $a \in L$ (this is correct since each $f \in F$, $f^{\mathbf{M}_a}$ are the same for $a \in L$). Furthermore, for a formula φ and $a \in L$ we define $||\varphi||^{a}_{\mathcal{C},v}$ as follows.

(i) for atomic formulas:

$$\begin{aligned} ||r(t_1,\ldots,t_n)||^{a}_{\mathcal{C},v} &= r^{\mathbf{M}_a}(||t_1||_{\mathcal{C},v},\ldots,||t_n||_{\mathcal{C},v}), \\ ||\mathbb{b}||^{a}_{\mathcal{C},v} &= 1 \quad \text{for } a \leq b \\ & 0 \quad \text{otherwise} \end{aligned}$$

($\mathbb{b}$ a truth constant);

(ii) if φ and ψ are formulas then

$$\begin{aligned} ||\varphi \mathbin{\&} \psi||^{a}_{\mathcal{C},v} &= ||\varphi||^{a}_{\mathcal{C},v} \wedge^{\mathbf{2}} ||\psi||^{a}_{\mathcal{C},v}, \\ ||\varphi \vee \psi||^{a}_{\mathcal{C},v} &= \bigvee^{\mathbf{2}}_{a \leq b \vee c} ||\varphi||^{b}_{\mathcal{C},v} \wedge^{\mathbf{2}} ||\psi||^{c}_{\mathcal{C},v}, \\ ||\varphi \otimes \psi||^{a}_{\mathcal{C},v} &= \bigvee^{\mathbf{2}}_{a \leq b \otimes c} ||\varphi||^{b}_{\mathcal{C},v} \wedge^{\mathbf{2}} ||\psi||^{c}_{\mathcal{C},v}, \\ ||\varphi \Rightarrow \psi||^{a}_{\mathcal{C},v} &= \bigwedge^{\mathbf{2}}_{b \in L} ||\varphi||^{b}_{\mathcal{C},v} \rightarrow^{\mathbf{2}} ||\psi||^{a \otimes b}_{\mathcal{C},v}; \end{aligned}$$

(iii) if φ is a formula and x a variable then

$$\begin{aligned} ||(\forall x)\varphi||^{a}_{\mathcal{C},v} &= \bigwedge^{\mathbf{2}}_{v' =_x v} ||\varphi||^{a}_{\mathcal{C},v'}, \\ ||(\exists x)\varphi||^{a}_{\mathcal{C},v} &= \bigvee^{\mathbf{2}}_{a \leq \bigvee_{v' =_x v} b_{v'}} \bigwedge^{\mathbf{2}}_{v' =_x v} ||\varphi||^{b_{v'}}_{\mathcal{C},v'}. \end{aligned}$$

First, we show that $\|\varphi\|^a_{\mathcal{C},v}$ is exactly the a-cut of $\|\varphi\|_{\mathbf{M}_\mathcal{C},v}$.

Lemma 3.139 *For an* $\mathbf{L}$*-structure* $\mathbf{M}$ *and the corresponding* $\mathbf{L}$*-indexed system* $\mathcal{C}_\mathbf{M}$ *we have*

$$^a\|\varphi\|_{\mathbf{M},v} = \|\varphi\|^a_{\mathcal{C}_\mathbf{M},v}.$$

Proof. We prove the assertion by induction over φ.

If φ is atomic (i.e. either $r(t_1,\dots,t_n)$ or a truth constant) then the assertion follows directly by definition and from $\|t_i\|_{\mathcal{C}_\mathbf{M},v} = \|t_i\|_{\mathbf{M},v}$.

Assume that the assertion is valid for φ and ψ.

We have $^a\|\varphi \mathbin{\&} \psi\|_{\mathbf{M},v} = 1$ iff $a \leq \|\varphi \mathbin{\&} \psi\|_{\mathbf{M},v}$ iff $a \leq \|\varphi\|_{\mathbf{M},v}$ and $a \leq \|\psi\|_{\mathbf{M},v}$ iff $^a\|\varphi\|_{\mathbf{M},v} = 1$ and $^a\|\psi\|_{\mathbf{M},v} = 1$ iff $\|\varphi\|^a_{\mathcal{C}_\mathbf{M},v} = 1$ and $\|\psi\|^a_{\mathcal{C}_\mathbf{M},v} = 1$ iff $\|\psi\|^a_{\mathcal{C}_\mathbf{M},v} \wedge^\mathbf{2} \|\psi\|^a_{\mathcal{C}_\mathbf{M},v} = 1$ iff $\|\varphi \mathbin{\&} \psi\|^a_{\mathcal{C}_\mathbf{M},v} = 1$, proving that the assertion is valid for $\varphi \mathbin{\&} \psi$.

For $\Cup$ we have $^a\|\varphi \Cup \psi\|_{\mathbf{M},v} = 1$ iff $a \leq \|\varphi\|_{\mathbf{M},v} \vee \|\psi\|_{\mathbf{M},v}$ iff there exist $b,c \in L$ with $a \leq b \vee c$ and $b \leq \|\varphi\|_{\mathbf{M},v}$, $c \leq \|\psi\|_{\mathbf{M},v}$ iff there exist $b,c \in L$ with $a \leq b \vee c$ and $^b\|\varphi\|_{\mathbf{M},v} = 1$, $^c\|\psi\|_{\mathbf{M},v} = 1$ iff there exist $b,c \in L$ with $a \leq b \vee c$ and $\|\varphi\|^b_{\mathcal{C}_\mathbf{M},v} = 1$, $\|\psi\|_{\mathcal{C}_\mathbf{M},v} = 1$ iff there exist $b,c \in L$ with $a \leq b \vee c$ and $\|\varphi\|^b_{\mathcal{C}_\mathbf{M},v} \wedge^\mathbf{2} \|\psi\|^c_{\mathcal{C}_\mathbf{M},v} = 1$ iff $\|\varphi \Cup \psi\|^a_{\mathcal{C}_\mathbf{M},v} = 1$, showing that the assertion holds true for $\varphi \Cup \psi$.

Consider $\otimes$: $^a\|\varphi \otimes \psi\|_{\mathbf{M},v} = 1$ iff $a \leq \|\varphi\|_{\mathbf{M},v} \otimes \|\psi\|_{\mathbf{M},v}$ iff there exist $b,c \in L$ such that $a \leq b \otimes c$ and $b \leq \|\varphi\|_{\mathbf{M},v}$, $c \leq \|\psi\|_{\mathbf{M},v}$ iff there exist $b,c \in L$ with $a \leq b \otimes c$ and $^b\|\varphi\|_{\mathbf{M},v} = 1$, $^c\|\psi\|_{\mathbf{M},v} = 1$ iff there exist $b,c \in L$ with $a \leq b \otimes c$ and $\|\varphi\|^b_{\mathcal{C}_\mathbf{M},v} = 1$, $\|\psi\|^c_{\mathcal{C}_\mathbf{M},v} = 1$ iff there exist $b,c \in L$ with $a \leq b \otimes c$ and $\|\varphi\|^b_{\mathcal{C}_\mathbf{M},v} \wedge^\mathbf{2} \|\psi\|^c_{\mathcal{C}_\mathbf{M},v} = 1$ iff $\|\varphi \otimes \psi\|^c_{\mathcal{C}_\mathbf{M},v} = 1$.

For $\Rightarrow$, we have $^a\|\varphi \Rightarrow \psi\|_{\mathbf{M},v} = 1$ iff $a \leq \|\varphi \Rightarrow \psi\|_{\mathbf{M},v}$ iff $a \otimes \|\varphi\|_{\mathbf{M},v} \leq \|\psi\|_{\mathbf{M},v}$. On the other hand, $\bigwedge^\mathbf{2}_{b\in L} \|\varphi\|^b_{\mathcal{C}_\mathbf{M},v} \to^\mathbf{2} \|\psi\|^{a\otimes b}_{\mathcal{C}_\mathbf{M},v} = 1$ iff for each $b \in L$, $\|\varphi\|^b_{\mathcal{C}_\mathbf{M},v} \to^\mathbf{2} \|\psi\|^{a\otimes b}_{\mathcal{C}_\mathbf{M},v} = 1$. Therefore, we have to show that $a \otimes \|\varphi\|_{\mathbf{M},v} \leq \|\psi\|_{\mathbf{M},v}$ is true iff for each $b \in L$, $\|\varphi\|^b_{\mathcal{C}_\mathbf{M},v} \to^\mathbf{2} \|\psi\|^{a\otimes b}_{\mathcal{C}_\mathbf{M},v} = 1$. Thus, let $a \otimes \|\varphi\|_{\mathbf{M},v} \leq \|\psi\|_{\mathbf{M},v}$ and take any $b \in L$. If $\|\varphi\|^b_{\mathcal{C}_\mathbf{M},v} = 1$ then, by assumption $b \leq \|\varphi\|_{\mathbf{M},v}$ and so $a \otimes b \leq a \otimes \|\varphi\|_{\mathbf{M},v} \leq \|\psi\|_{\mathbf{M},v}$, whence $^{a\otimes b}\|\psi\|_{\mathbf{M},v} = 1$, thus $\|\psi\|^{a\otimes b}_{\mathcal{C}_\mathbf{M},v} = 1$ by assumption. We thus have $\|\varphi\|^b_{\mathcal{C}_\mathbf{M},v} \to^\mathbf{2} \|\psi\|^{a\otimes b}_{\mathcal{C}_\mathbf{M},v} = 1$. Conversely, suppose that for each $b \in L$ we have $\|\varphi\|^b_{\mathcal{C}_\mathbf{M},v} \to^\mathbf{2} \|\psi\|^{a\otimes b}_{\mathcal{C}_\mathbf{M},v} = 1$. Then for $b = \|\varphi\|_{\mathbf{M},v}$ we particularly have $\|\varphi\|^{\|\varphi\|_{\mathbf{M},v}}_{\mathcal{C}_\mathbf{M},v} \to^\mathbf{2} \|\psi\|^{a\otimes\|\varphi\|_{\mathbf{M},v}}_{\mathcal{C}_\mathbf{M},v} = 1$. Now, since $\|\varphi\|_{\mathbf{M},v} \leq \|\varphi\|_{\mathbf{M},v}$, assumption yields $\|\varphi\|^{\|\varphi\|_{\mathbf{M},v}}_{\mathcal{C}_\mathbf{M},v} = 1$, and thus $\|\psi\|^{a\otimes\|\varphi\|_{\mathbf{M},v}}_{\mathcal{C}_\mathbf{M},v} = 1$. Invoking the assumption again we get $a \otimes \|\varphi\|_{\mathbf{M},v} \leq \|\psi\|_{\mathbf{M},v}$, proving that the assertion is true for $\varphi \Rightarrow \psi$.

For the general quantifier we have $^a\|(\forall x)\varphi\|_{\mathbf{M},v} = 1$ iff $a \leq \|(\forall x)\varphi\|_{\mathbf{M},v}$ iff for each valuation $v' =_x v$ we have $a \leq \|\varphi\|_{\mathbf{M},v'}$ iff for each valuation $v' =_x v$ we have $\|\varphi\|^a_{\mathcal{C}_{\mathbf{M}},v'} = 1$ iff $\bigwedge^{\mathbf{2}}{}_{v'=_x v} \|\varphi\|^a_{\mathcal{C}_{\mathbf{M}},v'} = 1$ iff $\|(\forall x)\varphi\|^a_{\mathcal{C}_{\mathbf{M}},v} = 1$.

Existential quantifier: ${}^a\|(\exists x)\varphi\|_{\mathbf{M},v} = 1$ iff $a \leq \bigvee_{v'=_x v} \|\varphi\|_{\mathbf{M},v'}$ iff there exist $b_{v'} \in L$ over all valuations $v' =_x v$ with $a \leq \bigvee_{v'=_x v} b_{v'}$ such that $b_{v'} \leq \|\varphi\|_{\mathbf{M},v'}$ iff there exist $b_{v'} \in L$ $(v' =_x v)$ with $a \leq \bigvee_{v'=_x v} b_{v'}$ such that $\|\varphi\|^{b_{v'}}_{\mathcal{C}_{\mathbf{M}},v'} = 1$ iff $\bigvee_{b_{v'} \in L, v'=_x v, a \leq \bigvee_{v'=_x v} b_{v'}} \bigwedge_{v'=_x v} \|\varphi\|^{b_{v'}}_{\mathcal{C}_{\mathbf{M}},v'} = 1$ iff $\|(\exists x)\varphi\|^a_{\mathcal{C}_{\mathbf{M}},v} = 1$.

The proof is complete. □

If **L** is linearly ordered, the situation for disjunction and the existential quantifier simplifies.

Lemma 3.140 *For a linearly ordered* **L** *we have*

$$\|\varphi \vee \psi\|^a_{\mathcal{C},v} = \|\varphi\|^a_{\mathcal{C},v} \vee^{\mathbf{2}} \|\psi\|^a_{\mathcal{C},v};$$

if M is, moreover, finite then

$$\|(\exists x)\varphi\|^a_{\mathcal{C},v} = \bigvee_{v'=_x v}^{\mathbf{2}} \|\varphi\|^a_{\mathcal{C},v'}.$$

Proof. Note that for $a \leq b$ we have $\|\varphi\|^a_{\mathcal{C},v} \geq \|\varphi\|^b_{\mathcal{C},v}$ (by Lemma 3.139 and by ${}^a\|\varphi\|_{\mathbf{M},v} \geq {}^b\|\varphi\|_{\mathbf{M},v}$ for $a \leq b$). Due to linearity of **L** we have $\|\varphi \vee \psi\|^a_{\mathcal{C},v} = \bigvee^{\mathbf{2}}{}_{a \leq b \vee c} \|\varphi\|^b_{\mathcal{C},v} \wedge^{\mathbf{2}} \|\psi\|^c_{\mathcal{C},v} = \bigvee^{\mathbf{2}}{}_{a \leq \max(b,c)} \|\varphi\|^b_{\mathcal{C},v} \wedge^{\mathbf{2}} \|\psi\|^c_{\mathcal{C},v} = \bigvee^{\mathbf{2}}{}_{a \leq b \text{ or } a \leq c} \|\varphi\|^b_{\mathcal{C},v} \wedge^{\mathbf{2}} \|\psi\|^c_{\mathcal{C},v} = \|\varphi\|^a_{\mathcal{C},v} \vee^{\mathbf{2}} \|\psi\|^a_{\mathcal{C},v}$. Furthermore, if the support M of $\mathbf{M}$ is finite then $\|(\exists x)\varphi\|^a_{\mathcal{C},v} = \bigvee^{\mathbf{2}}{}_{a \leq \max_{v'=_x v} b_{v'}} \bigwedge^{\mathbf{2}}{}_{v'=_x v} \|\varphi\|^{b_{v'}}_{\mathcal{C},v'} = \bigvee^{\mathbf{2}}{}_{v'=_x v} \|\varphi\|^a_{\mathcal{C},v'}$. □

Now, for an **L**-nested system $\mathcal{C}$ of structures, a valuation v, and a formula φ we define the degree $\|\varphi\|_{\mathcal{C},v}$ to which φ is true in $\mathcal{C}$ by

$$\|\varphi\|_{\mathcal{C},v} = \bigvee \{a \mid \|\varphi\|^a_{\mathcal{C},v} = 1\}.$$

Then we have

Theorem 3.141 *For an* **L**-*structure* **M** *and the corresponding* **L**-*indexed system $\mathcal{C}_{\mathbf{M}}$ we have*

$$\|\varphi\|_{\mathbf{M},v} = \|\varphi\|_{\mathcal{C}_{\mathbf{M}},v}.$$

Proof. The assertion follows directly by Lemma 3.139 using the obvious fact $\|\varphi\|_{\mathbf{M},v} = \bigvee \{a \mid {}^a\|\varphi\|_{\mathbf{M},v} = 1\}$. □

Remark 3.142 Note that Theorem 3.141 provides an alternative way to the definition of semantics. Instead of interpreting a formula φ directly in a fuzzy structure $\mathbf{M}$, it is, in principle, possible to consider subformulas of φ, interpret the subformulas in the crisp structures $\mathbf{M}_a$ to which $\mathbf{M}$ can be decomposed in a natural way (a-cuts), and then to use the (classical) truth values of the subformulas of φ to obtain the truth degree of φ in $\mathbf{M}$.

We will see applications of results elaborated in this section later on (Sections 5.3 and 7.2).

3.6 ALGEBRAS WITH FUZZY EQUALITIES

In this section we consider fuzzy structures with no other fuzzy relations than a fuzzy equality. That is, we consider $\mathbf{L}$-structures $\mathbf{M}$ of type $\langle R, F, \sigma\rangle$ where $R = \{\approx\}$. Moreover, we assume that $\approx^{\mathbf{M}}$ (i.e. the $\mathbf{L}$-equivalence relation interpreting symbol $\approx$) is an $\mathbf{L}$-equality. Such structures will be called *algebras with* $\mathbf{L}$*-equalities* or simply $\mathbf{L}$-algebras. As the type of an $\mathbf{L}$-algebra is uniquely given by $\langle F, \sigma\rangle$ ($R = \{\approx\}$, $\approx$ binary), we may, for simplicity, call a type of an $\mathbf{L}$-algebra the pair $\langle F, \sigma\rangle$. Therefore, an $\mathbf{L}$-algebra of type $\langle F, \sigma\rangle$ is a structure $\mathbf{M} = \langle M, \approx^{\mathbf{M}}, F^{\mathbf{M}}\rangle$ such that $\approx^{\mathbf{M}}$ is an $\mathbf{L}$-equality on M, $F^{\mathbf{M}} = \{f^{\mathbf{M}} \mid f \in F\}$ where $f^{\mathbf{M}}$ is a $\sigma(f)$-ary operation on M satisfying for $n = \sigma(f)$

$$\begin{aligned}&(m_1 \approx^{\mathbf{M}} m_1') \otimes \cdots \otimes (m_n \approx^{\mathbf{M}} m_n') \leq \\ \leq\ &(f^{\mathbf{M}}(m_1, \ldots, m_n) \approx^{\mathbf{M}} f^{\mathbf{M}}(m_1', \ldots, m_n'))\end{aligned}$$

for any $m_i, m_i' \in M$.

Clearly, for $\mathbf{L} = \mathbf{2}$, $\mathbf{L}$-algebras are precisely (universal) algebras. First we show that there are natural examples of $\mathbf{L}$-algebras.

Example 3.143 Let U be a set equipped with an $\mathbf{L}$-equality $\approx^U$. Let $M = S(U)$ be the set of all permutations of U (i.e. bijective mappings on U) which are compatible with $\approx^U$. The triple $\mathbf{M} = \langle M, \approx^{\mathbf{M}}, \circ^{\mathbf{M}}\rangle$ where $\pi \approx^{\mathbf{M}} \pi' = \bigwedge_{u\in U}(\pi(u) \approx^U \pi'(u))$ and $\circ^{\mathbf{M}}$ denotes the composition of permutations, is an $\mathbf{L}$-algebra. Indeed, Lemma 3.76 implies that $\approx^{\mathbf{M}}$ is an $\mathbf{L}$-equality on M. To verify that $\circ^{\mathbf{M}}$ is compatible with $\approx^{\mathbf{M}}$ take any $\pi, \pi', \rho, \rho' \in M$. We have $(\pi \approx^{\mathbf{M}} \pi') \otimes (\rho \approx^{\mathbf{M}} \rho') \leq (\pi \circ \rho \approx^{\mathbf{M}} \pi' \circ \rho')$ iff for each $u \in U$ we have $(\pi \approx^{\mathbf{M}} \pi') \otimes (\rho \approx^{\mathbf{M}} \rho') \leq (\rho(\pi(u)) \approx^U \rho'(\pi'(u)))$ which is true:

$$\begin{aligned}&(\pi \approx^{\mathbf{M}} \pi') \otimes (\rho \approx^{\mathbf{M}} \rho') \leq \\ \leq\ &(\pi(u) \approx^U \pi'(u)) \otimes (\rho(\pi'(u)) \approx^U \rho'(\pi'(u))) \leq\end{aligned}$$

$$\begin{aligned}&\leq (\rho(\pi(u)) \approx^U \rho(\pi'(u))) \otimes (\rho(\pi'(u)) \approx^U \rho'(\pi'(u))) \leq \\ &\leq (\rho(\pi(u)) \approx^U \rho'(\pi'(u))).\end{aligned}$$

Needless to say, since algebras with fuzzy equalities are special cases of fuzzy structures, all notions and results introduced earlier apply. We will be, however, interested in a particular case only. Namely, we will keep $\mathbf{L}$ fixed. More precisely, we take an arbitrary complete residuated lattice $\mathbf{L}$, fix it, and consider only algebras with fuzzy equalities with $\mathbf{L}$ as the structure of truth values. Moreover, if $h = \langle h_M, h_L \rangle \in \mathrm{Adm}_{\mathrm{hom}}(\mathbf{M}_1, \mathbf{M}_2)$ where $\mathbf{M}_1$ and $\mathbf{M}_2$ are $\mathbf{L}$-algebras, we suppose that $h_L : L \to L$ is the identity. Similarly, if $\theta = \langle \theta_M, \theta_L \rangle \in \mathrm{Adm}_{\mathrm{con}}(\mathbf{M}_1, \mathbf{M}_2)$, we suppose that θ_L is the identity on L.

We will be interested mainly in morphisms (and congruences), products, and substructures of algebras with fuzzy equalities. Since we consider only a particular case (fixed $\mathbf{L}$), our notions may be simplified: Thus, a morphism of an $\mathbf{L}$-algebra $\mathbf{M}_1$ to an $\mathbf{L}$-algebra $\mathbf{M}_2$ will be understood as a mapping $h : M_1 \to M_2$ such that $(m \approx^{\mathbf{M}_1} m') \leq (h(m) \approx^{\mathbf{M}_2} h(m'))$ and $h(f^{\mathbf{M}_1}(m_1, \dots, m_n)) = f^{\mathbf{M}_2}(h(m_1), \dots, h(m_n))$ for each n-ary $f \in F$. Recall that a congruence on an $\mathbf{L}$-algebra $\mathbf{M}$ is a pair $\langle \theta_M, \theta_L \rangle \in \mathrm{Adm}_{\mathrm{con}}(\mathbf{M}_1, \mathbf{M}_2)$ (θ_L is a complete congruence on $\mathbf{L}$ and θ_M is an $\mathbf{L}/\theta_L$-equivalence on M satisfying $\theta_M(m_1, m_2) \otimes [m_1 \approx^{\mathbf{M}} m_1']_{\theta_L} \otimes [m_2 \approx^{\mathbf{M}} m_2']_{\theta_L} \leq \theta_M(m_1', m_2')$) such that $\theta_M(m_1, m_1') \otimes \cdots \otimes \theta_M(m_n, m_n') \leq \theta_M(f^{\mathbf{M}}(m_1, \dots, m_n), f^{\mathbf{M}}(m_1', \dots, m_n'))$. A factor structure of an $\mathbf{L}$-algebra $\mathbf{M}$ by a congruence $\theta = \langle \theta_M, \theta_L \rangle$ is an $\mathbf{L}/\theta_L$-algebra as defined before. Now, since we suppose that θ_L is the identity on L, we modify the above notions, without changing the point, as follows: By a congruence on an $\mathbf{L}$-algebra $\mathbf{M}$ we understand an $\mathbf{L}$-equivalence relation θ on M satisfying $\theta(m_1, m_2) \otimes (m_1 \approx^{\mathbf{M}} m_1') \otimes (m_2 \approx^{\mathbf{M}} m_2') \leq \theta_M(m_1', m_2')$ for $m_i, m_i' \in M$, and $\theta_M(m_1, m_1') \otimes \cdots \otimes \theta_M(m_n, m_n') \leq \theta_M(f^{\mathbf{M}}(m_1, \dots, m_n), f^{\mathbf{M}}(m_1', \dots, m_n'))$ for each n-ary $f \in F$ and $m_i, m_i' \in M$. A factor structure of an $\mathbf{L}$-algebra $\mathbf{M}$ by a congruence θ on $\mathbf{M}$ will be understood to be an $\mathbf{L}$-structure $\mathbf{M}/\theta$ defined as follows: M/θ is $M/^1\theta$ (with elements $[m]_{^1\theta}$ denoted also simply by $[m]_\theta$ or even $[m]$); $([m]_\theta \approx^{\mathbf{M}} [m']_\theta) = (m \approx^{\mathbf{M}} m')$; $f^{\mathbf{M}/\theta}([m_1]_\theta, \dots, [m_2]_\theta) = [f^{\mathbf{M}}(m_1, \dots, m_n)]_\theta$. We can easily see that the thus-defined notions are indeed only accommodations of the general notions defined above to our particular case (fixed structure of truth values).

We are not going to investigate algebras with fuzzy equalities in the full breadth. Rather, we confine ourselves to a particular topic. Atomic formulas in the language of algebras with fuzzy equalities are (truth constants and) "identities" i.e. formulas $t_1 \approx t_2$ where t_1, t_2 are terms. In the following, we will be basically interested in two points:

(1) Given a fuzzy set Σ of identities, what is the class of all algebras $\mathbf{M}$ with fuzzy equalities satisfying each identity of Σ (i.e. $\mathbf{M}$ such that the degree to which any identity $t_1 \approx t_2$ is true in $\mathbf{M}$ is at least the degree to which $t_1 \approx t_2$ belongs to Σ)? We will see that such classes are exactly those ones closed under morphisms, substructures, and direct products.

(2) Given a fuzzy set Σ of identities, is there a simple set of syntactic rules for "fuzzy equational reasoning" that allow us to obtain the degree to which a given identity "follows" from Σ? We will see that there is such a set consisting of five elementary rules.

By X we will denote a set of variables, which may be a proper subset of the set of all variables. We denote by $T(X)$ the set of all terms over a set X of variables (i.e. a set of all terms t such that $\mathrm{var}(t) \subseteq X$). $T(X)$ can be naturally made into an $\mathbf{L}$-algebra $\mathbf{T}(X) = \langle T(X), \approx^{\mathbf{T}(X)}, F^{\mathbf{T}(X)} \rangle$: The support of $\mathbf{T}(X)$ is $T(X)$; functions are defined by

$$f^{\mathbf{T}(X)}(t_1, \ldots, t_n) = f(t_1, \ldots, t_n)$$

for any $t_1, \ldots, t_n \in T(X)$; $\mathbf{L}$-equality $\approx^{\mathbf{T}(X)}$ is defined by

$$(t_1 \approx^{\mathbf{T}(X)} t_2) = \begin{cases} 1 & \text{for } t_1 = t_2, \\ 0 & \text{for } t_1 \neq t_2. \end{cases}$$

Then $\approx^{\mathbf{T}(X)}$ is a crisp $\mathbf{L}$-equality and each $f^{\mathbf{T}(X)}$ is compatible with $\approx^{\mathbf{T}(X)}$. Note that $\mathbf{T}(X)$ exists whenever X is nonempty or there is some nullary $f \in F$; in the following we always assume that $\mathbf{T}(X)$ exists.

3.6.1 VARIETY THEOREM IN FUZZY SETTING

For a class $\mathcal{K}$ of $\mathbf{L}$-algebras of the same type we define four operators: $\mathrm{H}, \mathrm{I}, \mathrm{S}, \mathrm{P}$. $\mathrm{H}(\mathcal{K})$ is the class of all homomorphic images of $\mathbf{L}$-algebras from $\mathcal{K}$, i.e.

$$\mathrm{H}(\mathcal{K}) = \{h(\mathbf{M}) \mid \mathbf{M} \in \mathcal{K},\ h \text{ a morphism }\};$$

$\mathrm{I}(\mathcal{K})$ is the class of all $\mathbf{L}$-algebras isomorphic to some $\mathbf{M} \in \mathcal{K}$, i.e.

$$\mathrm{I}(\mathcal{K}) = \{\mathbf{N} \mid \mathbf{N} \text{ is isomorphic to some } \mathbf{M} \in \mathcal{K}\};$$

$\mathrm{S}(\mathcal{K})$ is the class of all substructures of $\mathbf{L}$-algebras from $\mathcal{K}$, i.e.

$$\mathrm{S}(\mathcal{K}) = \{\mathbf{N} \mid \mathbf{N} \text{ is a substructure of some } \mathbf{M} \in \mathcal{K}\};$$

$\mathrm{P}(\mathcal{K})$ is the class of all direct products of $\mathbf{L}$-algebras from $\mathcal{K}$, i.e.

$$\mathrm{P}(\mathcal{K}) = \{\times_{i\in I}\mathbf{M}_i \mid I \text{ an index set}, \mathbf{M}_i \in \mathcal{K}\}.$$

The operators may be composed, i.e. we may have $\mathrm{HS}(\mathcal{K})$, $\mathrm{HHPHS}(\mathcal{K})$, etc.

Definition 3.144 A class $\mathcal{K}$ of $\mathbf{L}$-algebras of the same type is called a *variety* iff it is closed under H, S, and P (i.e. if $\mathrm{H}(\mathcal{K}) \subseteq \mathcal{K}$, $\mathrm{S}(\mathcal{K}) \subseteq \mathcal{K}$, and $\mathrm{P}(\mathcal{K}) \subseteq \mathcal{K}$).

Let $\mathcal{K}$ be a class of $\mathbf{L}$-algebras of the same type, let $\mathbf{M}$ be an $\mathbf{L}$-algebra generated by $M' \subseteq M$ (i.e. $[M'] = M$). If for each $\mathbf{N} \in \mathcal{K}$ and for each mapping $g : M' \to N$ preserving $\approx$ (i.e. such that $(m \approx^{\mathbf{M}} m') \leq (g(m) \approx^{\mathbf{N}} g(n)))$ there exists a morphism $h : \mathbf{M} \to \mathbf{N}$ extending g (i.e. $h(m) = g(m)$ for each $m \in M'$), we say that $\mathbf{M}$ has a *universal mapping property for* $\mathcal{K}$ *over* M'; in this case M' is said to be a set of free generators of $\mathbf{M}$ over $\mathcal{K}$.

Lemma 3.145 *If* $\mathbf{M}$ *has a universal mapping property for* $\mathcal{K}$ *over* M' *and* $\mathbf{N} \in \mathcal{K}$ *then for any* $g : M' \to N$ *there exists a unique morphism* $h : M \to N$ *extending* g.

Proof. Follows directly from Remark 3.132 (3). □

Lemma 3.146 $\mathbf{T}(X)$ *has a universal mapping property for any class* $\mathcal{K}$ *of* $\mathbf{L}$*-algebras of a given type over* X.

Proof. For any $\mathbf{M} \in \mathcal{K}$ and any mapping $g : X \to M$, define $h : T(X) \to M$ inductively by $h(x) = g(x)$ for $x \in X$; $h(f(t_1, \dots, t_n)) = f^{\mathbf{M}}(h(t_1), \dots, h(t_n))$ for $f \in F$ and $t_i \in T(X)$. Trivially, h is a morphism. □

Let $\mathcal{K}$ be a class of $\mathbf{L}$-algebras of the same type, let X be a set of variables. Put

$$\Phi_{\mathcal{K}}(X) = \{\phi \in {}^1\mathrm{Con}(\mathbf{T}(X)) \mid \mathbf{T}(X)/\phi \in \mathrm{IS}(\mathcal{K})\},$$

i.e. $\Phi_{\mathcal{K}}$ is the set of all congruences ϕ on $\mathbf{T}(X)$ such that the factor $\mathbf{L}$-algebra $\mathbf{T}(X)/\phi$ is isomorphic to some substructure of some $\mathbf{M} \in \mathcal{K}$. Denote furthermore

$$\theta_{\mathcal{K}}(X) = \bigcap \Phi_{\mathcal{K}}(X)$$

the intersection of all congruences from $\Phi_{\mathcal{K}}(X)$.

By Theorem 3.113, $\theta_{\mathcal{K}}(X)$ is a congruence on $\mathbf{T}(X)$. We may thus form a factor $\mathbf{L}$-algebra $\mathbf{T}(X)/\theta_{\mathcal{K}}(X)$. For $x \in X$, denote by $\overline{x}$ the class $[x]_{^1\theta_{\mathcal{K}}(X)}$ and put $\overline{X} = \{\overline{x} \mid x \in X\}$. It is almost immediate that $\overline{X}$ is the set of generators of $\mathbf{T}(X)/\theta_{\mathcal{K}}(X)$. For convenience, we denote $\mathbf{T}(X)/\theta_{\mathcal{K}}(X)$ by $\mathcal{F}_{\mathcal{K}}(\overline{X})$.

Definition 3.147 $\mathcal{F}_{\mathcal{K}}(\overline{X})$ defined above is called a $\mathcal{K}$-free $\mathbf{L}$-algebra over $\overline{X}$.

Lemma 3.148 *$\mathcal{F}_{\mathcal{K}}(\overline{X})$ has a universal mapping property for $\mathcal{K}$ over $\overline{X}$.*

Proof. Let $\mathbf{M} \in \mathcal{K}$ and take a mapping $g : \overline{X} \to M$. Let $n : \mathbf{T}(X) \to \mathcal{F}_{\mathcal{K}}(\overline{X})$ denote the natural morphism (i.e. $n(t) = [t]_{^1\theta_{\mathcal{K}}(X)}$). Letting n_X denote the restriction of n to X, universal mapping property of $\mathbf{T}(X)$ implies that there is a morphism $k : \mathbf{T}(X) \to \mathbf{M}$ extending $n_X \circ g$. As $\mathbf{T}(X)/\theta_k$ is isomorphic to $k(\mathbf{T}(X))$ which is a substructure of $\mathbf{M}$, definition of $\theta_{\mathcal{K}}(X)$ implies that $\theta_{\mathcal{K}}(X) \subseteq \theta_k$.

We claim that there is a morphism $h : \mathcal{F}_{\mathcal{K}}(\overline{X}) \to \mathbf{M}$ such that $n \circ h = k$. Indeed, let $h = h_{\theta_k/\theta_{\mathcal{K}}(X)} \circ i_1 \circ i_2$ where $i_1 : [t]_{^1\theta_k/\theta_{\mathcal{K}}(X)} \mapsto [t]_{^1\theta_k}$ is an isomorphism due to Theorem 3.119 and $i_2 : \mathbf{T}(X)/\theta_k \to k(\mathbf{T}(X))$ is an isomorphism due to Theorem 3.118. Then $n \circ h(t) = h([t]_{^1\theta_{\mathcal{K}}(X)}) = i_2(i_1(h_{\theta_k/\theta_{\mathcal{K}}(X)}([t]_{^1\theta_{\mathcal{K}}(X)}))) = k(t)$.

Now, we have $h(\overline{x}) = h(n(x)) = n \circ h(x) = k(x) = n_X \circ g = g(\overline{x})$ showing that h extends g. □

Lemma 3.149 *If $\mathbf{T}(X)$ exists then for $\mathcal{K} \neq \emptyset$ we have $\mathcal{F}_{\mathcal{K}}(\overline{X}) \in$ ISP$(\mathcal{K})$. Thus, particularly, if $\mathcal{K}$ is a variety then $\mathcal{F}_{\mathcal{K}}(\overline{X}) \in \mathcal{K}$.*

Proof. First, we claim that $\mathcal{F}_{\mathcal{K}}(\overline{X})$ is isomorphic to a subalgebra of a direct product of $\mathbf{T}(X)/\phi$ for $\phi \in \Phi_{\mathcal{K}}(X)$, i.e. we claim that $\mathcal{F}_{\mathcal{K}}(\overline{X}) \in$ ISP$(\{\mathbf{T}(X)/\phi \mid \phi \in \Phi_{\mathcal{K}}\})$. Recall that $\mathcal{F}_{\mathcal{K}}(\overline{X}) = \mathbf{T}(X)/\theta_{\mathcal{K}}(X)$ where $\theta_{\mathcal{K}}(X) = \bigcap \Phi_{\mathcal{K}}(X)$. We verify that the mapping

$$i : \mathcal{F}_{\mathcal{K}}(\overline{X}) \to \times_{\phi \in \Phi_{\mathcal{K}}(X)} \mathbf{T}(X)/\phi$$

sending $[t]_{^1\theta_{\mathcal{K}}(X)}$ to $\langle \ldots, [t]_{^1\phi}, \ldots \rangle$ is an isomorphism of $\mathcal{F}_{\mathcal{K}}(\overline{X})$ to the subalgebra $i(\mathcal{F}_{\mathcal{K}}(\overline{X}))$ of $\times_{\phi \in \Phi_{\mathcal{K}}(X)} \mathbf{T}(X)/\phi$. i is an injection: if $i([t_1]_{^1\theta_{\mathcal{K}}(X)}) = i([t_2]_{^1\theta_{\mathcal{K}}(X)})$ then for each $\phi \in \Phi_{\mathcal{K}}(X)$ we have $[t_1]_{^1\phi} = [t_2]_{^1\phi}$, i.e. $\phi(t_1, t_2) = 1$ and thus also $(\theta_{\mathcal{K}}(X))(t_1, t_2) = 1$, i.e. $[t_1]_{^1\theta_{\mathcal{K}}(X)} = [t_2]_{^1\theta_{\mathcal{K}}(X)}$. Furthermore,

$$(\langle \ldots, [t_1]_{^1\phi}, \ldots \rangle \approx^{\times \mathbf{T}(X)/\phi} \langle \ldots, [t_2]_{^1\phi}, \ldots \rangle) =$$

$$= \bigwedge_{\phi \in \Phi_{\mathcal{K}}(X)} \phi(t_1, t_2) = (\bigwedge_{\phi \in \Phi_{\mathcal{K}}(X)} \phi)(t_1, t_2) = \theta_{\mathcal{K}}(X)(t_1, t_2) =$$
$$= ([t_1]_{^1\theta_{\mathcal{K}}(X)} \approx^{\mathcal{F}_{\mathcal{K}}(\overline{X})} [t_2]_{^1\theta_{\mathcal{K}}(X)}).$$

Obviously, $\mathrm{Hom}(i) = 1$. Therefore, i is an isomorphism of $\mathcal{F}_{\mathcal{K}}(\overline{X})$ to $i(\mathcal{F}_{\mathcal{K}}(\overline{X}))$.

Now, by definition, for each $\phi \in \Phi_{\mathcal{K}}(X)$, $\mathbf{T}(X)/\phi$ belongs to $\mathrm{IS}(\mathcal{K})$. To sum up, we have $\mathcal{F}_{\mathcal{K}}(\overline{X}) \in \mathrm{ISPIS}(\mathcal{K})$.

We need to show that $\mathrm{ISPIS}(\mathcal{K}) \subseteq \mathrm{ISP}(\mathcal{K})$. This is easy: we have $\mathrm{ISPIS}(\mathcal{K}) = \mathrm{ISPS}(\mathcal{K}) \subseteq \mathrm{ISSP}(\mathcal{K}) = \mathrm{ISP}(\mathcal{K})$. □

Recall that for a class $\mathcal{K}$ of $\mathbf{L}$-algebras, $\mathrm{Th}(\mathcal{K})$ is the $\mathbf{L}$-set of formulas where $(\mathrm{Th}(\mathcal{K}))(\varphi)$ is the truth degree to which φ is true in all $\mathbf{M} \in \mathcal{K}$. We put

$$\mathrm{Id}_X(\mathcal{K}) = \mathrm{Th}(\mathcal{K}) \cap \{t_1 \approx t_2 \mid t_1, t_2 \in T(X)\},$$

i.e. $\mathrm{Id}_X(\mathcal{K})$ is the $\mathbf{L}$-set of identities over X true in $\mathcal{K}$. If X is the set of all variables we also write only $\mathrm{Id}(\mathcal{K})$. Conversely, for an $\mathbf{L}$-set Σ of identities, ${}^1\mathrm{Mod}(\Sigma)$ is the set of all $\mathbf{L}$-algebras such that for each identity $t_1 \approx t_2$ we have $\Sigma(t_1 \approx t_2) \leq \|t_1 \approx t_2\|_{\mathbf{M}}$.

Lemma 3.150 *For a class $\mathcal{K}$ of $\mathbf{L}$-algebras and $t_1, t_2 \in T(X)$ we have*

$$\|t_1 \approx t_2\|_{\mathcal{K}} = \bigwedge_{\mathbf{M} \in \mathcal{K}} \bigwedge_{h: \mathbf{T}(X) \to \mathbf{M}} (h(t_1) \approx^{\mathbf{M}} h(t_2)).$$

Proof. By definition we have

$$\|t_1 \approx t_2\|_{\mathcal{K}} = \bigwedge_{\mathbf{M} \in \mathcal{K}} \bigwedge_{v: X \to M} (\|t_1\|_{\mathbf{M},v} \approx^{\mathbf{M}} \|t_2\|_{\mathbf{M},v}).$$

The assertion follows from the fact that there is a bijective correspondence between morphisms $h : \mathbf{T}(X) \to \mathbf{M}$ and valuations $v : X \to M$ (this follows from the universal mapping property of $\mathbf{T}(X)$, Lemma 3.146) and from the fact that for $h : \mathbf{T}(X) \to \mathbf{M}$ and v being the restriction of h to X we have $h(t_i) = \|t_i\|_{\mathbf{M},v}$. □

Lemma 3.151 *For a class $\mathcal{K}$ of $\mathbf{L}$-algebras we have*

$$\mathrm{Id}_X(\mathcal{K}) = \mathrm{Id}_X(\mathrm{I}(\mathcal{K})) = \mathrm{Id}_X(\mathrm{H}(\mathcal{K})) =$$
$$= \mathrm{Id}_X(\mathrm{S}(\mathcal{K})) = \mathrm{Id}_X(\mathrm{P}(\mathcal{K})).$$

Proof. First we show $\mathrm{Id}_X(\mathcal{K}) = \mathrm{Id}_X(\mathrm{I}(\mathcal{K}))$. Since $\mathcal{K} \subseteq \mathrm{I}(\mathcal{K})$ we have $\mathrm{Id}_X(\mathcal{K}) \supseteq \mathrm{Id}_X(\mathrm{I}(\mathcal{K}))$. Conversely, by Lemma 3.150, we have to show that for each $t_1, t_2 \in T(X)$, each isomorphism $g : \mathbf{M} \to \mathbf{N}$ where $\mathbf{M} \in \mathcal{K}$ and each morphism $h : \mathbf{T}(X) \to \mathbf{N}$ we have

$$(\mathrm{Id}_X(\mathcal{K}))(t_1 \approx t_2) \leq (h(t_1) \approx^{\mathbf{N}} h(t_2)).$$

However, this is true since morphisms $h : \mathbf{T}(X) \to \mathbf{N}$ are in a bijective correspondence with morphisms of $\mathbf{T}(X)$ to $\mathbf{M}$ (h corresponds to $h \circ g^{-1}$), we have $(h(t_1) \approx^{\mathbf{M}} h(t_2)) = (h \circ g^{-1}(t_1) \approx^{\mathbf{M}} h \circ g^{-1}(t_2))$, and one can apply Lemma 3.150.

Next, since $\mathcal{K} \subseteq \mathcal{O}(\mathcal{K})$ we have $\mathrm{Id}_X(\mathcal{K}) \supseteq \mathrm{Id}_X(\mathcal{O}(\mathcal{K}))$ for $\mathcal{O} = \mathrm{H}$, $\mathcal{O} = \mathrm{S}$, or $\mathcal{O} = \mathrm{P}$ (in fact, we have $\mathcal{K} \subseteq \mathrm{IP}(\mathcal{K})$; however, since $\mathrm{Id}_X(\mathcal{K}) = \mathrm{Id}_X(\mathrm{I}(\mathcal{K}))$, we may afford to neglect this). We thus need to establish the converse inclusions, i.e. to verify $(\mathrm{Id}_X(\mathcal{K}))(t_1 \approx t_2) \leq (\mathrm{Id}_X(\mathcal{O}(\mathcal{K})))(t_1 \approx t_2)$.

For $\mathrm{H}(\mathcal{K})$, we need to show that for each $\mathbf{M} \in \mathcal{K}$, morphism $h : \mathbf{M} \to \mathbf{N}$, $\mathbf{N} = h(\mathbf{M})$, and a valuation $v : X \to N$ we have $(\mathrm{Id}_X(\mathcal{K}))(t_1 \approx t_2) \leq (\|t_1\|_{\mathbf{N},v} \approx^{\mathbf{N}} \|t_2\|_{\mathbf{N},v})$. Due to surjectivity of h we may take $w : X \to M$ such that $h(w(x)) = v(x)$. Then we have

$$\begin{aligned}
&(\mathrm{Id}_X(\mathcal{K}))(t_1 \approx t_2) \leq (\|t_1\|_{\mathbf{M},w} \approx^{\mathbf{M}} \|t_2\|_{\mathbf{M},w}) \leq \\
\leq\ & (h(\|t_1\|_{\mathbf{M},w}) \approx^{\mathbf{N}} h(\|t_2\|_{\mathbf{M},w})) = (\|t_1\|_{\mathbf{M},w\circ h} \approx^{\mathbf{N}} \|t_2\|_{\mathbf{M},w\circ h}) = \\
=\ & (\|t_1\|_{\mathbf{M},v} \approx^{\mathbf{N}} \|t_2\|_{\mathbf{M},v}).
\end{aligned}$$

For $\mathrm{S}(\mathcal{K})$, we need to show that for each substructure $\mathbf{N}$ of some $\mathbf{M} \in \mathcal{K}$ and each valuation $v : X \to N$ we have $(\mathrm{Id}_X(\mathcal{K}))(t_1 \approx t_2) \leq (\|t_1\|_{\mathbf{N},v} \approx^{\mathbf{N}} \|t_2\|_{\mathbf{N},v})$. However, this is true since $(\|t_1\|_{\mathbf{N},v} \approx^{\mathbf{N}} \|t_2\|_{\mathbf{N},v}) = (\|t_1\|_{\mathbf{M},v} \approx^{\mathbf{M}} \|t_2\|_{\mathbf{M},v})$.

For $\mathrm{P}(\mathcal{K})$, we need to show that for each $\mathbf{N} = \times_{i\in I}\mathbf{M}_i$ ($\mathbf{M}_i \in \mathcal{K}$) and each valuation $v : X \to N$ we have $(\mathrm{Id}_X(\mathcal{K}))(t_1 \approx t_2) \leq (\|t_1\|_{\mathbf{N},v} \approx^{\mathbf{N}} \|t_2\|_{\mathbf{N},v})$ which is true iff for each $i \in I$ we have $(\mathrm{Id}_X(\mathcal{K}))(t_1 \approx t_2) \leq (\|t_1\|_{\mathbf{N},v(i)} \approx^{\mathbf{M}_i} \|t_2\|_{\mathbf{N},v(i)})$ which is true. □

Lemma 3.152 *For a class $\mathcal{K}$ of $\mathbf{L}$-algebras and $t_1, t_2 \in T(X)$ we have*

$$\|t_1 \approx t_2\|_{\mathcal{K}} = \|t_1 \approx t_2\|_{\mathcal{F}_{\mathcal{K}}(\overline{X})} = ([t_1]_{^1\theta_{\mathcal{K}}} \approx^{\mathcal{F}_{\mathcal{K}}(\overline{X})} [t_2]_{^1\theta_{\mathcal{K}}}) = \theta_{\mathcal{K}}(t_1, t_2).$$

Proof. For convenience, denote $\mathcal{F}_{\mathcal{K}}(\overline{X})$ by $\mathcal{F}$. We prove the assertion by showing $\|t_1 \approx t_2\|_{\mathcal{K}} \leq \|t_1 \approx t_2\|_{\mathcal{F}} \leq ([t_1]_{^1\theta_{\mathcal{K}}} \approx^{\mathcal{F}} [t_2]_{^1\theta_{\mathcal{K}}}) \leq \theta_{\mathcal{K}}(t_1, t_2) \leq \|t_1 \approx t_2\|_{\mathcal{K}}$.

$\|t_1 \approx t_2\|_{\mathcal{K}} \leq \|t_1 \approx t_2\|_{\mathcal{F}}$ follows from $\mathcal{F} \in \mathrm{ISP}(\mathcal{K})$ (Lemma 3.149) and Lemma 3.151.

$||t_1 \approx t_2||_{\mathcal{F}} \leq ([t_1]_{^1\theta_{\mathcal{K}}} \approx^{\mathcal{F}} [t_2]_{^1\theta_{\mathcal{K}}})$ is true since for $v' : X \to \mathcal{F}$ sending x to $[x]_{^1\theta_{\mathcal{K}}}$ we have

$$\begin{aligned} ||t_1 \approx t_2||_{\mathcal{F}} = & \bigwedge_{v:X\to\mathcal{F}} (||t_1||_{\mathcal{F},v} \approx^{\mathcal{F}} ||t_2||_{\mathcal{F},v}) \leq \\ \leq & (||t_1||_{\mathcal{F},v'} \approx^{\mathcal{F}} ||t_2||_{\mathcal{F},v'}) = ([t_1]_{^1\theta_{\mathcal{K}}} \approx^{\mathcal{F}} [t_2]_{^1\theta_{\mathcal{K}}}). \end{aligned}$$

$([t_1]_{^1\theta_{\mathcal{K}}} \approx^{\mathcal{F}} [t_2]_{^1\theta_{\mathcal{K}}}) \leq \theta_{\mathcal{K}}(t_1, t_2)$ holds true by definition since $\mathcal{F} = \mathbf{T}(X)/\theta_{\mathcal{K}}(X)$.

$\theta_{\mathcal{K}}(t_1, t_2) \leq ||t_1 \approx t_2||_{\mathcal{K}}$ is true iff for each $\mathbf{M} \in \mathcal{K}$ and each valuation $v : X \to M$ we have $\theta_{\mathcal{K}}(t_1, t_2) \leq (||t_1||_{\mathbf{M},v} \approx^{\mathbf{M}} ||t_2||_{\mathbf{M},v})$. However, due to the universal mapping property of $\mathbf{T}(X)$, v can be extended to a morphism $h : \mathbf{T}(X) \to \mathbf{M}$ for which we have $\theta_h(t_1, t_2) = (||t_1||_{\mathbf{M},v} \approx^{\mathbf{M}} ||t_2||_{\mathbf{M},v})$. As $\mathbf{T}(X)/\theta_h \in \mathrm{IS}(\mathcal{K})$, definition of $\theta_{\mathcal{K}}(X)$ gives $\theta_{\mathcal{K}}(X) \subseteq \theta_h$. The required inequality now readily follows. □

Lemma 3.153 *For an infinite set Y of variables we have* $\mathrm{Id}_X(\mathcal{K}) = \mathrm{Id}_X(\mathcal{F}_{\mathcal{K}})(\overline{Y})$.

Proof. Pick $Z \supseteq X$ such that $|Z| = |Y|$. Then, obviously, $\mathcal{F}_{\mathcal{X}}(\overline{Y})$ is isomorphic to $\mathcal{F}_{\mathcal{X}}(\overline{Z})$. Furthermore, for any identity $p \approx q$ where $p, q \in T(X)$ we have $p, q \in T(Z)$, whence by Lemma 3.152, $[\mathrm{Id}_X(\mathcal{K})](p \approx q) = ||p \approx q||_{\mathcal{K}} = ||p \approx q||_{\mathcal{F}_{\mathcal{K}}(\overline{Z})} = ||p \approx q||_{\mathcal{F}_{\mathcal{K}}(\overline{Z})}$. □

Definition 3.154 A class $\mathcal{K}$ of $\mathbf{L}$-algebras of the same type is called *equational* if there is an $\mathbf{L}$-set Σ of $\mathbf{L}$-identities such that $\mathcal{K} = {}^1\mathrm{Mod}(\Sigma)$.

Lemma 3.155 *If $\mathcal{V}$ is a variety of* $\mathbf{L}$*-algebras and X an infinite set of variables, then* $\mathcal{V} = {}^1\mathrm{Mod}(\mathrm{Id}_X(\mathcal{V}))$.

Proof. Denote $\mathcal{V}' = {}^1\mathrm{Mod}(\mathrm{Id}_X(\mathcal{V}))$. $\mathcal{V}'$ is a variety. Indeed, this follows from Lemma 3.151: Let $\mathcal{O}$ denote any of H, S, or P. For any identity $t_1 \approx t_2$ we have $||t_1 \approx t_2||_{\mathcal{O}(\mathcal{V}')} = ||t_1 \approx t_2||_{\mathcal{V}'} \geq (\mathrm{Id}_X(\mathcal{V}))(t_1 \approx t_2)$ which means that for any $\mathbf{M} \in \mathcal{O}(\mathcal{V}')$ we have $||t_1 \approx t_2||_{\mathbf{M}} \geq (\mathrm{Id}_X(\mathcal{V}))(t_1 \approx t_2)$ whence $\mathbf{M} \in \mathcal{V}'$. Obviously, $\mathcal{V} \subseteq \mathcal{V}'$. This implies $\mathrm{Id}_X(\mathcal{V}') \subseteq \mathrm{Id}_X(\mathcal{V})$ (cf. Remark 3.73). Conversely, $\mathrm{Id}_X(\mathcal{V}) \subseteq \mathrm{Id}_X(\mathcal{V}')$ is true iff for each $\mathbf{M} \in \mathcal{V}'$ we have $(\mathrm{Id}_X(\mathcal{V}))(t_1 \approx t_2) \leq ||t_1 \approx t_2||_{\mathbf{M}}$ which is true by definition of $\mathcal{V}'$. We thus have $\mathrm{Id}_X(\mathcal{V}') = \mathrm{Id}_X(\mathcal{V})$.

By Lemma 3.152, $\mathcal{F}_{\mathcal{V}}(\overline{X}) = \mathcal{F}_{\mathcal{V}'}(\overline{X})$. For each infinite set Y of variables we have by Lemma 3.153 $\mathrm{Id}_Y(\mathcal{V}') = \mathrm{Id}_Y(\mathcal{F}_{\mathcal{V}'})(\overline{X}) = \mathrm{Id}_Y(\mathcal{F}_{\mathcal{V}})(\overline{X}) = \mathrm{Id}_Y(\mathcal{V})$. Lemma 3.152 thus implies $\theta_{\mathcal{V}'}(Y) = \theta_{\mathcal{V}}(Y)$, i.e. $\mathcal{F}_{\mathcal{V}'}(Y) =$

$\mathcal{F}_{\mathcal{V}}(Y)$. Now, for each $\mathbf{M} \in \mathcal{V}'$ we have for some infinite Y that $\mathbf{M} \in \mathrm{H}(\mathcal{F}_{\mathcal{V}'}(\overline{Y}))$ (indeed, it suffices to have $|Y| \geq M$ and to consider a morphism induced by some surjection of Y to M) and thus $\mathbf{M} \in \mathrm{H}(\mathcal{F}_{\mathcal{V}}(\overline{Y}))$ whence $\mathbf{M} \in \mathcal{V}$. We thus established $\mathcal{V}' \subseteq \mathcal{V}$ and so $\mathcal{V} = \mathcal{V}'$. □

Theorem 3.156 (variety theorem) *A class $\mathcal{K}$ of $\mathbf{L}$-algebras of a given type is equational iff it is a variety.*

Proof. Let $\mathcal{K}$ be equational, i.e. $\mathcal{K} = {}^1\mathrm{Mod}(\Sigma)$ for some $\mathbf{L}$-set Σ of identities. Using Lemma 3.151, one can check that $\mathcal{K}$ is a variety (the same arguments as in the beginning of proof of Lemma 3.155).

Conversely, if $\mathcal{V}$ is a variety then Lemma 3.155 implies that for an infinite set X of variables and $\Sigma = \mathrm{Id}_X(\mathcal{V})$ we have $\mathcal{K} = {}^1\mathrm{Mod}(\Sigma)$, i.e. $\mathcal{K}$ is equational. □

Remark 3.157 One can immediately see that we get the corresponding classical bivalent notions result by setting $\mathbf{L} = \mathbf{2}$; particularly, Theorem 3.156 generalizes the well-known Birkhoff's theorem [40].

3.6.2 FUZZY EQUATIONAL LOGIC

Given an $\mathbf{L}$-set of identities, we may ask to what degree Σ entails a given identity $p \approx q$. Our semantical notions provide a natural definition of (semantic) entailment.

Definition 3.158 The degree $(\Sigma \models (p \approx q))$ to which an identity $p \approx q$ semantically follows from an $\mathbf{L}$-set Σ of identities is defined by

$$(\Sigma \models (p \approx q)) \quad = \quad \bigwedge_{\mathbf{M} \in {}^1\mathrm{Mod}(\Sigma)} \|p \approx q\|_{\mathbf{M}}.$$

That is, $(\Sigma \models (p \approx q))$ is the degree to which it is true that each $\mathbf{L}$-algebra $\mathbf{M}$ satisfying all identities from Σ satisfies $p \approx q$. Therefore, we have

$$(\Sigma \models (p \approx q)) \quad = \quad \bigwedge_{\mathbf{M}:(\forall r,s \in T(X))(\Sigma(r \approx s) \leq \|r \approx s\|_M)} \|p \approx q\|_M.$$

However, to obtain $(\Sigma \models (p \approx q))$, we need to go over an infinite number of $\mathbf{L}$-algebras. In practice, if we assume $p \approx q$ to be valid at least in degree a (i.e. $\Sigma(p \approx q) = a$) and $q \approx r$ to be valid at least in degree b (i.e. $\Sigma(q \approx r) = b$), then we may conclude that $q \approx p$ will be valid at least in degree a or that $p \approx r$ will be valid at least in degree $a \otimes b$. When making the above conclusions, we tacitly used elementary rules of "fuzzy equational reasoning". Namely, knowing a as a lower estimation of the validity of $p \approx q$ we inferred a as a lower estimation of validity of $q \approx p$; and knowing a and b as lower estimations of the validity of $p \approx q$ and $q \approx r$, respectively, we inferred $a \otimes b$ as a lower estimation of validity of $p \approx r$. A natural question therefore arises of whether there is some set of rules of inference that enable us to infer degrees of validity of identities from degrees of validity of identities, and a natural notion of a degree of provability of an identity $p \approx q$ from Σ (denoted $(\Sigma \vdash (p \approx q))$) in such a way that $(\Sigma \vdash (p \approx q)) = (\Sigma \models (p \approx q))$, i.e. the calculus of inference be in a sense complete. Note that $(\Sigma \vdash (p \approx q)) \leq (\Sigma \models (p \approx q))$ ensures that we cannot infer something which is actually not true; on the other hand, $(\Sigma \vdash (p \approx q)) \geq (\Sigma \models (p \approx q))$ ensures that $(\Sigma \models (p \approx q))$ can be verified using the calculus of inference.

In the following we devise a simple calculus of equational inference which is complete in the above sense. The calculus will be called *fuzzy equational logic*.

We are to define the notion of a syntactical consequence. A pair $\langle p \approx q, a \rangle$ where $p, q \in T(X)$ and $a \in L$ will be called an (**L**-)*evaluated identity*. Inference rules are defined on evaluated identities as follows (p, q, r, s refer to terms, x is a variable, a, b are from L; $t(u/w)$ denotes the term resulting from a term t by substitution of w for every occurrence of u):

(RE) (from the empty set of evaluated identities) infer $\langle p \approx p, 1 \rangle$ (reflexivity);

(SY) from $\langle p \approx q, a \rangle$ infer $\langle q \approx p, a \rangle$ (symmetry);

(TR) from $\langle p \approx q, a \rangle$ and $\langle q \approx r, b \rangle$ infer $\langle p \approx r, a \otimes b \rangle$ (transitivity);

(REP) from $\langle p \approx q, a \rangle$ infer $\langle r \approx s, a \rangle$ where r is a term containing p as a subterm and s results from r by replacing one occurrence of p by q (replacement);

(SUB) from $\langle p \approx q, a \rangle$ infer $\langle p(x/r) \approx q(x/r), a \rangle$ (substitution).

An (**L**-)*evaluated proof* of $p \approx q$ from an **L**-set Σ of identities is any sequence $\langle p_1 \approx q_1, a_1 \rangle, \ldots, \langle p_n \approx q_n, a_n \rangle$ of evaluated identities such that $p_n = p$, $q_n = q$, and for each $i = 1, \ldots, n$, either $a_i = \Sigma(p_i \approx q_i)$ or $\langle p_i \approx q_i, a_i \rangle$ follows from some $\langle p_j \approx q_j, a_j \rangle$('s), $j < i$, by some of the inference rules (RE)–(SUB). In such a case, a_n is called the *value* of the proof.

Definition 3.159 The *degree* $\Sigma\vdash(p \approx q)$ *of provability* of an identity $p \approx q$ from an **L**-set Σ of identities is defined by

$$\Sigma\vdash(p \approx q) \quad = \quad \bigvee\{a_n \mid \langle p_1 \approx q_1, a_1\rangle, \ldots, \langle p_n \approx q_n, a_n\rangle \text{ is an evaluated proof of } p \approx q \text{ from } \Sigma\}.$$

The balance of the degree of semantical consequence and the degree of provability is the subject of the following assertion.

Theorem 3.160 (completeness of fuzzy equational logic) *For any* **L**-*set* Σ *of identities and any* $p, q \in T(X)$ *we have* $\Sigma\models(p \approx q) = \Sigma\vdash(p \approx q)$.

In the following, we elaborate the proof. We call a congruence θ on **M** *fully invariant* if for any morphism h from **M** into **M** we have $\theta(m,n) \leq \theta(h(m), h(n))$. An **L**-set Σ of identities is called an *equational theory* iff $\Sigma = \mathrm{Id}(\mathcal{K})$ for some $\mathcal{K}$. For a congruence θ on **M**, we denote the class $[m]_{^1\theta}$ also by $[m]$ if there is no danger of confusion.

Lemma 3.161 $\mathrm{Id}(\mathcal{K})$ *is a fully invariant congruence on* $\mathbf{T}(X)$.

Proof. Reflexivity and symmetry of $\mathrm{Id}(\mathcal{K})$ are obvious. Transitivity follows by $(\mathrm{Id}(\mathcal{K}))(p,q) \otimes (\mathrm{Id}(\mathcal{K}))(q,r) = \bigwedge_{\mathbf{M}\in\mathcal{K}} \|p \approx q\|_{\mathbf{M}} \otimes \bigwedge_{\mathbf{M}\in\mathcal{K}} \|q \approx r\|_{\mathbf{M}} \leq \bigwedge_{\mathbf{M}\in\mathcal{K}}(\|p \approx q\|_{\mathbf{M}} \otimes \|q \approx r\|_{\mathbf{M}}) = \bigwedge_{\mathbf{M}\in K}(\bigwedge_{v:X\to M} \|p\|_{\mathbf{M},v} \approx^{\mathbf{M}} \|q\|_{\mathbf{M},v}) \otimes (\bigwedge_{v:X\to M} \|q\|_{\mathbf{M},v} \approx^{\mathbf{M}} \|r\|_{\mathbf{M},v}) \leq \bigwedge_{\mathbf{M}\in K}(\bigwedge_{v:X\to M}(\|p\|_{\mathbf{M},v} \approx^{\mathbf{M}} \|q\|_{\mathbf{M},v}) \otimes (\|q\|_{\mathbf{M},v} \approx^{\mathbf{M}} \|r\|_{\mathbf{M},v})) \leq \bigwedge_{\mathbf{M}\in K} \bigwedge_{v:X\to M}(\|p\|_{\mathbf{M},v} \approx^{\mathbf{M}} \|r\|_{\mathbf{M},v}) = (\mathrm{Id}(\mathcal{K}))(p,r)$. Since $\approx^{\mathbf{T}(X)}$ is the identity relation, $\approx^{\mathbf{T}(X)} \subseteq \mathrm{Id}(\mathcal{K})$ is trivial. Furthermore, $(\mathrm{Id}(\mathcal{K}))(f(p_1,\ldots,p_n), f(q_1,\ldots,q_n)) = \bigwedge_{\mathbf{M}\in\mathcal{K}} \bigwedge_{v:X\to M} \|f(p_1,\ldots,p_n)\|_{\mathbf{M},v} \approx^{\mathbf{M}} \|f(q_1,\ldots,q_n)\|_{\mathbf{M},v}$, thus $f^{\mathbf{T}(X)}$ is compatible w.r.t. $\mathrm{Id}(\mathcal{K})$ iff for any $\mathbf{M} \in \mathcal{K}$ and any $v : X \to M$ we have $\otimes_{i=1}^n(\mathrm{Id}(\mathcal{K}))(p_i,q_i) \leq \|f(p_1,\ldots,p_n)\|_{\mathbf{M},v} \approx^{\mathbf{M}} \|f(q_1,\ldots,q_n)\|_{\mathbf{M},v}$. The last inequality follows from compatibility of $f^{\mathbf{M}}$: $\otimes_{i=1}^n(\mathrm{Id}(\mathcal{K}))(p_i,q_i) \leq \otimes_{i=1}^n(\|p_i\|_{\mathbf{M},v} \approx^{\mathbf{M}} \|q_i\|_{\mathbf{M},v}) \leq f^{\mathbf{M}}(\|p_1\|_{\mathbf{M},v},\ldots,\|p_n\|_{\mathbf{M},v}) \approx^{\mathbf{M}} f^{\mathbf{M}}(\|q_1\|_{\mathbf{M},v},\ldots,\|q_n\|_{\mathbf{M},v})$. Finally, we check that $\mathrm{Id}(\mathcal{K})$ is fully invariant: Let $h : T(X) \to T(X)$ be a homomorphism. We have to check that $(\mathrm{Id}(\mathcal{K}))(p(x_1,\ldots,x_n), q(x_1,\ldots,x_n)) \leq (\mathrm{Id}(\mathcal{K}))(h(p(x_1,\ldots,x_n)), h(q(x_1,\ldots,x_n)))$. Since $h(r(x_1,\ldots,x_n)) = r(h(x_1),\ldots,h(x_n))$ is valid for any term $r \in T(X)$, the last inequality is true if for any $\mathbf{M} \in \mathcal{K}$ and any $v : X \to M$ there is some $v' : X \to M$ such that $\|p \approx q\|_{\mathbf{M},v'} \leq \|p(h(x_1),\ldots,h(x_n)) \approx q(h(x_1),\ldots,h(x_n))\|_{\mathbf{M},v}$. A moment's reflection shows that one may take any v' such that for any $i = 1,\ldots,n$, $v'(x_i) = \|h(x_i)\|_{\mathbf{M},v}$. □

Lemma 3.162 *For a fully invariant congruence* θ *on* $\mathbf{T}(X)$ *we have* $\|p \approx q\|_{\mathbf{T}(X)/\theta} = \theta(p,q)$ *for every* $p, q \in T(X)$.

Proof. "$\leq$": For $v : X \to T(X)/\theta$ such that $v(x) = [x]$ $(x \in X)$ we have $\|p \approx q\|_{\mathbf{T}(X)/\theta} \leq \|p\|_{\mathbf{T}(X)/\theta,v} \approx^{\mathbf{T}(X)/\theta} \|q\|_{\mathbf{T}(X)/\theta,v} = [p] \approx^{\mathbf{T}(X)/\theta} [q] = \theta(p,q)$.

"$\geq$": Take any $v : X \to T(X)/\theta$, let $v(x_1) = [r_1], \ldots, v(x_n) = [r_n]$. Consider any homomorphism $h : T(X) \to T(X)$ such that $h(x_i) = r_i$. We have $\theta(p,q) \leq \theta(h(p), h(q)) = \theta(p(h(x_1), \ldots, h(x_n)), q(h(x_1), \ldots, h(x_n))) = \theta(p(r_1, \ldots, r_n), q(r_1, \ldots, r_n)) = \|p\|_{\mathbf{T}(X)/\theta,v} \approx^{\mathbf{T}(X)/\theta} \|q\|_{\mathbf{T}(X)/\theta,v}$. Since v is chosen arbitrarily, we conclude $\theta(p,q) \leq \|p \approx q\|_{\mathbf{T}(X)/\theta}$. □

Corollary 3.163 *Let* Σ *be an* **L***-set of identities, put* $\theta(p,q) = \Sigma(p \approx q)$. *Then* θ *is a fully invariant congruence on* $\mathbf{T}(X)$ *iff* Σ *is an* **L***-equational theory.*

Proof. The "$\Leftarrow$" part is Lemma 3.161, the "$\Rightarrow$" part follows from Lemma 3.162 by putting $\mathcal{K} = \{\mathbf{T}(X)/\theta\}$. □

For an **L**-set Σ of identities denote by $\theta(\Sigma)$ the least fully invariant congruence on $\mathbf{T}(X)$ such that $\Sigma(p \approx q) \leq (\theta(\Sigma))(p,q)$ (its existence follows from the fact that the set of all fully invariant congruences on $\mathbf{T}(X)$ forms a complete lattice w.r.t. $\subseteq$).

Lemma 3.164 *For any* **L***-set* Σ *of identities, and every terms* $p, q \in T(X)$ *we have* $\Sigma \models (p \approx q) = (\theta(\Sigma))(p,q)$.

Proof. "$\leq$": We have $\Sigma(p \approx q) \leq (\theta(\Sigma))(p,q) = \|p \approx q\|_{\mathbf{T}(X)/\theta(\Sigma)}$, whence $\Sigma \models (p \approx q) \leq (\theta(\Sigma))(p,q)$.

"$\geq$": Take any $\mathbf{M}$ such that $\Sigma(r \approx s) \leq \|r \approx s\|_{\mathbf{M}}$. By Lemma 3.161, $\mathrm{Id}(\{\mathbf{M}\})$ is a fully invariant congruence on $\mathbf{T}(X)$ containing Σ. Therefore, $(\theta(\Sigma))(p,q) \leq \mathrm{Id}(\{\mathbf{M}\})(p,q) = \|p \approx q\|_{\mathbf{M}}$. The conclusion readily follows. □

The semantic consequence **L**-relation $\models$ defines naturally an $\mathbf{L}_{\{1\}}$-closure operator $\mathrm{C}^{\mathrm{sem}}$ on $T(X) \times T(X)$ (i.e. a mapping $\mathrm{C}^{\mathrm{sem}} : L^{T(X) \times T(X)} \to L^{T(X) \times T(X)}$ satisfying $\Sigma \subseteq \mathrm{C}^{\mathrm{sem}}(\Sigma)$, $\mathrm{C}^{\mathrm{sem}}(\Sigma_1) \subseteq \mathrm{C}^{\mathrm{sem}}(\Sigma_2)$ whenever $\Sigma_1 \subseteq \Sigma_2$, and $\mathrm{C}^{\mathrm{sem}}(\mathrm{C}^{\mathrm{sem}}(\Sigma)) = \mathrm{C}^{\mathrm{sem}}(\Sigma)$, see Section 7.1) by $(\mathrm{C}^{\mathrm{sem}}(\Sigma))(p,q) = \Sigma \models (p \approx q)$.

We define a syntactic consequence operator $\mathrm{C}^{\mathrm{syn}}$ by letting $\mathrm{C}^{\mathrm{syn}}(\Sigma)$ be the least **L**-set in $T(X) \times T(X)$ such that $\Sigma \subseteq \mathrm{C}^{\mathrm{syn}}(\Sigma)$ and

(RE') $(\mathrm{C}^{\mathrm{syn}}(\Sigma))(p, p) = 1$;

(SY') $(\mathrm{C}^{\mathrm{syn}}(\Sigma))(p, q) \leq (\mathrm{C}^{\mathrm{syn}}(\Sigma))(q, p)$;

(TR') $(\mathrm{C}^{\mathrm{syn}}(\Sigma))(p, q) \otimes (\mathrm{C}^{\mathrm{syn}}(\Sigma))(q, r) \leq (\mathrm{C}^{\mathrm{syn}}(\Sigma))(p, r)$;

(REP') $(\mathrm{C}^{\mathrm{syn}}(\Sigma))(p, q) \leq (\mathrm{C}^{\mathrm{syn}}(\Sigma))(r, s)$ where r is any term containing p as a subterm and s results from r by replacing one occurrence of p by q;

(SUB') $(\mathrm{C}^{\mathrm{syn}}(\Sigma))(p, q) \leq (\mathrm{C}^{\mathrm{syn}}(\Sigma))(p(x/r), q(x, r))$

holds for any $p, q, r \in T(X)$. Note that the existence of $\mathrm{C}^{\mathrm{syn}}$ follows from the fact that the set of all $\mathbf{L}$-sets in $T(X) \times T(X)$ which contain Σ and satisfy (RE')–(SUB') is not empty and is closed w.r.t. arbitrary intersections.

Lemma 3.165 *For any* $\mathbf{L}$*-set* Σ *of identities and every* $p, q \in T(X)$ *we have* $(\mathrm{C}^{\mathrm{sem}}(\Sigma))(p, q) = (\mathrm{C}^{\mathrm{syn}}(\Sigma))(p, q)$.

Proof. By Lemma 3.164, we have to prove $\mathrm{C}^{\mathrm{syn}}(\Sigma) = \theta(\Sigma)$. By (RE')–(TR'), $\mathrm{C}^{\mathrm{syn}}(\Sigma)$ is an $\mathbf{L}$-equivalence on $T(X)$ which contains Σ. To show that $\mathrm{C}^{\mathrm{syn}}(\Sigma)$ is a congruence on $\mathbf{T}(X)$, take any n-ary $f \in F$, any $p_1, q_1, \ldots, p_n, q_n \in T(X)$, and put $s_i = f(q_1, \ldots, q_i, p_{i+1}, \ldots, p_n)$ $(i = 0, \ldots, n)$. By (REP'), $(\mathrm{C}^{\mathrm{syn}}(\Sigma))(p_i, q_i) \leq (\mathrm{C}^{\mathrm{syn}}(\Sigma))(s_i, s_{i+1})$, therefore $(\mathrm{C}^{\mathrm{syn}}(\Sigma))(p_1, q_1) \otimes \cdots \otimes (\mathrm{C}^{\mathrm{syn}}(\Sigma))(p_n, q_n) \leq (\mathrm{C}^{\mathrm{syn}}(\Sigma))(s_1, s_2) \otimes \cdots \otimes (\mathrm{C}^{\mathrm{syn}}(\Sigma))(s_{n-1}, s_n) \leq (\mathrm{C}^{\mathrm{syn}}(\Sigma))(s_1, s_n) = (\mathrm{C}^{\mathrm{syn}}(\Sigma))(f(p_1, \ldots, p_n), f(q_1, \ldots, q_n))$, i.e. $\mathrm{C}^{\mathrm{syn}}(\Sigma)$ is a congruence. We show that $\mathrm{C}^{\mathrm{syn}}(\Sigma)$ is fully invariant. Take any homomorphism $h : T(X) \to T(X)$ and any variables $y_1, \ldots, y_n$ so that no y_i occurs in x_j or $h(x_j)$, $j = 1, \ldots, n$. (SUB') yields

$$
\begin{aligned}
& (\mathrm{C}^{\mathrm{syn}}(\Sigma))(p(x_1, \ldots, x_n), q(x_1, \ldots, x_n)) = \\
= \; & (\mathrm{C}^{\mathrm{syn}}(\Sigma))(p(y_1, \ldots, y_n), q(y_1, \ldots, y_n)) \leq \\
\leq \; & (\mathrm{C}^{\mathrm{syn}}(\Sigma))(p(h(x_1), \ldots, y_n), q(h(x_1), \ldots, y_n)) \leq \cdots \leq \\
\leq \; & (\mathrm{C}^{\mathrm{syn}}(\Sigma))(p(h(x_1), \ldots, h(x_n)), q(h(x_1), \ldots, h(x_n))) = \\
= \; & (\mathrm{C}^{\mathrm{syn}}(\Sigma))(h(p(x_1, \ldots, x_n)), h(q(x_1, \ldots, x_n))),
\end{aligned}
$$

completing the proof. □

Lemma 3.166 *For any* $\mathbf{L}$*-set* Σ *of identities and every* $p, q \in T(X)$ *we have* $\Sigma \vdash (p \approx q) = (\mathrm{C}^{\mathrm{syn}}(\Sigma))(p, q)$.

Proof. "$\leq$": Clearly, it suffices to check that if $\langle p_i \approx q_i, a_i\rangle$ is a member of some **L**-valuated proof, then $a_i \leq (\mathrm{C}^{\mathrm{syn}}(\Sigma))(p_i, q_i)$. This is obvious if $a_i = \Sigma(p_i \approx q_i)$. Otherwise (i.e. $\langle p_i \approx q_i, a_i\rangle$ is obtained by some inference rule), proceed by induction over i and suppose that the assertion is true for $j < i$. The following inference rules could have been used:

(RE): $p_i = q_i$ and $a_i = 1 = (\mathrm{C}^{\mathrm{syn}}(\Sigma))(p_i, q_i)$ by (RE');

(SY): $p_i = q_j$, $q_i = p_j$, and $a_i = a_j$ for some $j < i$ and thus $a_i \leq (\mathrm{C}^{\mathrm{syn}}(\Sigma))(p_j, q_j) \leq (\mathrm{C}^{\mathrm{syn}}(\Sigma))(p_i, q_i)$ by (SY');

(TR): $p_i = p_j$, $q_j = p_k$, $q_k = q_i$, and $a_i = a_j \otimes a_k$ for some $j, k < i$ and thus $a_i \leq (\mathrm{C}^{\mathrm{syn}}(\Sigma))(p_j, q_j) \otimes (\mathrm{C}^{\mathrm{syn}}(\Sigma))(p_k, q_k) \leq (\mathrm{C}^{\mathrm{syn}}(\Sigma))(p_i, q_i)$ by (TR');

(REP): q_i is obtained from p_i by replacement of some occurrence of p_j by q_j, and $a_i = a_j$ for some $j < i$, whence $a_i = (\mathrm{C}^{\mathrm{syn}}(\Sigma))(p_j, q_j) \leq (\mathrm{C}^{\mathrm{syn}}(\Sigma))(p_i, q_i)$ by (REP');

(SUB): $p_i = p_j(x/r)$, $q_i = q_j(x/r)$, and $a_i = a_j$ for some $r \in T(X)$ and $j < i$, thus $a_i = (\mathrm{C}^{\mathrm{syn}}(\Sigma))(p_j, q_j) \leq (\mathrm{C}^{\mathrm{syn}}(\Sigma))(p_i, q_i)$ by (SUB').

"$\geq$": It suffices to prove that the **L**-set D in $T(X) \times T(X)$ defined by $D(p, q) = \Sigma\vdash(p \approx q)$ contains Σ and satisfies (RE')–(SUB') (the inequality then follows from the fact that $\mathrm{C}^{\mathrm{syn}}(\Sigma)$ is the least **L**-set with these properties). Since $\langle p \approx q, \Sigma(p \approx q)\rangle$ is a proof from Σ, we have $\Sigma \subseteq D$.

(RE') follows from the fact that $\langle p \approx p, 1\rangle$ is a proof.

(SY') follows from the fact that if $\ldots, \langle p \approx q, a\rangle$ is a proof from Σ, then $\ldots, \langle p \approx q, a\rangle, \langle q \approx p, a\rangle$ is as well.

(TR'): Let $u_i, \langle p \approx q, a_i\rangle$ $(i \in I)$ and $v_j, \langle q \approx r, b_j\rangle$ $(j \in J)$ be all proofs of $p \approx q$ and $q \approx r$ from Σ, respectively. Since each $u_i, v_j, \langle p \approx q, a_i\rangle, \langle q \approx r, b_j\rangle, \langle p \approx r, a_i \otimes a_i \otimes b_j\rangle$ is a proof of $p \approx r$ from Σ, we have $D(p, q) \otimes D(q, r) = (\bigvee_{i\in I} a_i) \otimes (\bigvee_{j\in J} b_j) = \bigvee_{i\in I}(a_i \otimes (\bigvee_{j\in J} b_j)) = \bigvee_{i\in I, j\in J} a_i \otimes b_j \leq \bigvee\{c \mid w, \langle p \approx r, c\rangle$ is a proof from $\Sigma\} = D(p, r)$.

(REP') and (SUB') follow from the fact that if $\ldots, \langle p \approx q, a\rangle$ is a proof from Σ and $r \approx s$ is obtained from $p \approx q$ by replacement or substitution, then $\ldots, \langle p \approx q, a\rangle, \langle r \approx s, a\rangle$ is a proof from Σ as well. □

Theorem 3.160 is now a direct consequence of Lemmas 3.165 and 3.166.

Remark 3.167 The provability degree may be strictly greater than the value of any proof. Indeed, let $F = \{\circ\}$, $\circ$ be binary, denote by x^n the n-th power of x w.r.t. $\circ$, i.e. $x^3 = (x \circ x) \circ x$ etc. Let **L** be the standard Łukasiewicz algebra on $[0, 1]$. Define Σ by $\Sigma(x \circ x \approx x) = 1$, $\Sigma(x^n \approx y^n) = 1 - 1/n$, and $\Sigma(p \approx q) = 0$ otherwise. Clearly, $\langle x \circ x \approx x, 1\rangle$, $\langle y \circ y \approx y, 1\rangle$, $\langle x^n \approx y^n, 1 - 1/n\rangle$, $\langle x^{n-1} \approx y^n, 1 - 1/n\rangle$, $\langle x^{n-1} \approx y^{n-1},$

$1-1/n\rangle$,..., $\langle x \approx y, 1-1/n\rangle$ is an **L**-valuated proof of $x \approx y$ from Σ for any n. Therefore, $\Sigma\vdash(x \approx y) = 1$. On the other hand, there is no proof of $x \approx y$ from Σ the value of which is 1: Such a proof must not use any $\langle x^n \approx y^n, 1-1/n\rangle$, and therefore uses only $\langle x \circ x \approx x, 1\rangle$ which is impossible (any **L**-algebra satisfying $x \circ x \approx x$ in degree 1 but $x \approx y$ in degree less than 1, e.g. an idempotent monoid with at least two elements and the classical identity as the **L**-equality, serves as a counterexample).

Remark 3.168 For readers familiar with Pavelka-style fuzzy logic: As observed by Pavelka [158], we cannot have graded style completeness for arbitrary complete residuated lattice even in the case of propositional logic (the less so for the first-order case [105, 154]). However, since (RE)–(SUB) can be used in Pavelka-style first-order fuzzy logic as derived rules (more precisely: derived rules in first-order fuzzy logic with the usual inference rules where the predicate symbol $\approx$ is confined in an obvious sense by axioms of reflexivity, symmetry, transitivity, and compatibility), our result implies that the equational fragment (i.e. restriction to formulas of the form of identities) is completely axiomatizable (in the graded style) in first-order fuzzy logic using any complete residuated lattice as the structure of truth values.

3.7 MANY-SORTED CASE

The language of fuzzy logic we used so far is well-suited for situations where there are elements of one sort. We have at our disposal a set of object variables which have the same role: they (are supposed to) refer to elements of a universe that is specified when it comes to interpretation. In real situations, however, more often than not, we deal with elements of different sorts. For instance, we have people, real numbers, cities, cars, etc. In order to formulate formally assertions about a universe that is naturally many-sorted, we have basically two ways to do it. First, we can add unary relation symbols for distinguishing the sorts. Second, we can have different sets of variables for different sorts. Since the first way is a bit cumbersome, we will not follow it. Instead, we elaborate the second way. In this respect, the situation we have been interested in so far is a special case (one-sorted) of the more general one (many-sorted). Except for having more sorts and the resulting need for distinguishing sorts of respective arguments of fuzzy relations and functions, there is nothing substantially new, all results we obtained have their many-sorted generalizations that are easy to formulate. This enables us to proceed relatively quickly: We define the basic syntactic and semantic notions and show some selected results obtained before, now in their many-sorted formulation. The reader is left to formulate and prove other theorems.

Syntax. Suppose we have a nonempty set S whose elements s will be called *sorts*. *Many-sorted language* $\mathcal{J}$ given by a set S of sorts (also S-sorted language) is given by the following: a nonempty set R of relation symbols, each $r \in R$ with a (possibly empty) string $s_1s_2\dots s_n$ of sorts $s_i \in S$, $s_1\dots s_n$ is called the arity of r; a (possibly empty) set F of function symbols each $f \in F$ with a nonempty string $s_1\dots s_ns_{n+1}$ called arity of f; R and F are disjoint; object variables, each variable x given with its sort $s \in S$, for each sort s there is a denumerable number of variables of sort s; logical connectives, truth constants and quantifiers, and auxiliary symbols as for the ordinary one-sorted language; the language is called a language with equality if R contains a relation symbol $\approx_s$ of sort ss for each $s \in S$ (we almost always assume a language with equality).

The meaning of arities: If the arity of $r \in R$ is $s_1\dots s_n$, then r is to denote an n-ary fuzzy relation between sets of sorts $s_1,\dots,s_n$, respectively; i.e. the first argument is supposed to be of sort s_1 etc. Similarly, if the arity of $f \in F$ is $s_1\dots s_ns_{n+1}$, then f is to denote an n-ary function with the i-th argument from a set of sort s_i and the result is of sort s_{n+1}.

We denote the arity of $r \in R$ by $\sigma(r)$, the arity of $f \in F$ by $\sigma(f)$, and the sort of a variable x by $\sigma(x)$. Thus, each S-sorted language $\mathcal{J}$ is fully specified by the triple $\langle R, F, \sigma\rangle$, called a *type* of $\mathcal{J}$.

Terms are defined as follows: each variable x of sort s is a term of sort s; if $t_1,\dots,t_n$ are terms of sorts $s_1,\dots,s_n$, respectively, and $f \in F$ has arity $s_1\dots s_ns_{n+1}$, then $f(t_1,\dots,t_n)$ is a term of sort s_{n+1}.

Formulas: atomic formulas are of the form $r(t_1,\dots,t_n)$ where $r \in R$ has arity $s_1\dots s_n$ and t_i are terms of sorts s_i, and truth constants; composed formulas are defined from atomic ones using logical connectives and quantifiers as before.

Subterms and subformulas are defined obviously; so are the degrees $|t|_x$, $|t|_f$, $|\varphi|_r$, $|\varphi|_f$, $|\varphi|_x$ where x is a variable, t a term, φ a formula, $r \in R$, and $f \in F$. Similarly obvious are the modifications of other notions (var(t), var(φ), free(φ), bound(φ), etc.). As for substitutions, if a term t is being substituted for a variable x, both t and x are required to be of the same sort.

Semantics. An **L**-structure for an S-sorted language of type $\langle R, F, \sigma\rangle$ is a triple $\mathbf{M} = \langle M, R^{\mathbf{M}}, F^{\mathbf{M}}\rangle$ where: $M = \{M_s \mid s \in S\}$ is a system of nonempty sets M_s (M_s is called a universe of sort s); $R^{\mathbf{M}} =$

$\{r^{\mathbf{M}} \in L^{M_{s_1}\times\cdots\times M_{s_n}} \mid r \in R,\, \sigma(r) = s_1 \ldots s_n\}$ of **L**-relations, and a set $F^{\mathbf{M}} = \{f^{\mathbf{M}} : M_{s_1} \times \cdots \times M_{s_n} \to M_{s_{n+1}} \mid f \in F,\, \sigma(r) = s_1 \ldots s_{n+1}\}$ of functions such that each $\approx_s^{\mathbf{M}}$ is an **L**-equivalence relation on M_s, each $r^{\mathbf{M}} \in R^{\mathbf{M}}$ with $\sigma(r) = s_1 \ldots s_n$ is compatible with $\approx_{s_1}^{\mathbf{K}}, \ldots, \approx_{s_n}^{\mathbf{K}}$, and each $f^{\mathbf{M}} \in F^{\mathbf{M}}$ with $\sigma(f) = s_1 \ldots s_n s_{n+1}$ is compatible with $\approx_{s_1}^{\mathbf{K}}, \ldots, \approx_{s_n}^{\mathbf{K}}$, $\approx_{s_{n+1}}^{\mathbf{K}}$. In other words, an **L**-structure **M** for a many-sorted language is a system of **L**-relations and functions taking as arguments elements of the corresponding sorts.

Recall that the compatibility conditions for $r^{\mathbf{M}}$ and $f^{\mathbf{M}}$ say that

$$\bigotimes_{i=1}^{n}(m_i \approx_{s_i}^{\mathbf{M}} m_i') \leq r^{\mathbf{M}}(m_1, \ldots, m_n) \leftrightarrow r^{\mathbf{M}}(m_1', \ldots, m_n')$$

and

$$\bigotimes_{i=1}^{n}(m_i \approx_{s_i}^{\mathbf{M}} m_i') \leq f^{\mathbf{M}}(m_1, \ldots, m_n) \approx_{s_{n+1}}^{\mathbf{M}} f^{\mathbf{M}}(m_1', \ldots, m_n')$$

for any $m_i, m_i' \in M_{s_i}$.

Values of terms and formulas: Given an **L**-structure **M** for an S-sorted language, a valuation is a mapping assigning to any variable x an element $v(x) \in M_{\sigma(x)}$ (i.e. an element of sort s to a variable of sort s). Values of terms and formulas are defined as in Definitions 3.64 and 3.67 (only that the symbols in the definition have different meaning).

One easily verifies that all results obtained in the previous text for the one-sorted case have their generalizations for the many-sorted case. We will feel free to use these generalizations in the subsequent chapters. We just take one assertion for illustration.

Theorem 3.169 (preservation of similarity, many-sorted case) *For many-sorted* **L***-structures* $\mathbf{M}_1, \mathbf{M}_2$ *of type* $\langle R, F, \sigma\rangle$ *with* $M_1 = M_2$, *a formula* φ, *and valuations* v_1 *and* v_2 *we have*

$$\begin{aligned} &\bigotimes_{x \in \mathrm{free}(\varphi)} (v_1(x) \approx_{\sigma(s)}^{\mathbf{M}_i} v_2(x))^{|\varphi|_x} \otimes \\ &\otimes \bigotimes_{r\in R}(r^{\mathbf{M}_1} \approx r^{\mathbf{M}_2})^{|\varphi|_r} \otimes \bigotimes_{f\in F}(f^{\mathbf{M}_1} \approx f^{\mathbf{M}_2})^{|\varphi|_f} \leq \\ &\leq \|\varphi\|_{\mathbf{M}_1,v_1} \leftrightarrow \|\varphi\|_{\mathbf{M}_2,v_2} \end{aligned}$$

for $i = 1, 2$.

Note that for $r \in R$, $\sigma(r) = s_1 \ldots s_n$, and $f \in F$, $\sigma(f) = s_1 \ldots s_n s_{n+1}$, we have

$$(r^{\mathbf{M}_1} \approx r^{\mathbf{M}_2}) =$$

$$= \bigwedge_{m_i, m'_i \in M_{s_i}} r^{\mathbf{M}_1}(m_1, \ldots, m_n) \leftrightarrow r^{\mathbf{M}_2}(m'_1, \ldots, m'_n)$$

and

$$(f^{\mathbf{M}_1} \approx f^{\mathbf{M}_2}) =$$
$$= \bigwedge_{m_i, m'_i \in M_{s_i}} f^{\mathbf{M}_1}(m_1, \ldots, m_n)(\approx^{\mathbf{M}}_{s_{n+1}} \wedge \approx^{\mathbf{M}}_{s_{n+1}}) f^{\mathbf{M}_2}(m'_1, \ldots, m'_n).$$

3.8 BIBLIOGRAPHICAL REMARKS

Section 3.1: Zadeh's [216] is the first paper on fuzzy sets. There are, however, independent attempts to formally describe vaguely delineated collections of objects. The first paper systematically devoted to vagueness is Black's [42]; Black analyzed vagueness from the point of view of philosophy of science and devised a formalism for treating vaguely delineated collections of objects (so-called consistency profiles). Many-valued relations are considered by Menger [142, 143, 144]; however, Menger gives the degrees of membership a probabilistic interpretation (i.e. $R(x, y) = 0.8$ means that the probability of x and y being in the relation R is 0.8). Many-valued set theory was studied by Klaua, see e.g. [121]. Rough set theory introduced by Pawlak [159] provides a formalism for representing collections with imprecise boundaries where the imprecision results from an indistinguishability of elements of the universe. We give the basic definition: Let E be an equivalence relation (representing indistinguishability which is due to some limitations, e.g. measurement precision) on a universe set X. A rough set $R(A)$ representing a given subset $A \subseteq X$ is given by two subsets of X—the lower approximation $\underline{R}(A) = \cup\{[x]_E \mid [x]_E \subseteq A, x \in X\}$ and the upper approximation $\overline{R}(A) = \cup\{[x]_E \mid [x]_E \cap A \neq \emptyset, x \in X\}$. Vopěnka's Alternative set theory [199, 200, 201] provides a deeply founded attempt to treat nonsharply delineated collections of objects. Supervaluations and supertruth [206] represent an account of vagueness popular in philosophical community. The basic idea is that a possibly vague proposition may be evaluated by several truth valuations—in some of them it is true, in some of them it is false. If a proposition is true (false) in all such valuations, it is considered to be supertrue (superfalse), otherwise it is considered to be vague. Now, a negation of a vague proposition is vague. However, a conjuction of a proposition and its negation is always superfalse (even for a vague proposition).

Our approach to fuzzy sets was a "naive one"; axiomatic treatments can be found in [121, 192]. Fuzzy sets and their applications are well covered e.g. in [67, 96, 99, 127, 150, 196] (except for [196], the set of truth values is usually confined to $[0,1]$). The concept of an **L**-set is due to Goguen [90] (Goguen uses the term L-fuzzy set). Representation of fuzzy sets by a-cuts is a kind of folklore; however, only the fact that a fuzzy set can be reconstructed from the collection of its a-cuts is mentioned mostly. A characterization of systems of a-cuts of a fuzzy set with truth degrees from $[0,1]$ appeared the first time probably in [149]: Modifying a bit the original formulation, the result says that a system $\mathcal{S} = \{A_a \subseteq X \mid a \in L\}$ is a system of a-cuts of some **L**-set in X with $L = [0,1]$ iff it satisfies (a) $a \leq b$ implies $A_b \subseteq A_a$, (b) $A_0 = X$, (c) $A_a = \bigcap_{b<a} A_b$. Such a characterization is, however, not applicable for general structure **L** (e.g. for finite, at least three-element chain). Representation of fuzzy sets by representative subsets of $X \times L$ was used in [23]. The notion of a fuzzy equivalence is due to Zadeh [218] (for more comments on bibliography of fuzzy equivalences see Bibliographical remarks to Chapter 4).

There are also more general approaches to fuzzy sets: membership degrees are intervals in $[0,1]$ (interval-valued fuzzy sets), fuzzy sets in $[0,1]$ (fuzzy sets of type 2), elements of a lattice (lattice-valued fuzzy sets), see e.g. [127].

We defined the equality degree of fuzzy sets A and B by $(A \approx B) = \bigwedge_x (A(x) \leftrightarrow B(x)) = S(A,B) \wedge S(B,A)$. Note that Gottwald defines equality degree by $S(A,B) \otimes S(B,A)$ which in general differs from our definition. Other alternative approaches to some concepts can be found in [74, 75, 114, 210].

Section 3.2: Good books on mathematical logic are [12, 71, 146, 188]; [193] is an introduction to logic. Classical relational and functional systems are covered in [137]; [59] is a monograph on model theory (for classical logic). We also mention [58] devoted to nonclassical model theory where truth values form a special topological space; this book is surely worth reconsidering from the point of view of fuzzy approach.

Informative overview on historical development of many-valued and fuzzy logic (in narrow sense) can be found in [99, 105]. Although there is evidence about earlier approaches, the pioneering work on many-valued logic is due to Łukasiewicz [134, 135]. The first paper on fuzzy logic (with a clear interpretation of truth values: truth values are degrees of truth) is probably Goguen's [91] that appeared very shortly after Zadeh's seminal paper on fuzzy sets [216]. However, research on many-valued logics was quite fruitful even before Zadeh's and Goguen's papers (see [99, 105]). Pavelka's [158] (heavily inspired by [91]) is an important contribution to fuzzy logic; this paper introduces fuzzy approach to metalevel (degrees of provability, fuzzy inference rules, fuzzy set of axioms, etc.); Pavelka deals with propositional logic, Novák [151] extended this to the first-order case (see also [154]). Another contribution important for us is Höhle's papers [115, 116, 117] where a first-order fuzzy logic (so-called monoidal logic, see Section 7.3) with truth degrees in a complete residuated lattice is presented together with a syntactico-semantical completeness theorem. Lastly, we mention Hájek for his important contributions and unifying previous approaches. Recent monographs on fuzzy logic (in narrow sense) are [85, 99, 105, 154].

Section 3.3: For the role of similarity in approximate reasoning see Bibliographical remarks to Section 4.2.

Similarity issues in the general context of fuzzy predicate logic have been so far considered only in a restricted form: Hájek's [105] contains a section on similarity (i.e. fuzzy equivalence) relations, mainly as a preparation for an analysis of approximate inference (carried on in Chapter 7 of [105]): Lemma 5.6.8 provides (on syntax level) a rougher estimation than our Lemma 3.83 (this is because Hájek defines a so-called syntactic degree of a formula which is in a sense a rougher approach than our degree of a variable, a function symbol, and a relation symbol in a formula). A statement analogously related to our Lemma 3.81 is [107, Lemma 1] (Hájek does not consider function symbols in [105], he does so in [107]).

Section 3.4: Morphisms (and congruences), subsystems, and direct products count for basic algebraic constructions. The idea that morphisms in fuzzy setting can be regarded as pairs composed of a mapping between universe sets and of a morphism between the structures of truth values (and similarly for congruences) appeared for the first time in [64] ([64] uses a very general structure of truth values which has an arbitrary number of connectives and quantifiers); independently, it was used in a special form in [24] to factorize so-called fuzzy concept lattices. In [64], direct products of fuzzy structures with different structures of truth values are introduced basically the same way as our $\Pi\mathbf{M}_i$. Morphisms of fuzzy relations are the subject of [9].

A short section on model theory is contained in [154]. Ying [213, 214] studies ultraproduct theorem, compactness, and Löwenheim-Skolem property for fuzzy logic; direct products of structures of truth values are also discussed there. A section containing some model-theoretical results of fuzzy logic is presented in [99]; this covers ultraproduct theorem, compactness, and Löwenheim-Skolem property. Note that another version of the ultraproduct theorem for fuzzy logic with standard Łukasiewicz structure on $[0,1]$ is contained in [154] (based on Novák's [152]); however, this result is essentially vacuous since it results to a trivial case: Namely, the result assumes that for an index set I, a $\bigwedge$-complete ultrafilter (i.e. ultrafilter closed w.r.t. arbitrary infima) in $\mathbf{L}^I$ is given using which the ultraproduct is constructed. Now, it can be shown that each $\bigwedge$-complete ultrafilter G in $\mathbf{L}^I$ is of the form $G = G_j$ (for some $j \in J$) where $G_j = \{a \in L^I \mid a(j) = 1\}$, and that that the ultraproduct of $\mathbf{M}_i$'s by G_j is isomorphic to $\mathbf{M}_j$ and gives thus nothing new.

Fuzzy subalgebras started with [174], for a general approach see [65]. Fuzzy subalgebras (and fuzzified notions of further algebraic concepts like ideals etc.) have been since studied for several particular types of algebras. There are natural examples of fuzzy subalgebras for some types of algebras (see Remark 3.135); however, sometimes fuzzy subalgebras seem only a generalization without purpose, and thus too artificial.

Section 3.5: This section is based on [31]; it is an attempt to give a so-called cutworthiness firm foundations (see Section 7.2.2).

Section 3.6: This section is based on [26, 32], further information can be found in [34]. Note that the notion of a fuzzy congruence on an algebra appears frequently in the literature, however, with [0, 1] equipped with min as the structure of truth values; see e.g. [177] for such an approach to general algebras.

Section 3.7: Extension to many-sorted case is a standard way to have means for describing situations where elements are of different natural kinds (sorts); for many-sorted fuzzy logic see [105].

Chapter 4

BINARY FUZZY RELATIONS

It is impossible to describe a practical fact without attenuating by the use of the word "approximately" or "nearly"; on the other hand, all the elements constituting the theoretical fact are defined with rigorous exactness.

— Pierre Duhem (The Aim and Structure of Physical Theory, 1954)

Everything is vague to a degree you do not realize till you have tried to make it precise.

—Bertrand Russell (The Philosophy of Logical Atomism, 1918)

This chapter is devoted to binary fuzzy relations. For quite a variety of problems, binary relations turn out to be a simple but very powerful tool. Recall that a binary **L**-relation between (nonempty) sets X and Y is any mapping $R : X \times Y \to L$ (L is the support set of the complete residuated lattice **L**; for $x \in X$ and $y \in Y$, $R(x, y) \in L$ is the truth degree to which x and y are in the relation R). Thus, $L^{X \times Y}$ is the set of all binary **L**-relations between X and Y. If, moreover, the sets X and Y are equipped with **L**-equivalences $\approx^X$ and $\approx^Y$, respectively, we call R compatible w.r.t. $\approx^X$ and $\approx^Y$ if for any $x_1, x_2 \in X$ and $y_1, y_2 \in Y$ we have

$$(x_1 \approx^X x_2) \otimes (y_1 \approx^Y y_2) \otimes R(x_1, y_1) \leq R(x_2, y_2).$$

Through this chapter we leave x, y, z, etc. to denote elements of universe(s) of discourse and denote variables of a language of fuzzy logic by $\mathbb{x}$, $\mathbb{y}$, $\mathbb{z}$ (possibly with indices).

4.1 OPERATIONS AND PROPERTIES OF BINARY FUZZY RELATIONS

4.1.1 OPERATIONS WITH BINARY FUZZY RELATIONS

Binary **L**-relations between X and Y are but **L**-sets in $X \times Y$. Therefore, any operations with **L**-sets can be applied to binary **L**-relations. For instance, the intersection $R \cap S$ of two **L**-relations R and S between X and Y is defined by $(R \cap S)(x,y) = R(x,y) \wedge S(x,y)$. However, in case of binary relations, there are several other operations. The first operation we are interested in is the inversion of a binary **L**-relation.

Definition 4.1 The *inverse relation* of a binary **L**-relation R between $\langle X, \approx^X \rangle$ and $\langle Y, \approx^Y \rangle$ is the **L**-relation R^{-1} between $\langle Y, \approx^Y \rangle$ and $\langle X, \approx^X \rangle$ defined for any $x \in X$ and $y \in Y$ by

$$R^{-1}(y,x) = R(x,y).$$

Remark 4.2 Note that the compatibility of R^{-1} follows directly from the compatibility of R: $(y_1 \approx y_2) \otimes (x_1 \approx x_2) \otimes R^{-1}(y_1, x_1) \leq R(x_2, y_2) = R^{-1}(y_2, x_2)$.

Taking inverse relations has the following basic properties.

Lemma 4.3 *For $R, R_1, R_2 \in L^{X \times Y}$ we have*

$$\begin{aligned}
&R = (R^{-1})^{-1}, \\
&(\bigcap_i R_i)^{-1} = \bigcap_i R_i^{-1}, \\
&(\bigcup_i R_i)^{-1} = \bigcup_i R_i^{-1}, \\
&(R_1 \otimes R_2)^{-1} = R_1^{-1} \otimes R_2^{-1}, \\
&(R_1 \to R_2)^{-1} = R_1^{-1} \to R_2^{-1}, \\
&S(R_1, R_2) = S(R_1^{-1}, R_2^{-1}), \\
&(R_1 \approx R_2) = (R_1^{-1} \approx R_2^{-1}).
\end{aligned}$$

Proof. The proofs are trivial. As an illustration, we prove the third equality: $(R_1 \otimes R_2)^{-1}(y,x) = (R_1 \otimes R_2)(x,y) = R_1(x,y) \otimes R_2(x,y) = R_1^{-1}(y,x) \otimes R_2^{-1}(y,x) = (R_1^{-1} \otimes R_2^{-1})(y,x)$. □

The next operation we are going to introduce is the composition of binary fuzzy relations. This problem will be investigated in detail in Chapter 6. Here, we introduce only a special kind of composition called circlet composition (and will freely refer to properties of circlet composition that are proved in Chapter 6).

Definition 4.4 A *(circlet) composition* of an **L**-relation R between X and Y and an **L**-relation S between Y and Z (X, Y, Z equipped possibly with **L**-equivalences $\approx^X, \approx^Y, \approx^Z$, respectively) is the **L**-relation $R \circ S$ between X and Y defined by

$$(R \circ S)(x, z) = \bigvee_{y \in Y} R(x, y) \otimes S(y, z).$$

Remark 4.5 (1) As follows from results of Chapter 6, if R and S are compatible with $\approx^X, \approx^Y$ and $\approx^Y, \approx^Z$, respectively, then $R \circ S$ is compatible with $\approx^X, \approx^Z$. Moreover, $\circ$ is associative, i.e. we have $R \circ (S \circ T) = (R \circ S) \circ T$; and the relation to taking inverse relations is $(R \circ S)^{-1} = S^{-1} \circ R^{-1}$.

(2) The verbal description of circlet product is this one: the degree to which $\langle x, z\rangle$ belongs to $R \circ S$ equals the degree to which it is true that there exists $y \in Y$ such that $\langle x, y\rangle$ belongs to R and $\langle y, z\rangle$ belongs to S. Clearly, this generalizes the well-known composition of ordinary relations.

4.1.2 PROPERTIES OF BINARY FUZZY RELATIONS BETWEEN X AND Y

We are now going to introduce several important properties of binary fuzzy relations. We will be interested in properties that can be expressed by logical formulas. Such properties have a natural verbal description.

In this section, we will be interested in properties of binary fuzzy relations between possibly different sets. In subsequent sections, we will investigate binary fuzzy relations on a set. We start from a general point of view and define a general notion of a property given by a formula. This enables us to obtain some general results about properties of binary fuzzy relations at once by simple applications of results of Chapter 3. Then we define several particular properties by giving particular formulas. For each such property we list the corresponding expression describing the property in the way which is usual in the literature on fuzzy sets. This makes it possible to proceed also for those who read only Section 3.1 of Chapter 3.

The general notion of a property of a binary relation is introduced as follows. Consider a two-sorted language $\mathcal{J}_{\text{bin}}$ with sorts $\mathcal{X}, \mathcal{Y}$ of type $\langle R, F, \sigma\rangle$ with binary $r \in R$ such that $\sigma(r) = \mathcal{X}\mathcal{Y}$. Let φ be a formula that contains no relation or function symbols except for r, $\approx_{\mathcal{X}}$, and $\approx_{\mathcal{Y}}$, and has no free variables; denote by Prop the set of all such formulas.

The property of binary fuzzy relations that is determined by φ is defined as follows: Let R be an $\mathbf{L}$-relation between $\langle X, \approx^X \rangle$ and $\langle Y, \approx^Y \rangle$. Consider an $\mathbf{L}$-structure $\mathbf{M}$ given by $r^{\mathbf{M}} = R$, $\approx^{\mathbf{M}}_{\mathcal{X}}$ be $\approx^X$, $\approx^{\mathbf{M}}_{\mathcal{Y}}$ be $\approx^Y$.

Definition 4.6 The degree to which R satisfies the property given by φ, denoted by $\varphi(R)$, is defined by

$$\varphi(R) \; = \; \|\varphi\|_{\mathbf{M}}.$$

If R is an $\mathbf{L}$-relation on $\langle X, \approx^X \rangle$, one can consider $\mathcal{J}_{\text{bin}}$ a one-sorted language in an obvious way.

Remark 4.7 A property of a binary $\mathbf{L}$-relation is therefore represented by a closed formula of a two-sorted language with one binary relation symbol (and equality symbols). Given $\mathbf{L}$, X, and Y, the formula φ induces an $\mathbf{L}$-set in the set $L^{X \times Y}$ of all binary $\mathbf{L}$-relations between X and Y. A membership degree of any $R \in L^{X \times Y}$ to this $\mathbf{L}$-set is $\varphi(R)$. This $\mathbf{L}$-set can thus be interpreted as a (fuzzy) property: the truth degree to which R obeys this property is $\varphi(R)$.

The following result shows to what extent it is true that if R_1 has the property given by φ and if R_2 is similar to R_1, then R_2 has the property as well.

Lemma 4.8 *For* $\varphi \in \text{Prop}$ *we have*

$$(R_1 \approx R_2)^{|\varphi|_r} \otimes \varphi(R_1) \leq \varphi(R_2).$$

Proof. The assertion is a direct consequence of Lemma 3.85 and the above definitions. □

We are going to present the first group of properties.

Definition 4.9 Formulas Tot, Sur, Fun, Inj of $\mathcal{J}_{\text{bin}}$ are defined as follows:

$$\text{Tot is } (\forall \mathtt{x})(\exists \mathtt{y}) r(\mathtt{x}, \mathtt{y});$$

$$\text{Sur is } (\forall \mathtt{y})(\exists \mathtt{x}) r(\mathtt{x}, \mathtt{y});$$

$$\text{Fun is } (\forall \mathtt{x})(\forall \mathtt{y}_1, \mathtt{y}_2)((r(\mathtt{x}, \mathtt{y}_1) \otimes r(\mathtt{x}, \mathtt{y}_2)) \Rightarrow (\mathtt{y}_1 \approx_{\mathcal{Y}} \mathtt{y}_2));$$

$$\text{Inj is } (\forall \mathtt{x}_1, \mathtt{x}_2)(\forall \mathtt{y})((r(\mathtt{x}_1, \mathtt{y}) \otimes r(\mathtt{x}_2, \mathtt{y})) \Rightarrow (\mathtt{x}_1 \approx_{\mathcal{X}} \mathtt{x}_2)).$$

Formulas Tot, Sur, Fun, and Inj express properties that are called totality, surjectivity, functionality, and injectivity, respectively.

Remark 4.10 (1) For a binary **L**-relation R, $\mathrm{Tot}(R)$, $\mathrm{Sur}(R)$, $\mathrm{Fun}(R)$, and $\mathrm{Inj}(R)$ are called the degree to which R is total, surjective, functional, and injective, respectively.

(2) One can easily see that for $\mathbf{L} = \mathbf{2}$, the above properties generalize the corresponding well-known properties. For instance, for a binary relation R between X and Y it holds that R is a function of X into Y iff $\mathrm{Tot}(R) = 1$ and $\mathrm{Fun}(R) = 1$.

(3) We have

$$\begin{aligned}
\mathrm{Tot}(R) &= \bigwedge_{x\in X}\bigvee_{y\in Y} R(x,y),\\
\mathrm{Sur}(R) &= \bigwedge_{y\in Y}\bigvee_{x\in X} R(x,y),\\
\mathrm{Fun}(R) &= \bigwedge_{x\in X, y_1,y_2\in Y} (R(x,y_1)\otimes R(x,y_2)) \to (y_1 \approx^Y y_2),\\
\mathrm{Inj}(R) &= \bigwedge_{x_1,x_2\in X, y\in Y} (R(x_1,y)\otimes R(x_2,y)) \to (x_1 \approx^X x_2).
\end{aligned}$$

Theorem 4.11 (properties and similar fuzzy relations) *We have*

$$\begin{aligned}
(R_1 \approx R_2)\otimes \mathrm{Tot}(R_1) &\le \mathrm{Tot}(R_2),\\
(R_1 \approx R_2)\otimes \mathrm{Sur}(R_1) &\le \mathrm{Sur}(R_2),\\
(R_1 \approx R_2)^2\otimes \mathrm{Fun}(R_1) &\le \mathrm{Fun}(R_2),\\
(R_1 \approx R_2)^2\otimes \mathrm{Inj}(R_1) &\le \mathrm{Inj}(R_2).
\end{aligned}$$

Proof. The proof follows by Lemma 4.8 observing that $|\mathrm{Tot}|_r = 1$, $|\mathrm{Sur}|_r = 1$, $|\mathrm{Fun}|_r = 2$, and $|\mathrm{Inj}|_r = 2$. □

Lemma 4.12 *For a binary* **L**-*relation* R *between* $\langle X, \approx^X\rangle$ *and* $\langle Y, \approx^Y\rangle$ *we have* $\mathrm{Tot}(R) = \mathrm{Sur}(R^{-1})$ *and* $\mathrm{Fun}(R) = \mathrm{Inj}(R^{-1})$.

Proof. We have

$$\mathrm{Tot}(R) = \bigwedge_{x\in X}\bigvee_{y\in Y} R(x,y) = \bigwedge_{x\in X}\bigvee_{y\in Y} R^{-1}(y,x) = \mathrm{Sur}(R^{-1})$$

and

$$\begin{aligned}
\mathrm{Fun}(R) &= \bigwedge_{x\in X}\bigwedge_{y_1,y_2\in Y} ((R(x,y_1)\otimes R(x,y_2)) \to (y_1 \approx^Y y_2)) =\\
&= \bigwedge_{x\in X}\bigwedge_{y_1,y_2\in Y} ((R^{-1}(y_1,x)\otimes R^{-1}(y_2,x)) \to (y_1 \approx^Y y_2)) = \mathrm{Inj}(R^{-1}).
\end{aligned}$$

□

Remark 4.13 Clearly, the foregoing lemma implies that whatever is true about Tot and Fun, is true also for Sur and Inj. More precisely, if we replace in a true assertion Tot by Sur, and Fun by Inj, we get a true assertion.

We turn to compositions of fuzzy relations and the above properties.

Theorem 4.14 (closedness under composition) *For R between $\langle X, \approx^X \rangle$ and $\langle Y, \approx^Y \rangle$, and S between $\langle Y, \approx^Y \rangle$ and $\langle Z, \approx^Z \rangle$ we have*

$$\begin{aligned}
&\mathrm{Tot}(R) \otimes \mathrm{Tot}(S) \leq \mathrm{Tot}(R \circ S), \\
&\mathrm{Sur}(R) \otimes \mathrm{Sur}(S) \leq \mathrm{Sur}(R \circ S), \\
&\mathrm{Fun}(R) \otimes \mathrm{Fun}(S) \leq \mathrm{Fun}(R \circ S), \\
&\mathrm{Inj}(R) \otimes \mathrm{Inj}(S) \leq \mathrm{Inj}(R \circ S).
\end{aligned}$$

Proof. Due to Remark 4.13, it suffices to prove the first and the third inequality. The first one is true iff for each $x \in X$ we have $\mathrm{Tot}(R) \otimes \mathrm{Tot}(S) \leq \bigvee_{z \in Z}(R \circ S)(x, z)$. We have $\mathrm{Tot}(R) \otimes \mathrm{Tot}(S) = (\bigwedge_x \bigvee_y R(x,y)) \otimes (\bigwedge_y \bigvee_z S(y,z)) \leq (\bigvee_y R(x,y)) \otimes (\bigwedge_y \bigvee_z S(y,z)) = \bigvee_y (R(x,y) \otimes (\bigwedge_y \bigvee_z S(y,z))) \leq \bigvee_y (R(x,y) \otimes (\bigvee_z S(y,z))) = \bigvee_{z \in Z}(R \circ S)(x,z)$.

The third inequality is true iff for each $x \in X$, $z_1, z_2 \in Y$ we have $(R \circ S)(x, z_1) \otimes (R \circ S)(x, z_2) \otimes \mathrm{Fun}(R) \otimes \mathrm{Fun}(S) \leq (z_1 \approx^Z z_2)$. We have

$$\begin{aligned}
& (R \circ S)(x, z_1) \otimes (R \circ S)(x, z_2) \otimes \mathrm{Fun}(R) \otimes \mathrm{Fun}(S) = \\
= \; & (\bigvee_{y_1} R(x, y_1) \otimes S(y_1, z_1)) \otimes (\bigvee_{y_2} R(x, y_2) \otimes S(y_2, z_2)) \otimes \mathrm{Fun}(R) \otimes \\
& \otimes \mathrm{Fun}(S) \\
\leq \; & \bigvee_{y_1, y_2} (R(x, y_1) \otimes R(x, y_2) \otimes ((R(x, y_1) \otimes R(x, y_2)) \to (y_1 \approx^Y y_2)) \\
& \otimes S(y_1, z_1) \otimes S(y_2, z_2) \otimes \mathrm{Fun}(S)) \\
\leq \; & \bigvee_{y_1, y_2} ((y_1 \approx^Y y_2) \otimes S(y_1, z_1) \otimes S(y_2, z_2) \otimes \mathrm{Fun}(S)) \leq \\
\leq \; & \bigvee_{y_1, y_2} (S(y_2, z_1) \otimes S(y_2, z_2) \otimes \mathrm{Fun}(S)) \leq \cdots \leq (z_1 \approx^Z z_2).
\end{aligned}$$

The proof is complete. □

Verbal description of Theorem 4.14: composition of total (surjective, functional, injective) fuzzy relations is a total (surjective, functional, injective) fuzzy relation.

Notice that $\mathrm{Tot}(R) = a$ does not imply that for each $x \in X$ there is some $y \in Y$ such that $R(x, y) \geq a$; similarly for $\mathrm{Sur}(R)$. As a counterexample, consider $L = [0, 1]$, $X = Y = \mathbf{N}$, and R being defined by

$$R(x, y) = \frac{y-1}{y}.$$

Then $\mathrm{Tot}(R) = 1$ but there is no pair x, y for which $R(x, y) = 1$. Therefore, we put

$$\mathrm{Tot}_w(R) = \begin{cases} \mathrm{Tot}(R) & \text{if for each } x \in X \text{ there is } y \in Y \\ & \quad \text{such that } R(x, y) \geq \mathrm{Tot}(R) \\ 0 & \text{otherwise} \end{cases}$$

and

$$\mathrm{Sur}_w(R) = \begin{cases} \mathrm{Sur}(R) & \text{if for each } y \in Y \text{ there is } x \in X \\ & \quad \text{such that } R(x, y) \geq \mathrm{Sur}(R) \\ 0 & \text{otherwise.} \end{cases}$$

The meaning of $\mathrm{Tot}_w(R)$ is that R is total and has "witnesses"; similarly for $\mathrm{Sur}_w(R)$. If there is no danger of confusion we denote (some) witness by $w(x)$ (i.e. $R(x, w(x)) \geq \mathrm{Tot}(R)$) and $w(y)$ (i.e. $R(w(y), y) \geq \mathrm{Sur}(R)$).

Remark 4.15 Note that in the bivalent case, saying that that R is total and saying that R is total with witnesses are equivalent; similarly, saying that R is surjective and saying that R is surjective with witnesses are equivalent.

Theorem 4.16 (cancellation by fuzzy relations) *For R, R_1, R_2 between $\langle X, \approx^X\rangle$ and $\langle Y, \approx^Y\rangle$, and S, S_1, S_2 between $\langle Y, \approx^Y\rangle$ and $\langle Z, \approx^Z\rangle$ we have that $\mathrm{Sur}_w(R) = 1$ implies*

$$\mathrm{Fun}(R) \otimes (R \circ S_1 \approx R \circ S_2) \leq (S_1 \approx S_2)$$

and $\mathrm{Tot}_w(S) = 1$ implies

$$\mathrm{Inj}(S) \otimes (R_1 \circ S \approx R_2 \circ S) \leq (R_1 \approx R_2).$$

Proof. We prove only the first inequality; this is sufficient due to Remark 4.13. Thus, let $\mathrm{Sur}_w(R) = 1$. A moment's reflection shows that we have to check that for each $\langle y, z\rangle \in Y \times Z$ we have $S_1(y, z) \otimes \mathrm{Fun}(R) \otimes (R \circ S_1 \approx R \circ S_2) \leq S_2(y, z)$. We have $S_1(y, z) \otimes \mathrm{Fun}(R) \otimes (R \circ S_1 \approx R \circ S_2) \leq \bigwedge_x (S_1(y, z) \otimes (S_1(y, z) \to (R(x, y) \to (R \circ S_2)(x, z)))) \otimes \mathrm{Fun}(R) \leq$

$\bigwedge_x(R(x,y) \to (R \circ S_2)(x,z)) \otimes \mathrm{Fun}(R) \leq (R(w(y),y) \to (R \circ S_2)(w(y),z)) \otimes \mathrm{Fun}(R) = (1 \to (R \circ S_2)(w(y),z)) \otimes \mathrm{Fun}(R) = (R \circ S_2)(w(y),z) \otimes \mathrm{Fun}(R) = \bigvee_{y'}(R(w(y),y') \otimes S_2(y',z) \otimes \mathrm{Fun}(R)) = \bigvee_{y'}(S_2(y',z) \otimes R(w(y),y') \otimes R(w(y),y) \otimes \mathrm{Fun}(R)) \leq \bigvee_{y'}(S_2(y',z) \otimes (y' \approx^Y y)) \leq S_2(y,z)$, completing the proof. □

Our aim in the following is to discover conditions under which it is possible to define a function from a fuzzy relation between X and Y. If $f : X \to Y$ is a function compatible with $\approx^X$ and $\approx^Y$, one can consider an **L**-relation R_f between X and Y defined by

$$R_f(x,y) = (f(x) \approx^Y y).$$

R_f is compatible with $\approx^X$ and $\approx^Y$. Indeed, $R_f(x,y) \otimes (x \approx^X x') \otimes (y \approx^Y y') = (f(x) \approx^Y y) \otimes (x \approx^X x') \otimes (y \approx^Y y') \leq (f(x) \approx^Y y) \otimes (f(x) \approx^Y f(x')) \otimes (y \approx^Y y') \leq (f(x') \approx^X y') = R_f(x',y')$. Clearly, $\mathrm{Tot}_w(R_f) = 1$ (namely, one can take $w(x) = f(x)$ for the witness). Furthermore, $\mathrm{Fun}(R_f) = 1$. Indeed, $\mathrm{Fun}(R_f) = 1$ means that for each $x \in X$, y_1, y_2, we have $R_f(x,y_1) \otimes R_f(x,y_2) \leq (y_1 \approx^Y y_2)$, i.e. $(f(x) \approx^Y y_1) \otimes (f(x) \approx^Y y_2) \leq (y_1 \approx^Y y_2)$ which is true by transitivity of $\approx^Y$.

Conversely, if R is an **L**-relation between $\langle X, \approx^X\rangle$ and $\langle Y, \approx^Y\rangle$ such that $\mathrm{Tot}_w(R) = 1$, one can define a function $f_R : X \to Y$ such that $R(x, f_R(x)) = 1$. Indeed, $f_R(x)$ can be any witness of x. If, moreover, $\mathrm{Fun}(R) = 1$ and $\approx^Y$ is an **L**-equality, $f_R(x)$ is determined uniquely. For if $R(x,y_1) = 1$ and $R(x,y_2) = 1$, then $1 = 1 \otimes 1 = R(x,y_1) \otimes R(x,y_2) \leq (y_1 \approx^Y y_2)$ and thus $y_1 = y_2$.

Theorem 4.17 (functions vs. fuzzy relations) *Let $\approx^Y$ be an* **L***-equality on Y. Then the above-defined mappings sending f to R_f, and R to f_R, are mutually inverse bijections between the set of all functions of X to Y which are compatible with $\approx^X$ and $\approx^Y$ and the set of all* **L***-relations R between X and Y which are compatible with $\approx^X$ and $\approx^Y$, and for which* $\mathrm{Tot}_w(R) = 1$ *and* $\mathrm{Fun}(R) = 1$*, i.e. one has $f = f_{R_f}$ and $R = R_{f_R}$.*

Proof. Due to the above considerations we have to verify that f_R is compatible with $\approx^X$ and $\approx^Y$ whenever R is, and that $f = f_{R_f}$ and $R = R_{f_R}$.

Compatibility of f_R means $(x_1 \approx^X x_2) \leq (f_R(x_1) \approx^Y f_R(x_2))$. Since $R(x_i, f_R(x_i)) = 1$, $\mathrm{Fun}(R) = 1$ and compatibility of R yield $(x_1 \approx^X x_2) = (x_1 \approx^X x_2) \otimes R(x_1, f_R(x_1)) \otimes R(x_2, f_R(x_2)) \leq R(x_2, f_R(x_1)) \otimes R(x_2, f_R(x_2)) \otimes \mathrm{Fun}(R) \leq (f_R(x_1) \approx^Y f_R(x_2))$, verifying compatibility of f_R.

$f_{R_f}(x)$ is the unique element for which $R_f(x, f_{R_f}(x)) = 1$ and thus $f_{R_f}(x) = f(x)$, proving $f = f_{R_f}$. Moreover, $R_{f_R}(x,y) = (f_R(x) \approx^Y y) = 1 \otimes (f_R(x) \approx^Y y) = R(x, f_R(x)) \otimes (f_R(x) \approx^Y y) \leq R(x,y)$. Conversely, $R(x,y) = R(x,y) \otimes R(x, f_R(x)) \leq (y \approx^Y f_R(x)) = R_{f_R}(x,y)$, proving $R = R_{f_R}$. □

4.1.3 BINARY FUZZY RELATIONS ON A SET

We turn to binary fuzzy relations on a set X. This is in fact a special situation of that one of the previous section (namely, here we have $\langle Y, \approx^Y \rangle = \langle X, \approx^X \rangle$). Therefore, results of the previous section apply also here. We will be interested in fuzzy relations that are typically defined on a single set, and in their properties. As before, we introduce the properties via formulas of first-order fuzzy logic; a fuzzy relation satisfies a given property in degree to which the corresponding formula is true (in a fuzzy structure given by the fuzzy relation). We thus proceed exactly as in Definition 4.6, only that we assume for simplicity that $\mathcal{J}_{\text{bin}}$ is a one-sorted language. The properties we are interested in are represented by the following formulas.

Definition 4.18 We introduce the following formulas of $\mathcal{J}_{\text{bin}}$:

$$\begin{aligned}
&\text{Ref is } (\forall x)(r(x,x)),\\
&\text{Sym is } (\forall x,y)(r(x,y) \Rightarrow r(y,x)),\\
&\text{Tra is } (\forall x,y,z)((r(x,y) \otimes r(y,z)) \Rightarrow r(x,z)),\\
&\text{Ans is } (\forall x,y)((r(x,y) \wedge r(y,x)) \Rightarrow (x \approx y)),\\
&\text{Irr is } (\forall x) \neg r(x,x),\\
&\text{Asy is } (\forall x,y)(r(x,y) \Rightarrow \neg r(y,x)).
\end{aligned}$$

The properties represented by Ref, Sym, Tra, Ans, Irr, and Asy are called *reflexivity*, *symmetry*, *transitivity*, *antisymmetry*, *irreflexivity*, and *asymmetry*, respectively.

Thus, for a binary **L**-relation R on X (which is compatible with $\approx^X$), the degree Ref(R), ..., Asy(R) is called the degree to which R is reflexive, ..., asymmetric, respectively. Fixing **L**, X (and $\approx^X$), the above formulas Ref, ... thus define **L**-sets in the universe of all binary **L**-relations R on X (compatible with $\approx^X$), Ref(R), ... being the degree to which R belongs to the **L**-set Ref of reflexive **L**-relations, ... (since there is no danger of confusion, we denote the **L**-sets the same way as the corresponding formulas, e.g. Ref is the **L**-set of **L**-relations given by formula Ref).

Note that $\text{Ref}(R)$ is by definition equal to $\|\text{Ref}\|_{\mathbf{M}}$ where $\mathbf{M}$ is the $\mathbf{L}$-structure given by R, i.e. $M = X$ and $r^{\mathbf{M}} = R$ (and $\approx^{\mathbf{M}} = \approx^X$). The rules of evaluating truth degrees of formulas thus give

$$\begin{aligned}
\text{Ref}(R) &= \bigwedge_{x \in X} R(x,x), \\
\text{Sym}(R) &= \bigwedge_{x,y \in X} (R(x,y) \to R(y,x)), \\
\text{Tra}(R) &= \bigwedge_{x,y,z \in X} (R(x,y) \otimes R(y,z)) \to R(x,z), \\
\text{Ans}(R) &= \bigwedge_{x,y \in X} (R(x,y) \wedge R(y,x)) \to (x \approx^X y), \\
\text{Irr}(R) &= \bigwedge_{x \in X} \neg R(x,x), \\
\text{Asy}(R) &= \bigwedge_{x,y \in X} R(x,y) \to \neg R(y,x).
\end{aligned}$$

We say that R is reflexive, ..., asymmetric if $\text{Ref}(R) = 1$, ..., $\text{Asy}(R) = 1$ which means (easy observation) that for each $x, y, z \in X$,

$$\begin{aligned}
&R(x,x) = 1, \\
&R(x,y) = R(y,x), \\
&R(x,y) \otimes R(y,z) \leq R(x,z), \\
&R(x,y) \wedge R(y,x) \leq (x \approx^X y), \\
&R(x,x) = 0, \\
&R(x,y) \otimes R(y,x) = 0.
\end{aligned}$$

Remark 4.19 We thus see that the definition of reflexivity, symmetry, and transitivity of Section 3.1 is a particular case of the approach used here.

More generally, we say that R is a-reflexive, ..., a-asymmetric ($a \in L$), if $a \leq \text{Ref}(R)$, ..., $a \leq \text{Asy}(R)$. Using a-cuts, this can be equivalently expressed by saying that R belongs to ${}^a\text{Ref}$, ..., ${}^a\text{Asy}$.

Theorem 4.20 (properties and similar fuzzy relations) *For* $\mathbf{L}$*-relations* R_1, R_2 *in* $\langle X, \approx^X \rangle$ *we have*

$$\begin{aligned}
(R_1 \approx R_2) \otimes \text{Ref}(R_1) &\leq \text{Ref}(R_2), \\
(R_1 \approx R_2)^2 \otimes \text{Sym}(R_1) &\leq \text{Sym}(R_2), \\
(R_1 \approx R_2)^3 \otimes \text{Tra}(R_1) &\leq \text{Tra}(R_2),
\end{aligned}$$

$$(R_1 \approx R_2) \otimes \mathrm{Ans}(R_1) \leq \mathrm{Ans}(R_2),$$
$$(R_1 \approx R_2) \otimes \mathrm{Irr}(R_1) \leq \mathrm{Irr}(R_2),$$
$$(R_1 \approx R_2)^2 \otimes \mathrm{Asy}(R_1) \leq \mathrm{Asy}(R_2).$$

Proof. The proof follows by Lemma 4.8 observing that $|\mathrm{Ref}|_r = 1$, $|\mathrm{Sym}|_r = 2$, $|\mathrm{Tra}|_r = 3$, $|\mathrm{Ans}|_r = 1$, $|\mathrm{Irr}|_r = 1$, and $|\mathrm{Asy}|_r = 2$. □

We have an alternative definition of the properties using operations and relations on fuzzy sets as follows.

Lemma 4.21 *We have*

$$\begin{aligned}
\mathrm{Ref}(R) &= S(X, R),\\
\mathrm{Sym}(R) &= (R \approx R^{-1}),\\
\mathrm{Tra}(R) &= S(R \circ R, R),\\
\mathrm{Ans}(R) &= S(R \cap R^{-1}, \approx^X),\\
\mathrm{Irr}(R) &= (R \cap \mathrm{id}_X \approx \emptyset),\\
\mathrm{Asy}(R) &= (R \otimes R^{-1} \approx \emptyset).
\end{aligned}$$

Proof. The assertions are easy reformulations of the above semantic definitions of $\mathrm{Ref}(R), \ldots, \mathrm{Asy}(R)$. As an illustration we show $\mathrm{Tra}(R) = S(R \circ R, R)$. We have $\mathrm{Tra}(R) = \bigwedge_{x,y,z \in X}(R(x,y) \otimes R(y,z)) \to R(x,z)$ and $S(R \circ R, R) = \bigwedge_{x,z}(\bigvee_y R(x,y) \otimes R(y,z)) \to R(x,z)$. Clearly, $S(R \circ R, R) \leq \mathrm{Tra}(R)$. To show the converse, we have to check that for each $x, z \in X$ we have $(\bigvee_y R(x,y) \otimes R(y,z)) \otimes \mathrm{Tra}(R) \leq R(x,z)$. However, $(\bigvee_y R(x,y) \otimes R(y,z)) \otimes \mathrm{Tra}(R) = \bigvee_y (R(x,y) \otimes R(y,z) \otimes \mathrm{Tra}(R)) \leq \bigvee_y (R(x,y) \otimes R(y,z) \otimes (R(x,y) \otimes R(y,z) \to R(x,z))) \leq R(x,z)$. □

Closedness of the above properties under intersection is the subject of the following assertion.

Theorem 4.22 *For* **L***-relations R_i $(i \in I)$ on X we have*

$$\bigwedge_i \mathrm{Ref}(R_i) = \mathrm{Ref}(\bigcap_i R_i),$$
$$\bigwedge_i \mathrm{Sym}(R_i) \leq \mathrm{Sym}(\bigcap_i R_i),$$
$$\bigwedge_i \mathrm{Tra}(R_i) \leq \mathrm{Tra}(\bigcap_i R_i),$$
$$\bigwedge_i \mathrm{Ans}(R_i) \leq \mathrm{Ans}(\bigcap_i R_i),$$

$$\bigwedge_i \mathrm{Irr}(R_i) \le \mathrm{Irr}(\bigcap_i R_i),$$
$$\bigwedge_i \mathrm{Asy}(R_i) \le \mathrm{Asy}(\bigcap_i R_i).$$

Proof. The proofs are easy; as an illustration we prove the third inequality. $\bigwedge_i \mathrm{Tra}(R_i) \le \mathrm{Tra}(\bigcap_i R_i)$ is true iff for each $x, y, z \in X$ and each $i \in I$ we have $R_i(x,y) \otimes R_i(y,z) \otimes \bigwedge_i \mathrm{Tra}(R_i) \le R_i(x,z)$. This is true since $R_i(x,y) \otimes R_i(y,z) \otimes \bigwedge_i \mathrm{Tra}(R_i) \le R_i(x,y) \otimes R_i(y,z) \otimes ((R_i(x,y) \otimes R_i(y,z)) \to R_i(x,z)) \le R_i(x,z)$. □

Remark 4.23 (1) Note that Lemma 3.75 showing that an intersection of fuzzy equivalences is a fuzzy equivalence is an immediate consequence of the first three inequalities. Namely, if R_i are fuzzy equivalences, then $\mathrm{Ref}(R_i) = \mathrm{Sym}(R_i) = \mathrm{Tra}(R_i) = 1$ and thus also $\mathrm{Ref}(\bigcap_i R_i) = 1$, $\mathrm{Sym}(\bigcap_i R_i) = 1$, $\mathrm{Tra}(\bigcap_i R_i) = 1$, whence $\bigcap_i R_i$ is reflexive, symmetric, and transitive, i.e. it is a fuzzy equivalence.

(2) An easy consequence of Theorem 4.22 is that the set of all a-reflexive (a-symmetric, ...,a-transitive) fuzzy relations on X is a complete lattice w.r.t. $\subseteq$ where the meet coincides with intersection.

Closures w.r.t. properties. Given an **L**-relation R on X it may be desirable to have the least **L**-relation R' on X that contains R (i.e. $R \subseteq R'$) and satisfies some prescribed property. Such R' is called a *closure* of R w.r.t. the property but it does not exist in general. However, if the full **L**-relation $X \times X$ satisfies the property and if the property is closed w.r.t. intersection (i.e. $\bigcap_i R_i$ satisfies the property whenever each of R_i does), then R' always exists; namely R' is the intersection of all R'' such that R'' has the property and contains R.

In the following, we investigate closures of binary fuzzy relations w.r.t. a-reflexivity, a-symmetry, and a-transitivity; we denote these closures by $\mathrm{C}_{a\mathrm{Ref}}$, $\mathrm{C}_{a\mathrm{Sym}}$, and $\mathrm{C}_{a\mathrm{Tra}}$, respectively. Since $X \times X$ is a-reflexive, a-symmetric, and a-transitive, and since each of these properties is closed under intersection (Theorem 4.22), the above consideration implies that

$$\mathrm{C}_{a\mathrm{Ref}} = \bigcap\{R'' \mid R \subseteq R'',\ a \le \mathrm{Ref}(R'')\},$$
$$\mathrm{C}_{a\mathrm{Sym}} = \bigcap\{R'' \mid R \subseteq R'',\ a \le \mathrm{Sym}(R'')\},$$
$$\mathrm{C}_{a\mathrm{Tra}} = \bigcap\{R'' \mid R \subseteq R'',\ a \le \mathrm{Tra}(R'')\}.$$

We now present explicit expressions for the three closures. To this end let for R and $n \in \mathbf{N}$, $R^{\circ n}$ be defined by $R^{\circ 1} = R$; $R^{\circ(n+1)} = R \circ R^{\circ n}$

(this is correct on account of associativity of $\circ$). That is, $R^{\circ 2} = R \circ R$, $R^{\circ 3} = R \circ R \circ R$.

Theorem 4.24 (closures) *For a binary* $\mathbf{L}$*-relation* R *on* X *and* $a \in L$ *we have*

$$\begin{aligned} \mathrm{C}_{^a\mathrm{Ref}} &= R \cup a \otimes \mathrm{id}_X, \\ \mathrm{C}_{^a\mathrm{Sym}} &= R \cup a \otimes R^{-1}, \\ \mathrm{C}_{^a\mathrm{Tra}} &= R \cup \bigcup_{n \in \mathbf{N}} a^n \otimes R^{\circ(n+1)}. \end{aligned}$$

Proof. Put $R' = R \cup a \otimes \mathrm{id}_X$. Clearly, R' is a-reflexive because $(a \otimes \mathrm{id}_X)(x,x) = a$. If R'' is a-reflexive and contains R then clearly $a \otimes \mathrm{id}_X \subseteq R''$ and $R \subseteq R''$ whence also $R' \subseteq R''$ proving $R' = \mathrm{C}_{^a\mathrm{Ref}}$.

Put $R' = R \cup a \otimes R^{-1}$. R' is a-symmetric: $a \leq \bigwedge_{x,y}(R'(x,y) \to R'(y,x))$ iff for each x, y we have $a \otimes R'(x,y) \leq R'(y,x)$, i.e. iff $a \otimes (R(x,y) \vee a \otimes R^{-1}(x,y)) \leq R(y,x) \vee a \otimes R^{-1}(y,x)$. This is indeed true: $a \otimes (R(x,y) \vee a \otimes R^{-1}(x,y)) = a \otimes R(x,y) \vee a^2 \otimes R(y,x) \leq R(y,x) \vee a \otimes R(x,y)$. If R'' is a-symmetric and contains R then, since $a \leq R''(x,y) \to R''(y,x)$, clearly $a \otimes R^{-1} \subseteq a \otimes (R'')^{-1} \subseteq R''$ and $R \subseteq R''$ whence also $R' \subseteq R''$ proving $R' = \mathrm{C}_{^a\mathrm{Sym}}$.

Finally, put $R' = R \cup \bigcup_{n \in \mathbf{N}} a^n \otimes R^{\circ(n+1)}$. To show that R' is a-transitive we have to check $a \leq \mathrm{Tra}(R')$ which holds true iff $a \leq S(R' \circ R', R')$ which is true iff for each x, z we have $a \otimes (R' \circ R')(x,z) \leq R'(x,z)$. We have $a \otimes (R' \circ R')(x,z) = a \otimes \bigvee_y (R'(x,y) \otimes R'(y,z)) = \bigvee_y (a \otimes \bigvee_{n \geq 0} a^n \otimes R^{\circ(n+1)}(x,y) \otimes \bigvee_{m \geq 0} a^m \otimes R^{\circ(m+1)}(y,z)) \leq R'(x,z)$, proving that R' is a-transitive. If R'' is a-transitive and contains R then, since $a \otimes R''(x,y) \otimes R''(y,z) \leq R''(x,z)$, and thus $a^n \otimes R''^{\circ(n+1)} \subseteq R''$ (by induction on n). From $R \subseteq R''$ we thus have $a^n \otimes R^{\circ(n+1)} \subseteq a^n \otimes R''^{\circ(n+1)}$ for each n, whence $R' \subseteq R''$ proving $R' = \mathrm{C}_{^a\mathrm{Tra}}$. □

Remark 4.25 (1) For $\mathbf{L} = \mathbf{2}$, the closures (for $a = 1$) are the well-known reflexive, symmetric, and transitive closure of a binary relation.

(2) If there exists $n_0 \in \mathbf{N}$ such that $a^{n_0} = 0$ then, obviously, $\mathrm{C}_{^a\mathrm{Tra}} = R \cup \bigcup_{n \leq n_0} a^n \otimes R^{\circ(n+1)}$.

4.2 SIMILARITY: FUZZY EQUIVALENCES

This section is devoted to fuzzy equivalences (also called similarity relations) and fuzzy tolerances, i.e. to binary fuzzy relations that serve to model the graded similarity phenomenon. Note that we already presented definitions and basic properties of a fuzzy equivalence in Section 3.1.

Note also that some further properties of fuzzy equivalences will be given in the last part of Section 7.1.

Recall that *fuzzy equivalence* is a binary fuzzy relation which is reflexive, symmetric, and transitive. In this section we consider **L**-equivalences defined on a set X which is in general equipped with an **L**-equality relation $\approx^X$ (written simply also $\approx$); furthermore, we assume that the **L**-equivalences are compatible with $\approx^X$. If $\approx^X$ is the crisp identity on X (i.e. $(x \approx^X y) = 1$ for $x = y$ and $(x \approx^X y) = 0$ for $x \neq y$), this in fact means that we consider an **L**-equivalence on a set X. The **L**-equivalences on X will usually be denoted by $\theta, \phi, \ldots, E$ (possibly with indices), etc.

The primary intended meaning of the notion of a fuzzy equivalence is (gradual) similarity (thereof the term similarity used for fuzzy equivalences). That is, for a fuzzy equivalence θ, $\theta(x, y)$ is interpreted as the truth degree to which x and y are similar. There is an important difference between fuzzy equivalences and ordinary equivalences w.r.t. the question of whether (fuzzy) equivalence is an appropriate concept for modeling of similarity.

To elucidate this, recall a classical argument against the claim that ordinary equivalence captures similarity: If similarity is well-formalized by axioms of ordinary equivalence then being similar is a transitive relationship. However, this is not the case since it may happen that x is similar to y, and y is similar to z, still x is not similar to z. The argument goes further by pointing to an example where the universe is the set of real numbers and being similar for x and y means that the distance $|x - y|$ is less than or equal to some fixed q (threshold). It is argued that this is just an example of a similarity relation which is (reflexive, symmetric but) not transitive. In spite of this, one still feels that if x and y are similar and y and z are similar, then x is somewhat similar to z. Fuzzy approach solves this situation in a very natural way. The core is that similarity is a graded notion. Now, suppose that the degree to which x is similar to y is 0.8 and that the degree to which y is similar to z is 0.7. Intuitively, it may happen that the degree to which x is similar to z somewhat depends on 0.8 and 0.7, however, it may be strictly less than 0.8 and 0.7. Suppose that the t-norm $\otimes$ we use is the product. Then transitivity says that the degree to which x is similar to z is at least $0.7 \cdot 0.8 = 0.56$ (indeed, if E is the fuzzy equivalence then we have $0.56 = E(x, y) \otimes E(y, z) \leq E(x, z)$). The exact values do not matter for the argument, one may use other t-norms.

Therefore, using fuzzy approach, it is not true that transitivity is not natural. On the other hand, it captures well the intuition that graded similarity may decrease as one goes through a chain of pairwise similar elements.

A binary fuzzy relation is called a *fuzzy tolerance* (**L**-tolerance) if it is both reflexive and symmetric. Contrary to the bivalent case, if a fuzzy tolerance is not transitive, it may still have a property close to transitivity simply because transitivity is a graded property in fuzzy setting. Obviously, the notions of a fuzzy tolerance and a 0-transitive, reflexive, and symmetric fuzzy relation coincide.

For a set X and an **L**-equality $\approx$ on X we denote by $\mathrm{Tol}(X,\approx)$ (or simply by $\mathrm{Tol}(X)$) the set of all **L**-tolerances on $\langle X,\approx\rangle$, and by $\mathrm{Eqv}(X,\approx)$ (or simply by $\mathrm{Eqv}(X)$) the set of all **L**-equivalences on $\langle X,\approx\rangle$. Furthermore, $\mathrm{Sim}(X,\approx)$ (or simply $\mathrm{Sim}(X)$) will denote the **L**-set in $\mathrm{Tol}(X)$ defined by

$$(\mathrm{Sim}(X))(R) \;=\; \mathrm{Tra}(R),$$

i.e. $(\mathrm{Sim}(X))(R)$ is the degree to which an **L**-tolerance R is transitive. Then, ${}^a\mathrm{Sim}(X)$ is the set of all **L**-tolerances which are a-transitive. Clearly, we have

$$\mathrm{Eqv}(X) = {}^1\mathrm{Sim}(X) \subseteq {}^a\mathrm{Sim}(X) \subseteq \mathrm{Tol}(X)$$

for each $a \in L$.

Theorem 4.22 (see also Remark 4.23) implies that ${}^a\mathrm{Sim}(X)$ equipped with $\subseteq$ is a complete lattice where infimum coincides with intersection (note that $\theta_1 \subseteq \theta_2$ means that for each $x, y \in X$ we have $\theta_1(x,y) \leq \theta_2(x,y)$). The following theorem also provides description of suprema.

Theorem 4.26 (lattices of similarities) *For each* $a \in L$, $\langle {}^a\mathrm{Sim}(X), \subseteq\rangle$ *is a complete lattice where infima are given by*

$$\bigwedge_{i\in I} \theta_i = \bigcap_{i\in I} \theta_i$$

and for suprema we have

$$\bigvee_{i\in I} \theta_i = \bigcup_{n\geq 0, i_1,\ldots,i_{n+1}\in I} a^n \otimes \theta_{i_1} \circ \cdots \circ \theta_{i_{n+1}}.$$

Proof. Due to the above remarks, we only have to verify the formula for suprema. Denote the **L**-relation defined by the right-hand side by θ. Clearly, $\theta_i \subseteq \theta$ for each $i \in I$ ($\theta_i = a^0 \otimes \theta_i$ is a member of the union). Clearly, θ is reflexive and symmetric. That θ is a-transitive may be shown analogously as in the proof of Theorem 4.24 where we showed that R' is a-transitive (the situation here is more tedious but straightforward). Now, it remains to show that each a-transitive **L**-equivalence θ' that contains each θ_i, also contains θ, i.e. that θ' contains each member $a^n \otimes \theta_{i_1} \circ \cdots \circ \theta_{i_{n+1}}$ of the union. Since $\theta_i \subseteq \theta'$, we have

$a^n \otimes \theta_{i_1} \circ \cdots \circ \theta_{i_{n+1}} \subseteq a^n \otimes \theta'^{\circ(n+1)}$. Therefore, it suffices to show that θ' contains $a^n \otimes \theta'^{\circ(n+1)}$. This follows by induction: for $n = 0$ the assertion is trivial. Let the assertion be valid for $n-1$, i.e. suppose $a^{n-1} \otimes \theta'^{\circ n} \subseteq \theta'$. Recall that a-transitivity of θ' is equivalent to $a \leq S(\theta' \circ \theta', \theta')$, i.e. to $a \otimes (\theta' \circ \theta') \subseteq \theta'$. Therefore, we have $a^n \otimes \theta'^{\circ(n+1)} = a \otimes a^{n-1} \otimes \theta'^{\circ n} \circ \theta' \subseteq a \otimes (\theta' \circ \theta') \subseteq \theta'$ completing the induction step. The proof is complete. □

Corollary 4.27 *For suprema in* Tol(X) *we have*

$$\bigvee_{i \in I} \theta_i = \bigcup_{i \in I} \theta_i,$$

for suprema in Eqv(X) *we have*

$$\bigvee_{i \in I} \theta_i = \bigcup_{n \geq 1, i_1, \ldots, i_n \in I} \theta_{i_1} \circ \cdots \circ \theta_{i_n}.$$

In the following we show ways to represent (and induce) fuzzy equivalences.

Fuzzy equivalences and fuzzy partitions. Recall that for a fuzzy equivalence θ on X, $[x]_\theta$ is the **L**-set in X defined by $[x]_\theta(y) = \theta(x, y)$.

Definition 4.28 An **L**-*partition* of a set $\langle X, \approx \rangle$ is any set Π of **L**-sets in $\langle X, \approx \rangle$ such that

(i) for each $x \in X$ there is $A \in \Pi$ such that $A(x) = 1$;

(ii) for each $A \in \Pi$ there is $x \in X$ such that $A(x) = 1$;

(iii) for every $A, B \in \Pi$, $x \in X$ we have $A(x) \otimes B(x) \leq (A \approx B)$.

Remark 4.29 (1) It is easy easy to see that for $\mathbf{L} = \mathbf{2}$, the concept of an **L**-partition coincides with the concept of a partition ((i) says that Π covers X; (ii) says that each $A \in \Pi$ is nonempty; (iii) says that if x belongs to A and B then A and B are equal).

(2) We immediately have some consequences: (iii) implies that if $A(x) = 1$ and $B(x) = 1$ for some $A, B \in \Pi$, then $A = B$.

Theorem 4.30 (fuzzy equivalences and partitions) *Let θ be an* **L**-*equivalence in $\langle X, \approx \rangle$, Π be an* **L**-*partition in $\langle X, \approx \rangle$. Define a binary* **L**-*relation θ_Π in X by*

$$\theta_\Pi(x, y) = A_x(y)$$

where $A_x \in \Pi$ *is such that* $A_x(x) = 1$, *and put* $\Pi_\theta = \{[x]_\theta \mid x \in X\}$. *Then (1)* Π_θ *is an* $\mathbf{L}$*-partition of* $\langle X, \approx\rangle$; *(2)* θ_Π *is an* $\mathbf{L}$*-equivalence in* $\langle X, \approx\rangle$; *and (3)* $\theta = \theta_{\Pi_\theta}$ *and* $\Pi = \Pi_{\theta_\Pi}$.

Proof. (1): We verify that Π_θ is an $\mathbf{L}$-partition of $\langle X, \approx\rangle$: Since

$$[x]_\theta(y) \otimes (y \approx z) = \theta(x,y) \otimes (y \approx z) \leq \theta(x,z) = [x]_\theta(z),$$

each $[x]_\theta \in \Pi$ is compatible w.r.t. $\approx$. By reflexivity of θ, $[x]_\theta(x) = 1$ verifying (i) and (ii). To verify (iii) we have to show that for every $x, y, z \in X$ we have $[x]_\theta(z) \otimes [y]_\theta(z) \leq ([x]_\theta \approx [y]_\theta)$ which holds iff for each $u \in X$ we have both

$$[x]_\theta(z) \otimes [y]_\theta(z) \leq [x]_\theta(u) \to [y]_\theta(u)$$

and

$$[x]_\theta(z) \otimes [y]_\theta(z) \leq [y]_\theta(u) \to [x]_\theta(u).$$

Due to symmetry of both of the cases we check only the first inequality which is equivalent to

$$[x]_\theta(u) \otimes [x]_\theta(z) \otimes [y]_\theta(z) \leq [y]_\theta(u),$$

i.e. to

$$\theta(u,x) \otimes \theta(x,z) \otimes \theta(z,y) \leq \theta(u,y),$$

which is true by transitivity of θ. Therefore, (1) is proved.

(2): We verify that θ_Π is an $\mathbf{L}$-equivalence on $\langle X, \approx\rangle$. First, notice that θ_Π is defined correctly. Indeed, by Remark 4.29 (2), there is just one $A_x \in \Pi$ such that $A_x(x) = 1$. Reflexivity of θ_Π is obvious. Symmetry: We have to show that if $A(x) = 1$ and $B(y) = 1$ for $x, y \in X$ and $A, B \in \Pi$, then $A(y) = B(x)$. By definition we have $A(y) \otimes B(y) \leq (A \approx B)$, hence $A(y) = A(y) \otimes 1 = A(y) \otimes B(y) \leq A(x) \to B(x) = 1 \to B(x) = B(x)$, i.e. $A(y) \leq B(x)$; similarly, one gets $B(x) \leq A(y)$, proving the required equality. Transitivity: We have to show $A_x(y) \otimes A_y(z) \leq A_x(z)$, i.e. $A_x(y) \leq A_y(z) \to A_x(z)$. The last inequality is true since $A_x(y) = A_x(y) \otimes A_y(y) \leq (A_x \approx A_y) \leq A_y(z) \to A_x(z)$.

(3): We show $\theta = \theta_{\Pi_\theta}$: $\theta_{\Pi_\theta}(x,y) = A(y)$ for $A \in \Pi_\theta$ such that $A(x) = 1$. Since $A = [x]_\theta$, we have $\theta_{\Pi_\theta}(x,y) = [x]_\theta(y) = \theta(x,y)$. We show $\Pi = \Pi_{\theta_\Pi}$: We have $A \in \Pi_{\theta_\Pi}$ iff $A = [x]_{\theta_\Pi}$. Now, $A(y) = [x]_{\theta_\Pi}(y) = \theta_\Pi(x,y) = A_x(y)$ for such an $A_x \in \Pi$ for which $A_x(x) = 1$. Thus, $A = A_x \in \Pi$. □

Fuzzy equivalences and compatible mappings. In Section 3.4 we introduced the notion of a factor fuzzy structure by a congruence. If the type of the structure has no function and relation symbol except for the equality symbol, we get the following notion of a factor set modulo a fuzzy equivalence.

Let θ be an $\mathbf{L}$-equivalence on a set $\langle X, \approx \rangle$. The factor set of $\langle X, \approx \rangle$ modulo θ, denoted by $\langle X, \approx \rangle/\theta$ or simply X/θ, is a set X/θ equipped with an $\mathbf{L}$-equality $\approx^{X/\theta}$ defined as follows: X/θ is $X/{}^1\theta$, i.e. the ordinary factor set of X modulo the ordinary equivalence ${}^1\theta$; $\approx^{X/\theta}$ is defined by

$$([x]_{^1\theta} \approx^{X/\theta} [y]_{^1\theta}) = \theta(x, y).$$

For a mapping $f : X \to Y$ compatible with $\mathbf{L}$-equalities $\approx^X$ on X and $\approx^Y$ on Y, θ_f denotes the $\mathbf{L}$-relation on X defined by $\theta_f(x, y) = (f(x) \approx^Y f(y))$. For an $\mathbf{L}$-equivalence θ on $\langle X, \approx^X \rangle$, f_θ denotes a mapping of X to X/θ defined by $f_\theta(x) = [x]_{^1\theta}$. Results of Section 3.4 imply the following assertion.

Theorem 4.31 (similarities as kernels of compatible mappings) *For $f : X \to Y$ compatible with $\mathbf{L}$-equalities $\approx^X$ and $\approx^Y$, θ_f is an $\mathbf{L}$-equivalence on $\langle X, \approx^X \rangle$. For an $\mathbf{L}$-equivalence θ on $\langle X, \approx^X \rangle$, $f_\theta : X \to X/\theta$ is compatible with $\approx^X$ and $\approx^{X/\theta}$ and we have $\theta = \theta_{f_\theta}$.*

Remark 4.32 One can easily see that for an $\mathbf{L}$-equivalence θ on $\langle X, \approx^X \rangle$, $\langle X, \approx^X \rangle/\theta$ and Π_θ (the corresponding $\mathbf{L}$-partition) are in fact isomorphic structures. Indeed, the mapping sending $[x]_{^1\theta}$ to $[x]_\theta$ is a bijection of X/θ to Π_θ. Furthermore, $([x]_\theta \approx [y]_\theta) = ([x]_{^1\theta} \approx^{x/\theta} [y]_{^1\theta})$ and thus $\langle X/\theta, \approx^{X/\theta} \rangle$ and $\langle \Pi_\theta, \approx \rangle$ are isomorphic. In the bivalent case ($\mathbf{L} = \mathbf{2}$), X/θ and Π_θ coincide.

Leibniz identity rule. Recall from Section 3.1 that according to Leibniz, two objects that have the same properties are the same. Because of this, we called $\approx_S$ of Lemma 3.30 a Leibniz equivalence. In this part we investigate the Leibniz rule a bit further. Note first that the converse part to the above Leibniz rule, i.e. that the same objects have the same properties, is held for evident. We may thus give the Leibniz rule a clear verbal form:

x and y are the same iff for each property A: x has A iff y has A.

If we loosen a little bit, we have

x and y are the same iff
for each *relevant* property A: x has A iff y has A.

Now, a property of elements from a set X can be understood as a fuzzy set A, $A(x)$ being the degree to which x has the property A. Furthermore, being a relevant property may be considered a fuzzy set $\mathcal{S}$ in L^X, i.e. a fuzzy set in the set of all fuzzy sets in X (properties), $\mathcal{S}(A)$ being the degree to which $A \in L^X$ is a relevant property. Therefore, letting $x \approx_{\mathcal{S}} y$ denote the degree to which x and y are the same, the above rule gives

$$(x \approx_{\mathcal{S}} y) = \bigwedge_{A \in L^X} \mathcal{S}(A) \to (A(x) \leftrightarrow A(y)). \tag{4.1}$$

If $\mathcal{S}$ is crisp, i.e. $\mathcal{S}(A)$ is either 0 or 1 for each A, then since $0 \to a = 1$ and $1 \to a = a$, (3.27) from Section 3.1 is a special case of (4.1). We are interested in what properties the general fuzzy relation $\approx_{\mathcal{S}}$ has.

Example 4.33 In general, $\approx_{\mathcal{S}}$ is not transitive: Consider the Łukasiewicz structure on $L = \{0, \frac{1}{2}, 1\}$, $X = \{x, y, z\}$, $A(x) = 0$, $A(y) = \frac{1}{2}$, $A(z) = 1$, $\mathcal{S} = \{\,{}^{1}\!/\!_{2}/A\}$. Then $(x \approx_{\mathcal{S}} y) \otimes (y \approx_{\mathcal{S}} z) = 1 \otimes 1 = 1 \not\leq \frac{1}{2} = (x \approx_{\mathcal{S}} z)$.

We need the following: For $a \in L$, the *degree of idempotency* of a, denoted $\mathrm{idm}(a)$, is defined by

$$\mathrm{idm}(a) \;=\; a \leftrightarrow a \otimes a.$$

If a is idempotent then $a \otimes a = a$ and thus $\mathrm{idm}(a) = 1$. More generally, for $A \subseteq L$, we put

$$\mathrm{idm}(A) \;=\; \bigwedge_{a \in A} \mathrm{idm}(a);$$

intuitively, $\mathrm{idm}(A)$ is in the degree to which each $a \in A$ is idempotent. Furthermore, we say that an **L**-set $\mathcal{S}$ in L^X is compatible with a binary **L**-relation E in X if

$$\mathcal{S}(A) \otimes A(x) \otimes E(x, y) \leq A(y)$$

for each $A \in L^X$, $x, y \in X$. Under the above interpretation, this means: if x has a relevant property A and x is in a relation E to y, then y has A as well. Clearly, if $\mathcal{S}$ is crisp then it is compatible with E iff each A from $\mathcal{S}$ is compatible with E (as introduced earlier).

Theorem 4.34 (Leibniz rule) *For an* **L**-*set* $\mathcal{S} \in X$, *letting* $\mathrm{idm}(\mathcal{S}) = \mathrm{idm}(\{\mathcal{S}(A) \mid A \in L^X\})$, *we have that* $\approx_{\mathcal{S}} \in {}^{\mathrm{idm}(\mathcal{S})}\mathrm{Sim}$ *(i.e.* $\approx_{\mathcal{S}}$ *is an* $\mathrm{idm}(\mathcal{S})$*-transitive* **L**-*tolerance).* $\approx_{\mathcal{S}}$ *is the largest one from* ${}^{\mathrm{idm}(\mathcal{S})}\mathrm{Sim}$ *which is compatible with* $\mathcal{S}$. *Moreover,* $A_i(x) = 1$ *implies* $\mathcal{S}(A) \otimes [x]_{\approx_{\mathcal{S}}} \subseteq A_i$.

Proof. First we show that $\approx_{\mathcal{S}} \in {}^{\mathrm{idm}(\mathcal{S})}\mathrm{Sim}$: Reflexivity and symmetry of $\approx_{\mathcal{S}}$ are obvious. $\approx_{\mathcal{S}}$ is $\mathrm{idm}(\mathcal{S})$-transitive iff for each $x, y, z \in X$ we have $\mathrm{idm}(\mathcal{S}) \otimes (x \approx_{\mathcal{S}} y) \otimes (y \approx_{\mathcal{S}} z) \leq (x \approx_{\mathcal{S}} z)$ which is true iff for each $A \in L^X$ we have

$$\mathcal{S}(A) \otimes \mathrm{idm}(\mathcal{S}) \otimes (x \approx_{\mathcal{S}} y) \otimes (y \approx_{\mathcal{S}} z) \leq A(x) \leftrightarrow A(z).$$

Due to symmetry, it is sufficient to show $\mathcal{S}(A) \otimes \mathrm{idm}(\mathcal{S}) \otimes (x \approx_{\mathcal{S}} y) \otimes (y \approx_{\mathcal{S}} z) \leq A(x) \rightarrow A(z)$, i.e.

$$A(x) \otimes \mathcal{S}(A) \otimes \mathrm{idm}(\mathcal{S}) \otimes (x \approx_{\mathcal{S}} y) \otimes (y \approx_{\mathcal{S}} z) \leq A(z).$$

We have

$$\begin{aligned} & A(x) \otimes \mathcal{S}(A) \otimes \mathrm{idm}(\mathcal{S}) \otimes (x \approx_{\mathcal{S}} y) \otimes (y \approx_{\mathcal{S}} z) \leq \\ \leq\; & A(x) \otimes \mathcal{S}(A) \otimes (\mathcal{S}(A) \rightarrow \mathcal{S}(A) \otimes \mathcal{S}(A)) \otimes (x \approx_{\mathcal{S}} y) \otimes (y \approx_{\mathcal{S}} z) \leq \\ \leq\; & A(x) \otimes \mathcal{S}(A) \otimes \mathcal{S}(A) \otimes (x \approx_{\mathcal{S}} y) \otimes (y \approx_{\mathcal{S}} z) \leq \\ \leq\; & A(x) \otimes \mathcal{S}(A) \otimes \mathcal{S}(A) \otimes \\ & \otimes (\mathcal{S}(A) \rightarrow (A(x) \rightarrow A(y))) \otimes (\mathcal{S}(A) \rightarrow (A(y) \rightarrow A(z))) \leq \\ \leq\; & \cdots \leq A(z). \end{aligned}$$

We proved $\approx_{\mathcal{S}} \in {}^{\mathrm{idm}(\mathcal{S})}\mathrm{Sim}$.

Next we show that $\mathcal{S}$ is compatible with $\approx_{\mathcal{S}}$: $\mathcal{S}(A) \otimes A(x) \otimes (x \approx_{\mathcal{S}} y) \leq \mathcal{S}(A) \otimes A(x) \otimes (\mathcal{S}(A) \rightarrow (A(x) \leftrightarrow A(y))) \leq A(x) \otimes (A(x) \rightarrow A(y)) \leq A(y)$. If $E \in {}^{\mathrm{idm}(\mathcal{S})}\mathrm{Sim}$ is compatible with $\mathcal{S}$ then for each $x, y \in X$ and each $A \in L^X$ we have $\mathcal{S}(A) \otimes A(x) \otimes E(x,y) \leq A(y)$ and thus $\mathcal{S}(A) \otimes E(x,y) \leq A(x) \rightarrow A(y)$; similarly, $\mathcal{S}(A) \otimes E(x,y) \leq A(y) \rightarrow A(x)$. Therefore, $\mathcal{S}(A) \otimes E(x,y) \leq A(x) \leftrightarrow A(y)$ and thus $E(x,y) \leq \mathcal{S}(A) \rightarrow (A(x) \leftrightarrow A(y))$ for each $A \in L^X$. We thus conclude $E(x,y) \leq \bigwedge_{A \in L^X} \mathcal{S}(A) \rightarrow (A(x) \leftrightarrow A(y)) = (x \approx_{\mathcal{S}} y)$ proving that $\approx_{\mathcal{S}}$ is the largest one.

Finally, if $A_i(x) = 1$ then $\mathcal{S}(A) \otimes [x]_{\approx_{\mathcal{S}}}(y) = \mathcal{S}(A) \otimes (x \approx_{\mathcal{S}} y) \leq \mathcal{S}(A) \otimes (\mathcal{S}(A) \rightarrow (A_i(x) \leftrightarrow A_i(y))) \leq A_i(x) \rightarrow A_i(y) = A_i(y)$. The proof is finished. □

Remark 4.35 (1) Lemma 3.30 is a consequence of Theorem 4.34; just assume that $\mathcal{S}$ in Theorem 4.34 is crisp.

(2) In an analogous way it is possible to generalize Corollary 3.32; we omit details.

We are now going to elaborate a bit more on the degree of idempotency.

Lemma 4.36 *For each $a \in L$ we have*

(1) $\mathrm{idm}(0) = \mathrm{idm}(1) = 1$ *and* $a \vee \neg a \leq \mathrm{idm}(a)$*;*

(2) divisibility implies $a \otimes \mathrm{idm}(a) = a \otimes a$*;*

(3) in a Gödel algebra, $\mathrm{idm}(a) = 1$*;*

(4) in an MV-algebra and a product algebra, $\mathrm{idm}(a) = a \vee \neg a$*;*

(5) in a linear product algebra, $\mathrm{idm}(a) = a$ *for* $a > 0$*.*

Proof. (1): We have to verify $a \leq \mathrm{idm}(a)$ and $\neg a \leq \mathrm{idm}(a)$. $a \leq \mathrm{idm}(a)$ is equivalent to $a \leq a \to a \otimes a$ which follows by adjointness; $\neg a \leq \mathrm{idm}(a)$ is equivalent to $\neg a \leq a \to a \otimes a$ which is equivalent to $a \otimes \neg a \leq a \otimes a$ which follows from $a \otimes \neg a = 0$.

(2): By divisibility, $a \otimes \mathrm{idm}(a) = a \otimes (a \to (a \otimes a)) = a \wedge (a \otimes a) = a \otimes a$.

(3) follows from $a = a \otimes a$ which is true in Gödel algebras.

(4): For an MV-algebra: Due to (1) we have to show $\mathrm{idm}(a) \leq a \vee \neg a = (a \to \neg a) \to \neg a$ which is equivalent to $a \otimes (a \to \neg a) \otimes (a \to a \otimes a) \leq 0$. However, $a \otimes (a \to \neg a) \otimes (a \to a \otimes a) \leq a \otimes a \otimes (a \to \neg a) \leq a \otimes \neg a = 0$.

For a product algebra: Due to the subdirect representation (Theorem 2.69) we may safely assume that $\mathbf{L}$ is linearly ordered. For $a = 0$ we have $\mathrm{idm}(a) = 1 = a \vee \neg a$. Since $\mathbf{L}$ is linearly ordered, $a > 0$ implies $\neg a = 0$ and thus $\neg\neg a = 1$. Therefore, putting $x = a \to (a \otimes a)$, $y = a$, $z = a$, $\neg\neg z \leq ((z \otimes x) \to (z \otimes y)) \to (x \to y)$ yields $(z \otimes x) \to (z \otimes y) \leq x \to y$ whence $(a \otimes x) \to (a \otimes a) \leq x \to a$. Now, $(a \otimes x) \to (a \otimes a) = (a \otimes (a \to a \otimes a)) \to (a \otimes a) = (a \wedge a \otimes a) \to (a \otimes a) = (a \otimes a) \to (a \otimes a) = 1$ and thus $1 \leq x \to a$ whence $x \leq a$, i.e. $\mathrm{idm}(a) = a \to (a \otimes a) \leq a$. On account of (1), $\mathrm{idm}(a) = a = a \vee \neg a$.

(5): Follows from (1) and (4) and the fact that in a linearly ordered product algebra we have $\neg a = 0$ for $a > 0$. □

Remark 4.37 For the standard algebras on $[0,1]$ we have $\mathrm{idm}(a) = 1$ for a Gödel algebra, $\mathrm{idm}(a) = \max(a, 1 - a)$ for a Łukasiewicz algebra, $\mathrm{idm}(0) = 1$ and $\mathrm{idm}(a) = a$ for $a > 0$ in product algebra.

Corollary 4.38 *If $\mathcal{S}$ is crisp then $\approx_{\mathcal{S}}$ is an $\mathbf{L}$-equivalence. If $\mathbf{L}$ is a Gödel algebra then $\approx_{\mathcal{S}}$ is an $\mathbf{L}$-equivalence. If $\mathbf{L}$ is the standard Łukasiewicz algebra on $[0,1]$ then $\approx_{\mathcal{S}} \in {}^{0.5}\mathrm{Sim}$, i.e. $\approx_{\mathcal{S}}$ is 0.5-transitive.*

The next theorem shows that fuzzy equivalences are in fact exactly the fuzzy equivalences generated by a Leibniz rule from a crisp $\mathcal{S}$.

Theorem 4.39 (fuzzy equivalences vs. Leibniz equivalences) *For each set $\mathcal{S} \subseteq L^{\langle X,\approx\rangle}$, $\approx_{\mathcal{S}}$ is an* **L**-*equivalence on $\langle X,\approx\rangle$. For each* **L**-*equivalence θ on $\langle X,\approx\rangle$, there is some $\mathcal{S} \subseteq L^{\langle X,\approx\rangle}$ such that $\theta =\approx_{\mathcal{S}}$.*

Proof. The fact that $\approx_{\mathcal{S}}$ is an **L**-equivalence follows from Lemma 3.30. It remains to show that $\approx_{\mathcal{S}}$ is compatible with $\approx$, i.e. $(x \approx_{\mathcal{S}} y) \otimes (x \approx x') \otimes (y \approx y') \leq (x' \approx_{\mathcal{S}} y')$. Clearly, this is true iff for each $A \in \mathcal{S}$ we have $(x \approx_{\mathcal{S}} y) \otimes (x \approx x') \otimes (y \approx y') \leq (A(x') \leftrightarrow A(y'))$. Due to symmetry, it suffices to verify $(x \approx_{\mathcal{S}} y) \otimes (x \approx x') \otimes (y \approx y') \leq A(x') \to A(y')$, i.e. $A(x') \otimes (x \approx_{\mathcal{S}} y) \otimes (x \approx x') \otimes (y \approx y') \leq A(y')$. We have $A(x') \otimes (x \approx_{\mathcal{S}} y) \otimes (x \approx x') \otimes (y \approx y') \leq A(x) \otimes (A(x) \to A(y)) \otimes (y \approx y') \leq A(y')$.

Conversely, let θ be an **L**-equivalence on $\langle X,\approx\rangle$. Put $\mathcal{S} = \Pi_\theta$, i.e. $\mathcal{S} = \{[x]_\theta \mid x \in X\}$. Due to Theorem 4.30, it suffices to show that $(x \approx_{\mathcal{S}} y) = [x]_\theta(y)$. On the one hand, $(x \approx_{\mathcal{S}} y) = \bigwedge_{z\in X}[z]_\theta(x) \leftrightarrow [z]_\theta(y) \leq [x]_\theta(x) \leftrightarrow [x]_\theta(y) = [x]_\theta(y)$. On the other hand, $[x]_\theta(y) \leq (x \approx_{\mathcal{S}} y)$ iff for each $z \in X$ we have $[x]_\theta(y) \leq [z]_\theta(x) \leftrightarrow [z]_\theta(y)$. Due to symmetry, it is enough to verify $[x]_\theta(y) \leq [z]_\theta(x) \to [z]_\theta(y)$ which is equivalent to $[z]_\theta(x) \otimes [x]_\theta(y) \leq [z]_\theta(y)$ which follows from transitivity: $[z]_\theta(x) \otimes [x]_\theta(y) = \theta(z,x) \otimes \theta(x,y) \leq \theta(z,y) = [z]_\theta(y)$, completing the proof. □

Similarity of systems of elements with fuzzy equivalence. Suppose we are given **L**-sets A and B in a set X that is equipped with an **L**-equivalence $\approx$. In Section 3.1 we defined the degree $(A \approx B)$ of equality of A and B by $(A \approx B) = \bigwedge_{x\in X}(A(x) \leftrightarrow B(x))$. However, there is still another natural way to define the concept of a degree to which A and B are similar, namely one may put

$$(A \sim B) = \bigwedge_{x\in X}(A(x) \to \bigvee_{y\in X}(x \approx y) \otimes B(y)) \wedge \wedge \bigwedge_{x\in X}(B(x) \to \bigvee_{y\in X}(x \approx y) \otimes A(y)). \tag{4.2}$$

Remark 4.40 (1) $A \sim B$ is verbally described as the truth degree to which for each x of A there is a y in B which is similar to x and, conversely, for each x of B there is a y in A which is similar to x.

(2) As we will see in Section 7.1, if A and B are compatible with $\approx$, we have $B(x) = \bigvee_{y\in X}(x \approx y) \otimes B(y)$ and $A(x) = \bigvee_{y\in X}(x \approx y) \otimes A(y)$. Therefore, in this case we have $(A \sim B) = (A \approx B)$.

(3) The situation is thus interesting if A and B are not compatible with $\approx$. For instance, this naturally happens if A and B are crisp and $X = L^Y$, i.e. elements of X are themselves fuzzy sets. Then, X is

naturally equipped with $\approx$. This approach is used in Section 5.1 where we define similarity of two fuzzy concept lattices.

Theorem 4.41 *The* **L**-*relation* $\sim$ *defined by (4.2) is an* **L**-*equivalence.*

Proof. Reflexivity and symmetry are obvious. Transitivity: $(A \sim B) \otimes (B \sim C) \leq (A \sim C)$ holds iff if both

$$(A \sim B) \otimes (B \sim C) \leq \bigwedge_{x \in X} (A(x) \rightarrow \bigvee_{y \in X} (x \approx y) \otimes C(y))$$

and

$$(A \sim B) \otimes (B \sim C) \leq \bigwedge_{x \in X} (C(x) \rightarrow \bigvee_{y \in X} (x \approx y) \otimes A(y))$$

are true. Due to symmetry of both of the cases, we verify only the first one. The required inequality is true iff for each $x \in X$ we have

$$A(x) \otimes (A \sim B) \otimes (B \sim C) \leq \bigvee_{y \in X} (x \approx y) \otimes C(y).$$

We have

$$\begin{aligned}
& A(x) \otimes (A \sim B) \otimes (B \sim C) \leq \\
\leq\ & A(x) \otimes (A(x) \rightarrow \bigvee_{z \in X} (x \approx z) \otimes B(z)) \otimes (B \sim C) \leq \\
\leq\ & (\bigvee_{z \in X} (x \approx z) \otimes B(z)) \otimes (B \sim C) = \\
=\ & (\bigvee_{z \in X} (x \approx z) \otimes B(z) \otimes (B \sim C)) \leq \\
\leq\ & (\bigvee_{z \in X} (x \approx z) \otimes B(z) \otimes (B(z) \rightarrow \bigvee_{y \in X} (z \approx y) \otimes C(y))) \leq \\
\leq\ & \bigvee_{z \in X} (x \approx z) \otimes (\bigvee_{y \in X} (z \approx y) \otimes C(y)) \leq \\
\leq\ & \bigvee_{y \in X} (x \approx y) \otimes C(y),
\end{aligned}$$

completing the proof. □

4.3 HIERARCHY: FUZZY ORDER

The primary interpretation of order (i.e. the ordinary relation which is reflexive, antisymmetric, and transitive) is hierarchy. Our main aim in this section is to present a general definition of a fuzzy order and to show some of its properties. We concentrate on fuzzy order which is of lattice type. A "representative example" of a lattice-type fuzzy ordered sets are so-called fuzzy concept lattices introduced in Chapter 5.

4.3.1 L-ORDER AND LATTICE L-ORDER

Definition 4.42 An $\mathbf{L}$-*order* on a set X with an $\mathbf{L}$-equality relation $\approx$ is a binary $\mathbf{L}$-relation $\preceq$ which is compatible w.r.t. $\approx$ and satisfies

$$\begin{aligned} x \preceq x &= 1 \quad \text{(reflexivity)}, \\ (x \preceq y) \wedge (y \preceq x) &\leq x \approx y \quad \text{(antisymmetry)}, \\ (x \preceq y) \otimes (y \preceq z) &\leq x \preceq z \quad \text{(transitivity)}. \end{aligned}$$

If $\preceq$ is an $\mathbf{L}$-order on a set X with an $\mathbf{L}$-equality $\approx$, we call the pair $\mathbf{X} = \langle\langle X, \approx\rangle, \preceq\rangle$ an $\mathbf{L}$-ordered set.

Remark 4.43 (1) Clearly, if $\mathbf{L} = \mathbf{2}$, the two-element Boolean algebra, the notion of $\mathbf{L}$-order coincides with the usual notion of (partial) order.

(2) Consider a language $\mathcal{J}_{\preceq}$ with binary relation symbols $\preceq$ and $\approx$. For an $\mathbf{L}$-structure $\mathbf{M}$ for $\mathcal{J}_{\preceq}$, $\langle\langle M, \approx^{\mathbf{M}}\rangle, \preceq^{\mathbf{M}}\rangle$ is an $\mathbf{L}$-ordered set iff formulas Ref, Ans, and Tra are true in $\mathbf{M}$ (truth degree equals 1).

We say that $\mathbf{L}$-ordered sets $\langle\langle X, \approx_X\rangle, \preceq_X\rangle$ and $\langle\langle Y, \approx_Y\rangle, \preceq_Y\rangle$ are isomorphic if there is a bijective mapping $h : X \to Y$ such that $(x \approx_X x') = (h(x) \approx_Y h(x'))$ and $(x \preceq_X x') = (h(x) \preceq_Y h(x'))$ holds for all $x, x' \in X$.

Remark 4.44 This is just a simplified version of the notion of an isomorphism between fuzzy structures (see Section 3.4) with the same structures of truth values.

Lemma 4.45 *In an* $\mathbf{L}$*-ordered set* $\langle\langle X, \approx\rangle, \preceq\rangle$ *we have* $(x \preceq y) \wedge (y \preceq x) = (x \approx y)$.

Proof. The "$\leq$" part of the equality is the antisymmetry condition. The "$\geq$" part follows from compatibility of $\preceq$: $(x \approx y) = (x \preceq x) \otimes (x \approx y) \leq (x \preceq y)$, and similarly $(x \approx y) \leq (y \preceq x)$, whence the conclusion follows. □

Lemma 4.46 *If* $\mathbf{X} = \langle\langle X, \approx_X\rangle, \preceq_X\rangle$ *and* $\mathbf{Y} = \langle\langle Y, \approx_Y\rangle, \preceq_Y\rangle$ *are* $\mathbf{L}$*-ordered sets and* $h : X \to Y$ *is a mapping satisfying* $(x \preceq_X x') = (h(x) \preceq_Y h(x'))$*, then* $\mathbf{X}$ *and* $\mathbf{Y}$ *are isomorphic.*

Proof. By Lemma 4.45, $(h(x) \approx_Y h(x')) = ((h(x) \preceq_Y h(x')) \wedge (h(x') \preceq_Y h(x))) = (x \preceq_X x') \wedge (x' \preceq_X x) = (x \approx_X x')$, verifying the remaining condition of the definition of isomorphic **L**-ordered sets. □

Example 4.47 For any set $X \neq \emptyset$ and any subset $\emptyset \neq M \subseteq L^X$, $\langle\langle M, \approx\rangle, S\rangle$ is an **L**-ordered set. Indeed, reflexivity and antisymmetry are trivial. Transitivity, i.e. $S(A,B) \otimes S(B,C) \leq S(A,C)$, and compatibility, i.e. $S(A,B) \otimes (A \approx A') \otimes (B \approx B') \leq S(A',B')$, follow from Theorems 3.12 and 3.14.

For an **L**-ordered set $\langle\langle X, \approx\rangle, \preceq\rangle$ and $A \in L^X$ we define **L**-sets $\mathcal{U}(A) \in L^X$ and $\mathcal{L}(A) \in L^X$ by

$$\mathcal{U}(A)(x) = \bigwedge_{y \in X} A(y) \rightarrow (y \preceq x)$$

and

$$\mathcal{L}(A)(x) = \bigwedge_{y \in X} A(y) \rightarrow (x \preceq y).$$

We will see in Section 5.3 that $\langle \mathcal{U}, \mathcal{L}\rangle$ is just the so-called **L**-Galois connection between $\langle X, \approx\rangle$ and $\langle X, \approx\rangle$ which is induced by $\preceq$. We will use this fact and also use properties of **L**-Galois connections established in Chapter 5. Observe that $\mathcal{U}(A)(x)$ is the truth degree of a formula $(\forall \mathtt{y})(r_A(\mathtt{y}) \Rightarrow (\mathtt{y} \preceq \mathtt{x}))$ of the corresponding language $\mathcal{J}_\preceq$ (cf. Remark 4.43) in the structure **M** corresponding to the **L**-order $\preceq$. Clearly, for an **L**-set A in X, $\mathcal{U}(A)$ ($\mathcal{L}(A)$) is the **L**-set of elements which are greater (smaller) than all elements of A. Therefore, we call $\mathcal{U}(A)$ and $\mathcal{L}(A)$ the *upper cone* and the *lower cone* of A, respectively. We write $\mathcal{UL}(A)$ instead of $\mathcal{U}(\mathcal{L}(A))$ etc.

Now, we introduce the notion of infimum, supremum in an **L**-ordered set, and the notion of a completely lattice **L**-ordered set.

Definition 4.48 For an **L**-ordered set $\langle\langle X, \approx\rangle, \preceq\rangle$ and $A \in L^X$ we define the **L**-sets $\inf(A)$ and $\sup(A)$ in X by

$$\begin{aligned}(\inf(A))(x) &= (\mathcal{L}(A))(x) \wedge (\mathcal{UL}(A))(x),\\ (\sup(A))(x) &= (\mathcal{U}(A))(x) \wedge (\mathcal{LU}(A))(x).\end{aligned}$$

$\inf(A)$ and $\sup(A)$ are called the *infimum* and *supremum* of A, respectively.

Remark 4.49 The notions of infimum and supremum are generalizations of the classical notions. Indeed, if $\mathbf{L} = \mathbf{2}$, $(\inf(A))(x)$ is the truth value of the fact that x belongs to both the lower cone of A and the upper cone of the lower cone of A, i.e. x is the greatest lower bound of A; similarly for $\sup(A)$.

Lemma 4.50 *Let $\langle\langle X, \approx\rangle, \preceq\rangle$ be an* **L**-*ordered set, $A \in L^X$. If $(\inf(A))(x) = 1$ and $(\inf(A))(y) = 1$, then $x = y$ (and similarly for $\sup(A)$).*

Proof. $(\inf(A))(x) = 1$ and $(\inf(A))(y) = 1$ implies $(\mathcal{L}(A))(x) = 1$, $(\mathcal{L}(A))(y) = 1$, $(\mathcal{UL}(A))(x) = 1$, $(\mathcal{UL}(A))(y) = 1$. By definition of $\mathcal{L}$ and $\mathcal{U}$, $x \preceq y = 1$ and $y \preceq x = 1$, therefore, by antisymmetry, $1 = (x \preceq y) \wedge (y \preceq x) \leq x \approx y$. Since $\approx$ is an **L**-equality, we have $x = y$. □

Definition 4.51 An **L**-ordered set $\langle\langle X, \approx\rangle, \preceq\rangle$ is said to be *completely lattice* **L**-*ordered* if for any $A \in L^X$ both $\sup(A)$ and $\inf(A)$ are $\approx$-singletons.

Remark 4.52 Lemma 4.50 and Definition 4.51 imply that in a completely lattice **L**-ordered set **X**, supremum $\sup(A)$ of $A \in L^X$ is uniquely determined by the element $x \in X$ such that ${}^1\sup(A) = \{x\}$ (i.e. $(\sup(A))(x) = 1$).

Checking that an **L**-ordered set is completely lattice **L**-ordered may be simplified:

Lemma 4.53 *For an* **L**-*ordered set* **X** *and $A \in L^X$ we have: $\inf(A)$ is a $\approx$-singleton iff there is some $x \in X$ such that $(\inf(A))(x) = 1$. The same is true for suprema.*

Proof. Obviously, we have to show that if $(\inf(A))(x) = 1$ for some $x \in X$ then $\inf(A)$ is a $\approx$-singleton, i.e. $(\inf(A))(x') = (x \approx x')$ for all $x' \in X$. First, we show $(\inf(A))(x') \geq (x \approx x')$: We have to show that $(x \approx x') \leq (\mathcal{L}(A))(x')$ and $(x \approx x') \leq (\mathcal{UL}(A))(x')$. We show only the first inequality, the second one is dual. By definition of $\mathcal{L}$, $(x \approx x') \leq (\mathcal{L}(A))(x')$ holds iff $(x \approx x') \leq A(y) \to (x' \preceq y)$ for any $y \in X$ which is true iff $(x \approx x') \otimes A(y) \leq (x' \preceq y)$. Since $(\mathcal{L}(A))(x) = 1$, we have $(x \approx x') \otimes A(y) = (x \approx x') \otimes A(y) \otimes (\mathcal{L}(A))(x) \leq (x \approx x') \otimes A(y) \otimes (A(y) \to (x \preceq y)) \leq (x \approx x') \otimes (x \preceq y) \leq (x' \preceq y)$, verifying the required inequality.

Second, we show $(\inf(A))(x') \leq (x \approx x')$:

$$
\begin{aligned}
& (\inf(A))(x') = (\mathcal{L}(A))(x') \wedge (\mathcal{UL}(A))(x') = \\
= \; & (\mathcal{LUL}(A))(x') \wedge (\mathcal{UL}(A))(x') = \\
= \; & \bigwedge_{y \in X} ((\mathcal{UL}(A))(y) \to (x' \preceq y)) \wedge \bigwedge_{y \in X} ((\mathcal{L}(A))(y) \to (y \preceq x')) \leq \\
\leq \; & ((\mathcal{UL}(A))(x) \to (x' \preceq x)) \wedge ((\mathcal{L}(A))(x) \to (x \preceq x')) = \\
= \; & (1 \to (x' \preceq x)) \wedge (1 \to (x \preceq x')) = \\
= \; & (x' \preceq x) \wedge (x \preceq x') = (x \approx x').
\end{aligned}
$$

The case of suprema is dual. □

The following theorem generalizes the well-known fact that "infimum of a larger subset is smaller" and "supremum of a larger subset is bigger".

Lemma 4.54 *For an* **L**-*ordered set* **L**, $A, B \in L^X$, *and* $x, y \in X$ *we have*

$$
\begin{aligned}
S(A,B) \otimes (\inf(A))(x) \otimes (\inf(B))(y) & \leq (y \preceq x), \\
S(A,B) \otimes (\sup(A))(x) \otimes (\sup(B))(y) & \leq (x \preceq y).
\end{aligned}
$$

Proof. We have

$$
\begin{aligned}
& S(A,B) \otimes (\inf(A))(x) \otimes (\inf(B))(y) \leq \\
\leq \; & S(\mathcal{L}(B), \mathcal{L}(A)) \otimes (\inf(A))(x) \otimes (\inf(B))(y) \leq \\
\leq \; & (\mathcal{L}(B)(y) \to \mathcal{L}(A)(y)) \otimes (\inf(A))(x) \otimes (\mathcal{L}(B))(y) \leq \\
\leq \; & \mathcal{L}(A)(y) \otimes (\inf(A))(x) \leq (\mathcal{L}(A))(y) \otimes (\mathcal{UL}(A))(x) \leq \\
\leq \; & (\mathcal{L}(A))(y) \otimes \bigwedge_{x' \in X} ((\mathcal{L}(A))(x') \to (x' \preceq x)) \leq \\
\leq \; & (\mathcal{L}(A)(y)) \otimes (\mathcal{L}(A)(y) \to (y \preceq x)) \leq (y \preceq x)
\end{aligned}
$$

proving the first inequality. The second one is dual. □

Note that for an **L**-order $\preceq$, ${}^1\!\preceq$ (the 1-cut of $\preceq$, i.e. ${}^1\!\preceq \, = \{\langle x, y\rangle \in X \times X \mid (x \preceq y) = 1\}$) is a binary relation on X. $\langle x, y\rangle \in {}^1\!\preceq$ means that the fact that x is less than y is "fully true". The basic properties of the "fully true" part of an **L**-order are the subject of the following theorem.

Theorem 4.55 *For an* **L**-*ordered set* $\mathbf{X} = \langle\langle X, \approx\rangle, \preceq\rangle$, *the relation* ${}^1\!\preceq$ *is an order on* X. *Moreover, if* **X** *is completely lattice* **L**-*ordered then* ${}^1\!\preceq$ *is a complete lattice order on* X.

Proof. Denote ${}^1\preceq$ by $\subseteq$. Reflexivity of $\subseteq$ follows from reflexivity of $\preceq$. Antisymmetry of $\subseteq$: $x \subseteq y$ and $y \subseteq x$ implies $(x \preceq y) = 1$ and $(y \preceq x) = 1$. Antisymmetry of $\preceq$ thus yields $(x \approx y) = 1$. Since $\approx$ is an **L**-equality, we conclude $x = y$. If $x \subseteq y$ and $y \subseteq z$, then $(x \preceq y) = 1$ and $(y \preceq z) = 1$, therefore $1 = (x \preceq y) \otimes (y \preceq z) \leq (x \preceq z)$, whence $(x \preceq z) = 1$, i.e. $x \subseteq z$, by transitivity of $\preceq$.

Let $\mathbf{X}$ be completely lattice **L**-ordered and let $A \subseteq X$; denote by A' the **L**-set in X corresponding to A, i.e. $A'(x) = 1$ for $x \in A$ and $A'(x) = 0$ for $x \notin A$. We show that there exists a supremum $\bigvee A$ of A in $\langle X, \subseteq\rangle$ (the case of infimum is dual). Since $\mathbf{X}$ is completely lattice **L**-ordered, $\sup(\mathrm{C}_{\approx}(A'))$ is a $\approx$-singleton in $\langle X, \approx\rangle$ (for $\mathrm{C}_{\approx}$ see (3.28)). Denote by x^* the element of X such that $(\sup(\mathrm{C}_{\approx}(A')))(x^*) = 1$. Since $(\sup(\mathrm{C}_{\approx}(A')))(x^*) = (\mathcal{U}(\mathrm{C}_{\approx}(A')))(x^*) \wedge (\mathcal{LU}(\mathrm{C}_{\approx}(A')))(x^*)$ we have both $(\mathcal{U}(\mathrm{C}_{\approx}(A')))(x^*) = 1$ and $(\mathcal{LU}(\mathrm{C}_{\approx}(A')))(x^*) = 1$. From the former we have $\bigwedge_{x\in X}(\mathrm{C}_{\approx}(A'))(x) \to (x \preceq x^*) = 1$, i.e. $(\mathrm{C}_{\approx}(A'))(x) \leq (x \preceq x^*)$ by adjointness. Since $A'(x) \leq (\mathrm{C}_{\approx}(A'))(x)$ for any $x \in X$, we further conclude $(x \preceq x^*) = 1$ for any $x \in X$ such that $A'(x) = 1$ (i.e. $x \in A$). Therefore, x^* belongs to the upper cone (w.r.t. $\subseteq$) of A. In a similar way, using $\mathcal{U}(\mathrm{C}_{\approx}(A)) = \mathcal{U}(A)$, we can show that $(\mathcal{LU}(\mathrm{C}_{\approx}(A')))(x^*) = 1$ implies that x^* belongs to the lower cone of the upper cone (cones w.r.t. $\subseteq$) of A. Thus, x^* is the supremum of A w.r.t. $\subseteq$. □

Remark 4.56 Theorem 4.55 has the following consequence: If $\mathbf{X}$ is a completely lattice **L**-ordered set, we may talk about the infimum (supremum) of a (crisp) subset A of X w.r.t. ${}^1\preceq$. Unless otherwise specified, we adopt the following conventions to be used in what follows: whenever the context avoids possible confusion with the symbols related to the order on **L** (the structure of truth values), ${}^1\preceq$ will be denoted by $\leq$; infimum (supremum) of $A \subseteq X$ will be denoted by $\bigwedge A$ ($\bigvee A$) or any obvious modification of this notation. Due to the proof of Theorem 4.55, we have that for $A \subseteq X$ it holds ${}^1\inf(\mathrm{C}_{\approx}(A')) = \{\bigwedge A\}$ and ${}^1\sup(\mathrm{C}_{\approx}(A')) = \{\bigvee A\}$ where A' is the **L**-set in X corresponding to A (i.e. $A'(x) = 1$ for $x \in A$ and $A'(x) = 0$ for $x \notin A$). Therefore, in a sense, the infima of crisp subsets of X w.r.t. to $\preceq$ and w.r.t. to $\leq$ (i.e. ${}^1\preceq$) are consistent.

4.3.2 DEDEKIND–MACNEILLE COMPLETION

Let $\mathbf{X} = \langle\langle X, \approx_X\rangle, \preceq_X\rangle$ and $\mathbf{Y} = \langle\langle Y, \approx_Y\rangle, \preceq_Y\rangle$ be **L**-ordered sets. A mapping $g : X \to Y$ is called an embedding of $\mathbf{X}$ into $\mathbf{Y}$ if g is injective, $(x \preceq_X x') = (g(x) \preceq_Y g(x'))$, and $(x \approx_X x') = (g(x) \approx_Y g(x'))$ for every $x, x' \in X$. Therefore, the image of X under g is a "copy" of

X. We say that an embedding $g : X \to Y$ preserves infima (suprema) if for any $A \in L^X$ and $x \in X$ we have $(\mathcal{L}(A))(x) = (\mathcal{L}(g(A)))(g(x))$ and $(\mathcal{UL}(A))(x) = (\mathcal{UL}(g(A)))(g(x))$ $((\mathcal{U}(A))(x) = (\mathcal{U}(g(A)))(g(x))$ and $(\mathcal{LU}(A))(x) = (\mathcal{LU}(g(A)))(g(x)))$ where $g(A) \in L^Y$ is defined by $(g(A))(y) = A(x)$ if $y = g(x)$ and $(g(A))(y) = 0$ otherwise.

Remark 4.57 Clearly, the preservation of infima (suprema) implies that $(\inf(A))(x) = \inf(g(A))(g(x))$ $((\sup(A))(x) = \sup(g(A))(g(x)))$.

For an **L**-ordered set **X** and $x \in X$ we put $(x] := \mathcal{L}(\{\,^1/x\})$ and $[x) := \mathcal{U}(\{\,^1/x\})$. Therefore, $((x])(y) = (y \preceq x)$ and $([x))(y) = (x \preceq y)$ for each $y \in X$.

Obviously, the above-introduced notions generalize the well-known notions from the theory of ordered sets. Our aim in the following is a fundamental construction in the theory of ordered sets, so-called Dedekind–MacNeille completion (or completion by cuts). The objective is to describe a most economic completion of an ordered set which preserves infima and suprema, i.e. to describe "the least" completely lattice ordered set to which the original ordered set can be embedded in such a way that the embedding preserves infima and suprema.

Remark 4.58 For the bivalent case, the completion by cuts was probably exploited for the first time by Dedekind by the construction of real numbers from rational numbers. The construction has been generalized for arbitrary ordered sets by MacNeille [136].

For an **L**-ordered set $\langle\langle X, \approx\rangle, \preceq\rangle$ denote

$$\mathcal{B}(X, X, \preceq) = \{\langle A, B\rangle \in L^X \times L^X \mid \mathcal{U}(A) = B,\ \mathcal{L}(B) = A\}.$$

Putting

$$(\langle A_1, B_1\rangle \approx \langle A_2, B_2\rangle) = (A_1 \approx A_2)$$

and

$$(\langle A_1, B_1\rangle \preceq \langle A_2, B_2\rangle) = S(A_1, A_2),$$

Example 4.47 implies that $\langle\langle \mathcal{B}(X, X, \preceq), \approx\rangle, \preceq\rangle$ is an **L**-ordered set (note that we use $\preceq$ in two ways; however, there will be no danger of confusion). Furthermore, we will see in Chapter 5 that $(\langle A_1, B_1\rangle \approx \langle A_2, B_2\rangle) = (B_1 \approx B_2)$ and $(\langle A_1, B_1\rangle \preceq \langle A_2, B_2\rangle) = S(B_2, B_1)$. In the following we write simply $\mathcal{B}(X, X, \preceq)$ instead of $\langle\langle \mathcal{B}(X, X, \preceq), \approx\rangle, \preceq\rangle$.

Having prepared the necessary notions, we are ready to present the result.

Theorem 4.59 (Dedekind–MacNeille completion for L-order) *Let* **X** *be an* **L***-ordered set. Then* $g : x \mapsto \langle (x], [x)\rangle$ *is an embedding of*

$\mathbf{X}$ into a completely $\mathbf{L}$-ordered set $\mathcal{B}(X,X,\preceq)$ which preserves infima and suprema. Moreover, if f is an embedding of $\mathbf{X}$ into a completely lattice $\mathbf{L}$-ordered set $\mathbf{Y}$ which preserves infima and suprema, then there is an embedding h of $\mathcal{B}(X,X,\preceq)$ into $\mathbf{Y}$ such that $f = g \circ h$.

Proof. By Theorem 5.63, $\mathcal{B}(X,X,\preceq)$ is a completely lattice $\mathbf{L}$-ordered set. Furthermore, g is defined correctly since $\mathcal{U}((x]) = [x)$ and $\mathcal{L}([x)) = (x]$, i.e. $\langle (x],[x)\rangle \in \mathcal{B}(X,X,\preceq)$: We verify only $\mathcal{U}((x]) = [x)$, the second equality is symmetric. On the one hand, $\mathcal{U}((x])(y) = \bigwedge_{z\in X}((x](z) \to (z \preceq y)) \le (x](x) \to (x \preceq y) = (x \preceq y) = [x)(y)$. On the other hand, $[x)(y) \le \mathcal{U}((x])(y)$ holds iff $[x)(y) \le (x](z) \to (z \preceq y)$, i.e. $(x](z)\otimes[x)(y) \le (z \preceq y)$ for any $z \in X$. However, this is true since $(x](z)\otimes[x)(y) = (z \preceq x)\otimes(x \preceq y) \le (z \preceq y)$, by transitivity of $\preceq$.

We show that g is an embedding of $\mathbf{X}$ into $\mathcal{B}(X,X,\preceq)$. To this end it is clearly sufficient to show that $(x \preceq y) = (g(x) \preceq g(y))$. We prove both of the required inequalities:

"$\le$": As $(g(x) \preceq g(y)) = S((x],(y])$, the inequality holds iff $(x \preceq y) \le (x](z) \to (y](z)$ for each z which is equivalent to $(x](z)\otimes(x \preceq y) \le (y](z)$, i.e. $(z \preceq x)\otimes(x \preceq y) \le (z \preceq y)$ which holds by transitivity of $\preceq$.

"$\ge$": The inequality holds iff $\bigwedge_{z\in X}((x](z) \to (y](z)) \le (x \preceq y)$ which is true since $\bigwedge_{z\in X}((x](z) \to (y](z)) \le (x](x) \to (y](x) = 1 \to (x \preceq y) = (x \preceq y)$.

We now have to proved that g preserves infima and suprema. Due to the symmetry of both of the cases we proceed only for infima. By Remark 4.57, it is sufficient to show $(\mathcal{L}(M))(x) = (\mathcal{L}(g(M)))(g(x))$ and $(\mathcal{UL}(M))(x) = (\mathcal{UL}(g(M)))(g(x))$.

$(\mathcal{L}(M))(x) = (\mathcal{L}(g(M)))(g(x))$: Using the fact that for $\langle A,B\rangle \in \mathcal{B}(X,X,\preceq)$ we have $(g(M))(A,B) = 0$ if $\langle A,B\rangle \ne \langle (y],[y)\rangle$ for each $y \in X$, and $(g(M))(\langle (y],[y)\rangle) = (g(M))(g(y)) = M(y)$, we have $(\mathcal{L}(g(M)))(g(x)) = (\mathcal{L}(g(M)))(\langle (x],[x)\rangle) = \bigwedge_{\langle A,B\rangle\in\mathcal{B}(X,X,\preceq)}(g(M))(A,B) \to (\langle (x],[x)\rangle \preceq \langle A,B\rangle) = \bigwedge_{y\in X}(g(M))(\langle (y],[y)\rangle) \to (\langle (x],[x)\rangle \preceq \langle (y],[y)\rangle) = \bigwedge_{y\in X} M(y) \to S((x],(y]) = \bigwedge_{y\in X} M(y) \to (x \preceq y) = (\mathcal{L}(M))(x)$.

$(\mathcal{UL}(M))(x) = (\mathcal{UL}(g(M)))(g(x))$: On the one hand, $(\mathcal{UL}(g(M)))(g(x)) = (\mathcal{UL}(g(M)))(\langle (x],[x)\rangle) = \bigwedge_{\langle A,B\rangle\in\mathcal{B}(X,X,\preceq)}(\mathcal{L}(g(M)))(A,B) \to (\langle A,B\rangle \preceq \langle (x],[x)\rangle) = \bigwedge_{\langle A,B\rangle\in\mathcal{B}(X,X,\preceq)}(\mathcal{L}(g(M)))(A,B) \to S(A,(x]) \le \bigwedge_{\langle (y],[y)\rangle\in\mathcal{B}(X,X,\preceq)}(\mathcal{L}(g(M)))(\langle (y],[y)\rangle) \to S((y],(x]) = \bigwedge_{y\in Y}(\mathcal{L}(M))(y) \to (y \preceq x) = (\mathcal{UL}(M))(x)$.

On the other hand, we have $(\mathcal{UL}(M))(x) \le (\mathcal{UL}(g(M)))(g(x))$ iff for each $\langle A,B\rangle \in \mathcal{B}(X,X,\preceq)$ we have $(\mathcal{L}(g(M)))(A,B)\otimes(\mathcal{UL}(M))(x) \le S(A,(x])$ which holds (since $S(A,(x]) = \bigwedge_{y\in Y} A(y) \to (y \preceq x)$) iff for each $y \in Y$ we have $A(y) \otimes (\mathcal{L}(g(M)))(A, B) \otimes (\mathcal{UL}(M))(x) \leq (y \preceq x)$. Since $A(y) \otimes (\mathcal{L}(g(M)))(A, B) \otimes (\mathcal{UL}(M))(x) = A(y) \otimes (\mathcal{L}(g(M)))(A, B) \otimes \bigwedge_{u \in X}((\mathcal{L}(M))(u) \to (u \preceq x)) \leq A(y) \otimes (\mathcal{L}(g(M)))(A, B) \otimes ((\mathcal{L}(M))(y) \to (y \preceq x))$, it suffices to show that $A(y) \otimes (\mathcal{L}(g(M)))(A, B) \leq (\mathcal{L}(M))(y)$: To this end, observe that (A) $A(y) = S((y], A)$: on the one hand, $S((y], A) \leq (y](y) \to A(y) = A(y)$; on the other hand, $A(y) \leq S((y], A)$ iff for each $z \in X$ we have $(y](z) \otimes A(y) \leq A(z)$, i.e. $(z \preceq y) \otimes A(y) \leq A(z)$; using $A = \mathcal{LU}(A)$ we obtain that the last inequality holds iff $(z \preceq y) \otimes A(y) \otimes \mathcal{U}(A)(x) \leq (z \preceq x)$ for any $x \in X$ which holds since $(z \preceq y) \otimes A(y) \otimes \mathcal{U}(A)(x) \leq (z \preceq y) \otimes A(y) \otimes (A(y) \to (y \preceq x)) \leq (z \preceq y) \otimes (y \preceq x) \leq (z \preceq x)$. Now, $A(y) \otimes (\mathcal{L}(g(M)))(A, B) \leq (\mathcal{L}(M))(y)$ is true iff $A(y) \otimes M(z) \otimes (\mathcal{L}(g(M)))(A, B) \leq (y \preceq z)$ holds for each $z \in Z$. However, the last inequality is true: $A(y) \otimes M(z) \otimes (\mathcal{L}(g(M)))(A, B) \leq A(y) \otimes M(z) \otimes ((g(M))((z], [z)) \to (\langle A, B \rangle \preceq \langle (z], [z) \rangle)) = A(y) \otimes M(z) \otimes (M(z) \to S(A, (z])) \leq A(y) \otimes S(A, (z]) = S((y], A) \otimes S(A, (z]) \leq S((y], (z]) = (y \preceq z)$, using (A) and the fact that $x \mapsto \langle (x], [x) \rangle$ is an embedding. Thus, g preserves infima and suprema.

Let f be an embedding of $\mathbf{X}$ into $\mathbf{Y}$ which preserves infima and suprema. Define the mapping $h : \mathcal{B}(X, X, \preceq) \to Y$ as follows: for $\langle A, B \rangle \in \mathcal{B}(X, X, \preceq)$ let $h(A, B)$ be the (unique) element y of Y such that $(\sup(f(A)))(y) = 1$. We have to prove (1) $f = g \circ h$ and that (2) h is an embedding.

(1): Take any $x \in X$. Observe that $(\sup((x]))(x) = 1$: We have $(\mathcal{U}((x]))(x) = \bigwedge_{y \in X}(x](y) \to (y \preceq x) = \bigwedge_{y \in X}(y \preceq x) \to (y \preceq x) = 1$. Furthermore, $(\mathcal{LU}((x]))(x) = 1$ iff $(\mathcal{U}((x]))(y) \leq (x \preceq y)$ for any $y \in X$ which is true since $(\mathcal{U}((x]))(y) = \bigwedge_{z \in X}(x](z) \to (z \preceq y) \leq (x](x) \to (x \preceq y) = (x \preceq y)$. Therefore, $(\sup((x]))(x) = 1$. Since f preserves suprema, we get $1 = (\sup((x]))(x) = (\sup(f((x])))(f(x))$ which means that $(g \circ h)(x) = h((x], [x)) = f(x)$.

(2): Take $\langle A_1, B_1 \rangle, \langle A_2, B_2 \rangle \in \mathcal{B}(X, X, \preceq)$, denote $y_1 = h(A_1, B_1)$, $y_2 = h(A_2, B_2)$. We have to show $(\langle A_1, B_1 \rangle \preceq \langle A_2, B_2 \rangle) = (y_1 \preceq y_2)$, i.e. $S(A_1, A_2) = (y_1 \preceq y_2)$.

On the one hand, since $S(A_1, A_2) = S(f(A_1), f(A_2))$, Theorem 4.54 implies $S(A_1, A_2) = S(f(A_1), f(A_2)) \otimes (\sup(f(A_1)))(y_1) \otimes (\sup(f(A_2)))(y_2) \leq (y_1 \preceq y_2)$.

On the other hand, by $S(A_1, A_2) = S(f(A_1), f(A_2))$, we have to show $(y_1 \preceq y_2) \leq S(f(A_1), f(A_2))$. $(\sup(f(A_1)))(y_1) = 1$ implies $(\mathcal{U}(f(A_1)))(y_1) = 1$ from which one easily gets that $(f(A_1))(y) \leq (y \preceq y_1)$ for any $y \in X$, i.e. in particular $(f(A_1))(f(x)) \leq (f(x) \preceq y_1)$. Now, $(y_1 \preceq y_2) \leq S(f(A_1), f(A_2))$ holds iff for each $x \in X$ we have

$(f(A_1))(f(x))\otimes(y_1 \preceq y_2) \leq (f(A_2))(f(x))$. Since $(f(A_1))(f(x))\otimes(y_1 \preceq y_2) \leq (f(x) \preceq y_1)\otimes(y_1 \preceq y_2) \leq (f(x) \preceq y_2)$, it is sufficient to show that $(f(x) \preceq y_2) \leq (f(A_2))(f(x))$ for any $x \in X$. Since $[\mathcal{LU}(f(A_2))](f(x)) = [\mathcal{LU}(A_2)](x) = A_2(x) = (f(A_2))(f(x))$, it suffices to show that $(f(x) \preceq y_2) \leq (\mathcal{LU}(f(A_2)))(f(x))$: we have $(f(x) \preceq y_2) \leq (\mathcal{LU}(f(A_2)))(f(x))$ iff $(f(x) \preceq y_2)\otimes(\mathcal{U}(f(A_2)))(f(y)) \leq (f(x) \preceq y)$ for any $y \in Y$ which holds since $(f(x) \preceq y_2)\otimes(\mathcal{U}(f(A_2)))(f(y)) = (f(x) \preceq y_2)\otimes(\mathcal{ULU}(f(A_2)))(f(y)) \leq (f(x) \preceq y_2)\otimes((\mathcal{LU}(f(A_2)))(y_2) \to (y_2 \preceq y)) = (f(x) \preceq y_2)\otimes(1 \to (y_2 \preceq y)) = (f(x) \preceq y_2)\otimes(y_2 \preceq y) \leq (f(x) \preceq y)$. □

4.4 BIBLIOGRAPHICAL REMARKS

Binary relations are the basic tool for applications as well as for mathematics; a good text is [185], a good short overview is [172]. Binary fuzzy relations were already discussed in Zadeh's seminal paper [216]; what was later termed "compositional rule of inference" [220] was also introduced in [216]. Fuzzy equivalence (similarity) and fuzzy order were introduced by Zadeh in [218]. Binary fuzzy relations were investigated in numerous papers. Among important contributions we recall Yeh and Bang [212]; Ovchinikov [156, 157]; a comprehensive treatment is contained in [120]. Gottwald's [96] is a very good treatment of fuzzy relations and their applications approached systematically from the point of view of fuzzy logic (in the narrow sense); the structure of truth values is basically $[0,1]$ equipped with a left-continuous t-norm (thus a residuated lattice, cf. Section 2.4).

Section 4.1: Operations with binary fuzzy relations are a standard issue. Properties (and graded properties) of binary fuzzy relations (with $L = [0,1]$) are well-covered in [96]; further literature on this subject is [67, 120]. Functional fuzzy relations are treated e.g. in [105, 122]. Closures of binary fuzzy relations were studied by Bandler and Kohout, see e.g. [10] where the authors discuss also so-called local properties.

Section 4.2: Similarity relations in approximate reasoning are discussed e.g. in [39, 70, 105, 123, 124]. We already mentioned that similarity is another name for fuzzy equivalence; other names widely used are indistinguishability, likeness, and proximity. The crucial part of Theorem 4.39 stating that $\approx_{\mathcal{S}}$ is a fuzzy equivalence and that each fuzzy equivalence is of this form was proved for $L = [0,1]$ and a continuous t-norm $\otimes$ in [194]; in [123] one direction of this is proved for divisible complete residuated lattices. Compatibility of a fuzzy set with a fuzzy equivalence is sometimes called extensionality (of a fuzzy set w.r.t. the fuzzy equivalence, see e.g. [123]). A concept of a factor set modulo a fuzzy equivalence which is similar to our one is present in [156]. Alternative approaches and further results can be found in [54, 74, 92, 114, 118, 119, 197]. Note also that similarity phenomenon and its formalization are intensively discussed in cognitive science and psychology literature.

Section 4.3: This section is based on [27].

The classical readings on ordered sets and lattices are [41, 136]. As mentioned, the first paper on fuzzy order is Zadeh's [218] where $[0,1]$ with min is used for the structure of truth values (i.e. a standard Gödel algebra). Zadeh defined a fuzzy order as a binary relation R which is reflexive and transitive (in the standard way) and satisfies that $x = y$ whenever $R(x,y) > 0$ and $R(y,x) > 0$ (Zadeh calls this property antisymmetry). It is easy to see that Zadeh's antisymmetry is equivalent to our antisymmetry when the **L**-equality $\approx$ is a crisp identity; thus, Zadeh's fuzzy order is a special case of our general concept of an **L**-order. Antisymmetry is in fact a condition that is by some authors grasped by requiring $R(x,y) \otimes R(y,x) \leq (x \approx y)$: this is the case of Gottwald [96] (with $\approx$ being the crisp equality) and Bodenhofer [44, 45] ($\approx$ being a fuzzy equivalence); both Gottwald and Bodenhofer use $[0,1]$ equipped with a (left-continuous) t-norm. Condition $R(x,y) \otimes R(y,x) \leq (x \approx y)$ is, however, very often present only in a special form and it seems that the authors are unaware of the above simple form. To illustrate this, consider [198] where $L = [0,1]$, transitivity is expressed by $((R(x,y) + R(y,z) - 1) \vee 0) \leq R(x,z)$, and antisymmetry by $R(x,y) + R(y,x) > 1$ implies $x = y$. Clearly, these requirements are equivalent to saying that w.r.t. the standard Łukasiewicz algebra on $[0,1]$, R is transitive and satisfies $R(x,y) \otimes R(y,x) \leq (x \approx y)$ where $\approx$ is the crisp identity.

Other approaches to fuzzy order include [55, 198].

Chapter 5

OBJECT–ATTRIBUTE FUZZY RELATIONS AND FUZZY CONCEPT LATTICES

A concept is a basic unit for representing knowledge.

—John Sowa (Conceptual Structures, 1984)

Everything is a matter of degree.

—Anonymous

5.1 CONCEPTUAL GRANULATION: FUZZY CONTEXT, FUZZY CONCEPTS, AND FUZZY CONCEPT LATTICE

Elementary knowledge: relation between objects and attributes. When exploring an unknown domain of interest, the primarily observable data are in the form of a collection of relevant objects (minerals, animals, cities, documents, etc.), attributes that may apply to the objects (to be hard, to be warm-blooded, to be a capital city, to have many inhabitants, etc.), [1] and a relationship between objects and attributes specifying what objects have what attributes (some minerals are hard, some not; a city is either a capital or not; etc.). Most of empirical attributes are vague—a given attribute applies to a given object to a certain (truth) degree (mineral is hard to a certain degree).

[1]We shall confine ourselves to attributes of this kind, i.e. to qualitative attributes specifying whether the attribute applies to an object; we do not consider quantitative attributes (like speed) specifying a numerical value of an object (speed of an object).

Conceptual granulation. The form of elementary knowledge about a given domain is thus a triple consisting of a collection of objects, collection of attributes, and a relationship between objects and attributes. The data are naturally depicted in a table where rows correspond to objects, columns correspond to attributes, and the table items, i.e. intersections of rows and columns, specify the degrees to which objects have attributes. Except for describing the object–attribute relationship, the table does not directly carry any further information about the domain. Further information can be obtained by an appropriate analysis of the input data. The basic way people make the outer world comprehensible and mentally graspable is abstraction and formation of concepts. Concepts serve to granulate the domain of interest: a concept covers (applies to) several objects of the domain. As a result, concepts allow us to formulate facts and laws we observe in the domain concisely and in an intelligible way.

In the following, we confine ourselves just to the "conceptual analysis" of object–attribute data (and omit other forms of analysis of the data). Our aim is to provide formal means that enable us to speak of a collection of concepts that are hidden in an input data which is of the form of a binary relation between objects and attributes. What we will do is, in fact, a generalization of the approach to conceptual data analysis as pursued by Wille et al. [81]. This approach is based on the notion of a concept that is due to so-called traditional logic, particularly to Port-Royal logic [3].

Fuzzy context and fuzzy concepts. The input data which is in the form of a collection of objects, collection of attributes, and a binary relation (which is in general vague since attributes are vague) between objects and attributes, can be formalized as follows: We suppose that there is a nonempty set X whose elements will be called objects and a nonempty set Y whose elements will be called attributes. We choose a structure $\mathbf{L}$ of truth values ($\mathbf{L}$ is supposed to form a complete residuated lattice); $\mathbf{L}$ provides an appropriate scale of truth values with its structure; using L (the support of $\mathbf{L}$) one will evaluate propositions like "an object x has an attribute y". Furthermore, we suppose that there are $\mathbf{L}$-equivalence relations $\approx_X$ on X and $\approx_Y$ on Y ($\approx_X$ and $\approx_Y$ will also be denoted simply by $\approx$; there will be no danger of confusion). Finally, there is a binary $\mathbf{L}$-relation between $\langle X, \approx_X \rangle$ and $\langle Y, \approx_Y \rangle$; $I(x, y)$ being interpreted as the truth degree to which the object $x \in X$ has the attribute $y \in Y$.

Definition 5.1 The triple $\langle \langle X, \approx_X \rangle, \langle Y, \approx_Y \rangle, I \rangle$ as above is called an $\mathbf{L}$-*context* (or a fuzzy context).

We will also denote the **L**-context $\langle\langle X, \approx_X\rangle, \langle Y, \approx_Y\rangle, I\rangle$ simply by $\langle X, Y, I\rangle$.

Remark 5.2 (1) An **L**-context is in fact an **L**-structure of a simple many-sorted language. Namely, put $S = \{\mathcal{X}, \mathcal{Y}\}$ (sorts) and consider an S-sorted language $\mathcal{J}_{\text{Con}}$ given by $R = \{r_I, \approx_{\mathcal{X}}, \approx_{\mathcal{Y}}\}$, $F = \emptyset$, and $\sigma(r_I) = \mathcal{X}\mathcal{Y}$, $\sigma(\approx_{\mathcal{X}}) = \mathcal{X}\mathcal{X}$, $\sigma(\approx_{\mathcal{Y}}) = \mathcal{Y}\mathcal{Y}$. An **L**-structure for $\mathcal{J}_{\text{Con}}$ is of the form $\mathbf{M} = \langle M, R^{\mathbf{M}}, F^{\mathbf{M}}\rangle$ where $M = \{M_{\mathcal{X}}, M_{\mathcal{Y}}\}$, $R^{\mathbf{M}} = \{r_I^{\mathbf{M}}, \approx_{\mathcal{X}}^{\mathbf{M}}, \approx_{\mathcal{Y}}^{\mathbf{M}}\}$, $r_I^{\mathbf{M}}$ is a binary **L**-relation between $M_{\mathcal{X}}$ and $M_{\mathcal{Y}}$ that is compatible with **L**-equivalence relations $\approx_{\mathcal{X}}^{\mathbf{M}}$ on $M_{\mathcal{X}}$ and $\approx_{\mathcal{Y}}^{\mathbf{M}}$ on $M_{\mathcal{Y}}$. In other words, **M** is fully specified by the triple $\left\langle\langle M_{\mathcal{X}}, \approx_{\mathcal{X}}^{\mathbf{M}}\rangle, \langle M_{\mathcal{Y}}, \approx_{\mathcal{Y}}^{\mathbf{M}}\rangle, r_I^{\mathbf{M}}\right\rangle$, i.e. by an **L**-context.

(2) In the following we will several times refer to $\mathcal{J}_{\text{Con}}$. Unlike in Chapter 3, we will denote the object variables of sort $\mathcal{X}$ and $\mathcal{Y}$ by $\mathbb{x}$ (possibly with indices) and $\mathbb{y}$ (possibly with indices), respectively, and leave x and y for elements of X and Y, respectively (this is in order to be consistent with the existing literature). Furthermore, we will freely write $\mathbf{M} = \langle\langle X, \approx_X\rangle, \langle Y, \approx_Y\rangle, I\rangle$ or $\mathbf{M} = \langle X, Y, I\rangle$ and consider the **L**-context in question as an **L**-structure for $\mathcal{J}_{\text{Con}}$ as discussed above.

(3) We will use the following convention. Relation symbols of $\mathcal{J}_{\text{Con}}$ are r_I, r_A, r_{A_1}, r_B, etc. For the sake of brevity, their corresponding fuzzy relations and fuzzy sets $r_I^{\mathbf{M}}, r_A^{\mathbf{M}}, r_{A_1}^{\mathbf{M}}, r_B^{\mathbf{M}}$ will be denoted by I, A, A_1, B, respectively.

We need a natural definition of a concept. As mentioned, we are inspired by Port-Royal logic. By Port-Royal, a concept consists of its *extent* and its *intent*. The extent of a concept is the collection of all objects that are covered by the concept; the intent of a concept is the collection of all attributes covered by the concept. For example, the extent of a concept dog consists of all dogs while its intent consists of all attributes that apply to dogs (to bark, to have four extremities, etc.). Therefore, a concept is understood as a pair $\langle A, B\rangle$ consisting of the extent A and the intent B of a given concept. In order to qualify for a concept, a pair $\langle A, B\rangle$ has to satisfy some constraints. Needless to say, the choice of the constraints is crucially important. Port-Royal offers the following: for a collection A of objects and a collection B of attributes,

> $\langle A, B\rangle$ is a concept iff B is the collection of all attributes that are shared by all objects from A and, conversely, A is the collection of all objects sharing all attributes from B.

Naturally, when formalizing, we replace "collection of objects (attributes)" by "fuzzy set of objects (attributes)". What plays a role in the verbal definition of a concept are mappings $^{\uparrow}$ assigning to a collection A of objects a collection $A^{\uparrow}$ of all attributes common to all objects of A, and $^{\downarrow}$ assigning to a collection B of attributes a collection $B^{\downarrow}$ of all objects sharing all attributes of B. To formalize this, notice that for a fuzzy set A in X (fuzzy set of objects), the truth value $A^{\uparrow}(y)$ to which an attribute $y \in Y$ belongs to $A^{\uparrow}$ is the truth degree of the proposition (1): "for each object x from A, x has y"; similarly, for a fuzzy set B in Y (fuzzy set of attributes), the truth value $B^{\downarrow}(x)$ to which an object $x \in X$ belongs to $B^{\downarrow}$ is the truth degree of the proposition (2): "for each attribute y from B, x has y". Going to the language $\mathcal{J}_{\mathrm{Con}}$, the above propositions can be translated into corresponding formulas: proposition (1) translates into a formula $\varphi_{\uparrow}$ where

$$\varphi_{\uparrow} \quad \text{is} \quad (\forall \mathtt{x})(r_A(\mathtt{x}) \Rightarrow r_I(\mathtt{x}, \mathtt{y}));$$

proposition (2) translates to a formula $\varphi_{\downarrow}$ where

$$\varphi_{\downarrow} \quad \text{is} \quad (\forall \mathtt{y})(r_B(\mathtt{y}) \Rightarrow r_I(\mathtt{x}, \mathtt{y})).$$

Formulas $\varphi_{\uparrow}$ and $\varphi_{\downarrow}$ can be evaluated in an appropriate **L**-structure **M** for $\mathcal{J}_{\mathrm{Con}}$ which is given by an **L**-context $\langle\langle X, \approx_X\rangle, \langle Y, \approx_Y\rangle, I\rangle$ and **L**-sets A in X and B in Y (recall that in **M** we have $M_{\mathcal{X}} = X$, $M_{\mathcal{Y}} = Y$, $\approx_{\mathcal{X}}^{\mathbf{M}} = \approx_X$, $\approx_{\mathcal{Y}}^{\mathbf{M}} = \approx_Y$, $r_I^{\mathbf{M}} = I$, $r_A^{\mathbf{M}} = A$, $r_B^{\mathbf{M}} = B$). Using the basic semantical rules (cf. Section 3.2), we get for the truth degrees of $\varphi_{\uparrow}$ and $\varphi_{\downarrow}$ under a valuation v,

$$\|\varphi_{\uparrow}\|_{\mathbf{M},v} = \bigwedge_{x \in X} (r_A^{\mathbf{M}}(x) \to r_I^{\mathbf{M}}(x, v(\mathtt{y})))$$

and

$$\|\varphi_{\downarrow}\|_{\mathbf{M},v} = \bigwedge_{y \in Y} (r_B^{\mathbf{M}}(y) \to r_I^{\mathbf{M}}(v(\mathtt{x}), y)).$$

As discussed at the end of Section 3.2, $\varphi_{\uparrow}$ and $\varphi_{\downarrow}$ induce an **L**-set $\varphi_{\uparrow}^{\mathbf{M}}$ in Y and an **L**-set $\varphi_{\downarrow}^{\mathbf{M}}$ in X by

$$\varphi_{\uparrow}^{\mathbf{M}}(y) = \|\varphi_{\uparrow}\|_{\mathbf{M},v} \qquad \text{and} \qquad \varphi_{\downarrow}^{\mathbf{M}}(x) = \|\varphi_{\downarrow}\|_{\mathbf{M},w}$$

where the valuations v and w are such that $v(\mathtt{y}) = y$ and $w(\mathtt{x}) = x$. Moreover, A, I, and B, I can be thought of as arguments of binary functions $\varphi_{\uparrow}^{(\mathbf{L},X,Y)}$ and $\varphi_{\downarrow}^{(\mathbf{L},X,Y)}$ assigning to (A, I) and (B, I) an **L**-set $\varphi_{\uparrow}^{(\mathbf{L},X,Y)}(A, I)$ in Y and an **L**-set $\varphi_{\downarrow}^{(\mathbf{L},X,Y)}(B, I)$ in X by

$$[\varphi_{\uparrow}^{(\mathbf{L},X,Y)}(A, I)](y) = \|\varphi\|_{\mathbf{M},v}$$

where $v(\mathtt{y}) = y$, and

$$[\varphi_{\downarrow}^{(\mathbf{L},X,Y)}(B,I)](x) = \|\varphi\|_{\mathbf{M},v}$$

where $v(\mathtt{x}) = x$. In this case, we write also

$$A^{\uparrow_I} = \varphi_{\downarrow}^{(\mathbf{L},X,Y)}(A,I) \qquad \text{and} \qquad B^{\downarrow_I} = \varphi_{\uparrow}^{(\mathbf{L},X,Y)}(B,I).$$

If we think of I as fixed or if it is obvious, we write simply $A^{\uparrow}$ instead of $A^{\uparrow_I}$ and $B^{\downarrow}$ instead of $B^{\downarrow_I}$.

Thus, ${}^{\uparrow_I}$ and ${}^{\downarrow_I}$ may be thought of as mappings ${}^{\uparrow} : L^X \to L^Y$ and ${}^{\downarrow} : L^Y \to L^X$. From the above it follows that we have

$$A^{\uparrow_I}(y) = \bigwedge_{x \in X} A(x) \to I(x,y), \tag{5.1}$$

$$B^{\downarrow_I}(x) = \bigwedge_{y \in Y} B(y) \to I(x,y). \tag{5.2}$$

Remark 5.3 As we will see later on, if I is an **L**-relation between $\langle X, \approx\rangle$ and $\langle Y, \approx\rangle$, $A \in L^X$ and $B \in L^Y$ imply $A^{\uparrow_I} \in L^{\langle Y,\approx\rangle}$ and $B^{\downarrow_I} \in L^{\langle X,\approx\rangle}$, respectively.

Now, $A^{\uparrow}(y)$ is the truth degree to which each object of A has the attribute y and $B^{\downarrow}(x)$ is the truth degree to which each attribute of B is shared by the object x. Coming back to the verbally formulated definition of a concept provided by Port-Royal, we can define the notion of a fuzzy concept in a given fuzzy context.

Definition 5.4 An **L**-*concept* in a given **L**-context $\langle X, Y, I\rangle$ is any pair $\langle A, B\rangle$ of $A \in L^X$ and $B \in L^Y$ such that

$$A^{\uparrow} = B \qquad \text{and} \qquad B^{\downarrow} = A.$$

A is called the *extent*, B is the *intent* of $\langle A, B\rangle$. We immediately see that this definition formalizes properly the above verbal definition of Port-Royal concepts. The set of all **L**-concepts given by an **L**-concept is denoted by $\mathcal{B}(X,Y,I)$; that is,

$$\mathcal{B}(X,Y,I) = \{\langle A, B\rangle \in L^X \times L^Y \mid A^{\uparrow} = B,\ B^{\downarrow} = A\}.$$

Definition 5.5 $\mathcal{B}(X,Y,I)$ is called the **L**-*concept lattice* given by $\langle X, Y, I\rangle$ (the term "lattice" will be justified later).

We write only context, concept, and concept lattice if **L** is obvious.

Fuzzy concepts as maximal rectangles. We are going to show that $\mathbf{L}$-concepts of an $\mathbf{L}$-context $\langle\langle X,\approx\rangle,\langle Y,\approx\rangle,I\rangle$ correspond to the maximal rectangles contained in I. An $\mathbf{L}$-relation $I'\in L^{X\times Y}$ will be called a *rectangular relation* iff there are $A\in L^X$, $B\in L^Y$ such that $I'(x,y)=A(x)\otimes B(y)$ (for all $x\in X$, $y\in Y$), written $I'=A\otimes B$. In this case, the pair $\langle A,B\rangle$ is called a *rectangle.* A rectangle $\langle A,B\rangle$ is said to be *contained* in a relation $I''\in L^{X\times Y}$ if $A\otimes B$ is contained in I'', i.e. if $A\otimes B\subseteq I''$.

Lemma 5.6 *Let $A\in L^X$, $B\in L^Y$, $I\in L^{X\times Y}$ be such that $A\otimes B=I$. Let further $\approx_X$, $\approx_Y$ be $\mathbf{L}$-equivalences on X, Y, respectively. (1) If $A\in L^{\langle X,\approx_X\rangle}$, $B\in L^{\langle Y,\approx_Y\rangle}$, then I is an $\mathbf{L}$-relation between $\langle X,\approx_X\rangle$ and $\langle Y,\approx_Y\rangle$. (2) If I is an $\mathbf{L}$-relation between $\langle X,\approx_X\rangle$ and $\langle Y,\approx_Y\rangle$ and both A and B are normal (i.e. there are $x\in X$, $y\in Y$ with $A(x)=1$, $B(y)=1$), then $A\in L^{\langle X,\approx_X\rangle}$ and $B\in L^{\langle Y,\approx_Y\rangle}$.*

Proof. For any $x,x'\in X$, $y,y'\in Y$ we have

$$\begin{aligned}&I(x,y)\otimes(x\approx x')\otimes(y\approx y')=\\&\quad= A(x)\otimes B(y)\otimes(x\approx x')\otimes(y\approx y')\le A(x')\otimes B(y')=I(x',y'),\end{aligned}$$

i.e. (1) holds.

To obtain (2), let $x,x'\in X$, $y\in Y$ be such that $B(y)=1$. Then

$$\begin{aligned}&A(x)\otimes(x\approx x')=A(x)\otimes B(y)\otimes(x\approx x')\otimes(y\approx y)=\\&\quad= I(x,y)\otimes(x\approx x')\otimes(y\approx y)\le I(x',y)=A(x')\otimes B(y)=A(x'),\end{aligned}$$

proving (2). □

There is a naturally defined ordering $\trianglelefteq$ defined on the set of all rectangles by $\langle A_1,B_1\rangle\trianglelefteq\langle A_2,B_2\rangle$ iff for all $x\in X$, $y\in Y$ it holds $A_1(x)\le A_2(x)$ and $B_1(y)\le B_2(y)$. The following theorem has a simple interpretation for $\mathbf{L}=\mathbf{2}$: concepts are just maximal rectangles of I which are filled with 1's (if we consider the bivalent relation I as a matrix-table of 0's and 1's).

Theorem 5.7 *Let $\langle\langle X,\approx_X\rangle,\langle Y,\approx_Y\rangle,I\rangle$ be an $\mathbf{L}$-context. For each $A\in L^X$, $B\in L^Y$ it holds that $\langle A,B\rangle$ is an $\mathbf{L}$-concept iff it is a maximal rectangle contained in I.*

Proof. Let $\langle A,B\rangle$ be an $\mathbf{L}$-concept. If it were not maximal, there would be $\langle A',B'\rangle\in L^X\times L^Y$ such that $\langle A,B\rangle\triangleleft\langle A',B'\rangle$. Hence, there exists an $x\in X$ such that $A(x)<A'(x)$ or $y\in Y$ such that $B(y)<B'(y)$.

Suppose the former, i.e. $A(x) < A'(x)$ for some $x \in X$ (the latter may be handled analogously). By assumption, $B(y) \leq B'(y)$ holds for all $y \in Y$, therefore $A'(x) \otimes B(y) \leq A'(x) \otimes B'(y) \leq I(x,y)$, and thus $A'(x) \leq B(y) \rightarrow I(x,y)$ holds for each $y \in Y$. We conclude

$$A(x) < A'(x) \leq \bigwedge_{y \in Y} (B(y) \rightarrow I(x,y)) = B^{\downarrow}(x),$$

a contradiction to $A = B^{\downarrow}$ ($\langle A, B\rangle$ is a concept).

Conversely, let $\langle A, B\rangle$ be a maximal rectangle contained in I. We show $A = B^{\downarrow}$ and $B = A^{\uparrow}$. From $A(x) \otimes B(y) \leq I(x,y)$ it follows $A(x) \leq B(y) \rightarrow I(x,y)$ for all $x \in X$, $y \in Y$, i.e. $A(x) \leq B^{\downarrow}(x) = \bigwedge_{y\in Y}(B(y) \rightarrow I(x,y))$, thus $A \subseteq B^{\downarrow}$. $\left\langle B^{\downarrow}, B\right\rangle$ is contained in I, since $B^{\downarrow}(x) \otimes B(y) \leq I(x,y)$ is equivalent to $B^{\downarrow}(x) \leq B(y) \rightarrow I(x,y)$ which holds evidently. If $A \neq B^{\downarrow}$, i.e. $A(x) < B^{\downarrow}(x)$ for some $x \in X$, then $\langle A, B\rangle \lhd \left\langle B^{\downarrow}, B\right\rangle$, a contradiction to the maximality of $\langle A, B\rangle$ among rectangles contained in I. The condition $B = A^{\uparrow}$ may be shown analogously. From $A = B^{\downarrow}$ and $B = A^{\uparrow}$ we conclude $A \in L^{\langle X, \approx_X\rangle}$, $B \in L^{\langle Y, \approx_Y\rangle}$ by Remark 5.3. □

Lemma 5.6 and the second part of the preceding proof give the following corollary.

Corollary 5.8 *For each maximal rectangle $\langle A, B\rangle$ contained in a relation I between $\langle X, \approx_X\rangle$ and $\langle Y, \approx_Y\rangle$, $A \otimes B$ is an $\mathbf{L}$-relation between $\langle X, \approx_X\rangle$ and $\langle Y, \approx_Y\rangle$.*

Fuzzy concept lattices. We are now going to investigate the set of all **L**-concepts with its hierarchical structure. The hierarchical structure of $\mathcal{B}(\langle X, \approx\rangle, \langle Y, \approx\rangle, I)$ will be thoroughly investigated and characterized in Section 5.4. At this moment, we confine ourselves to a special case: we concentrate on "crisp" hierarchy of **L**-concepts and present a version of generalization of Wille's main theorem of concept lattices (later on, we will show two independent proofs; the first one making use of Wille's theorem, the second one making the assertion a consequence of much more general theorem concerning **L**-order on **L**-concepts).

So far, we considered the set $\mathcal{B}(X, Y, I)$ of all fuzzy concepts given by $\langle X, Y, I\rangle$ without any structure. The most natural structure on concepts is that of a conceptual hierarchy (subconcept–superconcept hierarchy); this was discussed already by Port-Royal: a concept c_1 is a subconcept of c_2 (c_2 is the superconcept of c_1) if the extent of c_1 is contained in the extent of c_2 (or, which is held to be equivalent, the intent of c_1 contains the intent of c_2). For instance, dog is a subconcept of mammal since each dog is a mammal (each attribute shared by mammals is shared by dogs). Thus, restricting ourselves to crisp subconcept–superconcept relation $\leq$ on $\mathcal{B}(\langle X, \approx\rangle, \langle Y, \approx\rangle, I)$, we can define

$$\langle A_1, B_1\rangle \leq \langle A_2, B_2\rangle \quad \text{iff} \quad A_1 \subseteq A_2 \quad (\text{ iff } B_1 \supseteq B_2). \tag{5.3}$$

The structure of $\mathcal{B}(\langle X, \approx\rangle, \langle Y, \approx\rangle, I)$ w.r.t. $\leq$ and its characterization is the subject of the following theorem; two proofs will be elaborated later.

Theorem 5.9 (main theorem of L-concept lattices, crisp order) *Let $\langle X, Y, I\rangle$ be an* **L***-context. (1) $\mathcal{B}(X, Y, I)$ is a complete lattice in which infima and suprema can be described as follows:*

$$\bigwedge_{j\in J} \langle A_j, B_j\rangle = \left\langle \bigcap_{j\in J} A_j, (\bigcap_{j\in J} A_j)^{\uparrow} \right\rangle = \left\langle \bigcap_{j\in J} A_j, (\bigcup_{j\in J} B_j)^{\downarrow\uparrow} \right\rangle, \tag{5.4}$$

$$\bigvee_{j\in J} \langle A_j, B_j\rangle = \left\langle (\bigcap_{j\in J} B_j)^{\downarrow}, \bigcap_{j\in J} B_j \right\rangle = \left\langle (\bigcup_{j\in J} A_j)^{\uparrow\downarrow}, \bigcap_{j\in J} B_j \right\rangle. \tag{5.5}$$

(2) Moreover, a complete lattice $\mathbf{V} = \langle V, \leq\rangle$ is isomorphic to $\mathcal{B}(X, Y, I)$ iff there are mappings $\gamma : X \times L \to V$, $\mu : Y \times L \to V$, such that $\gamma(X \times L)$ is supremally dense in $\mathbf{V}$, $\mu(Y \times L)$ is infimally dense in $\mathbf{V}$, and $a \otimes b \leq I(x, y)$ is equivalent to $\gamma(x, a) \leq \mu(y, b)$ for all $x \in X$, $y \in Y$, $a, b \in L$.

5.2 SIMILARITY AND LOGICAL PRECISION

5.2.1 SIMILARITY IN CONCEPT LATTICES

In this section we consider similarity relations in the framework of fuzzy concept lattices. Undoubtedly, gradual similarity of concepts is one of the fundamental features enabling a powerful human reasoning and communication. Similarity phenomenon is thus one of the most important ones accompanying conceptual structures. In our approach, similarity will be a fuzzy equivalence; however, because of the clear interpretation, we will speak of fuzzy similarity (**L**-similarity) instead of fuzzy equivalence.

We consider similarity on three natural levels: similarity of objects (and similarity of attributes), similarity of concepts, and similarity of concept lattices.

Similarity of objects (and attributes). We are going to propose a way to measure similarity of objects of a given fuzzy context; this similarity will be naturally induced by the structure of fuzzy concepts given by the fuzzy context. We prove that this similarity may be determined directly from the fuzzy context; this is relevant from the computational point of view.

Informally, two objects $x_1, x_2 \in X$ are similar if they cannot be separated by any concept, or more precisely, if for each concept c it holds that x_1 belongs to the extent of c iff x_2 belongs to the extent of c. This leads to the following definition of a relation $E^X_{\mathcal{B}(X,Y,I)} \in L^{X\times X}$:

$$E^X_{\mathcal{B}(X,Y,I)}(x_1,x_2) = \bigwedge_{\langle A,B\rangle\in\mathcal{B}(X,Y,I)} \Big(A(x_1) \leftrightarrow A(x_2)\Big) . \qquad (5.6)$$

The relation $E^X_{\mathcal{B}(X,Y,I)}$ will be called *induced* (by $\mathcal{B}(X,Y,I)$) *similarity on* X. By Theorem 4.34 we immediately get the following statement.

Theorem 5.10 *The relation $E^X_{\mathcal{B}(X,Y,I)}$ is the largest **L**-similarity relation on X compatible with the extents of all concepts of $\mathcal{B}(X,Y,I)$.*

From the computational point of view, the foregoing definition leads to the following algorithm for computing the similarity relation $E^X_{\mathcal{B}(X,Y,I)}$: Take an **L**-context, generate all the **L**-concepts of $\mathcal{B}(X,Y,I)$, and determine the similarity of each pair $\langle x_1, x_2\rangle \in X\times X$ by (5.6). The **L**-concept lattice may be, however, quite extensive. This poses the question whether the computational cost can be reduced. We give an (exact) solution which reduces the computational costs significantly. Define a relation $E^X_{\langle X,Y,I\rangle} \in L^{X\times X}$ by

$$E^X_{\langle X,Y,I\rangle}(x_1,x_2) = \bigwedge_{y\in Y} (I(x_1,y) \leftrightarrow I(x_2,y)) . \qquad (5.7)$$

$E^X_{\langle X,Y,I\rangle}(x_1,x_2)$ may be obtained from the L-context $\langle X,Y,I\rangle$ computing $|Y|$ times the operation $\leftrightarrow$. Using Theorem 4.34 we get

Theorem 5.11 *The relation $E^X_{\langle X,Y,I\rangle}$ is the largest **L**-similarity relation on X compatible with all $I(\_,y) \in L^X$, $y \in Y$.*

The following theorem solves the problem of finding an efficient procedure for computing the similarity relation $E^X_{\mathcal{B}(X,Y,I)}$.

Theorem 5.12 *Let $\langle X,Y,I\rangle$ be an **L**-context. Then for the similarity relations defined by (5.7) and (5.6) we have*

$$E^X_{\mathcal{B}(X,Y,I)} = E^X_{\langle X,Y,I\rangle} . \qquad (5.8)$$

Proof. We show $\bigwedge_{y\in Y}(I(x_1,y) \leftrightarrow I(x_2,y)) = \bigwedge_{\langle A,B\rangle\in\mathcal{B}(X,Y,I)}\big(A(x_1) \leftrightarrow A(x_2)\big)$ by checking both inequalities.

Consider the concept $\langle A_y, B_y\rangle = \langle \{\,{}^1\!/y\}^{\downarrow}, \{\,{}^1\!/y\}^{\downarrow\uparrow}\rangle$ for $y \in Y$. For each $x \in X$ we have

$$\begin{aligned} A_y(x) &= \bigwedge\{\{\,{}^1\!/y\}(y') \to I(x,y);\ y' \in Y\} = \\ &= 1 \to I(x,y) = I(x,y)\ . \end{aligned}$$

From $\{\langle A_y, B_y\rangle \mid y \in Y\} \subseteq \mathcal{B}(X,Y,I)$ and from the properties of infimum it follows that

$$\begin{aligned} &\bigwedge_{\langle A,B\rangle\in\mathcal{B}(X,Y,I)} \big(A(x_1) \leftrightarrow A(x_2)\big) \le \bigwedge_{y\in Y}\big(A_y(x_1) \leftrightarrow A_y(x_2)\big) = \\ &= \bigwedge_{y\in Y}(I(x_1,y) \leftrightarrow I(x_2,y)) \end{aligned}$$

for all $x_1, x_2 \in X$, proving the first inequality.

Conversely,

$$\bigwedge_{y\in Y}(I(x_1,y) \leftrightarrow I(x_2,y)) \le \bigwedge_{\langle A,B\rangle\in\mathcal{B}(X,Y,I)} \big(A(x_1) \leftrightarrow A(x_2)\big)$$

iff for each $\langle A, B\rangle \in \mathcal{B}(X,Y,I)$ we have

$$\begin{aligned} &\bigwedge_{y\in Y}(I(x_1,y) \leftrightarrow I(x_2,y)) \le \\ &\le\ A(x_1) \leftrightarrow A(x_2) = (A(x_1) \to A(x_2)) \wedge (A(x_2) \to A(x_1)) \end{aligned}$$

which is true iff both

$$\bigwedge_{y\in Y}(I(x_1,y) \leftrightarrow I(x_2,y)) \le A(x_1) \to A(x_2) \tag{5.9}$$

and

$$\bigwedge_{y\in Y}(I(x_1,y) \leftrightarrow I(x_2,y)) \le A(x_2) \to A(x_1) \tag{5.10}$$

hold. (5.9) is equivalent to

$$A(x_1) \otimes \bigwedge_{y\in Y}(I(x_1,y) \leftrightarrow I(x_2,y)) \le A(x_2).$$

Since $\langle A, B\rangle \in \mathcal{B}(X,Y,I)$, i.e. $A = A^{\uparrow\downarrow}$, the last inequality means

$$\begin{aligned} &\bigwedge_{y\in Y}(A^{\uparrow}(y) \to I(x_1,y)) \otimes \bigwedge_{y\in Y}(I(x_1,y) \leftrightarrow I(x_2,y)) \le \\ &\le\ \bigwedge_{y\in Y}(A^{\uparrow}(y) \to I(x_2,y)) \end{aligned}$$

which holds iff for each $y' \in Y$ we have

$$\bigwedge_{y \in Y} (A^{\uparrow}(y) \to I(x_1, y)) \otimes \bigwedge_{y \in Y} (I(x_1, y) \leftrightarrow I(x_2, y)) \leq$$
$$\leq A^{\uparrow}(y') \to I(x_2, y'),$$

i.e.

$$A^{\uparrow}(y') \otimes \bigwedge_{y \in Y} (A^{\uparrow}(y) \to I(x_1, y)) \otimes \bigwedge_{y \in Y} (I(x_1, y) \leftrightarrow I(x_2, y)) \leq$$
$$\leq I(x_2, y). \tag{5.11}$$

Evidently,

$$A^{\uparrow}(y') \otimes \bigwedge_{y \in Y} (A^{\uparrow}(y) \to I(x_1, y)) \otimes \bigwedge_{y \in Y} (I(x_1, y) \leftrightarrow I(x_2, y)) \leq$$
$$A^{\uparrow}(y') \otimes (A^{\uparrow}(y') \to I(x_1, y')) \otimes (I(x_1, y') \to I(x_2, y')) \leq I(x_2, y'),$$

thus (5.11) and therefore also (5.9) hold. (5.10) may be proved analogously. The second inequality and hence also the theorem is proved. □

The computation of $E^X_{\mathcal{B}(X,Y,I)}$ may thus be reduced to the computation of $E^X_{\langle X,Y,I \rangle}$ which is much more simple. From the foregoing theorem we also have the following consequence which is in accordance with our intuition: If we are given a family of (elementary) properties (attributes) of objects and consider the structure of concepts which is given by these properties, then the similarity among objects considered w.r.t. the structure of concepts is the same as the similarity w.r.t. to only the basic properties.

Considering the induced similarity relation (which will be denoted by E^X from now on) in question, one may ask: what are the similarity classes $[x]_{E^X}$ of the objects $x \in X$ with respect to the concepts of $\mathcal{B}(X, Y, I)$? Particularly, how are they related to (extents of) the elementary concepts $\left\langle \{{}^1/x\}^{\uparrow\downarrow}, \{{}^1/x\}^{\uparrow} \right\rangle$ generated by x? The following theorem shows that they are equal.

Theorem 5.13 *For every $\mathcal{B}(X, Y, I)$ and each $x \in X$ it holds $[x]_{E^X} = \{{}^1/x\}^{\uparrow\downarrow}$.*

Proof. Let $x, x' \in X$. On the one hand, $[x]_{E^X} \subseteq \{{}^1/x\}^{\uparrow\downarrow}$ follows by Theorem 4.34 since $\{{}^1/x\}^{\uparrow\downarrow}(x) = 1$. On the other hand,

$$\{{}^1/x\}^{\uparrow\downarrow}(x') = \bigwedge_{y \in Y} \{{}^1/x\}^{\uparrow}(y) \to I(x', y) = \bigwedge_{y \in Y} I(x, y) \to I(x', y)$$

$$\leq \bigwedge_{y \in Y} I(x,y) \leftrightarrow I(x',y) = E^X_{\langle X,Y,I\rangle}(x,x') = [x]_{E^X}(x'),$$

i.e. $\{{}^1/x\}^{\uparrow\downarrow} \subseteq [x]_{E^X}$. The proof is complete. □

In a completely analogous way we may get results for the similarity relations on Y.

Similarity of concepts. The next level on which we consider the similarity phenomenon is the level of concepts.

Consider first the relations E^{Ext} and E^{Int} on $\mathcal{B}(X,Y,I)$, call them *induced similarity* by extents and *induced similarity* by intents, respectively:

$$E^{Ext}(\langle A_1,B_1\rangle,\langle A_2,B_2\rangle) = (A_1 \approx A_2) = \bigwedge_{x \in X} (A_1(x) \leftrightarrow A_2(x)),$$

$$E^{Int}(\langle A_1,B_1\rangle,\langle A_2,B_2\rangle) = (B_1 \approx B_2) = \bigwedge_{y \in Y} (B_1(y) \leftrightarrow B_2(y)).$$

Corollary 3.32 mentioned in Remark 4.35 (2) gives immediately the following statement.

Theorem 5.14 *E^{Ext} and E^{Int} are the largest* **L***-similarity relations on $\mathcal{B}(X,Y,I)$ such that $A_1(x) \otimes E^{Ext}(\langle A_1,B_1\rangle,\langle A_2,B_2\rangle) \leq A_2(x)$ and $B_1(y) \otimes E^{Int}(\langle A_1,B_1\rangle,\langle A_2,B_2\rangle) \leq B_2(y)$ hold for every $x \in X$, $y \in Y$, $\langle A_1,B_1\rangle$, $\langle A_2,B_2\rangle \in \mathcal{B}(X,Y,I)$.*

To answer the question of how the relations E^{Ext} and E^{Int} are related, we derive some preliminary results. The next lemma states that the operators $^\uparrow$ and $^\downarrow$ preserve similarity.

Lemma 5.15 *Let $\langle X,Y,I\rangle$ be an* **L***-context, $A_1,A_2 \in L^X$, $B_1,B_2 \in L^Y$. Then we have $(A_1 \approx A_2) \leq (A_1^\uparrow \approx A_2^\uparrow)$ and $(B_1 \approx B_2) \leq (B_1^\downarrow \approx B_2^\downarrow)$.*

Proof. Consider the S-sorted language $\mathcal{J}_{\rm Con}$ from Remark 5.2. Let $\mathbf{M}_1$ be the **L**-structure for $\mathcal{J}_{\rm Con}$ given by $\langle X,Y,I\rangle$ with $r_A^{\mathbf{M}_1} = A_1$; $\mathbf{M}_2$ be the **L**-structure for $\mathcal{J}_{\rm Con}$ given by $\langle X,Y,I\rangle$ with $r_A^{\mathbf{M}_2} = A_2$ (thus $\mathbf{M}_1$ differs from $\mathbf{M}_2$ only by interpretation of r_A). Recall that the many-sorted version of Lemma 3.85 says

$$\bigotimes_{r \in R} (r^{\mathbf{M}_1} \approx r^{\mathbf{M}_2})^{|\varphi|_r} \otimes \bigotimes_{f \in F} (f^{\mathbf{M}_1} \approx f^{\mathbf{M}_2})^{|\varphi|_f} \leq \|\varphi\|_{\mathbf{M}_1,v} \leftrightarrow \|\varphi\|_{\mathbf{M}_2,v}.$$

Since $F = \emptyset$, $(r_I^{\mathbf{M}_1} \approx r_I^{\mathbf{M}_2}) = 1$, $|\varphi_\uparrow|_r = 1$ for $r = r_A$, $|\varphi_\uparrow|_r = 1$ for $r = r_I$, and $|\varphi_\uparrow|_r = 0$ otherwise, we have

$$(r_A^{\mathbf{M}_1} \approx r_A^{\mathbf{M}_2}) \leq \|\varphi_\uparrow\|_{\mathbf{M}_1,v} \leftrightarrow \|\varphi_\uparrow\|_{\mathbf{M}_2,v}.$$

As $r_A^{\mathbf{M}_1} = A_1$, $r_A^{\mathbf{M}_2} = A_2$, $\|\varphi_\uparrow\|_{\mathbf{M}_1,v} = A_1^\uparrow(v(\mathtt{y}))$, and $\|\varphi_\uparrow\|_{\mathbf{M}_2,v} = A_2^\uparrow(v(\mathtt{y}))$, we have

$$(A_1 \approx A_2) \leq (A_1^\uparrow(v(\mathtt{y})) \leftrightarrow A_2^\uparrow(v(\mathtt{y})))$$

for any valuation v. Recalling

$$(A_1^\uparrow \approx A_2^\uparrow) = \bigwedge_{v \text{ a valuation}} (A_1^\uparrow(v(\mathtt{y})) \approx A_2^\uparrow(v(\mathtt{y}))),$$

$(A_1 \approx A_2) \leq (A_1^\uparrow \approx A_2^\uparrow)$ follows. The second part is analogous (apply the same to $\varphi_\downarrow$). □

The following corollary is immediate.

Corollary 5.16 *Under the conditions of Lemma 5.15, we have* $(A_1 \approx A_2) \leq (A_1^{\uparrow\downarrow} \approx A_2^{\uparrow\downarrow})$ *and* $(B_1 \approx B_2) \leq (B_1^{\downarrow\uparrow} \approx B_2^{\downarrow\uparrow})$.

Note that even without Lemma 5.15 we have due to the properties of $\approx$ that $(A_1^{\uparrow\downarrow} \approx A_1) \otimes (A_1 \approx A_2) \otimes (A_2 \approx A_2^{\uparrow\downarrow}) \leq (A_1^{\uparrow\downarrow} \approx A_2^{\uparrow\downarrow})$, i.e. if $A_1^{\uparrow\downarrow}$ and A_1, A_1 and A_2, A_2 and $A_2^{\uparrow\downarrow}$ are pairwise similar, then also $A_1^{\uparrow\downarrow}$ and $A_2^{\uparrow\downarrow}$ are similar. Corollary 5.16 asserts a stronger result.

Corollary 5.17 *Under the conditions of Lemma 5.15 we have* $(A_1 \approx A_2) \otimes (A_1 \approx A_1^{\uparrow\downarrow}) \leq (A_1 \approx A_2^{\uparrow\downarrow})$ *and* $(B_1 \approx B_2) \otimes (B_1 \approx B_1^{\downarrow\uparrow}) \leq (B_1 \approx B_2^{\downarrow\uparrow})$.

Proof. $(A_1 \approx A_2) \otimes (A_1 \approx A_1^{\uparrow\downarrow}) \leq (A_1^{\uparrow\downarrow} \approx A_2^{\uparrow\downarrow}) \otimes (A_1 \approx A_1^{\uparrow\downarrow}) \leq (A_1 \approx A_2^{\uparrow\downarrow})$. The second part may be proved analogously. □

The following result shows that the similarities of concepts by extents and intents are equal.

Theorem 5.18 *For any* $\mathbf{L}$*-context* $\langle X, Y, I\rangle$, $E^{Ext} = E^{Int}$.

Proof. Let $\langle A_1, B_1\rangle, \langle A_2, B_2\rangle \in \mathcal{B}(X, Y, I)$, i.e. $A_i^\uparrow = B_i$ and $B_i^\downarrow = A_i$ for $i = 1, 2$. By Lemma 5.15 we get

$$\begin{aligned} E^{Ext}(\langle A_1, B_1\rangle, \langle A_2, B_2\rangle) &= (A_1 \approx A_2) \leq \\ \leq\ & (A_1^\uparrow \approx A_2^\uparrow) = (B_1 \approx B_2) = E^{Int}(\langle A_1, B_1\rangle, \langle A_2, B_2\rangle), \end{aligned}$$

and analogously, $E^{Int}(\langle A_1, B_1\rangle, \langle A_2, B_2\rangle) \leq E^{Ext}(\langle A_1, B_1\rangle, \langle A_2, B_2\rangle)$. To sum up, $E^{Ext}(\langle A_1, B_1\rangle, \langle A_2, B_2\rangle) = E^{Int}(\langle A_1, B_1\rangle, \langle A_2, B_2\rangle)$. □

We can therefore write E instead of E^{Ext} and E^{Int} and call it the *induced similarity* on concepts.

Compatible similarities and factorization. The primary importance of similarity relations in human reasoning is the *reduction of the complexity of the outer world at a reasonable price.* The complexity is reduced via considering the "collections of similar elements in concern" rather than the particular elements themselves. This is just what is known as the *abstraction process by factorization*: moving from a given level of abstraction (distinguishability) one level up where the elements are collections of elements of the lower level. Instead of the original system one therefore considers the "system modulo similarity". The price paid is the loss of precision.

Our concern in the following is the *reduction of the complexity of the concept lattice by factorization modulo similarity.* The concept lattice of a given context represents the overall conceptual structure which can be considerably intricate. To get an insight one has to look for methods for reducing the complexity of the structure. In fuzzy setting, one would expect methods for gradual reduction of complexity. The idea is to factorize the concept lattice by appropriate a-cut ${}^aE$ of the similarity E (recall that ${}^aE = \{\langle c_1, c_2\rangle \mid a \leq E(c_1, c_2)\}$), thus controlling the complexity by $a \in L$. Clearly, the lower $a \in L$, the coarser the factorization. The process of factorization of a system consists of two steps: first, specification of the elements, and, second, specification of the structure of the factor system. Since both of the steps are non-standard in our case we will describe them in more detail. In general, algebraic systems can be factorized by congruences, i.e. equivalences compatible with the structure of the system (see Chapter 1). We deal with conceptual structures which are complete lattices. The a-cut ${}^aE$ is clearly a tolerance relation (i.e. reflexive and symmetric), not transitive in general. In general, factorization of algebras by compatible tolerances is not possible (see [51]). Surprisingly, Czédli [62] showed a way to factorize lattices by compatible tolerance relations (see [204] for complete lattices). In the following we describe the construction of the factor lattice of an **L**-concept lattice by a compatible tolerance relation. Let $\langle X, Y, I\rangle$ be an **L**-context. A tolerance relation T on $\mathcal{B}(X, Y, I)$ is said to be *compatible* if it is preserved under arbitrary suprema and infima, i.e. if $\langle c_j, c'_j\rangle \in T$, $j \in J$, implies both $\langle \bigvee_{j\in J} c_j, \bigvee_{j\in J} c'_j\rangle \in T$ and $\langle \bigwedge_{j\in J} c_j, \bigwedge_{j\in J} c'_j\rangle \in T$ for any $c_j, c'_j \in \mathcal{B}(X, Y, I)$, $j \in J$. For a compatible tolerance relation T on $\mathcal{B}(X,Y,I)$ denote $c_T = \bigwedge_{\langle c,c'\rangle \in T} c'$ and $c^T = \bigvee_{\langle c,c'\rangle \in T} c'$; denote $[c]_T = [c_T, (c_T)^T] = \{c' \in \mathcal{B}(X,Y,I) \mid c_T \leq c' \leq (c_T)^T\}$ and $\mathcal{B}(X,Y,I)/T = \{[c]_T \mid c \in \mathcal{B}(X,Y,I)\}$. Introduce a relation $\leq_T$ on $\mathcal{B}(X,Y,I)/T$ by $[c]_T \leq_T [c']_T$ iff $\bigwedge[c]_T \leq \bigwedge[c']_T$ (iff $\bigvee[c]_T \leq \bigvee[c']_T$). The justification of the construction is given by the following statement which follows immediately from [204].

Theorem 5.19 *Let T be a compatible tolerance on $\mathcal{B}(X,Y,I)$. (1) $\mathcal{B}(X,Y,I)/T$ is the set of all maximal blocks of T (cf. Section 1.3). (2) $\langle \mathcal{B}(X,Y,I)/T, \leq_T\rangle$ is a complete lattice* (factor lattice) *where suprema and infima are described by*

$$\bigvee_{j\in J}[c_j]_T = [\bigvee_{j\in J} c_j]_T \quad \text{and} \quad \bigwedge_{j\in J}[c_j]_T = [(\bigwedge_{j\in J} c_j)^T]_T \tag{5.12}$$

for every $c_j \in \mathcal{B}(X,Y,I)$, $j \in J$.

Substituting (5.4) and (5.5) into (5.12) we get a more concrete description of the lattice operations.

Coming back to the induced similarity E on $\mathcal{B}(X,Y,I)$, the ultimate question is that of the compatibility of the a-cuts of E. Call an **L**-similarity relation F on $\mathcal{B}(X,Y,I)$ *compatible* if ${}^aE$ is a compatible tolerance relation on $\mathcal{B}(X,Y,I)$ for each $a \in L$. Notice that for the bivalent case the situation is completely uninteresting. Namely, the only cases are ${}^0E = \mathcal{B}(X,Y,I) \times \mathcal{B}(X,Y,I)$ and ${}^1E = \mathrm{id}_{\mathcal{B}(X,Y,I)} = \{\langle c,c\rangle \mid c \in \mathcal{B}(X,Y,I)\}$. In the first case, $\mathcal{B}(X,Y,I)/{}^0E = \{\mathcal{B}(X,Y,I)\}$, i.e. the factor lattice collapsed into a one-element lattice, while in the second case, $\mathcal{B}(X,Y,I)/{}^1E = \{\{\langle A,B\rangle\} \mid \langle A,B\rangle \in \mathcal{B}(X,Y,I)\}$, i.e. $\mathcal{B}(X,Y,I)$ and $\mathcal{B}(X,Y,I)/{}^1E$ are isomorphic.

Note that we have in no case to confine ourselves to the induced similarity E. On the other hand, taking into account only **L**-similarity relations F satisfying $A(x) \otimes F(\langle A,B\rangle, \langle A',B'\rangle) \leq A'(x)$ (which is quite natural—it reads "object belonging to the extent of some concept belongs also to the extent of any similar concept") for each $x \in X$, Theorem 5.14 tells us that E provides the most extensive reduction: for any other F and each $a \in L$, ${}^aE$ is coarser than ${}^aF$.

We will make use of the following lemma.

Lemma 5.20 *For every $\langle A_j, B_j\rangle, \left\langle A'_j, B'_j\right\rangle \in \mathcal{B}(X,Y,I)$, $j \in J$, we have*

$$\bigwedge_{j\in J} E(\langle A_j,B_j\rangle, \left\langle A'_j,B'_j\right\rangle) \leq E(\bigwedge_{j\in J}\langle A_j,B_j\rangle, \bigwedge_{j\in J}\left\langle A'_j,B'_j\right\rangle), \tag{5.13}$$

$$\bigwedge_{j\in J} E(\langle A_j,B_j\rangle, \left\langle A'_j,B'_j\right\rangle) \leq E(\bigvee_{j\in J}\langle A_j,B_j\rangle, \bigvee_{j\in J}\left\langle A'_j,B'_j\right\rangle). \tag{5.14}$$

Proof. By the above statements, (5.13) holds iff

$$\bigwedge_{j\in J}(A_j \approx A'_j) \leq (\bigwedge_{j\in J} A_j \approx \bigwedge_{j\in J} A'_j),$$

i.e.

$$\bigwedge_{j\in J}(\bigwedge_{x\in X} A_j(x) \leftrightarrow A'_j(x)) \leq \bigwedge_{x\in X}(\bigwedge_{j\in J} A_j(x)) \leftrightarrow (\bigwedge_{j\in J} A'_j(x))$$

which holds iff for each $x' \in X$ we have

$$\begin{aligned}\bigwedge_{j\in J}\bigwedge_{x\in X} A_j(x) \leftrightarrow A'_j(x) &\leq (\bigwedge_{j\in J} A_j(x')) \leftrightarrow (\bigwedge_{j\in J} A'_j(x')) = \\ &= (\bigwedge_{j\in J} A_j(x')) \to (\bigwedge_{j\in J} A'_j(x')) \wedge (\bigwedge_{j\in J} A'_j(x')) \to (\bigwedge_{j\in J} A_j(x')).\end{aligned}$$

The last inequality is true iff the left side is less than or equal to both of the conjuncts on the right side. As they can be handled analogously, we prove only

$$\bigwedge_{j\in J}\bigwedge_{x\in X}(A_j(x) \leftrightarrow A'_j(x)) \leq (\bigwedge_{j\in J} A_j(x')) \leftrightarrow (\bigwedge_{j\in J} A'_j(x'))$$

which is, by adjointness, equivalent to

$$(\bigwedge_{j\in J} A_j(x')) \otimes \bigwedge_{j\in J}\bigwedge_{x\in X}(A_j(x) \leftrightarrow A'_j(x)) \leq (\bigwedge_{j\in J} A'_j(x'))$$

which is true since

$$\begin{aligned}&(\bigwedge_{j\in J} A_j(x')) \otimes \bigwedge_{j\in J}\bigwedge_{x\in X}(A_j(x) \leftrightarrow A'_j(x)) \leq \\ &\leq (\bigwedge_{j\in J} A_j(x')) \otimes \bigwedge_{j\in J}(A_j(x') \to A'_j(x')) \leq \\ &\leq \bigwedge_{j\in J}\bigwedge_{j'\in J} A'_j(x') \otimes (A_j(x') \to A'_j(x')) \leq \\ &\leq \bigwedge_{j\in J} A_j(x') \otimes (A_j(x') \to A'_j(x')) \leq \bigwedge_{j\in J} A'_j(x').\end{aligned}$$

We have proved (5.13). (5.14) can be proved symmetrically using the fact $\bigvee_{j\in J}\langle A_j, B_j\rangle = \left\langle (\bigcup_{j\in J} A_j)^{\uparrow\downarrow}, \bigwedge_{j\in J} B_j\right\rangle$ of Theorem 5.9 and proceeding by the similarity E on the intents. □

Lemma 5.20 has an interesting corollary stating that the similarity of any two concepts is less than or equal to the similarity of any of them to their direct join or meet.

Corollary 5.21 *Let $\langle A_1, B_1\rangle, \langle A_2, B_2\rangle \in \mathcal{B}(X,Y,I)$. The following inequalities hold for $i = 1,2$:*

$$\begin{aligned} E(\langle A_1, B_1\rangle, \langle A_2, B_2\rangle) &\leq E(\langle A_i, B_i\rangle, \langle A_1, B_1\rangle \wedge \langle A_2, B_2\rangle), \\ E(\langle A_1, B_1\rangle, \langle A_2, B_2\rangle) &\leq E(\langle A_i, B_i\rangle, \langle A_1, B_1\rangle \vee \langle A_2, B_2\rangle). \end{aligned}$$

Proof. Put $J = x, y$, $\langle A_x, B_x\rangle = \langle A_y, B_y\rangle = \langle A_1, B_1\rangle$, $\langle A'_x, B'_x\rangle = \langle A_1, B_1\rangle$, $\left\langle A'_y, B'_y\right\rangle = \langle A_2, B_2\rangle$ and apply Lemma 5.20. □

Theorem 5.22 *The induced similarity E on $\mathcal{B}(X,Y,I)$ is compatible. If $a \in L$ is $\otimes$-idempotent (i.e. $a \otimes a = a$) then ${}^aE$ is, moreover, transitive, i.e. a congruence relation on $\mathcal{B}(X,Y,I)$.*

Proof. The first part follows immediately from (5.13) and (5.14) by the fact that if $a \leq E(\langle A_j, B_j\rangle, \left\langle A'_j, B'_j\right\rangle)$ for $j \in J$, then also $a \leq \bigwedge_{j\in J} E(\langle A_j, B_j\rangle, \left\langle A'_j, B'_j\right\rangle)$. The second part follows from the evident fact that if a is $\otimes$-idempotent and $a \leq b, c$, then $a \leq b \otimes c$. □

Remark 5.23 Theorem 5.22 and the above-described construction yield a method for factorizing any **L**-concept lattice $\mathcal{B}(X,Y,I)$ by any a-cut ${}^aE$ of the induced similarity E. It is worth noting that the similarity E is defined "internally", i.e. it is not supplied from the outside.

Next we formulate a statement concerning the relation of the similarity and the hierarchy of concepts which shows that the farther are the concepts in the hierarchy, the less similar they are.

Theorem 5.24 *Let $\langle A_i, B_i\rangle \in \mathcal{B}(X,Y,I)$, $i = 1,2,3$, be such that $\langle A_1, B_1\rangle \leq \langle A_2, B_2\rangle \leq \langle A_3, B_3\rangle$. Then*

$$\begin{aligned} E(\langle A_1, B_1\rangle, \langle A_3, B_3\rangle) &\leq E(\langle A_1, B_1\rangle, \langle A_2, B_2\rangle), \\ E(\langle A_1, B_1\rangle, \langle A_3, B_3\rangle) &\leq E(\langle A_2, B_2\rangle, \langle A_3, B_3\rangle). \end{aligned}$$

Proof. By the assumptions, i.e. $A_1(x) \leq A_2(x) \leq A_3(x)$ for all $x \in X$, and by the antitony of $\to$ in the first argument we have $E(\langle A_1, B_1\rangle, \langle A_3, B_3\rangle) = \bigwedge_{x\in X}(A_3(x) \to A_1(x)) \leq \bigwedge_{x\in X}(A_2(x) \to A_1(x)) = E(\langle A_1, B_1\rangle, \langle A_2, B_2\rangle)$. The second part may be obtained symmetrically. □

Similarity of concept lattices. Finally, we consider similarity of concept lattices. A natural way to define the similarity degree of two concept lattices over the sets X and Y is based on the following intuition. Concept lattices $\mathcal{B}(X,Y,I_1)$ and $\mathcal{B}(X,Y,I_2)$ are similar iff for each concept $c_1 \in \mathcal{B}(X,Y,I_1)$ there is a concept $c_2 \in \mathcal{B}(X,Y,I_2)$ such that c_1 and c_2 are similar and, conversely, for each concept $c_2 \in \mathcal{B}(X,Y,I_2)$ there is a concept $c_1 \in \mathcal{B}(X,Y,I_1)$ such that c_1 and c_2 are similar. In the following we write $\mathcal{B}_1$ and $\mathcal{B}_2$ instead of $\mathcal{B}(X,Y,I_1)$ and $\mathcal{B}(X,Y,I_2)$, respectively. According to how the similarity of concepts is measured we distinguish two rules for the definition of the similarity degree of two concept lattices:

$$\begin{aligned} & E^*(\mathcal{B}(X,Y,I_1),\mathcal{B}(X,Y,I_2)) = \\ & = \bigwedge_{\langle A_1,B_1\rangle\in\mathcal{B}_1}\ \bigvee_{\langle A_2,B_2\rangle\in\mathcal{B}_2} E^*(\langle A_1,B_1\rangle,\langle A_2,B_2\rangle)\wedge \\ & \quad \bigwedge_{\langle A_2,B_2\rangle\in\mathcal{B}_2}\ \bigvee_{\langle A_1,B_1\rangle\in\mathcal{B}_1} E^*(\langle A_1,B_1\rangle,\langle A_2,B_2\rangle), \end{aligned} \tag{5.15}$$

$* \in \{Ext, Int\}$, where we put

$$\begin{aligned} E^{Ext}(\langle A_1,B_1\rangle,\langle A_2,B_2\rangle) &= (A_1 \approx A_2),\\ E^{Int}(\langle A_1,B_1\rangle,\langle A_2,B_2\rangle) &= (B_1 \approx B_2). \end{aligned}$$

Note that E^{Ext} and E^{Int} correspond to the cases when the similarity of concepts is measured by extents and by intents of concepts, respectively. However, Theorem 5.18 cannot be applied to show $E^{Ext} = E^{Int}$ because concepts of different concept lattices are considered.

We are going to show that all of the above relations are in fact similarity relations, that they are equal, and that they are, moreover, equal to the similarity relation E defined on the set of all contexts by

$$E(\langle X,Y,I_1\rangle,\langle X,Y,I_2\rangle) = (I_1 \approx I_2) = \bigwedge_{\langle x,y\rangle\in X\times Y} I_1(x,y) \leftrightarrow I_2(x,y). \tag{5.16}$$

The following corollary follows directly by Theorem 4.34.

Corollary 5.25 *The relation E defined by (5.16) is the largest* $\mathbf{L}$*-similarity relation on* $\{\langle X,Y,I\rangle \mid I \in L^{X\times Y}\}$ *such that* $I_1(x,y)\otimes E(I_1,I_2) \leq I_2(x,y)$ *for every* $x\in X$, $y \in Y$.

We need the following lemmas.

Lemma 5.26 *Let $\langle X, Y, I_1\rangle$, $\langle X, Y, I_2\rangle$ be* **L**-*contexts*, $A \in L^X$, $B \in L^Y$. *Then* $E(I_1, I_2) \leq E(A^{\uparrow_{I_1}}, A^{\uparrow_{I_2}})$ *and* $E(I_1, I_2) \leq E(B^{\downarrow_{I_1}}, B^{\downarrow_{I_2}})$.

Proof. We proceed analogously as in the proof of Lemma 5.15. Consider $\mathcal{J}_{\mathrm{Con}}$ (Remark 5.2). Let $\mathbf{M}_1$ be the **L**-structure for $\mathcal{J}_{\mathrm{Con}}$ given by $\langle X, Y, I_1\rangle$ with $r_A^{\mathbf{M}_1} = A$; $\mathbf{M}_2$ be the **L**-structure for $\mathcal{J}_{\mathrm{Con}}$ given by $\langle X, Y, I_2\rangle$ with $r_A^{\mathbf{M}_2} = A$ (thus $\mathbf{M}_1$ differs from $\mathbf{M}_2$ only by interpretation of r_I). As $F = \emptyset$, $(r_A^{\mathbf{M}_1} \approx r_A^{\mathbf{M}_2})$, $|\varphi_\uparrow|_r = 1$ for $r = r_A$, $|\varphi_\uparrow|_r = 1$ for $r = r_I$, and $|\varphi_\uparrow|_r = 0$ otherwise, the many-sorted version of Lemma 3.85 yields

$$(r_I^{\mathbf{M}_1} \approx r_I^{\mathbf{M}_2}) \leq \|\varphi_\uparrow\|_{\mathbf{M}_1,v} \leftrightarrow \|\varphi_\uparrow\|_{\mathbf{M}_2,v}.$$

As $r_I^{\mathbf{M}_1} = I_1$, $r_I^{\mathbf{M}_2} = I_2$, $\|\varphi_\uparrow\|_{\mathbf{M}_1,v} = A^{\uparrow_{I_1}}(v(\mathtt{y}))$, and $\|\varphi_\uparrow\|_{\mathbf{M}_2,v} = A^{\uparrow_{I_2}}(v(\mathtt{y}))$, we have

$$(I_1 \approx I_2) \leq (A^{\uparrow_{I_1}}(v(\mathtt{y})) \leftrightarrow A^{\uparrow_{I_2}}(v(\mathtt{y})))$$

for any valuation v. Recalling

$$(A^{\uparrow_{I_1}} \approx A^{\uparrow_{I_2}}) = \bigwedge_{v \text{ a valuation}} (A^{\uparrow_{I_1}}(v(\mathtt{y})) \approx A^{\uparrow_{I_2}}(v(\mathtt{y}))),$$

$(I_1 \approx I_2) \leq (A^{\uparrow_{I_1}} \approx A^{\uparrow_{I_2}})$ follows. The second part is analogous (proceed for $\varphi_\downarrow$). □

Note that by Lemmas 5.15 and 5.26 we have $E(A_1, A_2) \otimes E(I_1, I_2) \leq E(A_1^{\uparrow_{I_1}}, A_2^{\uparrow_{I_2}})$ and $E(B_1, B_2) \otimes E(I_1, I_2) \leq E(B_1^{\downarrow_{I_1}}, B_2^{\downarrow_{I_2}})$.

Lemma 5.27 *For all* **L**-*contexts* $\langle X, Y, I_1\rangle$, $\langle X, Y, I_2\rangle$ *and* $* \in \{Ext, Int\}$ *we have*

$$E(\langle X, Y, I_1\rangle, \langle X, Y, I_2\rangle) \leq E^*(\mathcal{B}(X, Y, I_1), \mathcal{B}(X, Y, I_2)).$$

Moreover, both E^{Ext} *and* E^{Int} *are* **L**-*equivalence relations.*

Proof. We proceed only for E^{Ext}, the second case is symmetric. We have to prove $(I_1 \approx I_2) \leq E^{Ext}(\mathcal{B}_1, \mathcal{B}_2)$, which holds iff both

$$(I_1 \approx I_2) \leq \bigwedge_{\langle A_1, B_1\rangle \in \mathcal{B}_1} \bigvee_{\langle A_2, B_2\rangle \in \mathcal{B}_2} E^{Ext}(\langle A_1, B_1\rangle, \langle A_2, B_2\rangle)$$

and

$$(I_1 \approx I_2) \leq \bigwedge_{\langle A_2, B_2\rangle \in \mathcal{B}_2} \bigvee_{\langle A_1, B_1\rangle \in \mathcal{B}_1} E^{Ext}(\langle A_1, B_1\rangle, \langle A_2, B_2\rangle)$$

hold. Due to symmetry we prove only the first inequality which is true iff

$$(I_1 \approx I_2) \leq \bigvee_{\langle A_2, B_2\rangle \in \mathcal{B}_2} (A_1 \approx A_2)$$

holds for each $\langle A_1, B_1\rangle \in \mathcal{B}_1$. By Lemma 5.26 we have $(I_1 \approx I_2) \leq (B_1^{\downarrow_{I_1}} \approx B_1^{\downarrow_{I_2}}) = (A_1 \approx B_1^{\downarrow_{I_2}})$, i.e. putting $\langle A_2, B_2\rangle = \left\langle B_1^{\downarrow_{I_2}}, B_1^{\downarrow_{I_2}\uparrow_{I_2}}\right\rangle$ we have

$$(I_1 \approx I_2) \leq (A_1 \approx B_1^{\downarrow_{I_2}}) \leq \bigvee_{\langle A_2, B_2\rangle \in \mathcal{B}_2} (A_1 \approx A_2).$$

The fact that E^{Ext} and E^{Int} are similarity (i.e. **L**-equivalence) relations follows easily from Theorem 4.41. finishing the proof. □

5.2.2 LOGICAL PRECISION IN CONCEPT LATTICES

In this section we show how the general results on morphisms (and products) developed in Section 3.4. Particularly, we will show how the decrease of logical precision can be utilized to yield a method of factorization of fuzzy concept lattices.

Let $\langle X, Y, I\rangle$ be an **L**-context; let **M** be the corresponding **L**-structure for $\mathcal{J}_{\mathrm{Con}}$ (Remark 5.2). Recall from Section 3.4 that the idea of a decrease of logical precision is to identify those truth values from L which are for some reasons not necessary to distinguish (e.g. they are very close). Then, instead of $a, b \in L$ which are to be identified, we can consider another truth value $h_L(a) = h_L(b)$ (a common abstraction of a and b). In order for h_L to be a systematic replacement, h_L is supposed to be a complete morphism of **L** to another structure **L**′.

To decrease logical precision by h_L means to consider an **L**′-context $\langle X, Y, h_L(I)\rangle$ instead of an **L**-context $\langle X, Y, I\rangle$; more precisely, $\langle\langle X, h_L(\approx_X)\rangle, \langle Y, h_L(\approx_Y)\rangle, h_L(I)\rangle$ instead of $\langle\langle X, \approx_X\rangle, \langle Y, \approx_Y\rangle, I\rangle$, i.e. to consider $h_L(\mathbf{M})$ instead of **M** (everything is correct due to Lemma 3.93). Intuitively, changing the structure of truth values in a systematic way should result in a systematic change of the structure of information (conceptual structure in our case). The following two statements confirm the intuition. Recall that a lattice homomorphism $h : \mathbf{V}_1 \to \mathbf{V}_2$ between two complete lattices $\mathbf{V}_1$ and $\mathbf{V}_2$ is called *complete* if for each $K \subseteq V_1$ we have $h(\bigwedge_{k\in K}) = \bigwedge_{k\in K} h(k)$ and $h(\bigvee_{k\in K}) = \bigvee_{k\in K} h(k)$.

Lemma 5.28 *For a complete morphism $h_L : \mathbf{L} \to \mathbf{L}'$ and an **L**-context $\langle X, Y, I\rangle$ we have $h_L(A^{\uparrow_I}) = h_L(A)^{\uparrow_{h_L(I)}}$ and $h_L(B^{\downarrow_I}) = h_L(B)^{\downarrow_{h_L(I)}}$ for any $A \in L^X$ and $B \in L^Y$.*

Proof. Recall that for the language $\mathcal{J}_{\text{Con}}$, an $\mathbf{L}$-structure $\mathbf{M}$ corresponding to $\langle X, Y, I\rangle$, and a valuation v such that $v(\mathtt{y}) = y$ and $v(\mathtt{x}) = x$ we have $\|\varphi_{\uparrow}\|_{\mathbf{M},v} = A^{\uparrow}(y)$ and $\|\varphi_{\downarrow}\|_{\mathbf{M},v} = B^{\downarrow}(x)$. Therefore, the assertion follows directly from Theorem 3.94. □

Lemma 5.29 *Let $h_L : \mathbf{L} \to \mathbf{L}'$ be a complete onto morphism, $\langle X, Y, I\rangle$ be an $\mathbf{L}$-context. Then for $C \in L'^X$, $D \in L'^Y$, we have: $\langle C, D\rangle \in \mathcal{B}(X, Y, h_L(I))$ iff there are $A \in L^X$, $B \in L^Y$ such that $\langle A, B\rangle \in \mathcal{B}(X, Y, I)$, $h_L(A) = C$, and $h_L(B) = D$.*

Proof. Let $\langle A, B\rangle \in \mathcal{B}(X, Y, I)$, i.e. $A^{\uparrow} = B$, $B^{\downarrow} = A$. By Lemma 5.28, $h_L(A)^{\uparrow} = h_L(B)$ and $h_L(B)^{\downarrow} = h_L(A)$, whence $\langle h_L(A), h_L(B)\rangle \in \mathcal{B}(X, Y, I)$.

Conversely, let $\langle C, D\rangle \in \mathcal{B}(X, Y, h(I))$. Take arbitrary $A \in L^X$ such that $h_L(A(x)) = C(x)$ for each $x \in X$. Such an A always exists due to the surjectivity of h_L. By Lemma 5.28, $h(A^{\uparrow}) = (h(A))^{\uparrow} = C^{\uparrow} = D$ and $h(A^{\uparrow\downarrow}) = (h(A^{\uparrow}))^{\downarrow} = (h(A))^{\uparrow\downarrow} = C^{\uparrow\downarrow} = C$, and thus $\langle A^{\uparrow\downarrow}, A^{\uparrow}\rangle$ is the required concept from $\mathcal{B}(X, Y, I)$. □

Theorem 5.30 *Under the conditions of the preceding lemma, there is a complete homomorphism of $\mathcal{B}(X, Y, I)$ onto $\mathcal{B}(X, Y, h(I))$.*

Proof. Due to Lemma 5.29 (and its proof) it suffices to show that the mapping $h^* : \mathcal{B}(X, Y, I) \to \mathcal{B}(X, Y, h(I))$ defined by

$$h^*(A, B) = \langle h(A), h(B)\rangle$$

preserves arbitrary meets and joins. We proceed only for meets, the case of joins may be proved analogously. Let $\langle A_j, B_j\rangle \in \mathcal{B}(X, Y, I)$, $j \in J$. We have to prove

$$h^*(\bigwedge_{j\in J} \langle A_j, B_j\rangle) = \bigwedge_{j\in J} h^*(\langle A_1, B_1\rangle).$$

By Theorem 5.9 we have

$$\begin{aligned} & h^*(\bigwedge_{j\in J} \langle A_j, B_j\rangle) = \\ = \; & h^*(\left\langle \bigcap_{j\in J} A_j, (\bigcap_{j\in J} A_j)^{\uparrow} \right\rangle) = \left\langle h(\bigcap_{j\in J} A_j), h((\bigcap_{j\in J} A_j)^{\uparrow}) \right\rangle \end{aligned} \tag{5.17}$$

and

$$\bigwedge_{j\in J} h^*(\langle A_j, B_j\rangle) =$$

$$= \bigwedge_{j\in J} \langle h(A_j), h(B_j)\rangle = \left\langle \bigcap_{j\in J} h(A_j), (\bigcap_{j\in J} h(A_j))^{\uparrow} \right\rangle. \quad (5.18)$$

The concepts in (5.17) and (5.18) equal $(h(\bigcap_{j\in J} A_j))(x) = (\bigcap_{j\in J} h(A_j))(x)$. The proof is finished. □

Remark 5.31 The foregoing theorem is relevant also from the application point of view. Suppose we have a concept lattice with truth values from L. A further analysis on the level of L may be (for various reasons, e.g. computational ones) "too precise". We can then skip to the level of $L' = h(L)$ which is appropriate. Due to Theorem 5.30, the structure of concepts changes systematically, i.e. the structure of concepts in L is in a systematic way more precise than that in L'. An illustrative example is presented later.

Consider now another application of logical precision. The structure $\mathbf{L}$ of truth values in which our knowledge is formulated may be in some way composed of structures $\mathbf{L}_i$ $(i \in I)$. Since $\mathbf{L}_i$ are more simple, it is to be expected that if possible, the analysis and processing of knowledge formulated using $\mathbf{L}$ can be simplified by a suitable decomposition of the knowledge into parts which correspond to particular $\mathbf{L}_i$'s. In the following, we concentrate on the case where "$\mathbf{L}$ is composed of $\mathbf{L}_i$'s" means that $\mathbf{L}$ is a (sub)direct product of $\mathbf{L}_i$ $(i \in I)$. Note that if $\mathbf{L}$ is a complete subdirect product of $\mathbf{L}_i$ $(i \in I)$ then $\mathrm{pr}_i : L \to L_i$ is a complete morphism from $\mathbf{L}$ onto $\mathbf{L}_i$ for any $i \in I$.

Let $\mathbf{L}_1$ be a complete subalgebra of $\mathbf{L}_2$. If $\langle X, Y, I\rangle$ is an $\mathbf{L}_2$-context such that $I(x, y) \in L_1$, $(x \approx_X x') \in L_1$, $(y \approx_Y y') \in L_1$ for all $x, x' \in X$ and $y, y' \in Y$, then $\langle X, Y, I\rangle$ may be considered as an $\mathbf{L}_1$-context as well. In that case, we denote by $\mathcal{B}_{\mathbf{L}_1}(X, Y, I)$ and $\mathcal{B}_{\mathbf{L}_2}(X, Y, I)$ the $\mathbf{L}_1$-concept lattice and the $\mathbf{L}_2$-concept lattice, respectively, corresponding to $\langle X, Y, I\rangle$.

Lemma 5.32 *Under the above conditions, $\mathcal{B}_{\mathbf{L}_1}(X, Y, I)$ is a complete sublattice of $\mathcal{B}_{\mathbf{L}_2}(X, Y, I)$.*

Proof. Let $A \in L_1^X$. Denote by $^{\uparrow_1}$ and $^{\uparrow_2}$ the operators defined by I using the operations of $\mathbf{L}_1$ and $\mathbf{L}_2$, respectively. Since $\mathbf{L}_1$ is a complete residuated sublattice of $\mathbf{L}_2$, a moment's reflection shows that $A^{\uparrow_1} =$

$A^{\uparrow_2}$; similarly for $^{\downarrow_1}$ and $^{\downarrow_2}$. We immediately conclude $\mathcal{B}_{\mathbf{L}_1}(X, Y, I) \subseteq \mathcal{B}_{\mathbf{L}_2}(X, Y, I)$. Moreover, from (5.4) and (5.5) we infer that $\mathcal{B}_{\mathbf{L}_1}(X, Y, I)$ is a complete sublattice of $\mathcal{B}_{\mathbf{L}_2}(X, Y, I)$. □

Let $\mathbf{L}$ be a subalgebra of a direct product $\times_{j\in J}\mathbf{L}_j$ of complete residuated lattices. For an $\mathbf{L}$-relation I between X and Y, and each $j \in J$ we define an $\mathbf{L}_j$-relation between X and Y by

$$I_j(x, y) = \mathrm{pr}_j(I(x, y)).$$

For $\mathbf{L}_j$-contexts $\langle\langle X, \approx_j\rangle, \langle Y, \approx_j\rangle, I_j\rangle$ $(j \in J)$ we can define an $\times_{j\in J}\mathbf{L}_j$-context $\Pi_{j\in J}\langle\langle X, \approx_j\rangle, \langle Y, \approx_j\rangle, I_j\rangle$ by

$$\Pi_{j\in J}\langle\langle X, \approx_j\rangle, \langle Y, \approx_j\rangle, I_j\rangle = \langle\langle X, \Pi \approx_j\rangle, \langle Y, \Pi \approx_j\rangle, \Pi_{j\in J} I_j\rangle$$

where $\Pi_{j\in J} \approx_j$ and $\Pi_{j\in J} I_j$ are the products defined in Section 3.4. If $\mathbf{M}_j$ are the $\mathbf{L}_j$-structures corresponding to $\langle X, Y, I_j\rangle$ then the $\times\mathbf{L}_i$-structure corresponding to $\Pi_{j\in J}\langle\langle X, \approx_j\rangle, \langle Y, \approx_j\rangle, I_j\rangle$ is just $\Pi_j\mathbf{M}_j$ (cf. Definition 3.121). The definition is thus correct.

Theorem 5.33 *Let $\langle X, Y, I_j\rangle$ be $\mathbf{L}_j$-contexts $(j \in J)$. Then $\mathcal{B}(\Pi_{j\in J}\langle X, Y, I_j\rangle)$ is isomorphic to $\times_{j\in J}\mathcal{B}(X, Y, I_j)$ (direct product of lattices).*

Proof. We prove the theorem by showing that there is an isotone bijection $\varphi : \mathcal{B}(\Pi_{j\in J}\langle X, Y, I_j\rangle) \to \times_{j\in J}\mathcal{B}(X, Y, I_j)$. For $\langle A, B\rangle \in \mathcal{B}(\Pi_{j\in J}\langle X, Y, I_j\rangle)$ define $\varphi(A, B)$ by

$$\mathrm{pr}_j(\varphi(A, B)) = \langle A_j, B_j\rangle$$

where $A_j(x) = \mathrm{pr}_j(A(x))$, $B_j(x) = \mathrm{pr}_j(B(x))$. $pr_j : \times_{i\in J}\mathbf{L}_i \to \mathbf{L}_j$ $(j \in J)$ are complete onto morphisms. Furthermore, by the definition of $\Pi_{j\in J} I_j$ we have $\mathrm{pr}_j((\Pi_{i\in J} I_i)(x, y)) = I_j(x, y)$. Therefore, by Theorem 5.30, the mappings $\varphi_j : \langle A, B\rangle \mapsto \langle A_j, B_j\rangle$ $(j \in J)$, defined by $A_j(x) = \mathrm{pr}_j(A(x))$, $B_j(y) = \mathrm{pr}_j(B(y))$, are complete onto morphisms of $\mathcal{B}(\Pi_{j\in J}\langle X, Y, I_j\rangle)$ onto $\mathcal{B}(X, Y, I_j)$. This shows that φ is correctly defined and, moreover, is isotone. The injectivity of φ is immediate from the definition. It remains to show that φ is an onto mapping. Take therefore $\langle A_j, B_j\rangle \in \mathcal{B}(X, Y, I_j)$ for each $j \in J$. Take $\langle A, B\rangle \in L^X \times L^Y$ so that $\mathrm{pr}_j(A(x)) = A_j(x)$ and $\mathrm{pr}_j(B(y)) = B_j(y)$. Clearly, $\mathrm{pr}_j(\varphi(A, B)) = \langle A_j, B_j\rangle$. We need to show that $\langle A, B\rangle \in \mathcal{B}(\Pi_{j\in J}\langle X, Y, I_j\rangle)$, i.e. we need to verify $A^\uparrow = B$ and $B^\downarrow = A$. Due to the similarity of both of the cases we proceed for the first one only. Consider any $j \in J$.

Since $\mathrm{pr}_j : \times_{i\in J}\mathbf{L}_i \to \mathbf{L}_j$ is a complete morphism, Lemma 5.28 yields $\mathrm{pr}_j(A^{\uparrow}(y)) = \mathrm{pr}_j(A^{\uparrow}(y)) = \mathrm{pr}_j(B(y))$, i.e. $A^{\uparrow} = B$. The proof is complete. □

A combination of the preceding results yields the following assertion.

Theorem 5.34 *Let $\langle X, Y, I\rangle$ be an $\mathbf{L}$-context. If $\mathbf{L}$ is a complete subdirect product of $\mathbf{L}_j$ $(j \in J)$, then the $\mathbf{L}$-concept lattice $\mathcal{B}(X, Y, I)$ is isomorphic to a subdirect product of $\mathbf{L}_j$-concept lattices $\mathcal{B}\left(X, Y, \mathrm{pr}_j(I)\right)$. Each $\mathcal{B}\left(X, Y, \mathrm{pr}_j(I)\right)$ is a complete homomorphic image of $\mathcal{B}(X, Y, I)$.*

Proof. Since $I = \prod_{j\in J}\mathrm{pr}_j(I)$, Theorem 5.33 yields the existence of an isomorphism φ from $\mathcal{B}_{\times_{j\in J}\mathbf{L}_j}(X, Y, I)$ to $\Pi_{j\in J}\mathcal{B}_{\mathbf{L}_j}\left(X, Y, \mathrm{pr}_j(I)\right)$ (in what follows, φ refers to the isomorphism constructed in the proof of Theorem 5.33). Furthermore, by Lemma 5.32, $\mathcal{B}_{\mathbf{L}}(X, Y, I)$ is a complete sublattice of $\mathcal{B}_{\times_{j\in J}\mathbf{L}_j}(X, Y, I)$. Therefore, it is sufficient to prove that for each $j \in J$ and each $\langle A_j, B_j\rangle \in \mathcal{B}_{\mathbf{L}_j}\left(X, Y, \mathrm{pr}_j(I)\right)$ there is an $\langle A, B\rangle \in \mathcal{B}_{\mathbf{L}}(X, Y, I)$ such that $\mathrm{pr}_j(\varphi(A, B)) = \langle A_j, B_j\rangle$. Take any $\langle A_j, B_j\rangle \in \mathcal{B}_{\mathbf{L}_j}\left(X, Y, \mathrm{pr}_j(I)\right)$. Consider any $x \in X$. As $\mathbf{L}$ is a subdirect product of $\mathbf{L}_j$'s, there exists $a_x \in L$ such that $\mathrm{pr}_j(a_x) = A_j(x)$. Now, take $A' \in L^X$ such that $A'(x) = a_x$, and put $B = A'^{\uparrow}$, $A = B^{\downarrow}$. We have $\langle A, B\rangle \in \mathcal{B}_{\mathbf{L}}(X, Y, I)$. A moment's inspection shows that $\mathrm{pr}_j(A(x)) = A_j(x)$ and $\mathrm{pr}_j(B(y)) = B_j(y)$ for any $x \in X$ and $y \in Y$, i.e. $\mathrm{pr}_j(\varphi(A, B)) = \langle A_j, B_j\rangle$ completing the proof. □

5.2.3 ILLUSTRATIVE EXAMPLES

Consider the following illustrative example. We take $L = \{0, \frac{1}{2}, 1\}$. We will consider two structures defined on L, the Łukasiewicz one and the Gödel one. The context is given by Table 5.1. The set X contains nine elements (Mercury, ..., Pluto), the set Y contains four attributes ("size small", ..., "near to sun"). The corresponding fuzzy concept lattices are depicted in Fig. 5.1 (Łukasiewicz structure on L) and Fig. 5.2 (Gödel structure on L).

To get a deeper insight, the elements (i.e. concepts) of the lattice are identified in Table 5.2 (Łukasiewicz structure) and Table 5.3 (Gödel structure).

Consider the Gödel structure on L and the mapping $h : L \to \{0, 1\}$ defined by $h(1) = h(\frac{1}{2}) = 1$, $h(0) = 0$. It is easy to verify that h is a homomorphism. Then there is a complete onto homomorphism h^* :

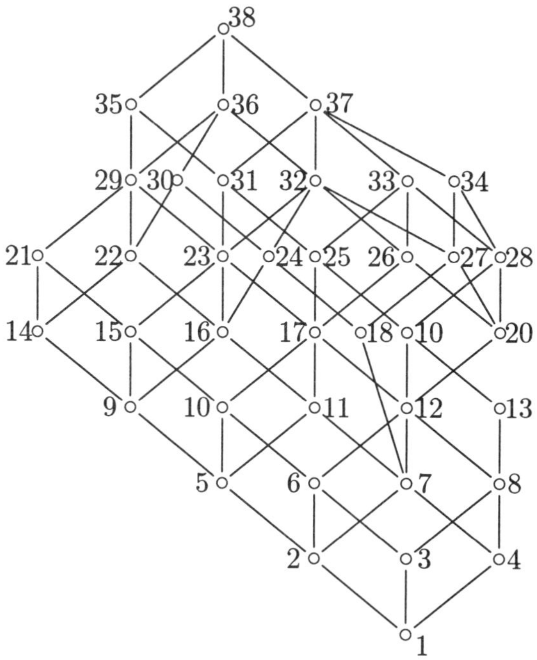

Figure 5.1. **L**-concept lattice of the context in Table 5.1 (Łukasiewicz structure).

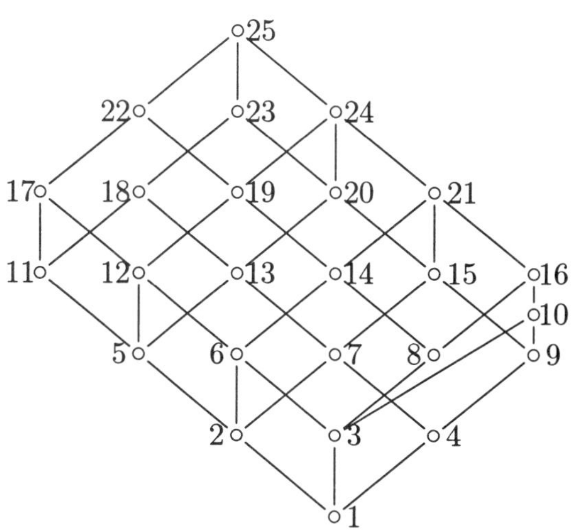

Figure 5.2. **L**-concept lattice of the context in Table 5.1 (Gödel structure).

	size		from sun	
	small (ss)	large (sl)	far (df)	near (dn)
Mercury (Me)	1	0	0	1
Venus (V)	1	0	0	1
Earth (E)	1	0	0	1
Mars (Ma)	1	0	$\frac{1}{2}$	1
Jupiter (J)	0	1	1	$\frac{1}{2}$
Saturn (S)	0	1	1	$\frac{1}{2}$
Uranus (U)	$\frac{1}{2}$	$\frac{1}{2}$	1	0
Neptune (N)	$\frac{1}{2}$	$\frac{1}{2}$	1	0
Pluto (P)	1	0	1	0

Table 5.1. **L**-context given by planets and their properties.

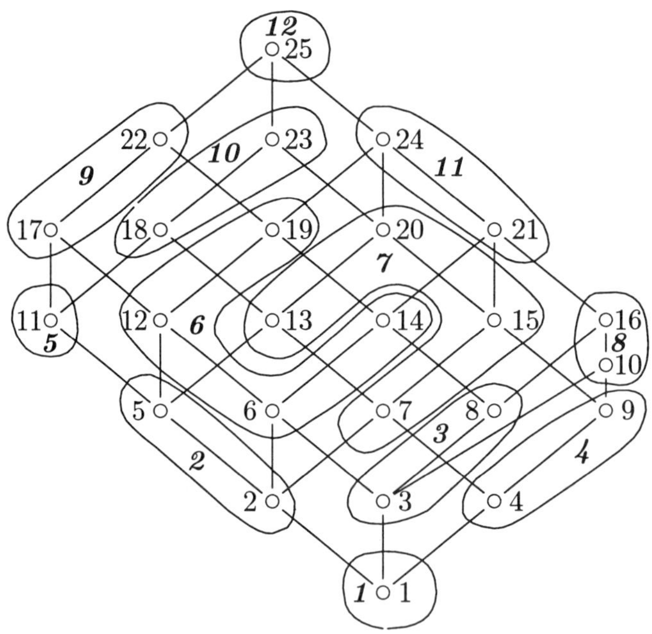

Figure 5.3. Classes of the congruence relation induced by h.

$\mathcal{B}(X,Y,I) \to \mathcal{B}(X,Y,h(I))$ (cf. Section 5.2.2). The lattice $\mathcal{B}(X,Y,h(I))$ is depicted in Fig. 5.5. The classes of the congruence relation induced by h^* are depicted in Fig. 5.3.

Consider now the Łukasiewicz structure and the a-cut of the induced similarity on $\mathcal{B}(X,Y,I)$ for $a = \frac{1}{2}$, i.e. ${}^{\frac{1}{2}}E$. The tolerance blocks (which are, in fact, complete sublattices) are depicted in Fig. 5.4. Note that

no.	extent									intent			
	Me	V	E	Ma	J	S	U	N	P	ss	sl	df	dn
1.	0	0	0	0	0	0	0	0	0	1	1	1	1
2.	0	0	0	$\frac{1}{2}$	0	0	0	0	0	1	$\frac{1}{2}$	1	1
3.	0	0	0	0	$\frac{1}{2}$	$\frac{1}{2}$	0	0	0	$\frac{1}{2}$	1	1	1
4.	0	0	0	0	0	0	$\frac{1}{2}$	$\frac{1}{2}$	0	1	1	1	$\frac{1}{2}$
5.	$\frac{1}{2}$	$\frac{1}{2}$	$\frac{1}{2}$	$\frac{1}{2}$	0	0	0	0	0	1	$\frac{1}{2}$	$\frac{1}{2}$	1
6.	0	0	0	$\frac{1}{2}$	$\frac{1}{2}$	$\frac{1}{2}$	0	0	0	$\frac{1}{2}$	$\frac{1}{2}$	1	1
7.	0	0	0	$\frac{1}{2}$	0	0	$\frac{1}{2}$	$\frac{1}{2}$	$\frac{1}{2}$	1	$\frac{1}{2}$	1	$\frac{1}{2}$
8.	0	0	0	0	$\frac{1}{2}$	$\frac{1}{2}$	$\frac{1}{2}$	$\frac{1}{2}$	0	$\frac{1}{2}$	1	1	$\frac{1}{2}$
9.	$\frac{1}{2}$	$\frac{1}{2}$	$\frac{1}{2}$	1	0	0	0	0	0	1	0	$\frac{1}{2}$	1
10.	$\frac{1}{2}$	$\frac{1}{2}$	$\frac{1}{2}$	$\frac{1}{2}$	$\frac{1}{2}$	$\frac{1}{2}$	0	0	0	$\frac{1}{2}$	$\frac{1}{2}$	$\frac{1}{2}$	1
11.	$\frac{1}{2}$	$\frac{1}{2}$	$\frac{1}{2}$	$\frac{1}{2}$	0	0	$\frac{1}{2}$	$\frac{1}{2}$	$\frac{1}{2}$	1	$\frac{1}{2}$	$\frac{1}{2}$	$\frac{1}{2}$
12.	0	0	0	$\frac{1}{2}$	$\frac{1}{2}$	$\frac{1}{2}$	$\frac{1}{2}$	$\frac{1}{2}$	$\frac{1}{2}$	$\frac{1}{2}$	$\frac{1}{2}$	1	$\frac{1}{2}$
13.	0	0	0	0	1	1	$\frac{1}{2}$	$\frac{1}{2}$	0	0	1	1	$\frac{1}{2}$
14.	1	1	1	1	0	0	0	0	0	1	0	0	1
15.	$\frac{1}{2}$	$\frac{1}{2}$	$\frac{1}{2}$	1	$\frac{1}{2}$	$\frac{1}{2}$	0	0	0	$\frac{1}{2}$	0	$\frac{1}{2}$	1
16.	$\frac{1}{2}$	$\frac{1}{2}$	$\frac{1}{2}$	1	0	0	$\frac{1}{2}$	$\frac{1}{2}$	$\frac{1}{2}$	1	0	$\frac{1}{2}$	$\frac{1}{2}$
17.	$\frac{1}{2}$	$\frac{1}{2}$	$\frac{1}{2}$	$\frac{1}{2}$	$\frac{1}{2}$	$\frac{1}{2}$	$\frac{1}{2}$	$\frac{1}{2}$	$\frac{1}{2}$	$\frac{1}{2}$	$\frac{1}{2}$	$\frac{1}{2}$	$\frac{1}{2}$
18.	0	0	0	$\frac{1}{2}$	0	0	$\frac{1}{2}$	$\frac{1}{2}$	1	1	0	1	0
19.	0	0	0	$\frac{1}{2}$	1	1	$\frac{1}{2}$	$\frac{1}{2}$	$\frac{1}{2}$	0	$\frac{1}{2}$	1	$\frac{1}{2}$
no.	extent									intent			
	Me	V	E	Ma	J	S	U	N	P	ss	sl	df	dn
20.	0	0	0	$\frac{1}{2}$	$\frac{1}{2}$	$\frac{1}{2}$	1	1	$\frac{1}{2}$	$\frac{1}{2}$	$\frac{1}{2}$	1	0
21.	1	1	1	1	$\frac{1}{2}$	$\frac{1}{2}$	0	0	0	$\frac{1}{2}$	0	0	1
22.	1	1	1	1	0	0	$\frac{1}{2}$	$\frac{1}{2}$	$\frac{1}{2}$	1	0	0	$\frac{1}{2}$
23.	$\frac{1}{2}$	$\frac{1}{2}$	$\frac{1}{2}$	1	$\frac{1}{2}$	$\frac{1}{2}$	$\frac{1}{2}$	$\frac{1}{2}$	$\frac{1}{2}$	$\frac{1}{2}$	0	$\frac{1}{2}$	$\frac{1}{2}$
24.	$\frac{1}{2}$	$\frac{1}{2}$	$\frac{1}{2}$	1	0	0	$\frac{1}{2}$	$\frac{1}{2}$	1	1	0	$\frac{1}{2}$	0
25.	$\frac{1}{2}$	$\frac{1}{2}$	$\frac{1}{2}$	$\frac{1}{2}$	1	1	$\frac{1}{2}$	$\frac{1}{2}$	$\frac{1}{2}$	0	$\frac{1}{2}$	$\frac{1}{2}$	$\frac{1}{2}$
26.	$\frac{1}{2}$	$\frac{1}{2}$	$\frac{1}{2}$	$\frac{1}{2}$	$\frac{1}{2}$	$\frac{1}{2}$	1	1	$\frac{1}{2}$	$\frac{1}{2}$	$\frac{1}{2}$	$\frac{1}{2}$	0
27.	0	0	0	$\frac{1}{2}$	$\frac{1}{2}$	$\frac{1}{2}$	1	1	1	$\frac{1}{2}$	0	1	0
28.	0	0	0	$\frac{1}{2}$	1	1	1	1	$\frac{1}{2}$	0	$\frac{1}{2}$	1	0
29.	1	1	1	1	$\frac{1}{2}$	$\frac{1}{2}$	$\frac{1}{2}$	$\frac{1}{2}$	$\frac{1}{2}$	$\frac{1}{2}$	0	0	$\frac{1}{2}$
30.	1	1	1	1	0	0	$\frac{1}{2}$	$\frac{1}{2}$	1	1	0	0	0
31.	$\frac{1}{2}$	$\frac{1}{2}$	$\frac{1}{2}$	1	1	1	$\frac{1}{2}$	$\frac{1}{2}$	$\frac{1}{2}$	0	0	$\frac{1}{2}$	$\frac{1}{2}$
32.	$\frac{1}{2}$	$\frac{1}{2}$	$\frac{1}{2}$	1	$\frac{1}{2}$	$\frac{1}{2}$	1	1	1	$\frac{1}{2}$	0	$\frac{1}{2}$	0
33.	$\frac{1}{2}$	$\frac{1}{2}$	$\frac{1}{2}$	$\frac{1}{2}$	1	1	1	1	$\frac{1}{2}$	0	$\frac{1}{2}$	$\frac{1}{2}$	0
34.	0	0	0	$\frac{1}{2}$	1	1	1	1	1	0	0	1	0
35.	1	1	1	1	1	1	$\frac{1}{2}$	$\frac{1}{2}$	$\frac{1}{2}$	0	0	0	$\frac{1}{2}$
36.	1	1	1	1	$\frac{1}{2}$	$\frac{1}{2}$	1	1	1	$\frac{1}{2}$	0	0	0
37.	$\frac{1}{2}$	$\frac{1}{2}$	$\frac{1}{2}$	1	1	1	1	1	1	0	0	$\frac{1}{2}$	0
38.	1	1	1	1	1	1	1	1	1	0	0	0	0

Table 5.2. **L**-concepts of the context of Table 5.1 (Łukasiewicz structure).

no.	extent									intent			
	Me	V	E	Ma	J	S	U	N	P	ss	sl	df	dn
1.	0	0	0	0	0	0	0	0	0	1	1	1	1
2.	0	0	0	$\frac{1}{2}$	0	0	0	0	0	1	0	1	1
3.	0	0	0	0	0	0	$\frac{1}{2}$	$\frac{1}{2}$	0	1	1	1	0
4.	0	0	0	0	$\frac{1}{2}$	$\frac{1}{2}$	0	0	0	0	1	1	1
5.	0	0	0	1	0	0	0	0	0	1	0	$\frac{1}{2}$	1
6.	0	0	0	$\frac{1}{2}$	0	0	$\frac{1}{2}$	$\frac{1}{2}$	1	1	0	1	0
7.	0	0	0	$\frac{1}{2}$	$\frac{1}{2}$	$\frac{1}{2}$	0	0	0	0	0	1	1
8.	0	0	0	0	0	0	1	1	0	$\frac{1}{2}$	$\frac{1}{2}$	1	0
9.	0	0	0	0	1	1	0	0	0	0	1	1	$\frac{1}{2}$
10.	0	0	0	0	1	1	$\frac{1}{2}$	$\frac{1}{2}$	0	0	1	1	0
11.	1	1	1	1	0	0	0	0	0	1	0	0	1
12.	0	0	0	1	0	0	$\frac{1}{2}$	$\frac{1}{2}$	1	1	0	$\frac{1}{2}$	0
13.	0	0	0	1	$\frac{1}{2}$	$\frac{1}{2}$	0	0	0	0	0	$\frac{1}{2}$	1
no.	extent									intent			
	Me	V	E	Ma	J	S	U	N	P	ss	sl	df	dn
14.	0	0	0	$\frac{1}{2}$	0	0	1	1	1	$\frac{1}{2}$	0	1	$\frac{1}{2}$
15.	0	0	0	$\frac{1}{2}$	1	1	0	0	0	0	0	1	$\frac{1}{2}$
16.	0	0	0	0	1	1	1	1	0	0	$\frac{1}{2}$	1	0
17.	1	1	1	1	0	0	$\frac{1}{2}$	$\frac{1}{2}$	1	1	0	0	0
18.	1	1	1	1	$\frac{1}{2}$	$\frac{1}{2}$	0	0	0	0	0	0	1
19.	0	0	0	1	0	0	1	1	1	$\frac{1}{2}$	0	$\frac{1}{2}$	0
20.	0	0	0	1	1	1	0	0	0	0	0	$\frac{1}{2}$	$\frac{1}{2}$
21.	0	0	0	$\frac{1}{2}$	1	1	1	1	1	0	0	1	0
22.	1	1	1	1	0	0	1	1	1	$\frac{1}{2}$	0	0	0
23.	1	1	1	1	1	1	0	0	0	0	0	0	$\frac{1}{2}$
24.	0	0	0	1	1	1	1	1	1	0	0	$\frac{1}{2}$	0
25.	1	1	1	1	1	1	1	1	1	0	0	0	0

Table 5.3. **L**-concepts of the context of Table 5.1 (Gödel structure).

each block is a maximal subset of **L**-concepts which are similar in the degree at least $\frac{1}{2}$. The corresponding factor lattice $\mathcal{B}(X,Y,I)/^{\frac{1}{2}}E$ is depicted in Fig. 5.5.

A few remarks are necessary on the examples. First, note that the concepts which were found depend on the fuzzy context. The fuzzy context is given by a subjective judgment (e.g. to what degree we consider "Mars is far from sun" to be true). The subjective judgment influences therefore the resulting structure of concepts.

Second, there are apparently natural concepts (consider, e.g., the Łukasiewicz structure) in the concept lattice (e.g. 14 ("small planet near to sun")), as well as concepts which "were found" (e.g. 26 ("a planet far from sun which is at least partially large")). Concept no. 1

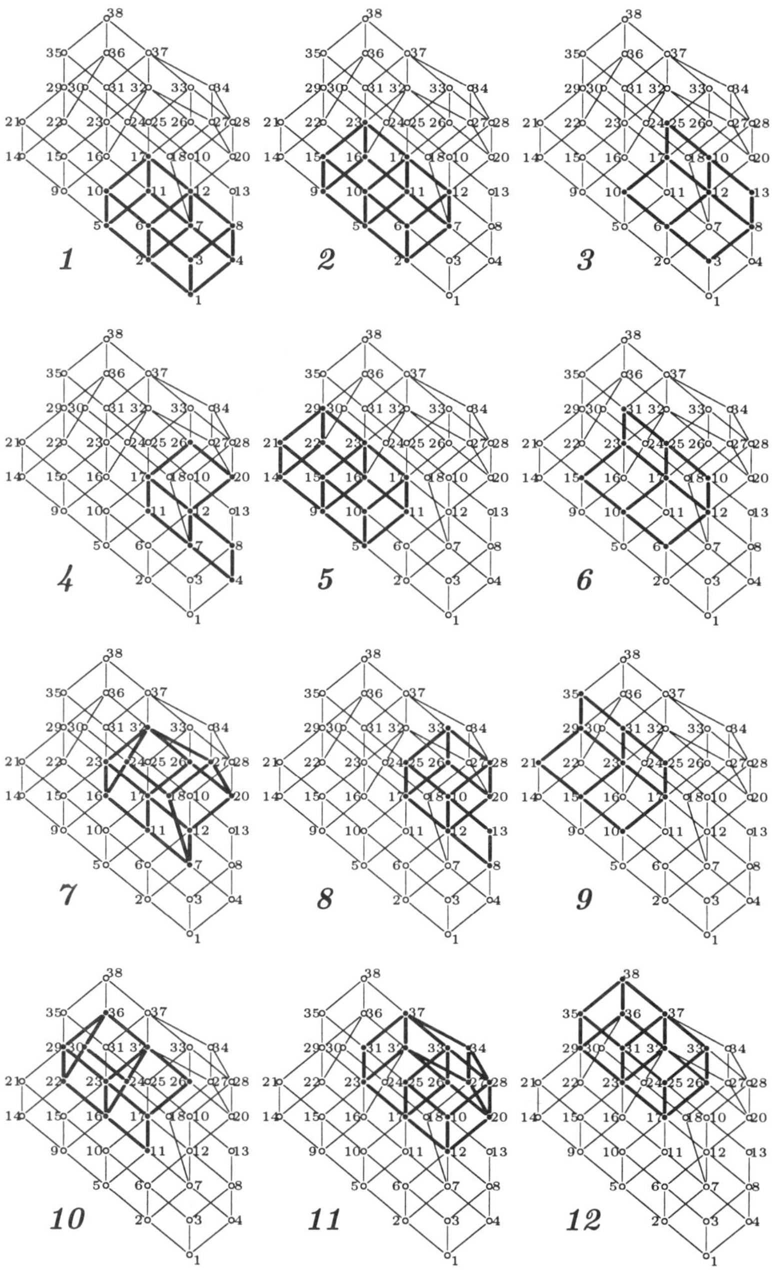

Figure 5.4. Blocks of the tolerance relation $\frac{1}{2}E$ on the concept lattice of Fig. 5.1.

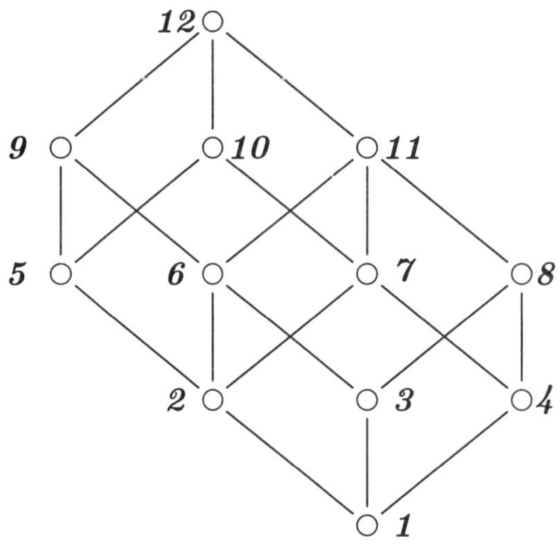

Figure 5.5. Lattice $\mathcal{B}(X, Y, h(I))$. Factor lattice $\mathcal{B}(X, Y, I)/\frac{1}{2}E$.

is an example of an empirically empty concept. Concepts which do not contain any element in the degree 1 in their extents (e.g. 1, 2, 3) could be called partially (empirically) empty. There is nothing wrong about these concepts just as there is nothing wrong about empirically empty concepts.

Note that both $\mathcal{B}(X, Y, h(I))$ obtained from $\mathcal{B}(X, Y, I)$ (Gödel structure) and $\mathcal{B}(X, Y, I)/\frac{1}{2}E$ obtained from $\mathcal{B}(X, Y, I)$ (Łukasiewicz structure) are isomorphic. Note also that taking the Gödel structure, $\frac{1}{2}E$ is just the partition induced by the homomorphism h^* induced above, i.e. $\mathcal{B}(X, Y, I)/\frac{1}{2}E$ again is isomorphic to the lattice in Fig. 5.5.

5.3 FUZZY GALOIS CONNECTIONS

In this section we investigate fuzzy Galois connections—the basic structures related to fuzzy concept lattices (other related structures, fuzzy closure operators, will be investigated in Section 7.1).

Unless otherwise stated, we adopt the following conventions: We denote by K (possibly with indices) a $\leq$-filter in a complete residuated lattice $\mathbf{L}$, i.e. a subset of L such that $a \in K$ and $a \leq b$ implies $b \in K$. A set $\langle X, \approx \rangle$ with an $\mathbf{L}$-equivalence will mostly be denoted just by X. Therefore, the set of all $\mathbf{L}$-sets in X compatible with $\approx$ will be denoted by L^X instead of $L^{\langle X, \approx \rangle}$ etc.

Definition 5.35 An $\mathbf{L}_K$-*Galois connection* (*fuzzy Galois connection*) between the sets $\langle X, \approx\rangle$ and $\langle Y, \approx\rangle$ is a pair $\langle {}^\uparrow, {}^\downarrow\rangle$ of mappings ${}^\uparrow : L^{\langle X,\approx\rangle} \to L^{\langle Y,\approx\rangle}$, ${}^\downarrow : L^{\langle Y,\approx\rangle} \to L^{\langle X,\approx\rangle}$, satisfying

$$\begin{aligned} S(A_1, A_2) &\leq S(A_2^\uparrow, A_1^\uparrow) \quad \text{whenever } S(A_1, A_2) \in K, && (5.19)\\ S(B_1, B_2) &\leq S(B_2^\downarrow, B_1^\downarrow) \quad \text{whenever } S(B_1, B_2) \in K, && (5.20)\\ A &\subseteq (A^\uparrow)^\downarrow, && (5.21)\\ B &\subseteq (B^\downarrow)^\uparrow, && (5.22) \end{aligned}$$

for every $A, A_1, A_2 \in L^{\langle X,\approx\rangle}$, $B, B_1, B_2 \in L^{\langle Y,\approx\rangle}$.

Remark 5.36 (1) It is easy to see that an $\mathbf{L}_K$-Galois connection between X and Y forms a Galois connection between the complete lattices $\langle L^X, \subseteq\rangle$ and $\langle L^Y, \subseteq\rangle$ (see Chapter 1). Also, Galois connections between sets (see Chapter 1) are (modulo obvious correspondence) just $\mathbf{L}$-Galois connections between sets with $\mathbf{L}$-equalities for $\mathbf{L} = \mathbf{2}$.

(2) There is a natural interpretation of (5.19)–(5.22): Think of X and Y as the set of objects and attributes, respectively. Consider the mappings ${}^\uparrow$ and ${}^\downarrow$ which assign to an $\mathbf{L}$-set of objects the corresponding $\mathbf{L}$-set of common attributes, and to an $\mathbf{L}$-set of attributes the corresponding $\mathbf{L}$-set of objects sharing all the attributes. Intuitively, ${}^\uparrow$ and ${}^\downarrow$ should satisfy (5.19)–(5.22).

If $K = L$, we omit the subscript K. The role of K is to control the sensitivity of the conditions "the more objects, the less common attributes" and "the more attributes, the less objects sharing them". For $K = \emptyset$ we have only the conditions (5.21) and (5.22), which say that the composed mappings ${}^{\uparrow\downarrow} : L^X \to L^X$, ${}^{\downarrow\uparrow} : L^Y \to L^Y$ are extensional.

Remark 5.37 Clearly, for $K_1 \subset K_2$, each $\mathbf{L}_{K_2}$-Galois connection is also an $\mathbf{L}_{K_1}$-Galois connection. We will be mostly concerned with $\mathbf{L}_K$-Galois connections where $K = L$ or $K = \{1\}$. We call $\mathbf{L}_L$-Galois connections just $\mathbf{L}$-*Galois connections*. The following example shows that there are $\mathbf{L}_{\{1\}}$-Galois connections which are not $\mathbf{L}$-Galois connections.

Consider $\mathbf{L}$ with $L = \{0, \frac{1}{2}, 1\}$ and the Łukasiewicz structure on L. Let further $X = \{x\}$, $Y = \{y\}$. Define the mappings ${}^\uparrow$ and ${}^\downarrow$ by $\left\{{}^0\!/x\right\}^\uparrow = \left\{{}^1\!/y\right\}$, $\left\{{}^{\frac{1}{2}}\!/x\right\}^\uparrow = \left\{{}^0\!/y\right\}$, $\left\{{}^1\!/x\right\}^\uparrow = \left\{{}^0\!/y\right\}$, $\left\{{}^0\!/y\right\}^\downarrow = \left\{{}^1\!/x\right\}$, $\left\{{}^{\frac{1}{2}}\!/y\right\}^\downarrow = \left\{{}^0\!/x\right\}$, $\left\{{}^1\!/y\right\}^\downarrow = \left\{{}^0\!/x\right\}$. One may easily verify that $\langle {}^\uparrow, {}^\downarrow\rangle$ is an $\mathbf{L}_{\{1\}}$-Galois connection which is not an $\mathbf{L}_L$-Galois connection for $S\left(\left\{{}^{\frac{1}{2}}\!/x\right\}, \left\{{}^0\!/x\right\}\right) = \frac{1}{2} \not\leq 0 = S\left(\left\{{}^1\!/y\right\}, \left\{{}^0\!/y\right\}\right)$.

The next theorem provides us with a simple characterization of $\mathbf{L}_K$-Galois connections.

Theorem 5.38 *A pair $\langle {}^\uparrow, {}^\downarrow \rangle$ forms an $\mathbf{L}_K$-Galois connection between X and Y iff $S\left(A, B^\downarrow\right) \in K$ or $S\left(B, A^\uparrow\right) \in K$ implies*

$$S\left(A, B^\downarrow\right) = S\left(B, A^\uparrow\right) \tag{5.23}$$

for all $A \in L^X$, $B \in L^Y$.

Proof. Let $\langle {}^\uparrow, {}^\downarrow \rangle$ be an $\mathbf{L}_K$-Galois connection. Suppose $S\left(A, B^\downarrow\right) \in K$. From $B \subseteq B^{\downarrow\uparrow}$ we get $S\left(B^{\downarrow\uparrow}, A^\uparrow\right) \le S\left(B, A^\uparrow\right)$, hence by (5.19)

$$S\left(A, B^\downarrow\right) \le S\left(B^{\downarrow\uparrow}, A^\uparrow\right) \le S\left(B, A^\uparrow\right).$$

From the property of K we get $S\left(B, A^\uparrow\right) \in K$ and repeating the arguments yields $S\left(B, A^\uparrow\right) \le S\left(A, B^\downarrow\right)$. If $S\left(B, A^\uparrow\right) \in K$, the argumentation is analogous.

Conversely, from $S\left(A^\uparrow, A^\uparrow\right) = 1$ we get $S\left(A, A^{\uparrow\downarrow}\right) = 1$, i.e. $A \subseteq A^{\uparrow\downarrow}$ proving (5.21). (5.22) may be proved symmetrically. Let $S(A_1, A_2) \in K$. From $A_2 \subseteq A_2^{\uparrow\downarrow}$ it follows by (5.23) that

$$S(A_1, A_2) \le S\left(A_1, A_2^{\uparrow\downarrow}\right) = S\left(A_2^\uparrow, A_1^\uparrow\right)$$

proving (5.19). The proof of (5.20) is symmetric. □

L-Galois connections vs. binary L-relations. We now turn our attention to **L**-Galois connections. It will turn out that **L**-Galois connections are precisely the mappings generated by an **L**-relation I by (5.1) and (5.2), i.e. the mappings that are at the core of the theory of fuzzy concept lattices. Note that $A^\uparrow$ (and $B^\downarrow$) were naturally interpreted as the fuzzy set of all attributes common to all objects from A (and the fuzzy set of all objects sharing all attributes from B). We are going to elaborate the details.

We will use some results of Section 7.1: $\langle L^{\langle X,\approx\rangle}, \subseteq \rangle$ is a complete sublattice of $\langle L^X, \subseteq \rangle$ (Corollary 7.65); recall from Section 3.1 that $(\mathrm{C}_\approx(A))(x') = \bigvee_{x\in X} A(x) \otimes (x \approx x')$ is the least **L**-set compatible with $\approx$ and containing A (this is proved in Lemma 7.61); $S(A_1, A_2) \le S(\mathrm{C}_\approx(A_1), \mathrm{C}_\approx(A_2))$ (Lemma 7.57).

Lemma 5.39 *For each* **L**-*relation* I *between* $\langle X, \approx_X \rangle$ *and* $\langle Y, \approx_Y \rangle$, *(5.1) and (5.2) define an* **L**-*Galois connection between* $\langle X, \approx_X \rangle$ *and* $\langle Y, \approx_Y \rangle$.

Proof. We prove (5.19) and (5.21); (5.20) and (5.22) may be proved analogously. Prove (5.19): We first show

$$S(A_1, A_2) \otimes (A_2(x) \to I(x,y)) \leq (A_1(x) \to I(x,y)) \tag{5.24}$$

for any $y \in Y$. We have $S(A_1, A_2) \leq A_1(x) \to A_2(x)$ for any $x \in X$ and thus

$$S(A_1, A_2) \otimes A_1(x) \leq A_1(x) \otimes (A_1(x) \to A_2(x)) \leq A_2(x) \ . \tag{5.25}$$

Together with

$$A_2(x) \leq (A_2(x) \to I(x,y)) \to I(x,y),$$

we therefore have

$$S(A_1, A_2) \otimes A_1(x) \leq (A_2(x) \to I(x,y)) \to I(x,y)$$

which is equivalent (applying the adjointness property twice) to (5.24). Since $a_i \leq b_i$ ($i \in I$) implies $\bigwedge_{i \in I} a_i \leq \bigwedge_{i \in I} b_i$, we conclude

$$\begin{aligned} & S(A_1, A_2) \otimes \bigwedge_{x \in X} A_2(x) \to I(x,y) \leq \\ & \leq \bigwedge_{x \in X} (S(A_1, A_2) \otimes (A_2(x) \to I(x,y))) \leq \bigwedge_{x \in X} A_1(x) \to I(x,y) \ . \end{aligned}$$

Applying the adjointness property we obtain

$$\begin{aligned} & S(A_1, A_2) \leq \\ & \leq \bigwedge_{x \in X} (A_2(x) \to I(x,y)) \to \bigwedge_{x \in X} (A_1(x) \to I(x,y)) = \\ & = A_2^{\uparrow}(y) \to A_1^{\uparrow}(y) \ . \end{aligned}$$

Since y is arbitrary,

$$S(A_1, A_2) \leq \bigwedge_{y \in Y} A_2^{\uparrow}(y) \to A_1^{\uparrow}(y) = S\left(A_2^{\uparrow}, A_1^{\uparrow}\right)$$

proving (5.19). Prove (5.21): Let $A \in L^X$, $x \in X$. We first prove that

$$A(x) \leq \left(\bigwedge_{x' \in X} (A(x') \to I(x', y)) \right) \to I(x,y) \tag{5.26}$$

for each $y \in Y$. By the adjointness property, (5.26) holds true iff

$$A(x) \otimes (\bigwedge_{x' \in X} (A(x') \to I(x', y))) \leq I(x, y)$$

which is due to commutativity of $\otimes$ and the adjointness property equivalent to

$$(\bigwedge_{x' \in X} (A(x') \to I(x', y))) \leq A(x) \to I(x, y)$$

which is evident, verifying (5.26). From (5.26) it follows

$$A(x) \leq \bigwedge_{y \in Y} (\bigwedge_{x' \in X} (A(x') \to I(x', y))) \to I(x, y) = A^{\uparrow\downarrow}(x) ,$$

i.e., $A \subseteq A^{\uparrow\downarrow}$ proving (5.21).

We now have to prove that if $A \in L^{\langle X, \approx_X \rangle}$, $B \in L^{\langle Y, \approx_Y \rangle}$, then also $A^{\uparrow_I} \in L^{\langle Y, \approx_Y \rangle}$ and $B^{\downarrow_I} \in L^{\langle X, \approx_X \rangle}$. Due to the symmetry of both of the cases we show only the first one. Let then $A \in L^{\langle X, \approx_X \rangle}$. We have to show that for each $y, y' \in Y$,

$$A^{\uparrow_I}(y) \otimes (y \approx_Y y') \leq A^{\uparrow_I}(y')$$

which is true iff for each $y, y' \in Y$,

$$(y \approx_Y y') \leq A^{\uparrow_I}(y) \leftrightarrow A^{\uparrow_I}(y').$$

Consider the language $\mathcal{J}_{\mathrm{Con}}$, the **L**-structure **M** for $\mathcal{J}_{\mathrm{Con}}$ that corresponds to $\langle X, Y, I \rangle$, and valuations v_1, v_2 such that $v_1(\mathtt{y}) = y$ and $v_2(\mathtt{y}) = y'$. Since $|\varphi_\uparrow|_{\mathtt{y}} = 1$, many-sorted version of Lemma 3.83 yields

$$(v_1(\mathtt{y}) \approx_{\mathtt{y}}^{\mathbf{M}} v_2(\mathtt{y})) \leq \|\varphi_\uparrow\|_{\mathbf{M}, v_1} \leftrightarrow \|\varphi_\uparrow\|_{\mathbf{M}, v_2}.$$

Recalling $\|\varphi_\uparrow\|_{\mathbf{M}, v} = A^\uparrow(v(\mathtt{y}))$, this inequality means

$$(y \approx_Y y') \leq A^\uparrow(y) \leftrightarrow A^\uparrow(y')$$

verifying the claim. □

Remark 5.40 As can be seen from the proof of Lemma 5.39, $A^{\uparrow_I} \in L^{\langle Y, \approx_Y \rangle}$ whenever $A \in L^X$ (the fact $A \in L^{\langle X, \approx_X \rangle}$ has not been used). Similarly for $B^{\downarrow_I}$.

We are going to prove that each **L**-Galois connection is in the above sense induced by some **L**-relation. For this purpose we will make use of the following lemma.

Lemma 5.41 *Let $\langle {}^{\uparrow}, {}^{\downarrow} \rangle$ be an* **L**-*Galois connection between $\langle X, \approx_X \rangle$ and $\langle Y, \approx_Y \rangle$. Then*

$$a \to \mathrm{C}(\left\{ {}^{b}/x \right\})^{\uparrow}(y) = b \to \mathrm{C}(\{ {}^{a}/y \})^{\downarrow}(x), \tag{5.27}$$

$$\mathrm{C}(\left\{ {}^{a \otimes b}/x \right\})^{\uparrow}(y) = a \to \mathrm{C}(\left\{ {}^{b}/y \right\})^{\downarrow}(x), \tag{5.28}$$

$$\mathrm{C}(\left\{ {}^{a \otimes b}/y \right\})^{\downarrow}(x) = a \to \mathrm{C}(\left\{ {}^{b}/x \right\})^{\uparrow}(y),$$

$$a \to \mathrm{C}(\left\{ {}^{b}/x \right\})^{\uparrow}(y) = b \to \mathrm{C}(\{ {}^{a}/x \})^{\uparrow}(y), \tag{5.29}$$

$$a \to \mathrm{C}(\left\{ {}^{b}/y \right\})^{\downarrow}(x) = b \to \mathrm{C}(\{ {}^{a}/y \})^{\downarrow}(x),$$

$$\mathrm{C}(\{ {}^{a}/x \})^{\uparrow}(y) = \mathrm{C}(\{ {}^{a}/y \})^{\downarrow}(x) \tag{5.30}$$

for all $a, b \in L$, $x \in X$, $y \in Y$.

Proof. We first show

Claim. $a \to \mathrm{C}(\left\{ {}^{b}/y \right\})^{\downarrow}(x) \leq b \to \mathrm{C}(\{ {}^{a}/x \})^{\uparrow}(y)$ *and* $a \to \mathrm{C}(\left\{ {}^{b}/x \right\})^{\uparrow}(y) \leq b \to \mathrm{C}(\{ {}^{a}/y \})^{\downarrow}(x)$.

Proof of Claim. By (5.19) and Lemma 5.39 we have

$$\begin{aligned} & S\left(\mathrm{C}(\left\{ {}^{\mathrm{C}(\{ {}^{b}/y \})^{\downarrow}(x)}/x \right\})^{\uparrow}, \mathrm{C}(\{ {}^{a}/x \})^{\uparrow}\right) \geq \\ \geq\ & S\left(\mathrm{C}(\{ {}^{a}/x \}), \mathrm{C}(\left\{ {}^{\mathrm{C}(\{ {}^{b}/y \})^{\downarrow}(x)}/x \right\})\right) \geq \\ \geq\ & S\left(\{ {}^{a}/x \}, \left\{ {}^{\mathrm{C}(\{ {}^{b}/y \})^{\downarrow}(x)}/x \right\}\right) = a \to \mathrm{C}(\left\{ {}^{b}/y \right\})^{\downarrow}(x) \end{aligned} \tag{5.31}$$

On the other hand,

$$\begin{aligned} & S\left(\mathrm{C}(\left\{ {}^{\mathrm{C}(\{ {}^{b}/y \})^{\downarrow}(x)}/x \right\})^{\uparrow}, \mathrm{C}(\{ {}^{a}/x \})^{\uparrow}\right) = \\ =\ & \bigwedge_{y' \in Y} \mathrm{C}(\left\{ {}^{\mathrm{C}(\{ {}^{b}/y \})^{\downarrow}(x)}/x \right\})^{\uparrow}(y') \to \mathrm{C}(\{ {}^{a}/x \})^{\uparrow}(y') \leq \\ \leq\ & \mathrm{C}(\left\{ {}^{\mathrm{C}(\{ {}^{b}/y \})^{\downarrow}(x)}/x \right\})^{\uparrow}(y) \to \mathrm{C}(\{ {}^{a}/x \})^{\uparrow}(y) \leq \\ \leq\ & b \to \mathrm{C}(\{ {}^{a}/x \})^{\uparrow}(y) \end{aligned} \tag{5.32}$$

provided the last inequality holds true. To see this it is enough to show that $b \leq \mathrm{C}(\left\{ {}^{\mathrm{C}(\{ {}^{b}/y \})^{\downarrow}(x)}/x \right\})^{\uparrow}(y)$. This is indeed true, we reason as follows. Obviously, $b \leq \left\{ {}^{b}/y \right\}^{\downarrow\uparrow}(y) \leq \mathrm{C}(\left\{ {}^{b}/y \right\})^{\downarrow\uparrow}(y)$.

Furthermore, $\mathrm{C}(\left\{{}^{b}/{y}\right\})^{\downarrow\uparrow}(y) \le \mathrm{C}(\left\{\mathrm{C}(\left\{{}^{b}/{y}\right\})^{\downarrow}(x)/x\right\})^{\uparrow}(y)$, since $\mathrm{C}(\left\{{}^{b}/{y}\right\})^{\downarrow\uparrow} \subseteq \mathrm{C}(\left\{\mathrm{C}(\left\{{}^{b}/{y}\right\})^{\downarrow}(x)/x\right\})^{\uparrow}$ which follows from $\mathrm{C}(\left\{\mathrm{C}(\left\{{}^{b}/{y}\right\})^{\downarrow}(x)/x\right\}) \subseteq \mathrm{C}(\left\{{}^{b}/{y}\right\})^{\downarrow\uparrow}$. The last relationship holds because of $\left\{\mathrm{C}(\left\{{}^{b}/{y}\right\})^{\downarrow}(x)/x\right\} \subseteq \mathrm{C}(\left\{{}^{b}/{y}\right\})^{\downarrow\uparrow}$ by Lemma 5.39 and the fact that $\mathrm{C}(\left\{{}^{b}/{y}\right\})^{\downarrow\uparrow}$ is closed w.r.t. C. To conclude, (5.31) and (5.32) give the first part of Claim. The second part is symmetric. *Q.E.D.*

By Claim we immediately get

$$a \to \mathrm{C}(\left\{{}^{b}/{x}\right\})^{\uparrow}(y) \le b \to \mathrm{C}(\{{}^{a}/{y}\})^{\downarrow}(x) \le a \to (\left\{{}^{b}/{x}\right\})^{\uparrow}(y),$$

verifying (5.27). Using (5.27), (5.28) may be obtained as follows:

$$\begin{aligned}
&\mathrm{C}(\left\{{}^{a\otimes b}/{x}\right\})^{\uparrow}(y) = 1 \to \mathrm{C}(\left\{{}^{a\otimes b}/{x}\right\})^{\uparrow}(y) = \\
&= \ b\otimes a \to \mathrm{C}(\left\{{}^{1}/{y}\right\})^{\downarrow}(x) = b \to \left(a \to \mathrm{C}(\left\{{}^{1}/{y}\right\})^{\downarrow}(x)\right) = \\
&= \ b \to \left(1 \to \mathrm{C}(\{{}^{a}/{x}\})^{\uparrow}(y)\right) = b \to \mathrm{C}(\{{}^{a}/{x}\})^{\uparrow}(y) = \\
&= \ a \to \mathrm{C}(\left\{{}^{b}/{y}\right\})^{\downarrow}(x);
\end{aligned}$$

the second part of (5.28) may be obtained symmetrically. By (5.28) we get

$$a \to \mathrm{C}(\left\{{}^{b}/{x}\right\})^{\uparrow}(y) = \mathrm{C}(\left\{{}^{a\otimes b}/{y}\right\})^{\downarrow}(x) = b \to \mathrm{C}(\{{}^{a}/{x}\})^{\uparrow}(y),$$

the first part of (5.29); the second part is symmetric. Putting $a = 1$ and $b = a$ in (5.28) we immediately get (5.30). □

Lemma 5.42 *Let $\langle {}^{\uparrow}, {}^{\downarrow}\rangle$ be an $\mathbf{L}$-Galois connection between $\langle X, \approx_X\rangle$ and $\langle Y, \approx_Y\rangle$, and $A_i \in L^{\langle X, \approx_X\rangle}$, $B_i \in L^{\langle Y, \approx_Y\rangle}$ $(i \in I)$. Then*

$$\left(\bigcup_{i\in I} A_i\right)^{\uparrow} = \bigcap_{i\in I} A_i^{\uparrow}, \qquad \left(\bigcup_{i\in I} B_i\right)^{\downarrow} = \bigcap_{i\in I} B_i^{\downarrow}.$$

Proof. We prove only the first part. By (5.23), we have for any $C \in L^Y$

$$S\left(C, (\bigcup_{i\in I} A_i)^{\uparrow}\right) = S\left(\left(\bigcup_{i\in I} A_i\right), C^{\downarrow}\right) =$$

$$= \bigwedge_{x\in X}\left(\bigvee_{i\in I} A_i(x) \to C^{\downarrow}(x)\right) = \bigwedge_{x\in X}\bigwedge_{i\in I}\left(A_i(x) \to C^{\downarrow}(x)\right) =$$
$$= \bigwedge_{i\in I}\bigwedge_{x\in X}\left(A_i(x) \to C^{\downarrow}(x)\right) = \bigwedge_{i\in I} S\left(A_i, C^{\downarrow}\right) =$$
$$= \bigwedge_{i\in I} S\left(C, A_i^{\uparrow}\right) = \bigwedge_{i\in I}\bigwedge_{x\in X} (C(x) \to A_i^{\uparrow}(x)) =$$
$$= \bigwedge_{x\in X} (C(x) \to \bigwedge_{i\in I} A_i^{\uparrow}(x)) = S\left(C, \bigcap_{i\in I} A_i^{\uparrow}\right).$$

The rest follows from Corollary 7.65 and Lemma 5.39. □

Lemma 5.43 *Let $\langle {}^{\uparrow}, {}^{\downarrow}\rangle$ be an $\mathbf{L}$-Galois connection between $\langle X, \approx_X\rangle$ and $\langle Y, \approx_Y\rangle$. Then there is an $\mathbf{L}$-relation I between $\langle X, \approx_X\rangle$ and $\langle Y, \approx_Y\rangle$ such that for the induced mappings ${}^{\uparrow_I}$ and ${}^{\downarrow_I}$ we have $\langle {}^{\uparrow_I}, {}^{\downarrow_I}\rangle = \langle {}^{\uparrow}, {}^{\downarrow}\rangle$.*

Proof. Introduce I by

$$I(x,y) = \mathrm{C}(\left\{ {}^{1}/x \right\})^{\uparrow}(y) = \mathrm{C}(\left\{ {}^{1}/y \right\})^{\downarrow}(x)$$

which is correct due to Lemma 5.41. First, we check that $I \in L^{\langle X\times Y, \approx_X \otimes \approx_Y\rangle}$ as follows.

$$\begin{aligned}
&I(x,y) \otimes (x \approx x') \otimes (y \approx y') = \\
&= \mathrm{C}(\left\{ {}^{1}/x \right\})^{\uparrow}(y) \otimes (x \approx x') \otimes (y \approx y') \leq \mathrm{C}(\left\{ {}^{1}/x \right\})^{\uparrow}(y') \otimes (x \approx x') = \\
&= \mathrm{C}(\left\{ {}^{1}/y' \right\})^{\downarrow}(x) \otimes (x \approx x') \leq \mathrm{C}(\left\{ {}^{1}/y' \right\})^{\downarrow}(x') = I(x', y').
\end{aligned}$$

Second, take any $A \in L^{\langle X, \approx_X\rangle}$. Since $\left\{ {}^{A(x)}/x \right\} \subseteq A$, we have $\mathrm{C}(\left\{ {}^{A(x)}/x \right\}) \subseteq \mathrm{C}(A) = A$ and thus $\bigcup_{x\in X} \mathrm{C}(\left\{ {}^{A(x)}/x \right\}) \subseteq A$. On the other hand, $A = \bigcup_{x\in X}\left\{ {}^{A(x)}/x \right\} \subseteq \bigcup_{x\in X} \mathrm{C}(\left\{ {}^{A(x)}/x \right\})$. To sum up, $A = \bigcup_{x\in X} \mathrm{C}(\left\{ {}^{A(x)}/x \right\})$. Using this fact, Lemma 5.41, and Lemma 5.42 we have

$$\begin{aligned}
A^{\uparrow}(y) &= (\bigcup_{x\in X} \mathrm{C}(\left\{ {}^{A(x)}/x \right\}))^{\uparrow}(y) = (\bigcap_{x\in X} \mathrm{C}(\left\{ {}^{A(x)}/x \right\})^{\uparrow})(y) = \\
&= \bigwedge_{x\in X} \mathrm{C}(\left\{ {}^{A(x)}/x \right\})^{\uparrow}(y) = \bigwedge_{x\in X} A(x) \to \mathrm{C}(\left\{ {}^{1}/x \right\})^{\uparrow}(y) = \\
&= \bigwedge_{x\in X} A(x) \to I(x,y) = A^{\uparrow_I}(y),
\end{aligned}$$

i.e. ${}^{\uparrow} = {}^{\uparrow_I}$. Using $\left\{ {}^1/x \right\}^{\uparrow}(y) = \left\{ {}^1/y \right\}^{\downarrow}(x)$, ${}^{\downarrow} = {}^{\downarrow_I}$ may be obtained symmetrically. □

Remark 5.44 Note also that if $\langle {}^{\uparrow}, {}^{\downarrow} \rangle$ is an **L**-Galois connection between $\langle X, \approx_X \rangle$ and $\langle Y, \approx_Y \rangle$, I is the induced relation from Lemma 5.43, and $\langle {}^{\uparrow_I}, {}^{\downarrow_I} \rangle$ is the pair defined by (5.1) and (5.2) considered as mappings taking arguments from L^X and L^Y, respectively, then for every $A \in L^X$, $B \in L^Y$ we have $A^{\uparrow_I} = \mathrm{C}(A)^{\uparrow_I}$ and $B^{\downarrow_I} = \mathrm{C}(B)^{\downarrow_I}$. We check only $A^{\uparrow_I} = \mathrm{C}(A)^{\uparrow_I}$. Since $A \subseteq \mathrm{C}(A)$ we have $\mathrm{C}(A)^{\uparrow_I} \subseteq A^{\uparrow_I}$. Conversely, we have to prove $A^{\uparrow_I} \subseteq \mathrm{C}(A)^{\uparrow_I}$, i.e. $A^{\uparrow_I}(y) \leq \mathrm{C}(A)(x) \rightarrow I(x, y)$ (for each $x \in X$), i.e. $A^{\uparrow_I}(y) \otimes \mathrm{C}(A)(x) \leq I(x, y)$. We have

$$
\begin{aligned}
A^{\uparrow_I}(y) \otimes \mathrm{C}(A)(x) &= A^{\uparrow_I}(y) \otimes \bigvee_{x' \in X} (A(x') \otimes (x \approx x')) = \\
&= \bigvee_{x' \in X} ((x \approx x') \otimes A(x') \otimes \bigwedge_{x'' \in X} (A(x'') \rightarrow I(x'', y))) \leq \\
&\leq \bigvee_{x' \in X} ((x \approx x') \otimes A(x') \otimes (A(x') \rightarrow I(x', y))) \leq \\
&\leq \bigvee_{x' \in X} ((x \approx x') \otimes I(x', y)) \leq \\
&\leq \bigvee_{x' \in X, y' \in Y} ((x \approx x') \otimes (y \approx y') \otimes I(x', y')) = \mathrm{C}(I)(x, y) = I(x, y)
\end{aligned}
$$

since I is a relation between $\langle X, \approx_X \rangle$and $\langle Y, \approx_Y \rangle$. We have proved the required inequality.

As a direct consequence we get the following theorem which shows a bijective correspondence between **L**-Galois connections and **L**-relations.

Theorem 5.45 *For a binary* **L**-*relation I between $\langle X, \approx_X \rangle$ and $\langle Y, \approx_Y \rangle$ denote $\langle {}^{\uparrow_I}, {}^{\downarrow_I} \rangle$ the mappings defined by (5.1) and (5.2). For an* **L**-*Galois connection $\langle {}^{\uparrow}, {}^{\downarrow} \rangle$ between $\langle X, \approx_X \rangle$ and $\langle Y, \approx_Y \rangle$ denote $I_{\langle \uparrow, \downarrow \rangle}$ the binary* **L**-*relation from Lemma 5.43. Then I is an* **L**-*relation between $\langle X, \approx_X \rangle$ and $\langle Y, \approx_Y \rangle$, and $\langle {}^{\uparrow_I}, {}^{\downarrow_I} \rangle$ is an* **L**-*Galois connection and we have*

$$
\langle {}^{\uparrow}, {}^{\downarrow} \rangle = \langle {}^{\uparrow_{I_{\langle \uparrow, \downarrow \rangle}}}, {}^{\downarrow_{I_{\langle \uparrow, \downarrow \rangle}}} \rangle \quad \text{and} \quad I = I_{\langle \uparrow_I, \downarrow_I \rangle}.
$$

Proof. By Lemmas 5.39 and 5.43 it suffices to prove $I = I_{\langle \uparrow_I, \downarrow_I \rangle}$. We have

$$
\begin{aligned}
I_{\langle \uparrow_I, \downarrow_I \rangle}(x, y) &= \mathrm{C}(\left\{ {}^1/x \right\})^{\uparrow}(y) = \bigwedge_{x' \in X} \mathrm{C}(\left\{ {}^1/x \right\})(x') \rightarrow I(x', y) \\
&= \bigwedge_{x' \in X} (x \approx x') \rightarrow I(x', y) = I(x, y).
\end{aligned}
$$

Indeed, the last equality holds true because, on the one hand, $\bigwedge_{x'\in X}(x \approx x') \to I(x',y) \leq (x \approx x) \to I(x,y) = I(x,y)$. On the other hand, $I(x,y) \leq \bigwedge_{x'\in X}(x \approx x') \to I(x',y)$ iff for each $x' \in X$, $I(x,y) \leq (x \approx x') \to I(x',y)$ iff $I(x,y) \otimes (x \approx x') \leq I(x',y)$ which is true by assumption ($I \in L^{\langle X\times Y, \approx_X \otimes \approx_Y\rangle}$). □

Representation of fuzzy Galois connections by crisp Galois connections. Our goal is to find a representation of fuzzy Galois connections by crisp Galois connections. We present two ways. The second one yields as its consequence a simple proof of Theorem 5.9 (the "crisp version" of the main theorem of **L**-concept lattices). In this section, we assume that the **L**-equivalence relations on X and Y are all classical identity relations (the more general case of **L**-equivalences would only yield complicated conditions not going into the point; moreover, the generalization for **L**-equivalences is straightforward).

First case: cutlike representation of L-Galois connections. Our aim now is to show that each **L**-Galois connection may be represented by a special system of **2**-Galois connections, in the cutlike manner. By Theorem 3.22, there is a one-to-one correspondence between the class of all **L**-sets in X and the class of all **L**-nested systems of subsets of X. The problem is to find an analogous representation theorem for **L**-Galois connections. The crucial observation is contained in the following lemma.

Lemma 5.46 *Let $I \in L^{X\times Y}$ be an **L**-relation, $\langle{}^\uparrow, {}^\downarrow\rangle$ be the **L**-Galois connection induced by I, and for $a \in L$ let $\langle{}^{\uparrow_a}, {}^{\downarrow_a}\rangle$ be the **2**-Galois connection induced by the **2**-relation ${}^aI$. Then for every **2**-set $A \in 2^X$, $B \in 2^Y$, $a \in L$, we have*

$$ {}^a(A^\uparrow) = A^{\uparrow_a}, \qquad {}^a(B^\downarrow) = B^{\downarrow_a}, \tag{5.33}$$

*and for every **L**-set $A \in L^X$, $B \in L^Y$, $a \in L$, we have*

$$ {}^a(A^\uparrow) = \bigcap_{b\in L} ({}^bA)^{\uparrow_{a\otimes b}}, \qquad {}^a(B^\downarrow) = \bigcap_{b\in L} ({}^bB)^{\downarrow_{a\otimes b}}. \tag{5.34}$$

Proof. Prove (5.33). Let $A \in 2^X$. By definition, $y \in {}^a(A^\uparrow)$ iff $\bigwedge_{x\in X}(A(x) \to I(x,y)) \geq a$. Since $A(x) \in \{0,1\}$ for each $x \in X$, this is equivalent to

$$\bigwedge_{x\in A}(A(x) \to I(x,y)) = \bigwedge_{x\in A}(1 \to I(x,y)) = \bigwedge_{x\in A} I(x,y) \geq a$$

which holds iff for each $x \in A$ it holds $\langle x, y\rangle \in {}^aI$. The last statement is equivalent to $y \in A^{\uparrow_a}$ proving the first part of (5.33). The second part can be proved analogously.

Prove (5.34). This is an easy consequence of Lemma 3.139: Recall language $\mathcal{J}_{\mathrm{Con}}$ from Remark 5.2 and the fact that for a formula $\varphi_\uparrow = (\forall \mathtt{x})(r_A(\mathtt{x}) \Rightarrow r_I(\mathtt{x}, \mathtt{y}))$, the **L**-structure **M** for $\mathcal{J}_{\mathrm{Con}}$ that corresponds to $\langle X, Y, I\rangle$ and A, and a valuation v such that $v(\mathtt{y}) = y$ we have

$$A^\uparrow(y) = \|\varphi_\uparrow\|_{\mathbf{M},v}.$$

Lemma 3.139 thus yields

$$\begin{aligned} ({}^aA^\uparrow)(y) &= {}^a\|\varphi_\uparrow\|_{\mathbf{M},v} = \\ &= \bigwedge^{\mathbf{2}}_{v'=_\mathtt{x}v}\bigwedge^{\mathbf{2}}_{b\in L} \|r_A\|^b_{\mathcal{C}_{\mathbf{M}},v'} \rightarrow^{\mathbf{2}} \|r_I\|^{a\otimes b}_{\mathcal{C}_{\mathbf{M}},v'} = \\ &= \bigwedge^{\mathbf{2}}_{b\in L}\bigwedge^{\mathbf{2}}_{x\in X} ({}^bA)(x) \rightarrow^{\mathbf{2}} ({}^{a\otimes b}I)(x,y) = \bigcap^{\mathbf{2}}_{b\in L} ({}^bA)^{\uparrow_{a\otimes b}} \end{aligned}$$

proving the claim. □

Definition 5.47 A system $\{\langle {}^{\uparrow_a}, {}^{\downarrow_a}\rangle \mid a \in L\}$ of **2**-Galois connections is called **L**-*nested* if (1) for each $a, b \in L$, $a \leq b$, $A \in 2^X$, $B \in 2^Y$, it holds $A^{\uparrow_a} \supseteq A^{\uparrow_b}$, $B^{\downarrow_a} \supseteq B^{\downarrow_b}$, (2) the set $\{a \in L \mid y \in \{x\}^{\uparrow_a}\}$ contains a greatest element.

Remark 5.48 Note that from (1) and (2) of Definition 5.47 it follows that $A^{\uparrow_0} = Y$ and $B^{\downarrow_0} = X$ for any $A \in 2^X$, $B \in 2^Y$. Indeed, by (2), $K = \{a \mid y \in \{x\}^{\uparrow_a}\}$ is non-empty and by (1), $0 \in K$ for any $x \in X$, $y \in Y$. Now, from $y \in \{x\}^{\uparrow_0}$ for any $x \in A$ we get $y \in A^{\uparrow_0}$ for any $y \in Y$, thus $A^{\uparrow_0} = Y$. For $B^{\downarrow_0} = X$ one can use the fact that $y \in \{x\}^{\uparrow_a}$ iff $x \in \{y\}^{\downarrow_a}$ and proceed in a similar way.

Lemma 5.49 *For $j = 1, 2$, let $\langle {}^{\uparrow_j}, {}^{\downarrow_j}\rangle$, and I_j be **L**-Galois connections and the corresponding **L**-relations between X and Y, i.e. $I_j = I_{\langle {}^{\uparrow_j}, {}^{\downarrow_j}\rangle}$ and $\langle {}^{\uparrow_j}, {}^{\downarrow_j}\rangle = \langle {}^{\uparrow_{I_j}}, {}^{\downarrow_{I_j}}\rangle$. Then we have*

$$I_1 \subseteq I_2 \quad \textit{iff} \quad \textit{for each } A \in L^X, B \in L^Y \textit{ it holds } A^{\uparrow_1} \subseteq A^{\uparrow_2},\ B^{\downarrow_1} \subseteq B^{\downarrow_2}.$$

Proof. The direct implication follows from (5.1) and (5.2). Conversely, from $A^{\uparrow_1} \subseteq A^{\uparrow_2}$ we conclude $I_1(x, y) = 1 \rightarrow I_1(x, y) = (\{{}^1/x\})^{\uparrow_1}(y) \leq (\{{}^1/x\})^{\uparrow_2}(y) = 1 \rightarrow I_2(x, y) = I_2(x, y)$ which means $I_1 \subseteq I_2$. □

Theorem 5.50 *For an* $\mathbf{L}$*-Galois connection* $\langle {}^{\uparrow}, {}^{\downarrow} \rangle$ *between* X *and* Y *denote* $\mathcal{C}_{\langle \uparrow,\downarrow \rangle} = \{\langle {}^{\uparrow_a}, {}^{\downarrow_a} \rangle \mid a \in L\}$ *where* ${}^{\uparrow_a} : 2^X \to 2^Y$ *and* ${}^{\downarrow_a} : 2^Y \to 2^X$ *are defined by* $A^{\uparrow_a} = {}^a(A^{\uparrow})$ *and* $B^{\downarrow_a} = {}^a(B^{\downarrow})$ *for* $A \in 2^X$, $B \in 2^Y$. *For an* $\mathbf{L}$*-nested system* $\mathcal{C} = \{\langle {}^{\uparrow_a}, {}^{\downarrow_a} \rangle \mid a \in L\}$ *of* $\mathbf{2}$*-Galois connections between* X *and* Y *denote* $\langle {}^{\uparrow_\mathcal{C}}, {}^{\downarrow_\mathcal{C}} \rangle$ *the pair of mappings* ${}^{\uparrow_\mathcal{C}} : L^X \to L^Y$ *and* ${}^{\downarrow_\mathcal{C}} : L^Y \to L^X$ *defined by*

$$A^{\uparrow_\mathcal{C}}(y) = \bigvee\{a \mid y \in \bigcap_{b \in L} ({}^bA)^{\uparrow_{a \otimes b}}\},$$

$$B^{\downarrow_\mathcal{C}}(x) = \bigvee\{a \mid x \in \bigcap_{b \in L} ({}^bB)^{\uparrow_{a \otimes b}}\}$$

for $A \in L^X$, $B \in L^Y$. *Then it holds*

(1) $\mathcal{C}_{\langle \uparrow,\downarrow \rangle}$ *is a nested system of* $\mathbf{L}$*-Galois connections between* X *and* Y,

(2) $\langle {}^{\uparrow_\mathcal{C}}, {}^{\downarrow_\mathcal{C}} \rangle$ *is an* $\mathbf{L}$*-Galois connection between* X *and* Y,

(3) $\mathcal{C} = \mathcal{C}_{\langle \uparrow_\mathcal{C}, \downarrow_\mathcal{C} \rangle}$ *and* $\langle {}^{\uparrow}, {}^{\downarrow} \rangle = \langle {}^{\uparrow_{\mathcal{C}_{\langle \uparrow,\downarrow \rangle}}}, {}^{\downarrow_{\mathcal{C}_{\langle \uparrow,\downarrow \rangle}}} \rangle$.

Proof. (1) follows by Lemma 5.49 from the fact that $a \leq b$ implies $I_a \supseteq I_b$. (2) follows by Lemma 5.46 from the fact that for any $A \in L^X$ it holds $A(x) = \bigvee\{a \mid x \in {}^aA\}$ (and similarly for $B \in L^Y$). (3) is a consequence of (1), (2), and Lemma 5.46. □

Remark 5.51 Theorem 5.50 (3) assures there is a one-to-one correspondence between the class of all **L**-Galois connections and the class of all **L**-nested systems of **2**-Galois connections between X and Y. The existence of the one-to-one correspondence is apparent from the one-to-one correspondence between **L**-Galois connections and binary **L**-relations and the representation of **L**-relations by systems of **2**-relations. However, Theorem 5.50 provides the direct way.

Second case: representation of $\mathbf{L}_K$-Galois connections and first proof of Theorem 5.9. Given an $\mathbf{L}_K$-Galois connection $\langle {}^{\uparrow}, {}^{\downarrow} \rangle$ between X and Y, we denote by $\mathcal{B}\left(X, Y, \langle {}^{\uparrow}, {}^{\downarrow} \rangle\right)$ the set of all fixed points of $\langle {}^{\uparrow}, {}^{\downarrow} \rangle$, i.e. $\mathcal{B}\left(X, Y, \langle {}^{\uparrow}, {}^{\downarrow} \rangle\right) = \{\langle A, B \rangle \in L^X \times L^Y \mid A^{\uparrow} = B, B^{\downarrow} = A\}$. Since $\langle {}^{\uparrow}, {}^{\downarrow} \rangle$ is a Galois connection between the complete lattices $\langle L^X, \subseteq \rangle$ and $\langle L^Y, \subseteq \rangle$, $\mathcal{B}\left(X, Y, \langle {}^{\uparrow}, {}^{\downarrow} \rangle\right)$ is a complete lattice w.r.t. $\leq$ defined by $\langle A_1, B_1 \rangle \leq \langle A_2, B_2 \rangle$ iff $A_1 \subseteq A_2$ (or, equivalently, iff $B_2 \subseteq B_1$), see Chapter 1.

Since $1 \in K$ holds for each $\leq$-filter K, each $\mathbf{L}_K$-Galois connection is also an $\mathbf{L}_{\{1\}}$-Galois connection. We are going to show that $\mathbf{L}_{\{1\}}$-Galois connections between X and Y are in one-to-one correspondence with special **2**-Galois connections between $X \times L$ and $Y \times L$. Recall that by Theorem 3.25, each **L**-set in X can be viewed as a (representative) subset of $X \times L$.

Definition 5.52 A **2**-Galois connection $\langle {}^\wedge, {}^\vee \rangle$ between $X \times L$ and $Y \times L$ is called *commutative w.r.t.* $\lfloor\lceil\ \rceil\rfloor$ if

$$\lfloor\lceil A\rceil\rfloor^\wedge = \lfloor\lceil A^\wedge\rceil\rfloor \quad \text{and} \quad \lfloor\lceil B\rceil\rfloor^\vee = \lfloor\lceil B^\vee\rceil\rfloor \tag{5.35}$$

holds for each $A \subseteq X \times L$, $B \subseteq Y \times L$.

Remark 5.53 Note that $\lfloor\lceil A\rceil\rfloor^\wedge \subseteq A^\wedge \subseteq \lfloor\lceil A^\wedge\rceil\rfloor$ holds true for any **2**-Galois connection $\langle {}^\wedge, {}^\vee \rangle$ between $X \times L$ and $Y \times L$. Indeed, $\lfloor\lceil A\rceil\rfloor^\wedge \subseteq A^\wedge$ follows from $A \subseteq \lfloor\lceil A\rceil\rfloor$ and antitony of ${}^\wedge$, whereas $A^\wedge \subseteq \lfloor\lceil A^\wedge\rceil\rfloor$ follows from the fact that $A \subseteq \lfloor\lceil A\rceil\rfloor$ is true for any $A \in X \times L$. It follows that the first condition of (5.35) is equivalent to $\lfloor\lceil A\rceil\rfloor^\wedge \supseteq \lfloor\lceil A^\wedge\rceil\rfloor$. Moreover, in this case we have $\lfloor\lceil A\rceil\rfloor^\wedge = A^\wedge = \lfloor\lceil A^\wedge\rceil\rfloor$. Similarly for the second condition of (5.35) which is, in fact, equivalent to $\lceil\lfloor B\rfloor\rceil^\vee \supseteq \lfloor\lceil B^\vee\rceil\rfloor$.

For a pair $\langle {}^\wedge, {}^\vee \rangle$ of mappings ${}^\wedge : X \times L \to Y \times L$, ${}^\vee : Y \times L \to X \times L$ introduce a pair $\langle {}^{\uparrow_{\langle\wedge,\vee\rangle}}, {}^{\downarrow_{\langle\wedge,\vee\rangle}} \rangle$ of mappings ${}^{\uparrow_{\langle\wedge,\vee\rangle}} : L^X \to L^Y$, ${}^{\downarrow_{\langle\wedge,\vee\rangle}} : L^Y \to L^X$ by

$$A^{\uparrow_{\langle\wedge,\vee\rangle}} = \lceil\lfloor A\rfloor^\wedge\rceil \quad \text{and} \quad B^{\downarrow_{\langle\wedge,\vee\rangle}} = \lceil\lfloor B\rfloor^\vee\rceil \tag{5.36}$$

for $A \in L^X$, $B \in L^Y$. For a pair $\langle {}^\uparrow, {}^\downarrow \rangle$ of mappings ${}^\uparrow : L^X \to L^Y$, ${}^\downarrow : L^Y \to L^X$ define a pair $\langle {}^{\wedge_{\langle\uparrow,\downarrow\rangle}}, {}^{\vee_{\langle\uparrow,\downarrow\rangle}} \rangle$ of mappings ${}^{\wedge_{\langle\uparrow,\downarrow\rangle}} : X \times L \to Y \times L$, ${}^{\vee_{\langle\uparrow,\downarrow\rangle}} : Y \times L \to X \times L$ by

$$A^{\wedge_{\langle\uparrow,\downarrow\rangle}} = \lfloor\lceil A\rceil^\uparrow\rfloor \quad \text{and} \quad B^{\vee_{\langle\uparrow,\downarrow\rangle}} = \lfloor\lceil B\rceil^\downarrow\rfloor \tag{5.37}$$

for $A \in X \times L$, $B \in Y \times L$.

Theorem 5.54 *Let $\langle {}^\uparrow, {}^\downarrow \rangle$ be an $\mathbf{L}_{\{1\}}$-Galois connection between X and Y and $\langle {}^\wedge, {}^\vee \rangle$ be a **2**-Galois connection between $X \times L$ and $Y \times L$ which is commutative w.r.t. $\lfloor\lceil\ \rceil\rfloor$. Then*

*(1) $\langle {}^{\wedge_{\langle\uparrow,\downarrow\rangle}}, {}^{\vee_{\langle\uparrow,\downarrow\rangle}} \rangle$ is a **2**-Galois connection between $X \times L$ and $Y \times L$ which is commutative w.r.t. $\lfloor\lceil\ \rceil\rfloor$;*

(2) $\langle {}^{\uparrow_{\langle\wedge,\vee\rangle}}, {}^{\downarrow_{\langle\wedge,\vee\rangle}} \rangle$ is an $\mathbf{L}_{\{1\}}$-Galois connection between X and Y;

(3) $\langle{}^{\wedge},{}^{\vee}\rangle = \langle {}^{\wedge_{\langle \uparrow_{\langle\wedge,\vee\rangle},\downarrow_{\langle\wedge,\vee\rangle}\rangle}}, {}^{\vee_{\langle \uparrow_{\langle\wedge,\vee\rangle},\downarrow_{\langle\wedge,\vee\rangle}\rangle}}\rangle$ *and*
$\langle{}^{\uparrow},{}^{\downarrow}\rangle = \langle {}^{\uparrow_{\langle \wedge_{\langle\uparrow,\downarrow\rangle},\vee_{\langle\uparrow,\downarrow\rangle}\rangle}}, {}^{\downarrow_{\langle \wedge_{\langle\uparrow,\downarrow\rangle},\vee_{\langle\uparrow,\downarrow\rangle}\rangle}}\rangle$.

Proof. (1) Let $A_1, A_2 \subseteq X \times L$, $A_1 \subseteq A_2$. We have $S(\lceil A_1\rceil, \lceil A_2\rceil) = 1$, thus $S(\lceil A_2\rceil^{\uparrow}, \lceil A_1\rceil^{\uparrow}) = 1$, and hence $A_2^{\wedge_{\langle\uparrow,\downarrow\rangle}} = \lfloor\lceil A_2\rceil^{\uparrow}\rfloor \subseteq \lfloor\lceil A_1\rceil^{\uparrow}\rfloor = A_1^{\wedge_{\langle\uparrow,\downarrow\rangle}}$. We have established that $A_1 \subseteq A_2$ implies $A_2^{\wedge_{\langle\uparrow,\downarrow\rangle}} \subseteq A_1^{\wedge_{\langle\uparrow,\downarrow\rangle}}$; similarly for $B_1, B_2 \in Y \times L$.

For $A \subseteq X \times L$ we have $A^{\wedge_{\langle\uparrow,\downarrow\rangle}\vee_{\langle\uparrow,\downarrow\rangle}} = \lfloor\lceil\lfloor\lceil A\rceil^{\uparrow}\rfloor\rceil^{\downarrow}\rfloor = \lfloor\lceil A\rceil^{\uparrow\downarrow}\rfloor$. If $\langle x, a\rangle \in A$ then $\lceil A\rceil(x) \geq a$, thus $\lceil A\rceil^{\uparrow\downarrow}(x) \geq \lceil A\rceil(x) \geq a$, therefore $\langle x, a\rangle \in \lfloor\lceil A\rceil^{\uparrow\downarrow}\rfloor$, i.e. $A \subseteq A^{\vee_{\langle\uparrow,\downarrow\rangle}\wedge_{\langle\uparrow,\downarrow\rangle}}$; similarly for $B \subseteq Y \times L$.

The commutativity follows by $\lfloor\lceil A\rceil\rfloor^{\wedge_{\langle\uparrow,\downarrow\rangle}} = \lfloor\lceil\lfloor\lceil A\rceil\rfloor\rceil^{\uparrow}\rfloor = \lfloor\lceil A\rceil^{\uparrow}\rfloor = \lfloor\lceil\lfloor\lceil A\rceil^{\uparrow}\rfloor\rceil\rfloor = \lfloor\lceil A^{\wedge_{\langle\uparrow,\downarrow\rangle}}\rceil\rfloor$.

(2) Let $A_1, A_2 \in L^X$, $S(A_1, A_2) = 1$. Then $\lfloor A_1\rfloor \subseteq \lfloor A_2\rfloor$, $\lfloor A_2\rfloor^{\wedge} \subseteq \lfloor A_1\rfloor^{\wedge}$ and $A_2^{\uparrow_{\langle\wedge,\vee\rangle}} = \lceil\lfloor A_2\rfloor^{\wedge}\rceil \subseteq \lceil\lfloor A_1\rfloor^{\wedge}\rceil = A_1^{\uparrow_{\langle\wedge,\vee\rangle}}$; similarly for $B_1, B_2 \in L^Y$.

For $A \in L^X$ we have $A^{\uparrow_{\langle\wedge,\vee\rangle}\downarrow_{\langle\wedge,\vee\rangle}} = \lceil\lfloor\lceil\lfloor A\rfloor^{\wedge}\rceil\rfloor^{\vee}\rceil = \lceil\lfloor\lceil\lfloor A\rfloor\rceil\rfloor^{\wedge\vee}\rceil = \lceil\lfloor A\rfloor^{\wedge\vee}\rceil \supseteq A$ due to the commutativity and the fact $\lfloor A\rfloor^{\wedge\vee} \supseteq \lfloor A\rfloor$; similarly for $B \in L^Y$.

(3) Due to Remark 5.53 and the fact that for $A' \in L^X$ it holds $\lceil\lfloor A'\rfloor\rceil = A'$ we have $A^{\wedge_{\langle \uparrow_{\langle\wedge,\vee\rangle},\downarrow_{\langle\wedge,\vee\rangle}\rangle}} = \lfloor\lceil A\rceil^{\uparrow_{\langle\wedge,\vee\rangle}}\rfloor = \lfloor\lceil\lfloor\lceil A\rceil\rfloor^{\wedge}\rceil\rfloor = \lfloor\lceil A^{\wedge}\rceil\rfloor = A^{\wedge}$ for any $A \subseteq X \times L$. For any $A \in L^X$, since $\lceil\lfloor A\rfloor\rceil = A$ holds, we have $A^{\uparrow_{\langle \wedge_{\langle\uparrow,\downarrow\rangle},\vee_{\langle\uparrow,\downarrow\rangle}\rangle}} = \lceil\lfloor A\rfloor^{\wedge_{\langle\uparrow,\downarrow\rangle}}\rceil = \lfloor\lceil\lfloor\lceil A\rceil\rfloor^{\wedge}\rceil\rfloor = A^{\uparrow}$. Similarly for $B \subseteq Y \times L$ and $B \in L^Y$. □

Remark 5.55 The condition that $\langle{}^{\wedge},{}^{\vee}\rangle$ is commutative is essential in the foregoing statement. Consider $X = \{x\}$, $Y = \{y\}$, $L = \{0, \frac{1}{2}, 1\}$, and $I \subseteq (X \times L) \times (Y \times L)$ given by $I = \{\langle\langle x, 0\rangle, \langle y, \frac{1}{2}\rangle\rangle, \langle\langle x, \frac{1}{2}\rangle, \langle y, 0\rangle\rangle, \langle\langle x, \frac{1}{2}\rangle, \langle y, \frac{1}{2}\rangle\rangle, \langle\langle x, 1\rangle, \langle y, \frac{1}{2}\rangle\rangle, \langle\langle x, 1\rangle, \langle y, 1\rangle\rangle\}$. Let further $\langle{}^{\wedge},{}^{\vee}\rangle$ be the Galois connection between $X \times L$ and $Y \times L$ induced by I. Note that it is not commutative since the commutativity fails for $A = \{\langle x, 1\rangle\}$. For the mappings $\langle{}^{\uparrow},{}^{\downarrow}\rangle$ induced by $\langle{}^{\wedge},{}^{\vee}\rangle$ by (5.36) we have $\{{}^{1}/x\}^{\uparrow\downarrow} = \{{}^{\frac{1}{2}}/x\} \not\supseteq \{{}^{1}/x\}$, i.e. $\langle{}^{\uparrow},{}^{\downarrow}\rangle$ is not an $\mathbf{L}_{\{1\}}$-Galois connection.

Note also that the $\mathbf{L}_{\{1\}}$-Galois connection $\langle{}^{\uparrow},{}^{\downarrow}\rangle$ induced by a Galois connection $\langle{}^{\wedge},{}^{\vee}\rangle$ need not be an $\mathbf{L}$-Galois connection (a counterexample is easy to get).

Theorem 5.56 *For any $\mathbf{L}_K$-Galois connection $\langle {}^{\uparrow}, {}^{\downarrow} \rangle$, $\mathcal{B}\left(X, Y, \langle {}^{\uparrow}, {}^{\downarrow} \rangle\right)$ and $\mathcal{B}(X \times L, Y \times L, \langle {}^{\wedge}, {}^{\vee} \rangle)$, where $\langle {}^{\wedge}, {}^{\vee} \rangle = \langle {}^{\wedge_{\langle \uparrow, \downarrow \rangle}}, {}^{\vee_{\langle \uparrow, \downarrow \rangle}} \rangle$ of Theorem 5.54, are isomorphic lattices. Moreover, $\mathcal{B}(X \times L, Y \times L, \langle {}^{\wedge}, {}^{\vee} \rangle) = \mathcal{B}(X \times L, Y \times L, I^{\times})$ where $\langle \langle x, a \rangle, \langle y, b \rangle \rangle \in I^{\times}$ iff $b \le \{ {}^{a}/x \}^{\uparrow}(y)$.*

Proof. We prove the assertion by showing that $h : \mathcal{B}\left(X, Y, \langle {}^{\uparrow}, {}^{\downarrow} \rangle\right) \to \mathcal{B}(X \times L, Y \times L, \langle {}^{\wedge}, {}^{\vee} \rangle)$ and $g : \mathcal{B}(X \times L, Y \times L, \langle {}^{\wedge}, {}^{\vee} \rangle) \to \mathcal{B}\left(X, Y, \langle {}^{\uparrow}, {}^{\downarrow} \rangle\right)$ defined by $h(\langle A, B \rangle) = \langle \lfloor A \rfloor, \lfloor B \rfloor \rangle$ and $g(\langle A', B' \rangle) = \langle \lceil A' \rceil, \lceil B' \rceil \rangle$ are mutually inverse order-preserving maps. First, we show that h and g are correctly defined, i.e. $h(\mathcal{B}\left(X, Y, \langle {}^{\uparrow}, {}^{\downarrow} \rangle\right)) \subseteq \mathcal{B}(X \times L, Y \times L, \langle {}^{\wedge}, {}^{\vee} \rangle)$ and $g(\mathcal{B}(X \times L, Y \times L, \langle {}^{\wedge}, {}^{\vee} \rangle)) \subseteq \mathcal{B}\left(X, Y, \langle {}^{\uparrow}, {}^{\downarrow} \rangle\right)$. If $\langle A, B \rangle \in \mathcal{B}\left(X, Y, \langle {}^{\uparrow}, {}^{\downarrow} \rangle\right)$ then $\lfloor A \rfloor^{\wedge} = \lfloor \lceil \lfloor A \rfloor \rceil^{\uparrow} \rfloor = \lfloor A^{\uparrow} \rfloor = \lfloor B \rfloor$ and, similarly, $\lfloor B \rfloor^{\vee} = \lfloor A \rfloor$, i.e. $h(\langle A, B \rangle) \in \mathcal{B}(X \times L, Y \times L, \langle {}^{\wedge}, {}^{\vee} \rangle)$. If $\langle A', B' \rangle \in \mathcal{B}(X \times L, Y \times L, \langle {}^{\wedge}, {}^{\vee} \rangle)$ then $\lceil A \rceil^{\uparrow} = \lceil \lfloor \lceil A' \rceil \rfloor^{\wedge} \rceil = \lceil A'^{\wedge} \rceil = \lceil B' \rceil$ and, similarly, $\lceil B' \rceil^{\vee} = \lceil A' \rceil$, i.e. $g(\langle A', B' \rangle) \in \mathcal{B}\left(X, Y, \langle {}^{\uparrow}, {}^{\downarrow} \rangle\right)$. By Theorem 3.25, h and g are mutually inverse. Finally, both h and g are clearly order-preserving.

To see that $\mathcal{B}(X \times L, Y \times L, \langle {}^{\wedge}, {}^{\vee} \rangle) = \mathcal{B}(X \times L, Y \times L, I^{\times})$, it is enough to show that $I^{\times}$ is the relation $I_{\langle \wedge, \vee \rangle}$ corresponding to $\langle {}^{\wedge}, {}^{\vee} \rangle$ by Theorem 5.45. This is indeed true since $\langle \langle x, a \rangle, \langle y, b \rangle \rangle \in I_{\langle \wedge, \vee \rangle}$ iff $\langle y, b \rangle \in \{ \langle x, a \rangle \}^{\wedge}$ which is by (5.37) equivalent to $b \le \{ {}^{a}/x \}^{\uparrow}(y)$, i.e. to $\langle \langle x, a \rangle, \langle y, b \rangle \rangle \in I^{\times}$. □

The following result shows that (modulo identifying **L**-sets with representative sets) **L**-concept lattices may be viewed as concept lattices.

Theorem 5.57 *Any **L**-concept lattice $\mathcal{B}(X, Y, I)$ is isomorphic to the concept lattice $\mathcal{B}(X \times L, Y \times L, I^{\times})$ where $\langle \langle x, a \rangle, \langle y, b \rangle \rangle \in I^{\times}$ iff $a \otimes b \le I(x, y)$.*

Proof. Let $\langle {}^{\uparrow}, {}^{\downarrow} \rangle$ be the **L**-Galois connection induced by I by Theorem 5.45. We have $\{ {}^{a}/x \}^{\uparrow}(y) = \bigwedge_{x' \in X} \{ {}^{a}/x \}(x') \to I(x, y) = a \to I(x, y)$, i.e. the adjunction property gives $b \le \{ {}^{a}/x \}^{\uparrow}(y)$ iff $a \otimes b \le I(x, y)$. The assertion now follows by Theorem 5.56. □

The following theorem gives a characterization of lattices of fixed points of $\mathbf{L}_K$-Galois connections. In particular, it contains Theorem 5.9, i.e. the crisp version of the main theorem of **L**-concept lattices, as a special case. Note that a subset K of a complete lattice $\mathbf{V}$ is called infimally (supremally) dense if each element of V is a supremum (infimum) of some elements of K.

Theorem 5.58 *Let $\langle{}^{\uparrow},{}^{\downarrow}\rangle$ be an $\mathbf{L}_K$-Galois connection between X and Y. (1) The set $\mathcal{B}\left(X,Y,\langle{}^{\uparrow},{}^{\downarrow}\rangle\right)$ is a complete lattice w.r.t. $\leq$ where the suprema and infima are given by*

$$\bigwedge_{j\in J}\langle A_j,B_j\rangle=\langle\bigcap_{j\in J}A_j,(\bigcup_{j\in J}B_j)^{\downarrow\uparrow}\rangle\ ,\ \bigvee_{j\in J}\langle A_j,B_j\rangle=\langle(\bigcup_{j\in J}A_j)^{\uparrow\downarrow},\bigcap_{j\in J}B_j\rangle\ .$$

(2) Moreover, an arbitrary complete lattice $\mathbf{V}=\langle V,\leq\rangle$ is isomorphic to $\mathcal{B}\left(X,Y,\langle{}^{\uparrow},{}^{\downarrow}\rangle\right)$ iff there are mappings $\gamma: X\times L\to V$, $\mu: Y\times L\to V$ such that

(i) $\gamma(X,L)$ is supremally dense in V, $\mu(Y,L)$ is infimally dense in V;

(ii) $\gamma(x,a)\leq\mu(y,b)$ iff $b\leq\{{}^a/x\}^{\uparrow}(y)$.

If $\langle{}^{\uparrow},{}^{\downarrow}\rangle$ is an $\mathbf{L}$-Galois connection, i.e. $\mathcal{B}\left(X,Y,\langle{}^{\uparrow},{}^{\downarrow}\rangle\right)=\mathcal{B}(X,Y,I)$ for some $I\in L^{X\times Y}$, then (ii) may be replaced by (ii') $\gamma(x,a)\leq\mu(y,b)$ iff $a\otimes b\leq I(x,y)$.

Proof. (1) follows from the description of infima and suprema of fixed points of Galois connection between complete lattices (see Chapter 1) and the fact that $\langle{}^{\uparrow},{}^{\downarrow}\rangle$ is a Galois connection between the complete lattices $\langle L^X,\subseteq\rangle$ and $\langle L^Y,\subseteq\rangle$.

(2): Note first that the so-called main theorem of (crisp) concept lattices (see [203]) says that for a relation I between sets C and D, a complete lattice $\mathbf{U}=\langle U,\leq\rangle$ is isomorphic to the (crisp) concept lattice $\mathcal{B}(C,D,I)$ iff there are mappings $g: C\to U$ and $m: D\to U$ such that $g(C)$ is supremally dense in U, $m(D)$ is infimally dense in U, and $g(c)\leq m(d)$ iff $\langle c,d\rangle\in I$. Now, (2) follows directly by Theorems 5.56 and 5.57. □

Remark 5.59 Let us remark that by [203], each complete lattice $\langle V,\leq\rangle$ is isomorphic to the crisp concept lattice $\mathcal{B}(V,V,\leq)$. Therefore, any $\mathbf{L}$-concept lattice $\mathcal{B}(X,Y,I)$ is isomorphic to the crisp concept lattice $\mathcal{B}(\mathcal{B}(X,Y,I),\mathcal{B}(X,Y,I),\leq)$. However, such a representation is not natural compared to that provided by Theorem 5.57. In Theorem 5.57, $\mathbf{L}$-concepts from $\mathcal{B}(X,Y,I)$ correspond to (ordinary) concepts from $\mathcal{B}(X\times L,Y\times L,I)$ in a very natural way: $\langle A,B\rangle$ corresponds to

$\langle \lfloor A \rfloor, \lfloor B \rfloor \rangle$. Moreover, our representation yields almost directly Theorem 5.58, i.e. the characterization theorem for $\mathbf{L}_K$-concept lattices.

Remark 5.60 Note also that the formulas for infima and suprema in Theorem 5.58 (1) can be derived directly from the assumption that they are valid for $\mathbf{L} = \mathbf{2}$. Namely, since, by Theorem 5.56, $\mathcal{B}\left(X, Y, \langle ^{\uparrow}, ^{\downarrow} \rangle\right)$ and $\mathcal{B}(X \times L, Y \times L, \langle ^{\wedge}, ^{\vee} \rangle)$ are isomorphic lattices with h, g (of Proof of Theorem 5.56) being the mutually inverse isomorphisms, we have $\bigwedge_{j \in J} \langle A_j, B_j \rangle = g(h(\bigwedge_{j \in J} \langle A_j, B_j \rangle))$. Furthermore, we have

$$
\begin{aligned}
g(h(\bigwedge_{j \in J} \langle A_j, B_j \rangle)) &= g(\bigwedge_{j \in J} h(\langle A_j, B_j \rangle)) = \\
&= g(\bigwedge_{j \in J} \langle \lfloor A_j \rfloor, \lfloor B_j \rfloor \rangle) = g(\langle \bigcap_{j \in J} \lfloor A_j \rfloor, (\bigcup_{j \in J} \lfloor B_j \rfloor)^{\vee\wedge} \rangle) = \\
&= \langle \lceil \bigcap_{j \in J} \lfloor A_j \rfloor \rceil, \lceil (\bigcup_{j \in J} \lfloor B_j \rfloor)^{\vee\wedge} \rceil \rangle = \langle \bigcap_{j \in J} \lceil \lfloor A_j \rfloor \rceil, \lceil \lfloor \lceil \bigcup_{j \in J} \lfloor B_j \rfloor \rceil^{\downarrow} \rfloor^{\wedge} \rceil \rangle = \\
&= \langle \bigcap_{j \in J} A_j, \lceil \lfloor \lceil \lfloor \lceil \bigcup_{j \in J} \lfloor B_j \rfloor \rceil^{\downarrow} \rfloor \rceil^{\uparrow} \rfloor \rceil \rangle = \langle \bigcap_{j \in J} A_j, \lceil \bigcup_{j \in J} \lfloor B_j \rfloor \rceil^{\downarrow\uparrow} \rangle = \\
&= \langle \bigcap_{j \in J} A_j, \bigcup_{j \in J} \lceil \lfloor B_j \rfloor \rceil^{\downarrow\uparrow} \rangle = \langle \bigcap_{j \in J} A_j, (\bigcup_{j \in J} B_j)^{\downarrow\uparrow} \rangle,
\end{aligned}
$$

i.e. the first part of (5.58) is true. We used the validity of (5.58) for $\mathbf{L} = \mathbf{2}$, Theorem 3.25, the equalities $\lceil \bigcap_{j \in J} \lfloor C_j \rfloor \rceil = \bigcap_{j \in J} \lceil \lfloor C_j \rfloor \rceil$, $\lceil \bigcup_{j \in J} \lfloor C_j \rfloor \rceil = \bigcup_{j \in J} \lceil \lfloor C_j \rfloor \rceil$ (which hold for $\mathbf{L}$-sets C_j), and the definitions of $^{\wedge}$ and $^{\vee}$. The second part of (5.58) can be obtained dually.

Remark 5.61 We have shown that each $\mathbf{L}$-concept lattice can be viewed as a $\mathbf{2}$-concept lattice (i.e. classical concept lattice). Our results are more general in that they concern lattices of fixed points of $\mathbf{L}_K$-Galois connections of which $\mathbf{L}$-concept lattices are a special case. The $\mathbf{L}$-concept lattice is interpreted as a structure of concepts determined by a given $\mathbf{L}$-context, i.e. a set of objects, a set of attributes, and an $\mathbf{L}$-relation "to have" between objects and attributes. In this respect, the context of the corresponding $\mathbf{2}$-concept lattice has no clear interpretation—each object of the new context is a pair $\langle x, a \rangle$ where x is an object of the original $\mathbf{L}$-context and a is a truth value, similarly for attributes. However, the result enables us to apply the results obtained for concept lattices to $\mathbf{L}$-concept lattices (a problem concerning an $\mathbf{L}$-concept lattice is to be "translated" into a problem concerning the corresponding concept lattice, the translated problem is to be solved by results available for concept lattices, and the solution obtained is to be translated back). On the other hand, there are several phenomena which are degenerate in the case of (**2**-)concept lattices (e.g. similarity or logical precision). The study of such phenomena cannot be directly "reduced" in the above-mentioned way.

5.4 MAIN THEOREM OF FUZZY CONCEPT LATTICES

The point of this section is to present a theorem characterizing the hierarchical structure of fuzzy concepts given by a fuzzy context. As we will see, the notion of a fuzzy concept lattice is a rather general one (for instance, Dedekind–MacNeille completion of an **L**-ordered set introduced in Chapter 4 is a particular case of a fuzzy concept lattice).

Recall that for an **L**-set A in U and $a \in L$, $a \otimes A$ and $a \to A$ denote the **L**-sets such that $(a \otimes A)(u) = a \otimes A(u)$ and $(a \to A)(u) = a \to A(u)$, respectively. Moreover, for an **L**-set $\mathcal{M}$ of **L**-sets in Y such that each $y \in Y$ is an **L**-set in X, **L**-sets $\bigcap \mathcal{M}$ and $\bigcup \mathcal{M}$ in X were defined in Section 3.1 by

$$(\bigcap \mathcal{M})(x) = \bigwedge_{A \in Y} \mathcal{M}(A) \to A(x),$$

$$(\bigcup \mathcal{M})(x) = \bigvee_{A \in Y} \mathcal{M}(A) \otimes A(x).$$

For an **L**-set $\mathcal{M}$ in $\mathcal{B}(X, Y, I)$, we put $\bigcap_X \mathcal{M} = \bigcap \mathrm{pr}_X(\mathcal{M})$, $\bigcup_X \mathcal{M} = \bigcup \mathrm{pr}_X(\mathcal{M})$, $\bigcap_Y \mathcal{M} = \bigcap \mathrm{pr}_Y(\mathcal{M})$, $\bigcup_Y \mathcal{M} = \bigcup \mathrm{pr}_Y(\mathcal{M})$, where $\mathrm{pr}_X(\mathcal{M})$ is an **L**-set in the set $\{A \in L^X \mid A = A^{\uparrow\downarrow}\}$ of all extents of $\mathcal{B}(X, Y, I)$ defined by $(\mathrm{pr}_X \mathcal{M})(A) = \mathcal{M}(A, A^{\uparrow})$ and, similarly, $\mathrm{pr}_Y(\mathcal{M})$ is an **L**-set in the set $\{B \in L^Y \mid B = B^{\uparrow\downarrow}\}$ of all intents of $\mathcal{B}(X, Y, I)$ defined by $(\mathrm{pr}_Y \mathcal{M})(B) = \mathcal{M}(B^{\downarrow}, B)$. Thus, $\bigcap_X \mathcal{M}$ is the "intersection of all extents from $\mathcal{M}$", etc.

Let **X** be a completely lattice **L**-ordered set, $L' \subseteq L$. A subset $K \subseteq X$ is called L'-infimally dense in **X** (L'-supremally dense in **X**) if for each $x \in X$ there is some $A \in {L'}^X$ such that $A(x) = 0$ for all $x \notin K$ and $(\inf(A))(x) = 1$ $((\sup(A))(x) = 1)$.

Remark 5.62 (1) Note that by Remark 4.56, K is $\{0, 1\}$-infimally dense ($\{0, 1\}$-supremally dense) in **X** if for each $x \in X$ there is some $K' \subseteq K$ such that $x = \bigwedge K'$ $(x = \bigvee K')$. Here, $\bigwedge$ $(\bigvee)$ refers to the infimum (supremum) w.r.t. $\leq$, the 1-cut of $\preceq$.

(2) For $\mathbf{L} = \mathbf{2}$ (the classical bivalent case), the above notions coincide with the usual notions of infimal and supremal density, see Chapter 1.

We are ready to present the main theorem of **L**-concept lattices (the general form for **L**-order). Recall by Remark 4.52 that in a complete lattice **L**-ordered set, $\inf(A)$ of $A \in L^X$ is uniquely given by ${}^1\inf(A) = \{x\}$; similarly for $\sup(A)$.

Theorem 5.63 (main theorem of L-concept lattices, fuzzy order) *Let $\langle X, Y, I\rangle$ be an **L**-context. (1) $\langle\langle \mathcal{B}(X,Y,I), \approx\rangle, \preceq\rangle$ is completely lattice **L**-ordered set in which infima and suprema can be described as follows: for an **L**-set $\mathcal{M}$ in $\mathcal{B}(X,Y,I)$ we have*

$$\begin{aligned} {}^1\inf(\mathcal{M}) &= \{\langle \bigcap_X \mathcal{M}, (\bigcap_X \mathcal{M})^{\uparrow}\rangle\} = \{\langle (\bigcup_Y \mathcal{M})^{\downarrow}, (\bigcup_Y \mathcal{M})^{\downarrow\uparrow}\rangle\}, &(5.38)\\ {}^1\sup(\mathcal{M}) &= \{\langle (\bigcap_Y \mathcal{M})^{\downarrow}, \bigcap_Y \mathcal{M}\rangle\} = \{\langle (\bigcup_X \mathcal{M})^{\uparrow\downarrow}, (\bigcup_X \mathcal{M})^{\uparrow}\rangle\} &(5.39)\end{aligned}$$

*(2) Moreover, a completely lattice **L**-ordered set $\mathbf{V} = \langle\langle V, \approx\rangle, \preceq\rangle$ is isomorphic to $\langle\langle \mathcal{B}(X,Y,I), \approx\rangle, \preceq\rangle$ iff there are mappings $\gamma : X \times L \to V$, $\mu : Y \times L \to V$, such that $\gamma(X \times L)$ is $\{0,1\}$-supremally dense in $\mathbf{V}$, $\mu(Y \times L)$ is $\{0,1\}$-infimally dense in $\mathbf{V}$, and $((a \otimes b) \to I(x,y)) = (\gamma(x,a) \preceq \mu(y,b))$ for all $x \in X$, $y \in Y$, $a, b \in L$. In particular, $\mathbf{V}$ is isomorphic to $\mathcal{B}(V, V, \preceq)$.*

Proof. For brevity, we write also $\mathcal{B}$ instead of $\mathcal{B}(X,Y,I)$.

Part (1): From Example 4.47 we know that $\langle\langle \mathcal{B}(X,Y,I), \approx\rangle, \preceq\rangle$ is an **L**-ordered set. We therefore have to show that $\inf(\mathcal{M})$ and $\sup(\mathcal{M})$ are $\approx$-singletons and that (5.38) and (5.39) hold for any **L**-set $\mathcal{M}$ in $\mathcal{B}(X,Y,I)$. We proceed only for suprema, the case of infima is symmetric. First, we show (5.39). To this end, denote $\langle A^*, B^*\rangle = \langle (\bigcup_X \mathcal{M})^{\uparrow\downarrow}, (\bigcup_X \mathcal{M})^{\uparrow}\rangle$. We start by proving $B^* = \bigcap_Y \mathcal{M}$, i.e. we have to prove

$$\bigwedge_{x \in X} ((\bigvee_{\langle A,B\rangle \in \mathcal{B}} \mathcal{M}(A,B) \otimes A(x)) \to I(x,y)) = \bigwedge_{\langle A,B\rangle \in \mathcal{B}} (\mathcal{M}(A,B) \to B(y)) \tag{5.40}$$

for any $y \in Y$. The "$\leq$" part of (5.40) holds iff $\bigwedge_{x\in X}((\bigvee_{\langle A,B\rangle \in \mathcal{B}} \mathcal{M}(A,B) \otimes A(x)) \to I(x,y)) \leq \mathcal{M}(A,B) \to B(y)$ is true for any $\langle A,B\rangle \in \mathcal{B}$ which holds iff $\mathcal{M}(A,B) \otimes \bigwedge_{x\in X}((\bigvee_{\langle A,B\rangle\in\mathcal{B}} \mathcal{M}(A,B) \otimes A(x)) \to I(x,y)) \leq B(y)$. The last inequality is true since

$$\begin{aligned} &\mathcal{M}(A,B) \otimes \bigwedge_{x\in X}((\bigvee_{\langle A,B\rangle \in \mathcal{B}} \mathcal{M}(A,B) \otimes A(x)) \to I(x,y)) \leq \\ \leq\ &\mathcal{M}(A,B) \otimes \bigwedge_{x\in X}((\mathcal{M}(A,B) \otimes A(x)) \to I(x,y)) \leq \\ \leq\ &\bigwedge_{x\in X}(\mathcal{M}(A,B) \otimes (\mathcal{M}(A,B) \to (A(x) \to I(x,y)))) \leq \end{aligned}$$

$$\leq \bigwedge_{x\in X} A(x) \to I(x,y) = B(y).$$

The "$\geq$" part of (5.40) holds iff $\bigwedge_{\langle A,B\rangle\in\mathcal{B}}(\mathcal{M}(A,B) \to B(y)) \leq (\bigvee_{\langle A,B\rangle\in\mathcal{B}}\mathcal{M}(A,B)\otimes A(x)) \to I(x,y)$ is valid for each $x \in X$ which holds iff $(\bigvee_{\langle A,B\rangle\in\mathcal{B}}\mathcal{M}(A,B)\otimes A(x))\otimes\bigwedge_{\langle A,B\rangle\in\mathcal{B}}(\mathcal{M}(A,B) \to B(y)) \leq I(x,y)$ which is true. Indeed,

$$\begin{aligned}
&(\bigvee_{\langle A,B\rangle\in\mathcal{B}} \mathcal{M}(A,B)\otimes A(x))\otimes \bigwedge_{\langle A,B\rangle\in\mathcal{B}} (\mathcal{M}(A,B)\to B(y)) \leq\\
\leq\ &\bigvee_{\langle A,B\rangle\in\mathcal{B}} (\mathcal{M}(A,B)\otimes A(x)\otimes \bigwedge_{\langle A,B\rangle\in\mathcal{B}} (\mathcal{M}(A,B)\to B(y))) \leq\\
\leq\ &\bigvee_{\langle A,B\rangle\in\mathcal{B}} (A(x)\otimes\mathcal{M}(A,B)\otimes(\mathcal{M}(A,B)\to B(y))) \leq\\
\leq\ &\bigvee_{\langle A,B\rangle\in\mathcal{B}} (A(x)\otimes B(y)) \leq I(x,y),
\end{aligned}$$

the last inequality being true by adjunction and the fact that $\langle A,B\rangle \in \mathcal{B}(X,Y,I)$. Therefore, (5.40) is established. Obviously, to prove (5.39) it is now sufficient to check ${}^1\text{sup}(\mathcal{M}) = \{\langle A^*,B^*\rangle\}$. Clearly, we have to show $(\mathcal{U}(\mathcal{M}))(A^*,B^*) \wedge (\mathcal{LU}(\mathcal{M}))(A^*,B^*) = 1$, i.e. $(\mathcal{U}(\mathcal{M}))(A^*,B^*) = 1$ and $(\mathcal{LU}(\mathcal{M}))(A^*,B^*) = 1$. We show $(\mathcal{U}(\mathcal{M}))(A^*,B^*) = 1$: By definition of $\mathcal{U}$ we have to show that $1 \leq \mathcal{M}(A,B) \to (\langle A,B\rangle \preceq \langle A^*,B^*\rangle)$, i.e. $1 \leq \mathcal{M}(A,B) \to S(A,A^*)$ which is equivalent to $\mathcal{M}(A,B) \leq S(A,A^*)$. The last inequality holds iff $\mathcal{M}(A,B) \leq A(x) \to A^*(x)$ holds for each $x \in X$, i.e. iff $\mathcal{M}(A,B)\otimes A(x) \leq A^*(x)$ which is true by definition of $\langle A^*,B^*\rangle$. We established $(\mathcal{U}(\mathcal{M}))(A^*,B^*) = 1$. We show $(\mathcal{LU}(\mathcal{M}))(A^*,B^*) = 1$: By definition of $\mathcal{L}$ we have to show $1 \leq (\mathcal{U}(\mathcal{M}))(A,B) \to (\langle A^*,B^*\rangle \preceq \langle A,B\rangle)$, i.e. $(\mathcal{U}(\mathcal{M}))(A,B) \leq (\langle A^*,B^*\rangle \preceq \langle A,B\rangle)$ for any $\langle A,B\rangle \in \mathcal{B}(X,Y,I)$. Since $(\langle A^*,B^*\rangle \preceq \langle A,B\rangle) = S(B,B^*)$, we have to prove $(\mathcal{U}(\mathcal{M}))(A,B) \leq B(y) \to B^*(y)$, i.e. $B(y)\otimes(\mathcal{U}(\mathcal{M}))(A,B) \leq B^*(y)$ for each $y \in Y$. As $B^* = A^{*\uparrow}$, we have to show $B(y)\otimes(\mathcal{U}(\mathcal{M}))(A,B) \leq A^*(x) \to I(x,y)$, i.e. $A^*\otimes B(y)\otimes(\mathcal{U}(\mathcal{M}))(A,B) \leq I(x,y)$ for each $x \in X$. By definition of A^*, this amounts to proving $(\bigvee_{\langle A',B'\rangle\in\mathcal{B}}\mathcal{M}(A',B')\otimes A'(x))\otimes B(y)\otimes(\mathcal{U}(\mathcal{M}))(A,B) \leq I(x,y)$, i.e. $\bigvee_{\langle A',B'\rangle\in\mathcal{B}}(\mathcal{M}(A',B')\otimes A'(x)\otimes B(y)\otimes(\mathcal{U}(\mathcal{M}))(A,B)) \leq I(x,y)$. Thus, we have to prove $\mathcal{M}(A',B')\otimes A'(x)\otimes B(y)\otimes(\mathcal{U}(\mathcal{M}))(A,B) \leq I(x,y)$ for any $\langle A',B'\rangle \in \mathcal{B}$. By definition of $\mathcal{U}(\mathcal{M})$, this is equivalent to $\mathcal{M}(A',B')\otimes A'(x)\otimes B(y)\otimes(\bigwedge_{\langle A',B'\rangle\in\mathcal{B}}\mathcal{M})(A',B') \to (\langle A',B'\rangle \preceq \langle A,B\rangle)) \leq I(x,y)$ which is true since

$$\mathcal{M}(A',B')\otimes A'(x)\otimes B(y)\otimes$$

$$\otimes\Big(\bigwedge_{\langle A'',B''\rangle\in\mathcal{B}} \mathcal{M}(A'',B'') \to (\langle A'',B''\rangle \preceq \langle A,B\rangle) \le$$

$$\le \mathcal{M}(A',B')\otimes A'(x)\otimes B(y)\otimes \mathcal{M}(A',B') \to S(A',A) \le$$

$$\le A'(x)\otimes\Big(\bigwedge_{x'\in X} A'(x') \to A(x')\Big)\otimes B(y) \le A(x)\otimes B(y) \le I(x,y)$$

the last inequality being true since $\langle A,B\rangle \in \mathcal{B}(X,Y,I)$. We proved $(\mathcal{LU}(\mathcal{M}))(A^*,B^*) = 1$.

It remains to show that $\sup(\mathcal{M})$ is an $\approx$-singleton in $\mathcal{B}(X,Y,I)$. This fact, however, follows by Lemma 4.53.

Part (2): Let $\mathcal{B}(X,Y,I)$ and $\mathbf{V}$ be isomorphic. We show the existence of γ, μ with the desired properties. It suffices to show the existence for $\mathbf{V} = \mathcal{B}(X,Y,I)$ because for the general case $\mathbf{V} \cong \mathcal{B}(X,Y,I)$ one can take $\gamma\circ\varphi : X\times L \to V$, $\mu\circ\varphi : Y\times L \to V$, where φ is the isomorphism of $\mathcal{B}(X,Y,I)$ onto $\mathbf{V}$. Let then $\gamma : X\times L \to \mathcal{B}(X,Y,I)$, $\mu : Y\times L \to \mathcal{B}(X,Y,I)$ be defined by

$$\gamma(x,a) = \Big\langle \{{}^{a}/x\}^{\uparrow\downarrow}, \{{}^{a}/x\}^{\uparrow} \Big\rangle,$$

$$\mu(y,b) = \Big\langle \Big\{{}^{b}/y\Big\}^{\downarrow}, \Big\{{}^{b}/y\Big\}^{\downarrow\uparrow} \Big\rangle$$

for every $x\in X$, $y\in Y$, $a,b\in L$. Since for each $\langle A,B\rangle \in \mathcal{B}(X,Y,I)$ it holds $A = \bigcup_{x\in X}\Big\{{}^{A(x)}/x\Big\}$, and $B = \bigcup_{y\in Y}\Big\{{}^{B(y)}/y\Big\}$, it follows from (5.38) and (5.39) that $\gamma(X\times L)$ and $\mu(Y\times L)$ are $\{0,1\}$-supremally dense and $\{0,1\}$-infimally dense in $\mathcal{B}(X,Y,I)$, respectively. We show that $((a\otimes b) \to I(x,y)) = (\gamma(x,a) \preceq \mu(y,b))$ is true for any $a,b\in L$, $x\in X$, and $y\in Y$: the equality easily follows by observing that $(\gamma(x,a) \preceq \mu(y,b)) = S(\{{}^{a}/x\}^{\uparrow\downarrow}, \{{}^{b}/y\}^{\downarrow}) = S(\{{}^{b}/y\}, \{{}^{a}/x\}^{\uparrow}) = b \to (a \to I(x,y))$ (we used Theorem 5.38) and the fact that $(a\otimes b) \to I(x,y) = b \to (a \to I(x,y))$.

Conversely, let γ and μ with the above properties exist. We prove the assertion by showing that there are mappings $\varphi : \mathcal{B}(X,Y,I) \to V$, $\psi : V \to \mathcal{B}(X,Y,I)$, such that $\varphi\circ\psi = id_{\mathcal{B}(X,Y,I)}$, $\psi\circ\varphi = id_V$, and $(\langle A_1,B_1\rangle \preceq \langle A_2,B_2\rangle) = (\varphi(A_1,B_1) \preceq \varphi(A_2,B_2))$. Then φ is a bijection and, by Lemma 4.46, $\mathcal{B}(X,Y,I)$ and $\mathbf{V}$ are isomorphic. We will need the following claims.

Claim A. $\gamma(x,\bigvee_{j\in J} a_j) = \bigvee_{j\in J}\gamma(x,a_j)$, $\mu(y,\bigvee_{j\in J} a_j) = \bigwedge_{j\in J}\mu(y,a_j)$ *for each* $x\in X$, $y\in Y$, $\{a_j \mid j\in J\} \subseteq L$, *i.e.* $\gamma(x,\_) : L \to V$ *are complete lattice* $\bigvee$*-morphisms and* $\mu(y,\_) : L \to V$ *are dual complete lattice* $\bigwedge$*-morphisms.*

Proof of Claim A. The $\{0,1\}$-infimal density of $\mu(Y\times L)$ implies that $\gamma(x,\bigvee_{j\in J} a_j) = \bigwedge_{\langle y,b\rangle\in X}\mu(y,b)$ for some $M \subseteq Y\times L$. Hence,

$\gamma(x, \bigvee_{j\in J} a_j) \leq \mu(y,b)$ which implies $(\bigvee_{j\in J} a_j) \otimes b \leq I(x,y)$, for each $\langle y,b\rangle \in M$. From $a_j \otimes b \leq (\bigvee_{j\in J} a_j) \otimes b$ we have $a_j \otimes b \leq I(x,y)$, i.e. $\gamma(x,a_j) \leq \mu(y,b)$ for every $j \in J$. This implies $\bigvee_{j\in J} \gamma(x,a_j) \leq \bigwedge_{\langle y,b\rangle\in M} \mu(y,b) = \gamma(x, \bigvee_{j\in J} a_j)$.

Conversely, the $\{0,1\}$-infimal density of $\mu(Y \times L)$ again implies the existence of some $M \subseteq Y \times L$ such that $\bigvee_{j\in J} \gamma(x,a_j) = \bigwedge_{\langle y,b\rangle\in M} \mu(y,b)$. That means that for each $j \in J$, $\langle y,b\rangle \in M$ we have $\gamma(x,a_j) \leq \mu(y,b)$, i.e. $a_j \otimes b \leq I(x,y)$. This implies $\bigvee_{j\in J}(a_j \otimes b) \leq I(x,y)$ and, by $\bigvee_{j\in J}(a_j \otimes b) = \bigvee_{j\in J} a_j \otimes b$, furthermore $\bigvee_{j\in J} a_j \otimes b \leq I(x,y)$, i.e. $\gamma(x, \bigvee_{j\in J} a_j) \leq \mu(y,b)$ for each $\langle y,b\rangle \in M$, thus $\gamma(x, \bigvee_{j\in J} a_j) \leq \bigwedge_{\langle y,b\rangle\in M} \mu(y,b) = \bigvee_{j\in J} \gamma(x,a_j)$, proving $\gamma(x, \bigvee_{j\in J} a_j) = \bigvee_{j\in J} \gamma(x,a_j)$.

$\mu(y, \bigvee_{j\in J} a_j) = \bigwedge_{j\in J} \mu(y,a_j)$ may be proved analogously using the $\{0,1\}$-supremal density of $\gamma(X \times L)$. *Q.E.D.*

Claim B. *For any $A \subseteq V$ and $u \in V$ we have $\bigwedge_{v\in A}(u \preceq v) = (u \preceq (\bigwedge_{v\in A} v))$ and $\bigwedge_{v\in A}(v \preceq u) = ((\bigvee_{v\in A} v) \preceq u)$.*

Proof of Claim B. Consider the crisp $\mathbf{L}$-set A' in V corresponding to A (i.e. $A'(v) = 1$ if $v \in A$ and $A'(v) = 0$ otherwise). We show $\bigwedge_{v\in A}(u \preceq v) = (u \preceq \bigwedge_{v\in A} v)$. Denote $v^* = \bigwedge_{v\in A} v$. Since $\bigwedge$ in the previous formula is derived from inf in $\mathbf{V}$, we have $(\inf(A'))(v^*) = ((\mathcal{L}(A'))(v^*)) \wedge ((\mathcal{UL}(A'))(v^*)) = 1$, therefore also $(\mathcal{UL}(A'))(v^*) = 1$. We have $(\mathcal{L}(A'))(u) = \bigwedge_{u'\in V} A'(u') \to (u \preceq u') = \bigwedge_{v\in A}(u \preceq v)$. On the other hand, $(\mathcal{LUL}(A'))(u) = \bigwedge_{u'\in V}(\mathcal{UL}(A'))(u') \to (u \preceq u') \leq (\mathcal{UL}(A'))(v^*) \to (u \preceq v^*) = (u \preceq v^*) = (u \preceq \bigwedge_{v\in A} v)$. Applying $\mathcal{L}(A') = \mathcal{LUL}(A')$ we get $\bigwedge_{v\in A}(u \preceq v) \leq (u \preceq \bigwedge_{v\in A} v)$. Conversely, $(u \preceq \bigwedge_{v\in A} v) \leq \bigwedge_{v\in A}(u \preceq v)$ holds iff $(u \preceq \bigwedge_{v\in A} v) \leq (u \preceq v)$ for any $v \in A$ which is true since $(u \preceq \bigwedge_{v\in A} v) = (u \preceq \bigwedge_{v\in A} v) \otimes (\bigwedge_{v\in A} v \preceq v) \leq (u \preceq v)$.

The second equality, i.e. $\bigwedge_{v\in A}(v \preceq u) = ((\bigvee_{v\in A} v) \preceq u)$ can be proved analogously: put $v^* = \bigvee_{v\in A} v$; one has $(\sup(A'))(v^*) = 1$, and so also $(\mathcal{LU}(A'))(v^*) = 1$; one can verify $(\mathcal{U}(A'))(u) = \bigwedge_{v\in A}(v \preceq u)$ and $(\mathcal{ULU}(A'))(u) \leq (\bigvee_{v\in A} v \preceq u)$; using $\mathcal{U}(A') = \mathcal{ULU}(A')$ we get $\bigwedge_{v\in A}(v \preceq u) \leq (\bigvee_{v\in A} v \preceq u)$; conversely, we have $(\bigvee_{v\in A} v \preceq u) = (v \preceq \bigvee_{v\in A} v) \otimes (\bigvee_{v\in A} v \preceq u) \leq (v \preceq u)$ for any $v \in A$ yielding $(\bigvee_{v\in A} v \preceq u) \leq \bigwedge_{v\in A}(v \preceq u)$. *Q.E.D.*

Claim C. $(a \to b) \leq (\gamma(x,a) \preceq \gamma(x,b))$ *for every $x \in X$ and $a, b \in L$.*

Proof of Claim C. From the $\{0,1\}$-infimal density of $\mu(Y \times L)$ it follows that $\gamma(x,b) = \bigwedge_{\gamma(x,b)\leq\mu(y,b')} \mu(y,b')$. Therefore, we have to show $(a \to b) \leq (\gamma(x,a) \preceq (\bigwedge_{\gamma(x,b)\leq\mu(y,b')} \mu(y,b')))$. Putting $A = \{\mu(y,b') \mid \gamma(x,b) \leq \mu(y,b')\}$, Claim B yields $(\gamma(x,a) \preceq (\bigwedge_{\gamma(x,b)\leq\mu(y,b')} \mu(y,b'))) = \bigwedge_{\gamma(x,b)\leq\mu(y,b')}(\gamma(x,a) \preceq (\mu(y,b')))$.

To prove $(a \to b) \leq (\gamma(x,a) \preceq \gamma(x,b))$ we therefore have to show $(a \to b) \leq \bigwedge_{\gamma(x,b)\leq\mu(y,b')}(\gamma(x,a) \preceq \mu(y,b'))$, i.e. we have to show that $(a \to b) \leq (\gamma(x,a) \preceq \mu(y,b'))$ is true for any y and b' such that $\gamma(x,b) \leq \mu(y,b')$. By assumption, $(\gamma(x,a) \preceq \mu(y,b')) = (a \otimes b') \to I(x,y)$. If $\gamma(x,b) \leq \mu(y,b')$ then using the assumption we get $1 = \gamma(x,b) \to \mu(y,b') = (b \otimes b') \to I(x,y)$, i.e. $b \otimes b' \leq I(x,y)$. We therefore have $a \to b \leq (\gamma(x,a) \preceq \mu(y,b'))$ iff $a \to b \leq (a \otimes b') \to I(x,y)$ iff (we now use $(a \otimes b') \to I(x,y) = a \to (b' \to I(x,y))$ and adjointness) $a \otimes (a \to b) \otimes b' \leq I(x,y)$ which is true since $a \otimes (a \to b) \otimes b' \leq b \otimes b'$ and $b \otimes b' \leq I(x,y)$ (by the above observation). *Q.E.D.*

Claim D. $I(x,y) = \bigvee_{\gamma(x,a)\leq\mu(y,b)} a \otimes b$.

Proof of Claim D. The inequality $I(x,y) \geq \bigvee_{\gamma(x,a)\leq\mu(y,b)} a \otimes b$ follows immediately. For $a = I(x,y)$, $b = 1$ we have $a \otimes b = I(x,y) \otimes 1 \leq I(x,y)$, hence $\gamma(x, I(x,y)) \leq \mu(y,1)$, thus the equality holds. *Q.E.D.*

Define the mapping $\varphi : \mathcal{B}(X,Y,I) \to V$ by

$$\varphi(A,B) = \bigvee_{x\in X} \gamma(x, A(x)) \tag{5.41}$$

for each $\langle A,B\rangle \in \mathcal{B}(X,Y,I)$.

We prove the existence of an inverse mapping ψ of φ. Define $\psi : V \to \mathcal{B}(X,Y,I)$ by

$$\psi(v) = \langle A,B\rangle, \qquad \text{where } A(x) = \bigvee_{\gamma(x,a)\leq v} a,\ B(y) = \bigwedge_{\mu(y,b)\geq v} b \tag{5.42}$$

for each $v \in V$, and every $x \in X$, $y \in Y$. First, we show that for each $v \in V$, $\psi(v)$ is a fixed point of $\mathcal{B}(X,Y,I)$, i.e. $A^{\uparrow} = B$ and $B^{\downarrow} = A$. We show only $B^{\downarrow} = A$, the second case may be proved symmetrically. By Claim D we have

$$B^{\downarrow}(x) = \bigwedge_{y\in Y} B(y) \to I(x,y) = \bigwedge_{y\in Y}\Big(\bigwedge_{\mu(y,b)\geq v} b \to \bigvee_{\gamma(x,a)\leq\mu(y,b)} a \otimes b\Big).$$

We show $A(x) \leq B^{\downarrow}(x)$. We have

$$\bigvee_{\gamma(x,a)\leq v} a \otimes \bigwedge_{\mu(y,b)\geq v} b \leq \bigvee_{\gamma(x,a)\leq v\leq\mu(y,b)} a \otimes b \leq \bigvee_{\gamma(x,a)\leq\mu(y,b)} a \otimes b,$$

i.e.

$$A(x) = \bigvee_{\gamma(x,a)\leq v} a \leq \bigwedge_{\mu(y,b)\geq v} \to \bigvee_{\gamma(x,a)\leq\mu(y,b)} a \otimes b$$

holds for each $y \in Y$, hence also

$$A(x) = \bigvee_{\gamma(x,a)\leq v} a \leq \bigwedge_{y\in Y}\Big(\bigwedge_{\mu(y,b)\geq v} b \to \bigvee_{\gamma(x,a)\leq\mu(y,b)} a\otimes b\Big) = B^{\downarrow}(x)$$

holds. We now show the equality $A(x) = B^{\downarrow}(x)$ as follows. Suppose there is an $a \in L$ such that for each $y \in Y$ it holds

$$a \leq \bigvee_{\mu(y,b)\geq v} b \to I(x,y) \tag{5.43}$$

(i.e. a is a lower bound) and show that $a \leq \bigvee_{\gamma(x,a)\leq v} a = A(x)$ (i.e. $A(x)$ is the infimum, i.e. $B^{\downarrow}(x)$). (5.43) holds iff

$$a \otimes \bigvee_{\mu(y,b)\geq v} b \leq I(x,y),$$

i.e. by $a\otimes\bigvee_{\mu(y,b)\geq v} b = \bigvee_{\mu(y,b)\geq v}(a\otimes b)$ we get that for each b such that $\mu(y,b) \geq v$ it holds $a\otimes b \leq I(x,y)$. The last fact implies that for each b such that $\mu(y,b) \geq v$ it holds $\gamma(x,a) \leq \mu(y,b)$ which holds for each $y \in Y$. From the $\{0,1\}$-infimal density of $\mu(Y \times L)$ it follows that $v = \bigwedge_{v\leq\mu(y,b)} \mu(y,b)$, and hence $\gamma(x,a) \leq v$ which implies $a \leq \bigvee_{\gamma(x,a')\leq v} a' = A(x)$. We have proved $A = B^{\downarrow}$.

Next we show that $\varphi\circ\psi = id_{\mathcal{B}(X,Y,I)}$ and $\psi\circ\varphi = id_V$. For each $v \in V$ we have by Claim A and the $\{0,1\}$-supremal density of $\gamma(X \times L)$

$$\begin{aligned}
\psi\circ\varphi(v) &= \varphi(A,B) = \bigvee_{x\in X}\gamma(x,A(x)) = \\
&= \bigvee_{x\in X}\gamma(x,\bigvee_{\gamma(x,a)\leq v} a) = \bigvee_{x\in X}\bigvee_{\gamma(x,a)\leq v}\gamma(x,a) = \\
&= \bigvee_{\gamma(x,a)\leq v}\gamma(x,a) = v,
\end{aligned}$$

i.e. $\psi\circ\varphi(v) = v$. Consider now $\varphi\circ\psi(A,B)$ for $\langle A,B\rangle \in \mathcal{B}(X,Y,I)$. First we show

$$\bigvee_{x\in X}\gamma(x,A(x)) = \bigwedge_{y\in Y}\mu(y,B(y)). \tag{5.44}$$

The inequality $\bigvee_{x\in X}\gamma(x,A(x)) \leq \bigwedge_{y\in Y}\mu(y,B(y))$ is inferred from the fact that for every $x \in X$, $y \in Y$ we have $\gamma(x,A(x)) \leq \mu(y,B(y))$ which follows from Claim D as here: $\gamma(x,A(x)) \leq \mu(y,B(y))$ hold-s iff $A(x)\otimes B(y) \leq I(x,y)$ iff $A(x) \leq B(y) \to I(x,y)$ which hold-s because of $A(x) = \bigwedge_{y'\in Y}(B(y') \to I(x,y)) \leq B(y) \to I(x,y)$. To get the equality (5.44), denote $v = \varphi(A,B) = \bigvee_{x\in X}\gamma(x,A(x))$.

We show that $\bigwedge_{y\in Y}\mu(y,B(y)) = v$. From the $\{0,1\}$-infimal density of $\mu(Y\times L)$ we have clearly $v = \bigwedge_{\mu(y,b)\geq v}\mu(y,b)$. We show that for each y,b such that $\mu(y,b)\geq v$ it holds $b\leq B(y)$. Indeed, if $\mu(y,b)\geq v$ then clearly $\mu(y,b)\geq\gamma(x,A(x))$ for all $x\in X$. If $b\leq B(y)$ is not the case then consider $b\vee B(y)$. For each $x\in X$ we have $\mu(y,b)\geq\gamma(x,A(x))$, $\mu(y,B(y))\geq\gamma(x,A(x))$, hence, by Claim A, $\mu(y,b\vee B(y)) = \mu(y,b)\wedge\mu(y,B(y))\geq\gamma(x,A(x))$. This implies $A(x)\otimes B(y)\leq A(x)\otimes(b\vee B(y))\leq I(x,y)$, i.e. $b\vee B(y)\leq A(x)\to I(x,y)$ for each $x\in X$, i.e. $b\vee B(y)\leq\bigwedge_{x\in X}A(x)\to I(x,y) = B(y)$, i.e. $b\leq B(y)$, a contradiction. Furthermore, from $b\leq B(y)$ it follows by Claim A that $\mu(y,B(y))\leq\mu(y,b)$. Thus, from $v\leq\mu(y,b)$ it follows $\mu(y,B(y))\leq\mu(y,b)$. We conclude

$$v = \bigvee_{x\in X}\gamma(x,A(x))\leq\bigwedge_{y\in Y}\mu(y,B(y))\leq\bigwedge_{\mu(y,b)\geq v}\mu(y,b) = v,$$

i.e. (5.44) holds. We therefore have

$$\begin{aligned}\varphi\circ\psi(A,B) &= \psi(\bigwedge_{y\in Y}\mu(y,B(y))) = \\ &= \Big\langle\{\langle x, \bigvee_{\gamma(x,a)\leq\bigwedge_{y\in Y}\mu(y,B(y))} a\rangle\mid x\in X\}, \\ &\quad \{\langle y, \bigvee_{\mu(y,b)\geq\bigwedge_{y\in Y}\mu(y,B(y))} b\rangle\mid y\in Y\}\Big\rangle.\end{aligned}$$

As $\varphi\circ\psi(A,B)\in\mathcal{B}(X,Y,I)$, it suffices to show that

$$\bigvee_{\gamma(x,a)\leq\bigwedge_{y\in Y}\mu(y,B(y))} a = A(x).$$

From (5.44) we have $\gamma(x,A(x))\leq\bigwedge_{y\in Y}\mu(y,B(y))$ and therefore

$$\bigvee_{\gamma(x,a)\leq\bigwedge_{y\in Y}\mu(y,B(y))} a \geq A(x).$$

Conversely, if $\gamma(x,a)\leq\bigwedge_{y\in Y}\mu(y,B(y))$, then $\gamma(x,a)\leq\mu(y,B(y))$ for each $y\in Y$, i.e. $a\otimes B(y)\leq I(x,y)$, which yields $a\leq B(y)\to I(x,y)$, for each $y\in Y$, and hence $a\leq\bigwedge_{y\in Y}B(y)\to I(x,y) = A(x)$ which implies $\bigvee_{\gamma(x,a)\leq\bigwedge_{y\in Y}\mu(y,B(y))}a\leq A(x)$. We have proved $\varphi\circ\psi(A,B) = \langle A,B\rangle$.

It now suffices to show that $(\langle A_1,B_1\rangle\preceq\langle A_2,B_2\rangle) = (\varphi(A_1,B_1)\preceq\varphi(A_2,B_2))$ for any $\langle A_1,B_1\rangle,\langle A_2,B_2\rangle\in\mathcal{B}(X,Y,I)$. We prove this fact by showing (a) $(\langle A_1, B_1\rangle \preceq \langle A_2, B_2\rangle) \leq (\varphi(A_1, B_1) \preceq \varphi(A_2, B_2))$ and (b) $(u \preceq v) \leq (\psi(u) \preceq \psi(v))$ (for any $u, v \in V$).

(a): By definitions, we have to prove $S(A_1, B_1) \leq (\bigvee_{x\in X} \gamma(x, A_1(x)) \preceq \bigvee_{x\in X} \gamma(x, A_2(x)))$. Denote $v^* = \bigvee_{x\in X} \gamma(x, A_2(x))$. The $\{0,1\}$-infimal density of $\mu(Y \times L)$ yields $v^* = \bigwedge_{v^*\leq\mu(y,b)} \mu(y, b)$. Therefore, we have to show $S(A_1, B_1) \leq (\bigvee_{x\in X} \gamma(x, A_1(x)) \preceq \bigwedge_{v^*\leq\mu(y,b)} \mu(y, b))$. Since, by Claim B, we have $(\bigvee_{x\in X} \gamma(x, A_1(x)) \preceq \bigwedge_{v^*\leq\mu(y,b)} \mu(y, b)) = \bigwedge_{x\in X} \bigwedge_{v^*\leq\mu(y,b)} (\gamma(x, A_1(x)) \preceq \mu(y, b))$, we have to show that $S(A_1, A_2) \leq (\gamma(x, A_1(x)) \preceq \mu(y, b))$ holds for any $x \in X$, $y \in Y$, $b \in L$ such that $v^* \leq \mu(y, b)$. Since $(\gamma(x, A_1(x)) \preceq \mu(y, b)) = (A_1(x) \otimes b) \rightarrow I(x, y))$, we have to show that $S(A_1, A_2) \leq (A_1(x) \otimes b) \rightarrow I(x, y))$, i.e. $(b \otimes A_1(x)) \otimes S(A_1, A_2) \leq I(x, y))$. Since $(b \otimes A_1(x)) \otimes S(A_1, A_2) = b \otimes A_1(x) \otimes \bigwedge_{x\in X}(A_1(x) \rightarrow A_2(x)) \leq b \otimes A_1(x) \otimes (A_1(x) \rightarrow A_2(x)) \leq b \otimes A_2(x)$ it is now sufficient to prove $b \otimes A_2(x) \leq I(x, y))$. However, since $\bigvee_{x\in X} \gamma(x, A_2(x)) \leq \mu(y, b)$, thus also $\gamma(x, A_2(x)) \leq \mu(y, b)$, we have $1 = (\gamma(x, A_2(x)) \preceq \mu(y, b)) = b \otimes A_2(x) \rightarrow I(x, y)$ from which the required inequality $b \otimes A_2(x) \leq I(x, y)$ directly follows.

(b): Denote $\psi(u) = \langle A_u, B_u\rangle$ and $\psi(v) = \langle A_v, B_v\rangle$. We have to show $(u \preceq v) \leq S(A_u, A_v)$, i.e. $(u \preceq v) \leq \bigwedge_{x\in X} A_u(x) \rightarrow A_v(x)$ which holds iff for each $x \in X$ we have $A_u(x) \otimes (u \preceq v) \leq A_v(x)$. By definition of ψ we thus have to show $(\bigvee_{\gamma(x,a)\leq u} a) \otimes (u \preceq v) \leq \bigvee_{\gamma(x,b)\leq v} b$, i.e. $\bigvee_{\gamma(x,a)\leq u} (a \otimes (u \preceq v)) \leq \bigvee_{\gamma(x,b)\leq v} b$ which holds iff $(a \otimes (u \preceq v)) \leq \bigvee_{\gamma(x,b)\leq v} b$ for each $a \in L$ such that $\gamma(x, a) \leq u$. The last inequality certainly holds provided $\gamma(x, a \otimes (u \preceq v)) \leq v$ which we are now going to prove: First, we show $\gamma(x, a \otimes (u \preceq v)) \leq \gamma(x, (\gamma(x, a) \preceq v))$. $\gamma(x, a) \leq u$ yields $(u \preceq v) \leq (\gamma(x, a) \preceq v)$, and so $a \otimes (u \preceq v) \leq a \otimes (\gamma(x, a) \preceq v)$. By Claim C we therefore have $1 = a \otimes (u \preceq v) \rightarrow a \otimes (\gamma(x, a) \preceq v) \leq (\gamma(x, a \otimes (u \preceq v)) \preceq \gamma(x, a \otimes (\gamma(x, a) \preceq v)))$, hence $\gamma(x, a \otimes (u \preceq v)) \leq \gamma(x, a \otimes (\gamma(x, a) \preceq v))$. Now, the $\{0,1\}$-infimal density of $\mu(Y \times L)$ implies that $v = \bigwedge_{v\leq\mu(y,b)} \mu(y, b)$. Therefore, to show $\gamma(x, a \otimes (u \preceq v)) \leq v$ it is sufficient to show that $\gamma(x, a \otimes (\gamma(x, a) \preceq v)) \leq \mu(y, b)$ for any y and b such that $v \leq \mu(y, b)$. The required inequality is equivalent to $(\gamma(x, a \otimes (\gamma(x, a) \preceq v)) \preceq \mu(y, b)) = 1$ which is equivalent (using $(\gamma(x, a \otimes (\gamma(x, a) \preceq v)) \preceq \mu(y, b)) = (a \otimes b \otimes (\gamma(x, a) \preceq v)) \rightarrow I(x, y))$ to $(a \otimes b \otimes (\gamma(x, a) \preceq v)) \rightarrow I(x, y) = 1$, i.e. to $(\gamma(x, a) \preceq v) \leq (a \otimes b) \rightarrow I(x, y)$. However, the last inequality is true: $v \leq \mu(y, b)$ implies $(\gamma(x, a) \preceq v) \leq (\gamma(x, a) \preceq \mu(y, b)) = (a \otimes b) \rightarrow I(x, y)$.

We proved that $\langle\langle \mathcal{B}(X, Y, I), \approx\rangle, \preceq\rangle$ and $\mathbf{V}$ are isomorphic.

To complete the proof we show that any completely lattice $\mathbf{L}$-ordered set $\mathbf{V}$ is isomorphic to $\mathcal{B}(V, V, \preceq)$. By what we just verified, it is enough to show that there are mappings $\gamma : V \times L \rightarrow V$ and $\mu : V \times L \rightarrow V$ with the required properties. For $a \in L$ and $x \in V$, let $\gamma(x,a)$ be the (unique) element $x^* \in V$ such that $(\sup(\{{}^{a}/x\}))(x^*) = 1$; $\mu(x,a)$ be the (unique) element $x^* \in V$ such that $(\inf(\{{}^{a}/x\}))(x^*) = 1$. Since $\gamma(x,1) = x$ and $\mu(x,1) = x$, we have $\gamma(V,L) = V$ and $\mu(V,L) = V$, therefore $\gamma(V,L) = V$ and $\mu(V,L) = V$ are $\{0,1\}$-supremally dense and $\{0,1\}$-infimally dense in $\mathbf{V}$. We show $((a \otimes b) \to (x \preceq y)) = (\gamma(x,a) \preceq \mu(y,b))$ by proving both of the inequalities.

"$\leq$": Denote $x^* = \gamma(x,a)$, $y^* = \mu(y,b)$. We have to show $((a \otimes b) \to (x \preceq y)) \leq (x^* \preceq y^*)$. By definition, $(\sup(\{{}^{a}/x\}))(x^*) = 1$ and $(\inf(\{{}^{b}/y\}))(y^*) = 1$. One easily verifies that $\mathcal{U}(\{{}^{a}/x\})(x') = a \to (x \preceq x')$. Therefore, $(\mathcal{LU}(\{{}^{a}/x\}))(x^*) = \bigwedge_{x' \in X}((a \to (x \preceq x')) \to (x^* \preceq x')) \leq (a \to (x \preceq y)) \to (x^* \preceq y)$. Since, by definition of sup, $(\mathcal{LU}(\{{}^{a}/x\}))(x^*) = 1$, we conclude $(a \to (x \preceq y)) \leq (x^* \preceq y)$. Similarly, $1 = (\mathcal{UL}(\{{}^{b}/y\}))(y^*) = \bigwedge_{x' \in X}((b \to (x' \preceq y)) \to (x' \preceq y^*)) \leq (b \to (x^* \preceq y)) \to (x^* \preceq y^*)$, thus $(b \to (x^* \preceq y)) \leq (x^* \preceq y^*)$. We thus have

$$\begin{aligned}(a \otimes b) \to (x \preceq y) &= b \to (a \to (x \preceq y)) \leq \\ \leq\ & b \to (x^* \preceq y) \leq (x^* \preceq y^*).\end{aligned}$$

"$\geq$": $(x^* \preceq y^*) \leq (a \otimes b) \to (x \preceq y)$ iff $a \otimes b \otimes (x^* \preceq y^*) \leq (x \preceq y)$. Now, $(\inf(\{{}^{b}/y\}))(y^*) = 1$ and $(\sup(\{{}^{a}/x\}))(x^*) = 1$ yield $(\mathcal{L}(\{{}^{b}/y\}))(y^*) = b \to (y^* \preceq y) = 1$ and $(\mathcal{U}(\{{}^{a}/x\}))(x^*) = a \to (x^* \preceq x) = 1$, respectively. Therefore,

$$\begin{aligned}& a \otimes b \otimes (x^* \preceq y^*) = \\ =\ & a \otimes (a \to (x \preceq x^*)) \otimes (x^* \preceq y^*) \otimes b \otimes (b \to (y^* \preceq y)) \leq \\ \leq\ & (x \preceq x^*) \otimes (x^* \preceq y^*) \otimes (y^* \preceq y) \leq (x \preceq y)\end{aligned}$$

proving the inequality.

The proof of Theorem 5.63 is complete. □

Remark 5.64 An inspection of the proof of Theorem 5.63 shows an alternative proof of Theorem 5.9: Part (1) of Theorem 5.9 follows directly from part (1) of Theorem 5.63 by Theorem 4.55 and Remark 4.56. Part (2) of Theorem 5.9: Consider any $\mathcal{B}(X,Y,I)$ and a complete lattice $\mathbf{V}$. If $\mathcal{B}(X,Y,I)$ and $\mathbf{V}$ are isomorphic as lattices then the order $\leq$ of $\mathbf{V}$ can clearly be (in an obvious way) extended to an $\mathbf{L}$-order $\preceq$ on V in such a way that $\mathcal{B}(X,Y,I)$ and $\langle\langle V, \approx\rangle, \preceq\rangle$ are isomorphic as $\mathbf{L}$-ordered sets (here, $\approx$ on V is uniquely determined by $(u \approx v) = (u \preceq v) \wedge (v \preceq u)$, see Lemma 4.45). Part (2) of Theorem 5.63 then yields the mappings γ and μ which satisfy the density conditions and $((a \otimes b) \to I(x,y)) = (\gamma(x,a) \preceq \mu(y,b))$. The required condition $a \otimes b \leq I(x,y)$ iff $\gamma(x,a) \leq \mu(y,b)$ now follows from

$((a \otimes b) \to I(x,y)) = (\gamma(x,a) \preceq \mu(y,b))$. Conversely, if γ and μ satisfying the conditions of part (2) of Theorem 5.9 exist, then the order preserving bijections $\varphi : \mathcal{B}(X,Y,I) \to V$ and $\psi : V \to \mathcal{B}(X,Y,I)$ can be constructed as in the proof of Theorem 5.63 (to verify this claim, just go through the proof).

5.5 BIBLIOGRAPHICAL REMARKS

Section 5.1: Mathematical theory of concept lattices and formal concept analysis goes back to the 1980s and is due to Rudolf Wille [203]; foundations of formal concept analysis can be found in [81]. Theory of concept lattices is based on the concept of a Galois connection which is a useful concept playing an important role in reasoning and mathematics (see e.g. [155, 41]); concept lattices are in a sense Galois connections plus the conceptual interpretation. Although Galois connections were studied long before concept lattices, the conceptual interpretation opened a new view. Nowadays, there is a large literature on concept lattices and their applications in data analysis.

The first paper on fuzzy concept lattices is [47]; the structure of truth values is a complete lattice L equipped with complementation $'$; however, the basic operators $^{\uparrow}$ and $^{\downarrow}$ do not satisfy $A^{\uparrow\downarrow\uparrow} = A^{\uparrow}$ and $B^{\downarrow\uparrow\downarrow} = B^{\downarrow}$ (this is because adjointness is missing). Independently, [167] and [13, 36] (which is a bit more general; [36] is a collection of journal papers [14, 15, 17, 23, 24, 28, 27]) are further approaches to fuzzy concept lattices. Both use completed residuated lattices as structures of truth values (note that they are called L-fuzzy algebras in [167]). Both [167] and [36] focus on different topics; [36] goes more into nonclassical issues. Main theorem of fuzzy concept lattices for crisp order appears in [167] (in a bit special case; **L**-equivalences on X and Y are not considered), and [15, 23] (two independent proofs). Further literature on fuzzy concept lattices is [48, 49, 73].

Section 5.2 is based on [17, 24].

Section 5.3 contains results from [14, 16, 23]; for $\mathbf{L} = \mathbf{2}$, the presented results yield classical results on Galois connections [41, 155].

Section 5.4 is based on [27]. Let us comment on the historical development of Theorem 5.63. It was Birkhoff (see e.g. [41]) who observed that for any (bivalent) relation I between X and Y, $\mathcal{B}(X,Y,I)$ is a complete lattice. Note that Birkhoff calls the induced mappings $^{\uparrow}$ and $^{\downarrow}$ polarities, cf. [41, p. 122]. The characterization of $\mathcal{B}(X,Y,I)$ for special cases (I is a quasiorder or an order) has been pursued by Banaschewski [5] and Schmidt [183]. The general case when I is an arbitrary binary relation is due to Wille [203]. Theorem 5.63 is then a further improvement of these results. A moment's inspection also shows that Wille's theorem is a special case of Theorem 5.63 for $\mathbf{L} = \mathbf{2}$ (i.e. for $\mathbf{L}$ being the two-element Boolean algebra).

Chapter 6

COMPOSITION AND DECOMPOSITION OF FUZZY RELATIONS

From causes which appear similar, we expect similar effects. This is the sum total of all our experimental conclusions.
—David Hume (An Enquiry Concerning Human Understanding, 1758)

6.1 COMPOSITION OF FUZZY RELATIONS

We are going to introduce is compositions of binary fuzzy relations. The basic situation is this: Given an **L**-relation R between X and Y and an **L**-relation S between Y and Z, it might be desirable to obtain from R and S a binary **L**-relation $R{*}S$ between X and Z. $R{*}S$ will be called the composition of R and S. In this general setting, a composition of binary **L**-relations between X and Y, and Y and Z is therefore a mapping $* : L^{X\times Y} \times L^{Y\times Z} \to L^{X\times Z}$, i.e. a mapping assigning to any $R \in L^{X\times Y}$ and $S \in L^{Y\times Z}$ their composition $R * S \in L^{X\times Z}$ (all X, Y, and Z may be equipped with **L**-equalities). We are interested in cases where $*$ can be defined by logical formulas, i.e. in cases where the composition can be described verbally.

Example 6.1 Let X be a set of patients, Y be a set of symptoms (of diseases), and Z be a set of diseases. Let R be a fuzzy relation between X and Y, S be a fuzzy relation between Y and Z. R may represent results of a medical examination, i.e. for a patient $x \in X$ and a symptom $y \in Y$ (e.g. a headache), $R(x, y)$ is the truth degree to which x has a headache (notice that typically, R is a noncrisp fuzzy relation). Similarly, S may represent expert knowledge (that can be found in medical literature), i.e. for a symptom $y \in Y$ and a disease $z \in Z$, $S(y, z)$ is the truth value to which y is a symptom of z (again, S is a typical fuzzy relation). A natural question is whether one can find out which patients have which diseases etc. From this point of view we have to consider "patient x has a disease z iff for each symptom y it holds that x has y iff y is a symptom of z". The quoted sentence gives rise to a fuzzy relation $R * S$ between X and Z which is a composition of R and S verbally described by the quoted sentence.

The situation can be approached formally as follows. Let $\mathcal{J}_{\text{comp}}$ be a many-sorted language with sorts $\mathcal{X}, \mathcal{Y}, \mathcal{Z}$, of type $\langle R, F, \sigma \rangle$ with binary $r, s \in R$ such that $\sigma(r) = \mathcal{X}\mathcal{Y}$ and $\sigma(s) = \mathcal{Y}\mathcal{Z}$, let $\mathtt{x}, \mathtt{y}, \mathtt{z}$ be variables of sorts $\mathcal{X}, \mathcal{Y}, \mathcal{Z}$, respectively. Let $\varphi(\mathtt{x}, \mathtt{z})$ be a formula (which will represent the verbal description of the composite relation and should thus contain r and s). The *composition* of **L**-relations defined by φ is defined as follows. For an **L**-relation R between $\langle X, \approx^X \rangle$ and $\langle Y, \approx^Y \rangle$, and an **L**-relation S between $\langle Y, \approx^Y \rangle$ and $\langle Z, \approx^Z \rangle$, consider an **L**-structure $\mathbf{M}$ given by $r^{\mathbf{M}} = R$, $s^{\mathbf{M}} = S$, $\approx^{\mathbf{M}}_{\mathcal{X}}$ be $\approx^X$, $\approx^{\mathbf{M}}_{\mathcal{Y}}$ be $\approx^Y$, and $\approx^{\mathbf{M}}_{\mathcal{Z}}$ be $\approx^Z$ (φ could in principle contain other symbols but we will not consider such a situation). Recalling the end of Section 3.2, φ induces a binary **L**-relation $\varphi^{\mathbf{M}}$ between X and Z. Note that $\varphi^{\mathbf{M}}$ is defined by

$$\varphi^{\mathbf{M}}(x, z) = \|\varphi\|_{\mathbf{M},v}$$

where $x \in X$, $z \in Z$, and v is a valuation such that $v(\mathtt{x}) = x$, $v(\mathtt{z}) = z$.

Definition 6.2 Binary **L**-relation $\varphi^{\mathbf{M}}$ defined above is called a *φ-composition* of R and S.

Remark 6.3 We will freely use the term φ-composition in two ways: either as referring to the particular fuzzy relation obtained from R and S, or as referring to the operation that assigns to any two fuzzy relations R and S the resulting fuzzy relation (their φ-composition in the first sense).

The question of compatibility of the φ-product is answered by the following lemma.

Lemma 6.4 *The φ-product of R and S is a $(|\varphi|_{\mathtt{x}}, |\varphi|_{\mathtt{z}})$-compatible* **L**-*relation between $\langle X, \approx^X \rangle$ and $\langle Z, \approx^Z \rangle$.*

Proof. The assertion follows directly from the many-sorted version of Lemma 3.83. □

In what follows we will be interested in the following compositions.

Definition 6.5 A φ-composition is called

a $\circ$-*composition* (circlet composition) if $\varphi = \varphi_\circ$ where

$$\varphi_\circ \text{ is } (\exists \mathrm{y})(r(\mathrm{x},\mathrm{y}) \otimes s(\mathrm{y},\mathrm{z}));$$

a $\lhd$-*composition* (subproduct composition) if $\varphi = \varphi_\lhd$ where

$$\varphi_\lhd \text{ is } (\forall \mathrm{y})(r(\mathrm{x},\mathrm{y}) \Rightarrow s(\mathrm{y},\mathrm{z}));$$

a $\rhd$-*composition* (superproduct composition) if $\varphi = \varphi_\rhd$ where

$$\varphi_\rhd \text{ is } (\forall \mathrm{y})(s(\mathrm{y},\mathrm{z}) \Rightarrow r(\mathrm{x},\mathrm{y}));$$

a $\Box$-*composition* (square composition) if $\varphi = \varphi_\Box$ where

$$\varphi_\Box \text{ is } (\forall \mathrm{y})(r(\mathrm{x},\mathrm{y}) \Leftrightarrow s(\mathrm{y},\mathrm{z})).$$

For $* \in \{\circ, \lhd, \rhd, \Box\}$ we will denote the $*$-composition of R and S by $R * S$, thus we have $R \circ S$, $R \lhd S$, $R \rhd S$, and $R\Box S$; $*$-composition is also called a $*$-product.

Remark 6.6 Let us show the direct formulas for $\circ$-, $\lhd$-, $\rhd$-, $\Box$-products. One can immediately see that we have

$$\begin{aligned}
(R \circ S)(x,z) &= \bigvee_{y \in Y} (R(x,y) \otimes S(y,z)),\\
(R \lhd S)(x,z) &= \bigwedge_{y \in Y} (R(x,y) \rightarrow S(y,z)),\\
(R \rhd S)(x,z) &= \bigwedge_{y \in Y} (S(y,z) \rightarrow R(x,y)),\\
(R \Box S)(x,z) &= \bigwedge_{y \in Y} (R(x,y) \leftrightarrow S(y,z))
\end{aligned}$$

for all $x \in X$, $z \in Z$.

Remark 6.7 Example 6.1 enables us to elucidate the natural meaning of the above four types of composition. $(R \circ S)(x,z)$ is the truth degree to which there is a symptom y such that patient x has symptom y and y is a symptom of disease z; $(R \lhd S)(x,z)$ is the truth degree to which it is true that if x has a symptom y then y is a symptom of z; $(R \rhd S)(x,z)$ is the truth degree to which it is true that patient x has all symptoms of disease z; $(R\Box S)(x,z)$ is the truth degree to which it is true that patient x has exactly the symptoms of disease z.

Theorem 6.8 (products and compatibility) *If R is an $\mathbf{L}$-relation between $\langle X, \approx^X\rangle$ and $\langle Y, \approx^Y\rangle$, and S is an $\mathbf{L}$-relation between $\langle Y, \approx^Y\rangle$ and $\langle Z, \approx^Z\rangle$, then each of $R \circ S$, $R \lhd S$, $R \rhd S$, and $R \Box S$ is an $\mathbf{L}$-relation between $\langle X, \approx^X\rangle$ and $\langle Z, \approx^Z\rangle$.*

Proof. By definition, we have to show that $R * S$ is $(1,1)$-compatible with $\langle X, \approx^X\rangle$ and $\langle Z, \approx^Z\rangle$. However, this follows immediately from Lemma 6.4 observing that for each φ_* $(* \in \{\circ, \lhd, \rhd, \Box\})$ we have $|\varphi_*|_x = 1$ and $|\varphi_*|_z = 1$. □

In the following we list several properties of the above four compositions. The first issue is that of preserving similarity.

Theorem 6.9 (products and similarity) *We have*

$$\begin{aligned}
(R_1 \approx R_2) \otimes (S_1 \approx S_2) &\leq (R_1 \circ S_1) \approx (R_2 \circ S_2),\\
(R_1 \approx R_2) \otimes (S_1 \approx S_2) &\leq (R_1 \lhd S_1) \approx (R_2 \lhd S_2),\\
(R_1 \approx R_2) \otimes (S_1 \approx S_2) &\leq (R_1 \rhd S_1) \approx (R_2 \rhd S_2),\\
(R_1 \approx R_2) \otimes (S_1 \approx S_2) &\leq (R_1 \Box S_1) \approx (R_2 \Box S_2).
\end{aligned}$$

Proof. The assertion is a direct consequence of (many-sorted version of) Lemma 3.85 observing that for each φ_* $(* \in \{\circ, \lhd, \rhd, \Box\})$ we have $|\varphi_*|_r = 1$ and $|\varphi_*|_s = 1$. □

Remark 6.10 Theorem 6.9 has the following verbal description: if R_1 is similar to R_2 and S_1 is similar to S_2, then $R_1 * S_1$ is similar to $R_2 * S_2$ for any $*$ of the above four compositions.

Theorem 6.11 (products and inverse relations) *We have*

$$\begin{aligned}
(R \circ S)^{-1} &= S^{-1} \circ R^{-1},\\
(R \lhd S)^{-1} &= S^{-1} \rhd R^{-1},\\
(R \rhd S)^{-1} &= S^{-1} \lhd R^{-1},\\
(R \Box S)^{-1} &= S^{-1} \Box R^{-1}.
\end{aligned}$$

Proof. The proof is very easy; we just illustrate the second equality: $(R \lhd S)^{-1}(z,x) = (R \lhd S)(x,z) = \bigwedge_{y \in Y} R(x,y) \rightarrow S(y,z) = \bigwedge_{y \in Y} R^{-1}(y,x) \rightarrow S^{-1}(z,x) = S^{-1} \rhd R^{-1}$. The proofs of the remaining assertions are analogous. □

Remark 6.12 We can easily see that $R\Box S = (R \lhd S) \cap (R \rhd S)$. Moreover, Theorem 6.11 implies that

$$(R \lhd S) = (S^{-1} \rhd R^{-1})^{-1}$$

and $(R \rhd S) = (S^{-1} \lhd R^{-1})^{-1}$. Therefore, one can consider as the basic relational products $\circ$-product and $\lhd$-product, or $\circ$-product and $\rhd$-product.

Theorem 6.13 (products and subsethood) *We have*

$$\begin{aligned} S(R_1, R_2) \otimes S(S_1, S_2) &\le S(R_1 \circ S_1, R_2 \circ S_2), \\ S(R_2, R_1) \otimes S(S_1, S_2) &\le S(R_1 \lhd S_1, R_2 \lhd S_2), \\ S(R_1, R_2) \otimes S(S_2, S_1) &\le S(R_1 \rhd S_1, R_2 \rhd S_2). \end{aligned}$$

Proof. The proof of the first inequality is a consequence of monotony of $\otimes$: The inequality is equivalent (after some immediate steps) to $(R_1 \circ S_1)(x,z) \otimes S(R_1, R_2) \otimes S(S_1, S_2) \le (R_2 \circ S_2)(x,z)$ which is true. Indeed,

$$\begin{aligned} & (R_1 \circ S_1)(x,z) \otimes S(R_1, R_2) \otimes S(S_1, S_2) = \\ = \ & (\bigvee_{y \in Y} R_1(x,y) \otimes S_1(y,z)) \otimes S(R_1, R_2) \otimes S(S_1, S_2) = \\ = \ & \bigvee_{y \in Y} (R_1(x,y) \otimes S_1(y,z) \otimes S(R_1, R_2) \otimes S(S_1, S_2)) \le \\ \le \ & \bigvee_{y \in Y} (R_1(x,y) \otimes (R_1(x,y) \to R_2(x,y)) \otimes S_1(y,z) \otimes \\ & \otimes (S_1(y,z) \to S_2(y,z))) \le \\ \le \ & \bigvee_{y \in Y} R_2(x,y) \otimes S_2(y,z) = (R_2 \circ S_2)(x,z). \end{aligned}$$

The second inequality follows by antitony and monotony of $\to$ in the first and the second argument, respectively; the proof is very similar to the above one and will thus be omitted. The third inequality follows by Remark 6.12 and Lemma 4.3. $\square$

Theorem 6.14 (products and distributivity) *We have*

$$\begin{aligned} (\bigcap_i R_i) \circ S &\subseteq \bigcap_i (R_i \circ S), \\ R \circ (\bigcap_i S_i) &\subseteq \bigcap_i (R \circ S_i), \end{aligned}$$

$$
\begin{aligned}
&(\bigcup_i R_i) \circ S = \bigcup_i (R_i \circ S),\\
&R \circ (\bigcup_i S_i) = \bigcup_i (R \circ S_i),\\
&\bigcup_i (R_i \lhd S) \subseteq (\bigcap_i R_i) \lhd S,\\
&R \lhd (\bigcap_i S_i) = \bigcap_i (R \lhd S_i),\\
&(\bigcup_i R_i) \lhd S = \bigcap_i (R_i \lhd S),\\
&\bigcup_i (R \lhd S_i) \subseteq R \lhd (\bigcup_i S_i),\\
&(\bigcap_i R_i) \rhd S = \bigcup_i (R_i \rhd S),\\
&\bigcup_i (R \rhd S_i) \subseteq R \rhd (\bigcap_i S_i),\\
&\bigcap_i (R_i \rhd S) \subseteq (\bigcup_i R_i) \rhd S,\\
&R \rhd (\bigcup_i S_i) = \bigcap_i (R \rhd S_i).
\end{aligned}
$$

Proof. The first assertion is a consequence of $a \otimes \bigwedge_i b_i \leq \bigwedge_i (a \otimes b_i)$: We have $[(\bigcap_i R_i) \circ S](x, z) = \bigvee_{y \in Y} (\bigwedge_i R_i(x, y)) \otimes S(y, z) \leq \bigvee_{y \in Y} \bigwedge_i (R_i(x, y) \otimes S(y, z)) \leq \bigwedge_i \bigvee_{y \in Y} (R_i(x, y) \otimes S(y, z)) = [\bigcap_i (R_i \circ S)](x, z)$. The proofs of the remaining assertions are similar: The second to fourth assertions are consequences of $a \otimes \bigwedge_i b_i \leq \bigwedge_i (a \otimes b_i)$, $(\bigvee_i a_i) \otimes b = \bigvee_i (a_i \otimes b)$, $a \otimes \bigvee_i b_i = \bigvee_i (a \otimes b_i)$, respectively. The fifth to eighth assertions are consequences of $\bigvee_i (a_i \to b) \leq (\bigwedge_i a_i) \to b$, $a \to (\bigwedge_i b_i) = \bigwedge_i (a \to b_i)$, $(\bigvee_i a_i) \to b = \bigwedge_i (a_i \to b)$, $\bigvee_i (a \to b_i) \leq a \to (\bigvee_i b_i)$, respectively. The last four assertions follow from the fifth to eighth ones by Remark 6.12 and Lemma 4.3. □

Theorem 6.15 (products and associativity) *We have*

$$
\begin{aligned}
&R \circ (S \circ T) = (R \circ S) \circ T,\\
&R \lhd (S \rhd T) = (R \lhd S) \rhd T,\\
&R \lhd (S \lhd T) = (R \circ S) \lhd T,\\
&(R \rhd S) \rhd T = R \rhd (S \circ T).
\end{aligned}
$$

Proof. The proofs are easy: The first equality is a consequence of associativity of $\otimes$; the second one is a consequence of $a \to (c \to b) = c \to$

$(a \to b)$; the third one is a consequence of $a \to (b \to c) = (a \otimes b) \to c$; the fourth one follows from the third one by Remark 6.12 and Lemma 4.3. □

Remark 6.16 (1) Besides the fact that the above four products are useful in system analysis and knowledge extraction as repeatedly demonstrated by Bandler and Kohout [6, 7, 8], there are well-known cases which are in fact instances of these relational products. As an example, let us mention Zadeh's *compositional rule of inference* [220] which is formulated as follows: Given a fuzzy set A in Y and a fuzzy relation S between Y and Z, the fuzzy set $A \circ' S$ in Z obtained by the compositional rule of inference is given by

$$(A \circ' S)(z) = \bigvee_{y \in Y} A(y) \otimes S(y, z).$$

Putting $X = \{x\}$ and defining a fuzzy relation R_A between X and Y by $R_A(x, y) = A(y)$ we have that $(A \circ' S)(z) = (R_A \circ S)(x, z)$ and thus the compositional rule of inference is a special case of a $\circ$-product. Theorem 6.9 implies that we have $(A \approx A') \leq ((A \circ' R) \approx (A' \circ' R))$ and $(R \approx R') \leq ((A \circ' R) \approx (A \circ' R'))$. Now, R represents a linguistic rule base describing a relationship between inputs (elements of Y) and outputs (elements of Z) [220], A represents an (approximately specified) input. The above inequalities say that similar inputs lead to similar outputs (the first inequality) and that similar rule bases have similar behavior (the second inequality).

(2) Another example of a particular case of relational products are the operators $^{\uparrow}$ and $^{\downarrow}$ induced by a binary fuzzy relation I between sets X and Y which are the core of so-called fuzzy concept lattices (see Chapter 5). If we put $Z = \{z\}$ and define for each fuzzy set B in Y a fuzzy relation S_B between Y and Z by $S_B(y, z) = B(y)$, then we immediately have that $B^{\downarrow}(x) = (I \triangleright S_B)(x, z)$. Therefore, $B^{\downarrow}$ is in fact obtained by the $\triangleright$-product; similarly, $A^{\uparrow}$ is obtained by the $\triangleleft$-product.

6.2 DECOMPOSITION: FUZZY RELATIONAL EQUATIONS

In the previous section we considered ways of composition of a binary fuzzy relation R between sets X and Y, and a binary fuzzy relation S between sets Y and Z. The result is a binary fuzzy relation $R * S$ between X and Z and its meaning depends on the type of composition $*$ being used. Each of the fuzzy relations R, S, and $R * S$ represents some knowledge about the domain of interest. It may well happen that instead of R and S we know R and $R * S$ (and $*$, i.e. the type of composition) and want to know S, or we know S and $R * S$ (and $*$) and want to know R. That is, we know the composed fuzzy relation $R * S$ and only one of the fuzzy relations it was composed of. In order to find the second fuzzy relation (S or R depending on whether R or S is known), we need to solve the problem of a decomposition of $R * S$.

In general we suppose that two out of three fuzzy relations R (between X and Y), S (between Y and Z), and T (between X and Z) are given and the remaining one is unknown w.r.t. the constraint

$$R * S \text{ is (at least approximately) equal to } T \tag{6.1}$$

where $*$ denotes an operator (composition operator) assigning to any fuzzy relation R between X and Y and a fuzzy relation S between Y and Z a fuzzy relation $R*S$ between X and Z. Relationships of type (6.1) are known as *fuzzy relational equations*. In the following we restrict ourselves to the case where $*$ is one of the relational compositions discussed in the previous section.

Since there is a large literature on fuzzy relational equations, the main purpose of this section is to present fuzzy relational equations from the point of view of general fuzzy relational systems, to discuss some general aspects, and to illustrate (analytical) methods of solution. References to the literature on fuzzy relational equations are given in the Bibliographical remarks section at the end of this chapter.

Given a fuzzy relational equation like (6.1), the problem is to find the unknown fuzzy relation so that (6.1) holds true; as usual we speak of solving of a fuzzy relational equation. That is, we want that $R * S$ is at least approximately equal to T, the more the better. This requires the use of a suitable measure of equality of $R * S$ and T. To this end we use the equality degree $\approx$, i.e. $R*S \approx T = \bigwedge_{\langle x,z\rangle \in X\times Z} R*S(x,z) \leftrightarrow T(x,z)$. Note that in this case we have that $(R * S \approx T) = 1$ iff $(R * S) = T$.

Remark 6.17 Note that considering also approximate solutions of fuzzy relational equations is only natural in the context of fuzzy relational modeling. Considering only exact solutions (i.e. solutions for which $(R * S \approx T) = 1$) and disregarding very good approximate solutions (e.g. solutions for which $(R * S \approx T) = 0.98$) would be somewhat artificial, especially in situations where the equation results from an expert's description of the situation.

The problem of solving a fuzzy relational equation has a good sense. We know from the previous section that starting from R and S that represent some (elementary) knowledge, composition of fuzzy relations allows us to obtain new knowledge which is linguistically described using R and S. Proceeding in an inverse way, decomposition allows us to obtain knowledge represented e.g. by R provided we know S and $R * S$.

Note that if the unknown fuzzy relation is T, the problem of solving the fuzzy relational equation is trivial: one just takes T to be $R * S$ and then the equality $R * S = T$ is trivially satisfed. Therefore, we concentrate on cases where the unknown is R or S. As usual, we denote a fuzzy relational equation (6.1) where R is unknown by

$$X * S = T$$

whereas by

$$R * X = T$$

we denote a fuzzy relational equation where S is unknown.

The case of a single fuzzy relational equation can be naturally extended to a system of fuzzy relational equations. The interesting situation in considering a system of fuzzy relational equations is of course the case where the single equations of the system are not independent. It is natural to assume that the unknown fuzzy relation is common to all of the fuzzy relational equations of the system, that the type of composition is common to all of the equations of the system, and that the unknown fuzzy relation occurs in the same place in the equations—this will be a general form of systems of fuzzy relational equations we will consider. For instance, we can consider a system consisting of two equations $R_1 \circ X = T_1$ and $R_2 \circ X = T_2$. In general, we consider systems of equations of the form $R_j * X = T_j$ $(j \in J)$ or of the form $X * S_j = T_j$ $(j \in J)$.

We will use the following notation. Fuzzy relational equations as well as systems of fuzzy relational equations will be denoted by $\mathcal{E}$ possibly with indices; the fact that $\mathcal{E}$ is a system consisting of single equations $\mathcal{E}_j$ $(j \in J)$ will be denoted by writing $\mathcal{E} = \{\mathcal{E}_j \mid j \in J\}$. For an equation $\mathcal{E}$ of type $X * S = T$ we denote by $\mathrm{Sol}(\mathcal{E})$ the **L**-set in $L^{X \times Y}$ defined for $R \in L^{X \times Y}$ by

$$[\mathrm{Sol}(\mathcal{E})](R) \;=\; (R * S \approx T);$$

furthermore, we denote

$$\mathrm{Solv}(\mathcal{E}) = \bigvee\{\mathrm{Sol}(\mathcal{E})(R) \mid R \in L^{X \times Y}\}.$$

Similarly, for equations $\mathcal{E}$ of type $R * X = T$ we denote by $\mathrm{Sol}(\mathcal{E})$ the **L**-set in $L^{Y \times Z}$ defined for $S \in L^{Y \times Z}$ by

$$[\mathrm{Sol}(\mathcal{E})](S) \;=\; (R * S \approx T);$$

and put

$$\mathrm{Solv}(\mathcal{E}) = \bigvee\{\mathrm{Sol}(\mathcal{E})(S) \mid R \in L^{Y\times Z}\}.$$

The degree $[\mathrm{Sol}(\mathcal{E})](R)$ is called the degree to which R is a solution of $\mathcal{E}$; $\mathrm{Solv}(\mathcal{E})$ is called the degree of solvability of $\mathcal{E}$; similarly for the type $R * X = T$.

For a system $\mathcal{E} = \{\mathcal{E}_j \mid j \in J\}$ of fuzzy relational equations we denote

$$[\mathrm{Sol}(\mathcal{E})](R) \;=\; \bigwedge_{j\in J}[\mathrm{Sol}(\mathcal{E}_j)](R)$$

and

$$\mathrm{Solv}(\mathcal{E}) \;=\; \bigvee_{R}[\mathrm{Sol}(\mathcal{E})](R).$$

$[\mathrm{Sol}(\mathcal{E})](R)$ and $\mathrm{Solv}(\mathcal{E})$ are the degree to which R is a solution of $\mathcal{E}$ and the degree of solvability of $\mathcal{E}$, respectively.

Thus, for $a \in L$, ${}^{a}\mathrm{Sol}(\mathcal{E})$ is the set of all fuzzy relations which are solutions of $\mathcal{E}$ in degree at least a; elements of ${}^{1}\mathrm{Sol}(\mathcal{E})$ will be called solutions to $\mathcal{E}$.

In the following we concentrate on equations and systems of equations with $\circ$-composition, $\lhd$-composition, and $\rhd$-composition. Since $R \circ S = (S^{-1} \circ R^{-1})^{-1}$, equation $R \circ X = T$ is equivalent to $X^{-1} \circ R^{-1} = T^{-1}$. Moreover, since $R \rhd S = (S^{-1} \lhd R^{-1})^{-1}$, $R \rhd X = T$ is equivalent to $X^{-1} \lhd R^{-1} = T^{-1}$ and $X \rhd S = T$ is equivalent to $S^{-1} \lhd X^{-1} = T^{-1}$. Therefore, the six possible types of equations $X \circ S = T$, $R \circ X = T$, $X \lhd S = T$, $R \lhd X = T$, $X \rhd S = T$, and $R \rhd X = T$ and their systems reduce to only three types $X \circ R = T$, $X \lhd S = T$, and $R \lhd X = T$.

6.2.1 SENSITIVITY ISSUES

The first problem we concentrate on is to check to what extent it is true that if R is a solution of a system of fuzzy relational equations and if R' is similar to R, then R' is also a solution. The following assertion answers this question for systems of fuzzy relational equations.

Lemma 6.18 *For systems $\mathcal{E}_1 = \{X \circ S_j = T_j \mid j \in J\}$, $\mathcal{E}_2 = \{X \lhd S_j = T_j \mid j \in J\}$, and $\mathcal{E}_3 = \{R_j \lhd X = T_j \mid j \in J\}$ of fuzzy relational equations we have*

$$[\mathrm{Sol}(\mathcal{E}_1)](R) \otimes (R \approx R') \le [\mathrm{Sol}(\mathcal{E}_1)](R'),$$
$$[\mathrm{Sol}(\mathcal{E}_2)](R) \otimes (R \approx R') \le [\mathrm{Sol}(\mathcal{E}_2)](R'),$$
$$[\mathrm{Sol}(\mathcal{E}_3)](S) \otimes (S \approx S') \le [\mathrm{Sol}(\mathcal{E}_3)](S').$$

Proof. By Theorem 6.9 we have

$$
\begin{aligned}
&[\mathrm{Sol}(\mathcal{E}_1)](R) \otimes (R \approx R') = \\
= \;& \bigwedge_{j \in J} [\mathrm{Sol}(X \circ S_j = T_j)](R) \otimes (R \approx R') \leq \\
\leq \;& \bigwedge_{j \in J} ((R \circ S_j \approx T_j) \otimes (R \approx R')) \leq \\
\leq \;& \bigwedge_{j \in J} ((R \circ S_j \approx T_j) \otimes (R \circ S_j \approx R' \circ S_j)) \leq \\
\leq \;& \bigwedge_{j \in J} (R' \circ S_j \approx T_j) = [\mathrm{Sol}(\mathcal{E}_1)](R').
\end{aligned}
$$

For $\mathcal{E}_2$ and $\mathcal{E}_3$, the proof is similar. □

The next question is to what extent it is true that for two systems of fuzzy relational equations that are similar (in the sense that the given fuzzy relations determining the single fuzzy relational equations are pairwise similar), if a fuzzy relation is a solution to one of the systems then it is also a solution to the other one. For systems $\mathcal{E}_1 = \{X \circ S_j = T_j \mid j \in J\}$, $\mathcal{E}_2 = \{X \lhd S_j = T_j \mid j \in J\}$, $\mathcal{E}_3 = \{R_j \lhd X = T_j \mid j \in J\}$, and $\mathcal{E}'_1 = \{X \circ S'_j = T'_j \mid j \in J\}$, $\mathcal{E}'_2 = \{X \lhd S'_j = T'_j \mid j \in J\}$, $\mathcal{E}'_3 = \{R'_j \lhd X = T'_j \mid j \in J\}$ we denote

$$
\begin{aligned}
(\mathcal{E}_1 \approx \mathcal{E}'_1) &= \bigwedge_{j \in J} ((S_j \approx S'_j) \otimes (T_j \approx T'_j)), \\
(\mathcal{E}_2 \approx \mathcal{E}'_2) &= \bigwedge_{j \in J} ((S_j \approx S'_j) \otimes (T_j \approx T'_j)), \\
(\mathcal{E}_3 \approx \mathcal{E}'_3) &= \bigwedge_{j \in J} ((R_j \approx R'_j) \otimes (T_j \approx T'_j));
\end{aligned}
$$

$(\mathcal{E}_i \approx \mathcal{E}'_i)$ is called the degree to which the systems $\mathcal{E}_i$ and $\mathcal{E}'_i$ are similar.

Lemma 6.19 *For systems $\mathcal{E}_1 = \{X \circ S_j = T_j \mid j \in J\}$, $\mathcal{E}_2 = \{X \lhd S_j = T_j \mid j \in J\}$ and $\mathcal{E}_3 = \{R_j \lhd X = T_j \mid j \in J\}$, and the corresponding systems $\mathcal{E}'_1$, $\mathcal{E}'_2$, $\mathcal{E}'_3$ of fuzzy relational equations we have*

$$
\begin{aligned}
(\mathcal{E}_1 \approx \mathcal{E}'_1) \otimes [\mathrm{Sol}(\mathcal{E}_1)](R) &\leq [\mathrm{Sol}(\mathcal{E}'_1)](R), \\
(\mathcal{E}_2 \approx \mathcal{E}'_2) \otimes [\mathrm{Sol}(\mathcal{E}_2)](R) &\leq [\mathrm{Sol}(\mathcal{E}'_2)](R), \\
(\mathcal{E}_3 \approx \mathcal{E}'_3) \otimes [\mathrm{Sol}(\mathcal{E}_3)](S) &\leq [\mathrm{Sol}(\mathcal{E}'_3)](S).
\end{aligned}
$$

Proof. By Theorem 6.9 we have

$$
(\mathcal{E}_1 \approx \mathcal{E}'_1) \otimes [\mathrm{Sol}(\mathcal{E}_1)](R) =
$$

$$
\begin{aligned}
&= \bigwedge_{j\in J}((S_j \approx S'_j)\otimes(T_j \approx T'_j))\otimes \bigwedge_{j\in J}(R\circ S_j \approx T_j) \leq \\
&\leq \bigwedge_{j\in J}((R\circ S_j \approx R\circ S'_j)\otimes(T_j \approx T'_j)\otimes(R\circ S_j \approx T_j)) \leq \\
&\leq \bigwedge_{j\in J}(R\circ S'_j \approx T'_j) = [\mathrm{Sol}(\mathcal{E}'_1)](R).
\end{aligned}
$$

For the other types, the proof is similar. □

To sum up, we obtain

Theorem 6.20 *For systems $\mathcal{E}_1 = \{X\circ S_j = T_j \mid j \in J\}$, $\mathcal{E}_2 = \{X\triangleleft S_j = T_j \mid j \in J\}$ and $\mathcal{E}_3 = \{R_j \triangleleft X = T_j \mid j \in J\}$, and the corresponding systems $\mathcal{E}'_1$, $\mathcal{E}'_2$, $\mathcal{E}'_3$ of fuzzy relational equations we have*

$$
\begin{aligned}
(\mathcal{E}_1 \approx \mathcal{E}'_1)\otimes[\mathrm{Sol}(\mathcal{E}_1)](R)\otimes(R\approx R') &\leq [\mathrm{Sol}(\mathcal{E}'_1)](R'),\\
(\mathcal{E}_2 \approx \mathcal{E}'_2)\otimes[\mathrm{Sol}(\mathcal{E}_2)](R)\otimes(R\approx R') &\leq [\mathrm{Sol}(\mathcal{E}'_2)](R'),\\
(\mathcal{E}_3 \approx \mathcal{E}'_3)\otimes[\mathrm{Sol}(\mathcal{E}_3)](S)\otimes(S\approx S') &\leq [\mathrm{Sol}(\mathcal{E}'_3)](S').
\end{aligned}
$$

Proof. Direct consequence of Lemmas 6.18 and 6.19. □

Interpreted verbally, Theorem 6.20 says that if R is a solution of a system $\mathcal{E}_1$ of fuzzy relational equations, if $\mathcal{E}'_1$ is a system of pairwise similar fuzzy relational equations, and in R' is similar to R then R' is a solution to $\mathcal{E}'_1$ (and similarly for $\mathcal{E}_2$ and $\mathcal{E}_3$).

6.2.2 STRUCTURES OF SOLUTIONS

It is well-known that a linear combination of solutions to a system of homogenous linear equations is again a solution to this system; the set of all solutions forms therefore a vector space. We will now consider analogous questions for systems of fuzzy relational equations.

Lemma 6.21 *For a system $\mathcal{E} = \{X\circ S_j = T_j \mid j \in J\}$ we have*

$$\bigwedge_{i\in I}[\mathrm{Sol}(\mathcal{E})](R_i) \leq [\mathrm{Sol}(\mathcal{E})](\bigcup_{i\in I} R_i).$$

Proof. By definition, we have to show $\bigwedge_{i\in I}[\mathrm{Sol}(\mathcal{E})](R_i) \leq \bigwedge_{j\in J}((\bigcup R_i)\circ S_j \approx T_j)$, which is true iff for each $j \in J$ and $\langle x,z\rangle \in X\times Z$ we have both (a) $\bigwedge_{i\in I}[\mathrm{Sol}(\mathcal{E})](R_i) \leq ((\bigcup R_i)\circ S_j)(x,z) \to T_j(x,z)$ and (b) $\bigwedge_{i\in I}[\mathrm{Sol}(\mathcal{E})](R_i) \leq T_j(x,z) \to ((\bigcup R_i)\circ S_j)(x,z)$. In order to verify (a) we have to show $\bigwedge_{i\in I}[\mathrm{Sol}(\mathcal{E})](R_i)\otimes((\bigcup R_i)\circ S_j)(x,z) \leq T_j(x,z)$. We have

$$\begin{aligned}
&\bigwedge_{i\in I}[\mathrm{Sol}(\mathcal{E})](R_i)\otimes((\bigcup R_i)\circ S_j)(x,z) \leq \\
\leq\ &\bigvee_{i\in I}((R_i\circ S_j)(x,z)\otimes((R_i\circ S_j)(x,z)\to T_j(x,z))) \leq T_j(x,z).
\end{aligned}$$

In order to verify (b) we have to show $\bigwedge_{i\in I}[\mathrm{Sol}(\mathcal{E})](R_i)\otimes T_j(x,z) \leq ((\bigcup R_i)\circ S_j)(x,z)$. We have

$$\begin{aligned}
&\bigwedge_{i\in I}[\mathrm{Sol}(\mathcal{E})](R_i)\otimes T_j(x,z) \leq \\
\leq\ &\bigwedge_{i\in I}((R_i\circ S_j)\approx T_j)\otimes T_j(x,z) \leq \\
\leq\ &\bigwedge_{i\in I}(R_i\circ S_j)(x,z) \leq ((\bigcup R_i)\circ S_j)(x,z).
\end{aligned}$$

□

Lemma 6.22 *For a system $\mathcal{E}=\{X\lhd S_j=T_j \mid j\in J\}$ we have*

$$\bigwedge_{i\in I}[\mathrm{Sol}(\mathcal{E})](R_i) \leq [\mathrm{Sol}(\mathcal{E})](\bigcup_{i\in I} R_i).$$

Proof. By definition, we have to show $\bigwedge_{i\in I}[\mathrm{Sol}(\mathcal{E})](R_i) \leq \bigwedge_{j\in J}((\bigcup R_i)\lhd S_j \approx T_j)$, which is true iff for each $j\in J$ and $\langle x,z\rangle \in X\times Z$ we have both (a) $\bigwedge_{i\in I}[\mathrm{Sol}(\mathcal{E})](R_i) \leq ((\bigcup R_i)\lhd S_j)(x,z)\to T_j(x,z)$ and (b) $\bigwedge_{i\in I}[\mathrm{Sol}(\mathcal{E})](R_i) \leq T_j(x,z)\to((\bigcup R_i)\lhd S_j)(x,z)$. In order to verify (a) we have to show $\bigwedge_{i\in I}[\mathrm{Sol}(\mathcal{E})](R_i)\otimes((\bigcup R_i)\lhd S_j)(x,z) \leq T_j(x,z)$. We have

$$\begin{aligned}
&\bigwedge_{i\in I}[\mathrm{Sol}(\mathcal{E})](R_i)\otimes((\bigcup R_i)\lhd S_j)(x,z) \leq \\
\leq\ &\bigwedge_{i\in I}((R_i\lhd S_j)(x,z)\otimes((R_i\lhd S_j)(x,z)\to T_j(x,z))) \leq T_j(x,z).
\end{aligned}$$

In order to verify (b) we have to show $\bigwedge_{i\in I}[\mathrm{Sol}(\mathcal{E})](R_i)\otimes T_j(x,z) \leq ((\bigcup R_i)\lhd S_j)(x,z)$. We have

$$\begin{aligned}
&\bigwedge_{i\in I}[\mathrm{Sol}(\mathcal{E})](R_i)\otimes T_j(x,z) \leq \\
\leq\ &\bigwedge_{i\in I}((R_i\lhd S)\approx T_j)\otimes T_j(x,z) \leq \\
\leq\ &\bigwedge_{i\in I}(R_i\lhd S_j)(x,z) = ((\bigcup R_i)\lhd S_j)(x,z).
\end{aligned}$$

□

Lemma 6.23 *For a system $\mathcal{E} = \{R_j \lhd X = T_j \mid j \in J\}$ we have*

$$\bigwedge_{i\in I}[\mathrm{Sol}(\mathcal{E})](S_i) \leq [\mathrm{Sol}(\mathcal{E})](\bigcap_{i\in I} S_i).$$

Proof. By definition, we have to show $\bigwedge_{i\in I}[\mathrm{Sol}(\mathcal{E})](S_i) \leq \bigwedge_{j\in J}(R_j \lhd (\bigcap S_i) \approx T_j)$, which is true iff for each $j \in J$ and $\langle x, z\rangle \in X \times Z$ we have both (a) $\bigwedge_{i\in I}[\mathrm{Sol}(\mathcal{E})](S_i) \leq (R_j \lhd (\bigcap S_i))(x,z) \to T_j(x,z)$ and (b) $\bigwedge_{i\in I}[\mathrm{Sol}(\mathcal{E})](S_i) \leq T_j(x,z) \to (R_j \lhd (\bigcap S_i))(x,z)$. In order to verify (a) we have to show $\bigwedge_{i\in I}[\mathrm{Sol}(\mathcal{E})](S_i) \otimes (R_j \lhd (\bigcap S_i))(x,z) \leq T_j(x,z)$. We have

$$\begin{aligned} &\bigwedge_{i\in I}[\mathrm{Sol}(\mathcal{E})](S_i) \otimes (R_j \lhd (\bigcap S_i))(x,z) \leq \\ \leq\ &\bigwedge_{i\in I}((R_j \lhd S_i)(x,z) \otimes ((R_j \lhd S_i)(x,z) \to T_j(x,z))) \leq T_j(x,z). \end{aligned}$$

In order to verify (b) we have to show $\bigwedge_{i\in I}[\mathrm{Sol}(\mathcal{E})](S_i) \otimes T_j(x,z) \leq (R_j \lhd (\bigcap S_i))(x,z)$. We have

$$\begin{aligned} &\bigwedge_{i\in I}[\mathrm{Sol}(\mathcal{E})](S_i) \otimes T_j(x,z) \leq \\ \leq\ &\bigwedge_{i\in I}((R_j \lhd S_i) \approx T_j) \otimes T_j(x,z) \leq \\ \leq\ &\bigwedge_{i\in I}(R_j \circ S_i)(x,z) \leq (R_j \lhd (\bigcap S_i))(x,z). \end{aligned}$$

□

Theorem 6.24 *For systems $\mathcal{E}_1 = \{X \circ S_j = T_j \mid j \in J\}$, $\mathcal{E}_2 = \{X \lhd S_j = T_j \mid j \in J\}$, and $\mathcal{E}_3 = \{R_j \lhd X = T_j \mid j \in J\}$ of fuzzy relational equations and each $a \in L$ we have*

$$R_i \in {}^a\mathrm{Sol}(\mathcal{E}_1) \textit{ for each } i \in I \textit{ implies } \bigcup_{i\in I} R_i \in {}^a\mathrm{Sol}(\mathcal{E}_1),$$
$$R_i \in {}^a\mathrm{Sol}(\mathcal{E}_2) \textit{ for each } i \in I \textit{ implies } \bigcup_{i\in I} R_i \in {}^a\mathrm{Sol}(\mathcal{E}_2),$$
$$S_i \in {}^a\mathrm{Sol}(\mathcal{E}_3) \textit{ for each } i \in I \textit{ implies } \bigcap_{i\in I} S_i \in {}^a\mathrm{Sol}(\mathcal{E}_3).$$

Thus, in particular, if R_i are solutions of $\mathcal{E}_1$ then so is the union $\bigcup R_i$; if R_i are solutions of $\mathcal{E}_2$ then so is the union $\bigcup R_i$; if S_i are solutions of $\mathcal{E}_3$ then so is the intersection $\bigcap R_i$.

Proof. Follows directly from Lemmas 6.21, 6.22, and 6.23. For instance, $R_i \in {}^a\mathrm{Sol}(\mathcal{E}_1))$, i.e. $a \leq [\mathrm{Sol}(\mathcal{E}_1)](R_i)$, implies $a \leq \bigwedge_{i\in I}[\mathrm{Sol}(\mathcal{E}_1)](R_i)$, Lemma 6.21 thus yields that $a \leq [\mathrm{Sol}(\mathcal{E}_1)](\bigcup_i R_i)$, i.e. $\bigcup_i R_i \in {}^a\mathrm{Sol}(\mathcal{E}_1))$. The proof of the other statements is similar. □

Remark 6.25 Theorem 6.24 shows that for systems $\mathcal{E}_1$, $\mathcal{E}_2$, and $\mathcal{E}_3$, defined above, ${}^a\mathrm{Sol}(\mathcal{E}_1)$ is a complete join-semilattice, ${}^a\mathrm{Sol}(\mathcal{E}_2)$ is a complete join-semilattice, and ${}^a\mathrm{Sol}(\mathcal{E}_3)$ is a complete meet-semilattice (whenever the respective ${}^a\mathrm{Sol}(\mathcal{E}_i)$ is non-empty), all with respect to the subsethood $\subseteq$.

6.2.3 SOLUTIONS AND APPROXIMATE SOLUTIONS

We now show criteria of solvability of (systems of) fuzzy relational equations and mention also the issue of finding approximate solutions.

Lemma 6.26 *A fuzzy relational equation $X \circ S = T$ has a solution iff $(S \lhd T^{-1})^{-1}$ is a solution.*

Proof. We have to show that if $X \circ S = T$ has a solution then $(S \lhd T^{-1})^{-1}$ is a solution. First, notice that $(S \lhd T^{-1})^{-1} \circ S \subseteq T$. Indeed,

$$
\begin{aligned}
&((S \lhd T^{-1})^{-1} \circ S)(x,z) = \\
= &\bigvee_{y\in Y} S(y,z) \otimes (\bigwedge_{z'\in Z} S(y,z') \to T(x,z')) \leq \\
\leq &\bigvee_{y\in Y} S(y,z) \otimes (S(y,z) \to T(x,z)) \leq T(x,z).
\end{aligned}
$$

Moreover, if R is a solution of $X \circ S = T$ then for each $x \in X$ and $y \in Y$,

$$\bigvee_{y\in Y} R(x,y) \otimes S(y,z) \leq T(x,z),$$

and thus, using adjointness, for each $y \in Y$ we have $R(x,y) \leq S(y,z) \to T(x,z)$, i.e. $R(x,y) \leq \bigwedge_{z\in Z} S(y,z) \to T^{-1}(z,x)$ which means $R(x,y) \leq (S \lhd T^{-1})(y,x)$, i.e. $R \subseteq (S \lhd T^{-1})^{-1}$. We thus have

$$T = R \circ S \subseteq (S \lhd T^{-1})^{-1} \circ S \subseteq T,$$

thus $(S \lhd T^{-1})^{-1} \circ S = T$ proving that $(S \lhd T^{-1})^{-1}$ is a solution. □

Lemma 6.27 *A fuzzy relational equation $X \lhd S = T$ has a solution iff $(T \lhd S^{-1})$ is a solution.*

Proof. We have to show that if $X \lhd S = T$ has a solution then $T \lhd S^{-1}$ is a solution. Notice first that $T \subseteq (T \lhd S^{-1}) \lhd S$. Indeed, by definition and using adjointness, $T \subseteq (T \lhd S^{-1}) \lhd S$ holds true iff for each x, y, z we have $T(x,z) \otimes (T \lhd S^{-1})(x,y) \leq S(y,z)$ which is true: $T(x,z) \otimes (T \lhd S^{-1})(x,y) \leq T(x,z) \otimes (T(x,z) \to S(y,z)) \leq S(y,z)$. Moreover, if R is a solution of $X \lhd S = T$ then for each $x \in X$ and $y \in Y$,

$$T(x,z) \leq \bigwedge_{y \in Y} R(x,y) \to S(y,z),$$

and thus, using adjointness, for each $y \in Y$ we have $R(x,y) \leq T(x,z) \to S(y,z)$, i.e. $R(x,y) \leq \bigwedge_{z \in Z} T(x,z) \to S^{-1}(z,y)$ thus $R \subseteq T \lhd S^{-1}$. We thus have

$$T \subseteq (T \lhd S^{-1}) \lhd S \subseteq R \lhd S = T$$

thus $(T \lhd S^{-1}) \lhd S = T$ proving that $T \lhd S^{-1}$ is a solution. □

Lemma 6.28 *A fuzzy relational equation $R \lhd X = T$ has a solution iff $R^{-1} \circ T$ is a solution.*

Proof. We have to show that if $R \lhd X = T$ has a solution then $R^{-1} \circ T$ is a solution. Notice first that $T \subseteq R \lhd (R^{-1} \circ T)$. Indeed, $T \subseteq R \lhd (R^{-1} \circ T)$ holds true iff for each x, y, z we have $T(x,y) \otimes R(x,y) \lhd (R^{-1} \circ T)(y,z)$ which is evidently satisfied. Moreover, if S is a solution of $R \lhd X = T$ then for each $x \in X$ and $y \in Y$,

$$T(x,z) \leq \bigwedge_{y \in Y} R(x,y) \to S(y,z),$$

and thus, using adjointness, for each $y \in Y$ we have $R(x,y) \otimes T(x,z) \leq S(y,z)$, i.e. $\bigvee_{x \in X} R^{-1}(y,x) \otimes T(x,z) \leq S(y,z)$ thus $R^{-1} \circ T \subseteq S$. We thus have

$$T \subseteq R \lhd (R^{-1} \circ T) \subseteq R \lhd S = T$$

and thus $R \lhd (R^{-1} \circ T) = T$ proving that $R^{-1} \circ T$ is a solution. □

Remark 6.29 From the proofs of the previous three lemmas we get that if $X \circ S = T$ has a solution then $(S \lhd T^{-1})^{-1}$ is the greatest one; if $X \lhd S = T$ has a solution then $T \lhd S^{-1}$ is the greatest one; if $R \lhd X = T$ has a solution then $R^{-1} \circ T$ is the least one.

We now turn to systems of fuzzy relational equations.

Theorem 6.30 *For systems* $\mathcal{E}_1 = \{X \circ S_j = T_j \mid j \in J\}$, $\mathcal{E}_2 = \{X \lhd S_j = T_j \mid j \in J\}$, *and* $\mathcal{E}_3 = \{R_j \lhd X = T_j \mid j \in J\}$ *of fuzzy relational equations we have*

(1) $\mathcal{E}_1$ *has a solution iff* $\bigcap_{j \in J}(S_j \lhd T_j^{-1})^{-1}$ *is a solution of* $\mathcal{E}_1$*;*

(2) $\mathcal{E}_2$ *has a solution iff* $\bigcap_{j \in J} T_j \lhd S_j^{-1}$ *is a solution of* $\mathcal{E}_2$*;*

(3) $\mathcal{E}_3$ *has a solution iff* $\bigcup_{j \in J} R_j^{-1} \circ T_j$ *is a solution of* $\mathcal{E}_3$*.*

Proof. We use Lemmas 6.26, 6.27, 6.28, and a key observation that for each of the systems, a fuzzy relation is a solution to the system iff it is a solution to each of the equations of the system.

(1): Suppose R is a solution to $\mathcal{E}_1$. Then R is a solution to each of $X \circ S_j = T_j$, $j \in J$. By Lemma 6.26, $(S_j \lhd T_j^{-1})^{-1}$ is a solution to $X \circ S_j = T_j$, and by Remark 6.29, it is the greatest one. Therefore, $R \subseteq (S_j \lhd T_j^{-1})^{-1}$ for each $j \in J$, hence $R \subseteq \bigcap_{j \in J}(S_j \lhd T_j^{-1})^{-1}$. By Theorem 6.13 we thus have

$$T_j = R \circ S_j \subseteq (\bigcap_{j' \in J} (S_{j'} \lhd T_{j'}^{-1})^{-1}) \circ S_j \subseteq (S_j \lhd T_j^{-1})^{-1} \circ S_j = T_j,$$

hence $(\bigcap_{j' \in J}(S_{j'} \lhd T_{j'}^{-1})^{-1}) \circ S_j = T_j$, i.e. $\bigcap_{j' \in J}(S_{j'} \lhd T_{j'}^{-1})^{-1}$ is a solution to each $X \circ S_j = T_j$ and hence a solution to $\mathcal{E}_1$.

(2): Suppose R is a solution to $\mathcal{E}_2$, i.e. R is a solution to each of $X \lhd S_j = T_j$, $j \in J$. By Lemma 6.27, $T_j \lhd S_j^{-1}$ is a solution to $X \lhd S_j = T_j$, and by Remark 6.29, it is the greatest one. Therefore, $R \subseteq T_j \lhd S_j^{-1}$ for each $j \in J$, hence $R \subseteq \bigcap_{j \in J} T_j \lhd S_j^{-1}$. Using Theorem 6.13 we get

$$T_j = R \lhd S_j \supseteq (\bigcap_{j' \in J} (T_{j'} \lhd S_{j'}^{-1})) \lhd S_j \supseteq (T_j \lhd S_j^{-1}) \circ S_j = T_j,$$

hence $(\bigcap_{j' \in J} T_{j'} \lhd S_{j'}^{-1}) \lhd S_j = T_j$, i.e. $\bigcap_{j' \in J} T_{j'} \lhd S_{j'}^{-1}$ is a solution to each $X \lhd S_j = T_j$ and hence a solution to $\mathcal{E}_2$.

(3): Suppose S is a solution to $\mathcal{E}_2$, i.e. S is a solution to each of $R_j \lhd X = T_j$, $j \in J$. By Lemma 6.28, $T_j \lhd S_j^{-1}$ is a solution to $X \lhd S_j = T_j$, and by Remark 6.29, it is the least one. Therefore, $R_j^{-1} \circ T_j \subseteq S$ for each $j \in J$, hence $\bigcup_{j \in J} R_j^{-1} \circ T_j \subseteq S$. Using Theorem 6.13 we get

$$T_j = R_j \lhd S \supseteq R_j \lhd (\bigcup_{j' \in J} R_{j'}^{-1} \circ T_{j'}) \supseteq R_j \lhd (R_j^{-1} \circ T_j) = T_j,$$

hence $R_j \lhd (\bigcup_{j' \in J} R_{j'}^{-1} \circ T_{j'}) = T_j$, i.e. $\bigcup_{j' \in J} R_{j'}^{-1} \circ T_{j'}$ is a solution to each $X \lhd S_j = T_j$ and hence a solution to $\mathcal{E}_3$. □

Corollary 6.31 *For systems $\mathcal{E}_1$, $\mathcal{E}_2$, $\mathcal{E}_3$ of Theorem 6.30 we have that if $\mathcal{E}_i$ ($i = 1,2,3$) has a solution (i.e. ${}^1\mathrm{Sol}(\mathcal{E}_1)$ is non-empty) then $\langle {}^1\mathrm{Sol}(\mathcal{E}_1), \subseteq\rangle$ is a complete join-semilattice with the greatest element $\bigcap_{j\in J}(S_j \lhd T_j^{-1})^{-1}$, $\langle {}^1\mathrm{Sol}(\mathcal{E}_2), \subseteq\rangle$ is a complete join-semilattice with the greatest element $\bigcap_{j\in J} T_j \lhd S_j^{-1}$, $\langle {}^1\mathrm{Sol}(\mathcal{E}_3), \subseteq\rangle$ is a complete meet-semilattice with the least element $\bigcup_{j\in J} R_j^{-1} \circ T_j$.*

Proof. Follows from Theorem 6.24 and the proof of Theorem 6.30. □

Approximate solutions. Theorem 6.30 gives criteria under which a system of fuzzy relational equations is solvable. Clearly, it can be used to find out whether a given system has a solution. If the system does not have any solution, i.e. a solution in degree 1, it might be desirable to have at least a fuzzy relation that is a solution of the system to some high degree. One may always use heuristics and take fuzzy relations from Theorem 6.30 even if the system has no solution in degree 1. In the following we show that if we restrict ourselves to structures of truth values with $\otimes$ being $\wedge$, fuzzy relations from Theorem 6.30 are the best ones to take even if the system has no solution in degree 1.

Theorem 6.32 *Let $\mathbf{L}$ be a structure of truth values with $\otimes = \wedge$. For systems $\mathcal{E}_1 = \{X \circ S_j = T_j \mid j \in J\}$, $\mathcal{E}_2 = \{X \lhd S_j = T_j \mid j \in J\}$, and $\mathcal{E}_3 = \{R_j \lhd X = T_j \mid j \in J\}$ of fuzzy relational equations we have*

(1) $\mathrm{Solv}(\mathcal{E}_1) = [\mathrm{Sol}(\mathcal{E}_1)](\bigcap_{j\in J}(S_j \lhd T_j^{-1})^{-1})$;

(2) $\mathrm{Solv}(\mathcal{E}_2) = [\mathrm{Sol}(\mathcal{E}_2)](\bigcap_{j\in J} T_j \lhd S_j^{-1})$;

(3) $\mathrm{Solv}(\mathcal{E}_3) = [\mathrm{Sol}(\mathcal{E}_3)](\bigcup_{j\in J} R_j^{-1} \circ T_j)$.

Proof. (1): The "$\geq$"-part is trivial. Conversely, putting $R^* = \bigcap_{j\in J}(S_j \lhd T_j^{-1})^{-1}$, we have to show

$$\bigvee_R \bigwedge_{j\in J} (R \circ S_j \approx T_j) \leq \bigwedge_{j\in J} (R^* \circ S_j \approx T_j).$$

Since

$$(R^* \circ S_j)(x,z) = \bigvee_{y\in Y} S_j(y,z) \otimes R^*(x,y) \leq$$

$$\leq \bigvee_{y\in Y} S_j(y,z) \otimes (S_j(y,z) \to T_j(x,z)) \leq T_j(x,z),$$

i.e. $R^* \circ S_j \subseteq T_j$, and thus $(R^* \circ S_j \approx T_j) = S(T_j, R^* \circ S_j)$. Therefore, we have to show

$$\bigvee_R \bigwedge_{j\in J}(R \circ S_j \approx T_j) \leq \bigwedge_{j\in J} S(T_j, R^* \circ S_j)$$

which holds true iff for each R and each i we have

$$\bigwedge_{j\in J}(R \circ S_j \approx T_j) \leq S(T_i, R^* \circ S_i).$$

We have

$$\begin{aligned}
&\bigwedge_{j\in J}(R \circ S_j \approx T_j) = \\
= \ & \bigwedge_{j\in J} S(R \circ S_j, T_j) \wedge \bigwedge_{j\in J} S(T_j, R \circ S_j) \leq \\
\leq \ & \bigwedge_{j\in J} S(R, (S_j \lhd T_j^{-1})^{-1}) \wedge \bigwedge_{j\in J} S(T_j, R \circ S_j) = \\
= \ & S(R, \bigcap_j (S_j \lhd T_j^{-1})^{-1}) \wedge \bigwedge_{j\in J} S(T_j, R \circ S_j) = \\
= \ & S(R, R^*) \wedge \bigwedge_{j\in J} S(T_j, R \circ S_j) \leq \\
\leq \ & S(R \circ S_j, R^* \circ S_j) \otimes \bigwedge_{j\in J} S(T_j, R \circ S_j) \leq S(T_i, R^* \circ S_i)
\end{aligned}$$

proving (1).

(2): Again, the "$\geq$"-part is trivial. For the "$\leq$"-part, putting $R^* = \bigcap_{j\in J}(T_j \lhd S_j^{-1})$, we have to show

$$\bigvee_R \bigwedge_{j\in J}(R \lhd S_j \approx T_j) \leq \bigwedge_{j\in J}(R^* \lhd S_j \approx T_j).$$

Since

$$\begin{aligned}
(R^* \lhd S_j)(x,z) = \bigwedge_{y\in Y}(\bigwedge_{j\in J}(\bigwedge_{z'\in Z} T_j(x,z') \to S_j(y,z')) \to S_j(y,z)) \geq \\
\geq \ \bigwedge_{y\in Y}(T_j(x,z) \to S_j(y,z)) \to S_j(y,z) \geq T_j(x,z),
\end{aligned}$$

i.e. $T_j \subseteq R^* \lhd S_j$, and thus $(R^* \lhd S_j \approx T_j) = S(R^* \lhd S_j, T_j)$. We thus have to show

$$\bigvee_R \bigwedge_{j\in J}(R \lhd S_j \approx T_j) \leq \bigwedge_{j\in J} S(R^* \lhd S_j, T_j)$$

which is true iff for each R and each i we have

$$\bigwedge_{j\in J}(R \lhd S_j \approx T_j) \le S(R^* \lhd S_i, T_i).$$

We have

$$\begin{aligned}
&\bigwedge_{j\in J}(R \lhd S_j \approx T_j) = \\
= \; & \bigwedge_{j\in J} S(R \lhd S_j, T_j) \wedge \bigwedge_{j\in J} S(T_j, R \lhd S_j) = \\
= \; & \bigwedge_{j\in J} S(R \lhd S_j, T_j) \wedge \bigwedge_{j\in J} S(R, T_j \lhd S_j^{-1}) \le \\
\le \; & S(R \lhd S_i, T_i) \wedge S(R, \bigcap_{j\in J} T_j \lhd S_j^{-1}) \le \\
\le \; & S(R \lhd S_i, T_i) \otimes S((\bigcap_{j\in J} T_j \lhd S_j^{-1}) \lhd S_i, R \lhd S_i) \le \\
\le \; & S(R^* \lhd S_i, T_i)
\end{aligned}$$

proving (2).

(3): The "$\ge$"-part is trivial. For the "$\le$"-part, putting $S^* = \bigcup_{j\in J}(R_j^{-1} \circ T_j)$, we have to show

$$\bigvee_S \bigwedge_{j\in J}(R_j \lhd S \approx T_j) \le \bigwedge_{j\in J}(R_j \lhd S^* \approx T_j).$$

Since

$$\begin{aligned}
&(R_j \lhd S^*)(x,z) = (R_j \lhd (\bigcup_i R_i^{-1} \circ T_i))(x,z) \ge \\
\ge \; & \bigwedge_y R_j(x,y) \to (R_j(x,y) \otimes T_j(x,z)) \ge T_j(x,z),
\end{aligned}$$

i.e. $T_j \subseteq R_j \lhd S^*$, and thus $(R_j \lhd S^* \approx T_j) = S(R_j \lhd S^*, T_j)$. We thus have to show

$$\bigvee_S \bigwedge_{j\in J}(R_j \lhd S^* \approx T_j) \le \bigwedge_{j\in J} S(R_j \lhd S^*, T_j)$$

which is true iff for each S and each j we have

$$\bigwedge_{j\in J}(R_j \lhd S \approx T_j) \le S(R_i \lhd S^*, T_i).$$

We have

$$
\begin{aligned}
&\bigwedge_{j\in J}(R_j \lhd S \approx T_j) = \\
= \quad & \bigwedge_{j\in J} S(R_j \lhd S, T_j) \wedge \bigwedge_{j\in J} S(T_j, R_j \lhd S) = \\
= \quad & S(R_i \lhd S, T_i) \wedge \bigwedge_{j\in J} S(R_j^{-1} \circ T_j, S) \leq \\
\leq \quad & S(R_i \lhd S, T_i) \wedge S(\bigcup_j R_j^{-1} \circ T_j, S) \leq \\
\leq \quad & S(R_i \lhd S, T_i) \otimes S(R_i \lhd (\bigcup_j R_j^{-1} \circ T_j), R_i \lhd S) \leq \\
& S(R_i \lhd S^*, T_i)
\end{aligned}
$$

proving (3). □

6.2.4 SOME PARTICULAR TYPES OF FUZZY RELATIONAL EQUATIONS

We are now going to consider some special types of fuzzy relational equations and discuss their relationship to those discussed earlier. For a fuzzy relation R between X and Y, a fuzzy set A in X, and a fuzzy set B in Y, let us define a fuzzy set $R_\circ(A)$ in Y, $R_\lhd(A)$ in Y, and a fuzzy set $R_\rhd(B)$ in X by

$$
\begin{aligned}
[R_\circ(A)](y) &= \bigvee_{x\in X} A(x) \otimes R(x,y), \\
[R_\lhd(A)](y) &= \bigwedge_{x\in X} A(x) \to R(x,y), \\
[R_\rhd(B)](x) &= \bigwedge_{y\in Y} B(y) \to R(x,y).
\end{aligned}
$$

It is easy to see that $R_\circ(A)$, $R_\lhd(A)$, and $R_\rhd(B)$ are definable in terms of $\circ$-, $\lhd$-, and $\rhd$-composition. Indeed, one can associate with A a fuzzy relation A' between a one-element set $U = \{u\}$ and X defined by $A'(u,x) = A(x)$. Then $[R_\circ(A)](y) = (A' \circ R)(u,y)$ and $[R_\lhd(A)](y) = (A' \lhd R)(u,y)$. Similarly, putting $Z = \{z\}$, $B'(y,z) = B(y)$, we have $[R_\rhd(B)](x) = (R \rhd B')(x,z)$. Therefore, it is a matter of routine to transform systems of fuzzy relational equations of the form $R_{j*}(X) = B_j$ $(j \in J)$ and $X_*(A_j) = B_j$ $(j \in J)$ where $*$ is $\circ$, $\lhd$, or $\rhd$, into the types discussed above. In the rest of this section we concentrate on two situations where fuzzy relational equations naturally occur.

Zadeh's compositional rule of inference. The so-called compositional rule of inference (CRI) is a simple rule to obtain a fuzzy set B in Y from a fuzzy set A using a binary fuzzy relation R between X and Y. In terms of fuzzy relational compositions, B is given by $B = R_{\circ}(A)$. CRI is used in approximate reasoning, and particularly as the main part of so-called fuzzy controllers of Mamdani type [127]. The situation is basically the following one. An expert formulates his knowledge in the form of statements like "if temperature is high and rainfall is low then the amount of sediment is big". The terms "high temperature", "low rainfall", "big amount of sediment" can be naturally modeled by fuzzy sets. Moreover, if we assume only one antecedent, the if–then statements are of the form IF A THEN B where A and B are appropriate fuzzy sets. In general, one may have several if–then rules IF A_j THEN B_j ($j \in J$). If one assumes that the outputs B_j should be obtained from the inputs A_j using CRI, the problem of finding an appropriate inference mechanism can formulated as finding a solution to a system

$$X_{\circ}(A_j) = B_j \quad (j \in J)$$

of fuzzy relational equations. It follows from Theorem 6.30 that such a system has a solution iff R defined by

$$R(x, y) = \bigwedge_{j \in J} (A_j(x) \to B_j(y))$$

is a solution to this system. Moreover, for $\otimes = \wedge$, R equals the degree of solvability of this system, i.e. R is the best approximate solution; this follows from Theorem 6.32.

On the other hand, if we assume that R is known, one may ask what input leads to a particular output B. This leads to a fuzzy relational equation

$$R_{\circ}(X) = B.$$

This equation has a solution iff $A \in L^X$ defined by

$$A(x) = R(x, y) \to B(y)$$

is a solution, by Theorem 6.30. Theorem 6.32 implies that for $\otimes = \wedge$, A is the best approximate solution.

For a thorough discussion on CRI and fuzzy relational equations see [96].

Concept equations. Recall from Chapter 5 that starting from a binary fuzzy relation I between a set X (of objects) and a set Y (of attributes), the set $\mathcal{B}(X, Y, I) = \{\langle A, B\rangle \in L^{X \times Y} \mid A^{\uparrow} = B, B^{\downarrow} = A\}$ is naturally interpreted as the set of all concepts (in the sense of traditional logic) which are hidden in the input data I (the object–attribute relation).

Conversely, one may ask the following. Given a system $T = \{\langle A_j, B_j\rangle \in L^{X\times Y} \mid j \in J\}$, is there some fuzzy relation I such that $T \subseteq \mathcal{B}(X,Y,I)$, i.e. can the system T be thought of as a set of concepts induced by some data, i.e. a set which is in a sense conceptually consistent? Noticing that in the present notation, $A^{\uparrow} = B$ and $B^{\downarrow} = A$, is equivalent to $I_{\triangleleft}(A) = B$ and $I_{\triangleright}(B) = A$, respectively, the problem is equivalent to the question of whether there is a solution to the system

$$X_{\triangleleft}(A_j) = B_j, \quad X_{\triangleright}(B_j) = A_j \quad (j \in J).$$

Theorem 6.30 applies here: $X_{\triangleleft}(A_j) = B_j$ is equivalent to $A'_j \triangleleft X = B'_j$ where $A'_j(u,x) = A_j(x)$ and $B'_j(u,y) = B_j(y)$ and $U = \{u\}$ (see above) and thus $X_{\triangleleft}(A_j) = B_j$ $(j \in J)$ has a solution iff $I \in L^{X\times Y}$ defined by

$$I(x,y) = (\bigcup_{j\in J} A'^{-1}_j \circ B_j)(x,y) = \bigvee_{j\in J} A_j(x) \otimes B_j(y) \qquad (6.2)$$

is a solution. In a similar way, one can see that $X_{\triangleright}(B_j) = A_j$ $(j \in J)$ has a solution iff $I \in L^{X\times Y}$ defined by $I(x,y) = \bigvee_{j\in J} A_j(x) \otimes B_j(y)$ is a solution. Therefore, for a system $T = \{\langle A_j, B_j\rangle \in L^{X\times Y} \mid j \in J\}$, there is some fuzzy relation I such that $T \subseteq \mathcal{B}(X,Y,I)$ iff for $I^T \in L^{X\times Y}$ defined by

$$I^T(x,y) = \bigvee_{j\in J} A_j(x) \otimes B_j(y) \qquad (6.3)$$

we have $T \subseteq \mathcal{B}\left(X,Y,I^T\right)$.

Example. We present an example of a system of fuzzy relational equations. Particularly, this is an example of concept equations discussed above.

Let $L = \{0, \frac{1}{2}, 1\}$ and consider the Gödel operations $\otimes, \rightarrow$ on L. Let $|X| = 9$, $|Y| = 25$. We will represent each pair $\langle A, B\rangle$ of $\mathbf{L}$-sets $A \in L^X$ and $B \in L^Y$ by a pair of 3-level grayscale bitmap pictures, a 3×3-bitmap for A and 5×5-bitmap for B. The membership degrees $A(x)$ and $B(y)$ are represented by grayscales of corresponding elements of the bitmaps (take and fix any one-to-one correspondence between elements of X and the pixels of the 3×3-bitmap, and elements of Y and the 5×5-bitmap). The values 0, $\frac{1}{2}$, and 1 are represented by white, gray, and black pixels, respectively. Consider the set T consisting of four pairs $\langle A, B\rangle$ represented by the pairs 13, 14, 16, and 25 of Fig. 6.1. One may verify that a solution to a system of concept equations given by T exists.

The concept lattice $\mathcal{B}(X, Y, I)$ determined by the least solution I given by (6.2) is depicted in Fig. 6.2; all **L**-concepts of $\mathcal{B}(X, Y, I)$ are depicted in Fig. 6.1.

6.3 BIBLIOGRAPHICAL REMARKS

Section 6.1: $\circ$-composition of fuzzy relations was introduced by Zadeh in [216] (for $L = [0, 1]$ and min as a t-norm). Other types (subproduct, superproduct, square composition) were introduced and systematically discussed in the context of fuzzy sets by Bandler and Kohout [6, 7, 8] (note, however, that compositions of fuzzy relations appeared implicitly in papers on fuzzy relational equations earlier); for bivalent relations, subproduct, superproduct, and square composition is treated in [172]. [20] shows that all of the four basic types of composition of fuzzy relations preserve similarity (Theorem 6.9).

Section 6.2: The origin of fuzzy relational equations goes back to Sanchez [178, 179, 180, 181] whose work was inspired by applications of fuzzy relations in medicine. Fuzzy relational equations have been discussed in connection with design problems of fuzzy controllers, decision-making, prediction, and identification of systems (see e.g. [160, 161, 96, 163] and also references in [127]). Gottwald [93] discussed systems of fuzzy relational equations. Approximate solutions of fuzzy relational equations were initiated by Pedrycz [161] and then investigated by Wu [209], Gottwald and Pedrycz [100, 162, 94, 97, 98]. Degrees of solvability were introduced and studied by Gottwald [94], see also [165]. [110] deals with finite case (and contains an algorithm for computing minimal solutions).

[66] is a comprehensive overview of fuzzy relational equations, a good overview is [96, Chapter 3]. Note also that methods of solving of fuzzy relation equations using artificial neural networks and so-called genetic algorithms were proposed (see [127] for further information).

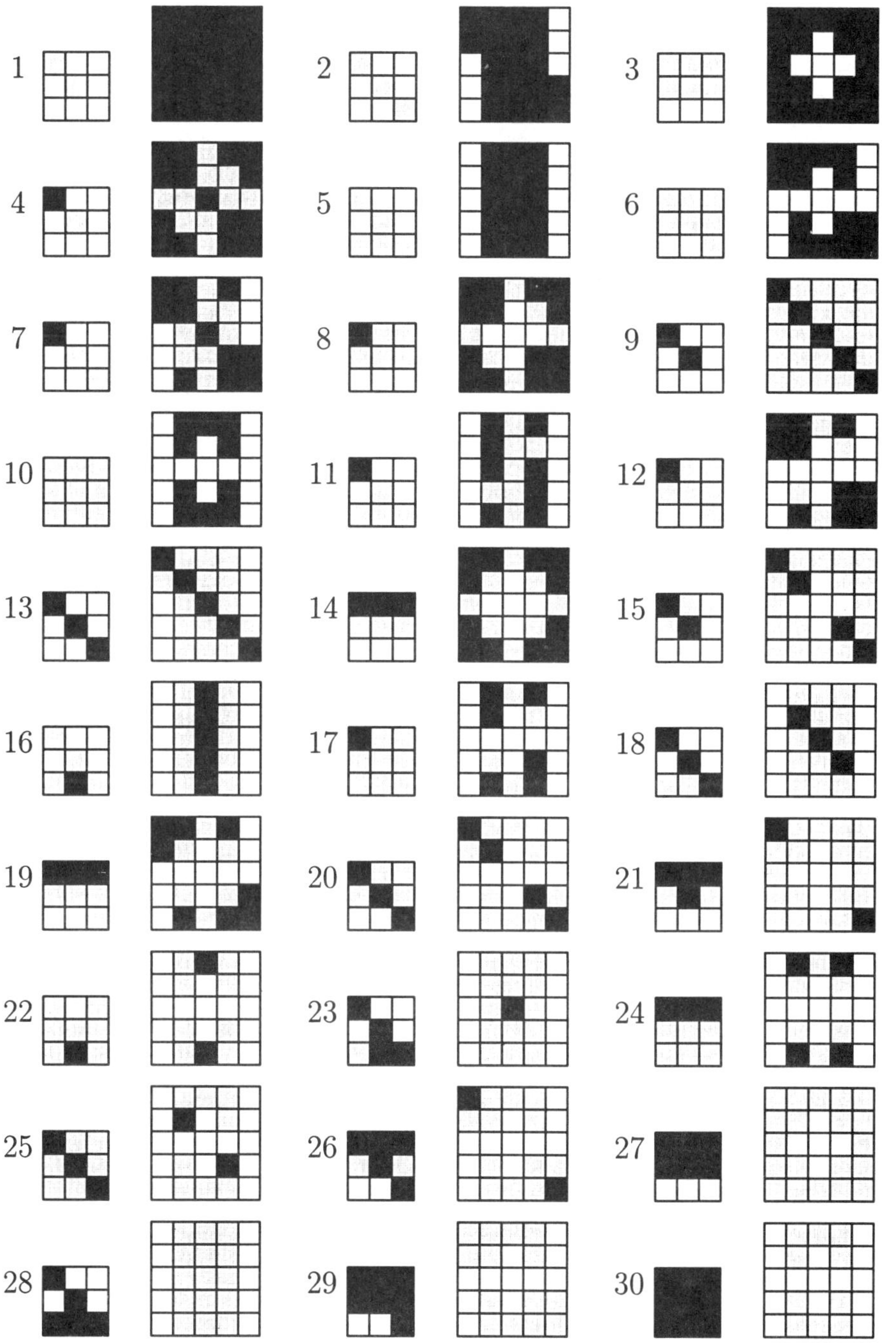

Figure 6.1. **L**-concepts of $\mathcal{B}\left(X, Y, I^T\right)$..

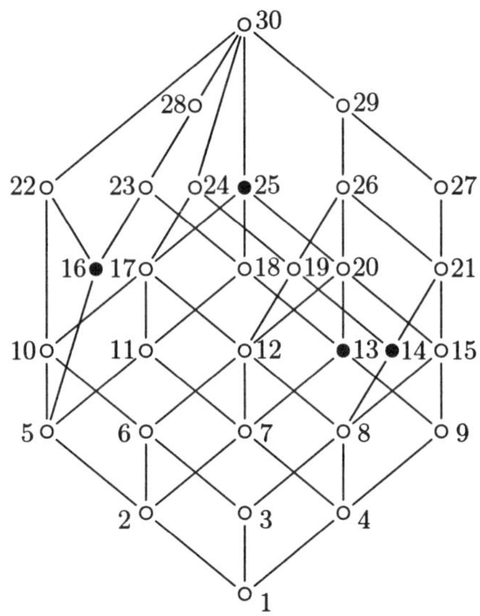

Figure 6.2. **L**-concept lattice $\mathcal{B}\left(X, Y, I^{T}\right)$. The elements of T correspond to filled circles.

Chapter 7

MISCELLANEA

Nothing in the world, no object or event, would be true or false if there were not thinking creatures.

— Donald Davidson

7.1 FUZZY CLOSURE OPERATORS AND SIMILARITY SPACES

Closure operators (and closure systems and related structures) play a significant role in both pure and applied mathematics. Once we adopt and use fuzzy approach to modeling, ordinary closure operators are no longer an appropriate tool and need to be replaced by a more general one. A general theory of closure operators and related structures from the point of view of fuzzy approach is the subject of this section.

Recall that a nonempty subset $K \subseteq L$ is called an *$\leq$-filter* if for every $a, b \in L$ such that $a \leq b$ we have that $b \in K$ whenever $a \in K$. Then, an $\leq$-filter K is a filter if $a, b \in K$ implies $a \otimes b \in K$. Unless otherwise stated, in what follows we denote by $\mathbf{L}$ a complete residuated lattice and by K an $\leq$-filter in $\mathbf{L}$ (both $\mathbf{L}$ and K possibly with indices).

7.1.1 L_K-CLOSURE OPERATORS AND L_K-CLOSURE SYSTEMS

Recall that a closure operator on a set X is a mapping $\mathrm{C} : 2^X \to 2^X$ satisfying the following conditions: $A \subseteq \mathrm{C}(A)$, if $A_1 \subseteq A_2$ then $\mathrm{C}(A_1) \subseteq \mathrm{C}(A_2)$, and $\mathrm{C}(A) = \mathrm{C}(\mathrm{C}(A))$, for any $A, A_1, A_2 \in 2^X$. More generally, if $\subseteq$ denotes a partial order, we get the notion of a closure operator in an ordered set (cf. Chapter 1).

Definition 7.1 An $\mathbf{L}_K$*-closure operator* (*fuzzy closure operator*) on the set X is a mapping $\mathrm{C} : L^X \to L^X$ satisfying

$$A \subseteq \mathrm{C}(A), \tag{7.1}$$
$$S(A_1, A_2) \leq S(\mathrm{C}(A_1), \mathrm{C}(A_2)) \quad \text{whenever } S(A_1, A_2) \in K, \tag{7.2}$$
$$\mathrm{C}(A) = \mathrm{C}(\mathrm{C}(A)) \tag{7.3}$$

for every $A, A_1, A_2 \in L^X$.

If K and $\mathbf{L}$ are obvious, we speak of a fuzzy closure operator. If $K = L$, we omit the subscript K and call C an $\mathbf{L}$-closure operator. The set K plays the role of the set of designated truth values: Condition (7.2) says that the closure preserves also partial subsethood whenever the subsethood degree is designated. Since K is an $\leq$-filter in $\mathbf{L}$, the designated truth values represent, in a sense, sufficiently high truth values. In this view, (7.2) reads "if A_1 is almost included in A_2 then $\mathrm{C}(A_1)$ is almost included in $\mathrm{C}(A_2)$". It is easily seen that each $\mathbf{L}_K$-closure operator on X is a closure operator on the complete lattice $\langle L^X, \subseteq \rangle$.

Remark 7.2 (1) One easily verifies that for $L = \{0, 1\}$, $\mathbf{L}_K$-closure operators are precisely the ordinary closure operators (no matter what K). Clearly, if $K_1 \subseteq K_2$ then each $\mathbf{L}_{K_2}$-closure operator is an $\mathbf{L}_{K_1}$-closure operator. As we will see, the converse is not true. Note also that for $L = [0, 1]$, $\mathbf{L}_{\{1\}}$-closure operators are what is usually known as fuzzy closure operators (see Bibliographical remarks).

(2) We show that for residuated lattices $\mathbf{L}$ where $L = [0, 1]$ with Łukasiewicz, Gödel, and product structures, K is relevant: Take $X = \{x_1, x_2\}$, and define C by $\mathrm{C}(A)(x_1) = 0, \mathrm{C}(A)(x_2) = 0.5$ for $A(x_1) = 0, A(x_2) \leq 0.5$, and $\mathrm{C}(A)(x_1) = \mathrm{C}(A)(x_2) = 1$ otherwise. An easy inspection shows that C is an $\mathbf{L}_{\{1\}}$-closure operator. However, for A_1, A_2 given by $A_1(x_1) = A_2(x_1) = 0$, $A_1(x_2) = 1$, $A_2(x_2) = 0.5$ it holds $S(A_1, A_2) = 0.5 > 0 = S(\mathrm{C}(A_1), \mathrm{C}(A_2))$ (for all of the three above-mentioned structures). Thus, C is not an $\mathbf{L}_{[0.5,1]}$-closure operator.

Theorem 7.3 $\mathrm{C} : L^X \to L^X$ *is an* $\mathbf{L}_K$*-closure operator on* X *iff it satisfies (7.1) and the following condition:*

$$S(A_1, \mathrm{C}(A_2)) \leq S(\mathrm{C}(A_1), \mathrm{C}(A_2)) \quad \textit{whenever } S(A_1, \mathrm{C}(A_2)) \in K. \tag{7.4}$$

Proof. Suppose (7.1)–(7.3) hold. If $S(A_1, \mathrm{C}(A_2)) \in K$ then by (7.2) and (7.3) we have $S(A_1, \mathrm{C}(A_2)) \leq S(\mathrm{C}(A_1), \mathrm{C}(\mathrm{C}(A_2))) = S(\mathrm{C}(A_1), \mathrm{C}(A_2))$, i.e. (7.4) holds.

Conversely, let (7.1) and (7.4) hold. Suppose $S(A_1, A_2) \in K$. By (7.1), $A_2 \subseteq \mathrm{C}(A_2)$, whence $S(A_1, A_2) \leq S(A_1, \mathrm{C}(A_2)) \in K$ (K is a $\leq$-filter). Furthermore, (7.4) implies $S(A_1, \mathrm{C}(A_2)) \leq S(\mathrm{C}(A_1), \mathrm{C}(A_2))$, hence $S(A_1, A_2) \leq S(\mathrm{C}(A_1), \mathrm{C}(A_2))$, proving (7.2). By (7.4), $1 = S(\mathrm{C}(A), \mathrm{C}(A)) \leq S(\mathrm{C}(\mathrm{C}(A)), \mathrm{C}(A))$, we have $\mathrm{C}(\mathrm{C}(A)) \subseteq \mathrm{C}(A)$. Since the converse inclusion holds by (7.1), we conclude (7.3). □

Definition 7.4 A system $\mathcal{S} = \{A_i \in L^X \mid i \in I\}$ is called *closed under S_K-intersections* iff for each $A \in L^X$ it holds that

$$\bigcap_{i \in I, S(A, A_i) \in K} (S(A, A_i) \to A_i) \in \mathcal{S}$$

where

$$\Big(\bigcap_{i \in I, S(A, A_i) \in K} S(A, A_i) \to A_i\Big)(x) = \bigwedge_{i \in I, S(A, A_i) \in K} (S(A, A_i) \to A_i(x))$$

for each $x \in X$. A system closed under S_K-intersections will be called an *$\mathbf{L}_K$-closure system.*

For $K = L$ the subscript will again be omitted.

Remark 7.5 (1) We have

$$\bigcap_{i \in I, S(A, A_i) \in \{1\}} S(A, A_i) \to A_i = \bigcap_{i \in I, A \subseteq A_i} A_i.$$

Therefore, $\mathcal{S}$ is a $\mathbf{2}$-closure system iff for each $A \subseteq X$ it holds $\bigcap_{A \subseteq A_i} A_i \in \mathcal{S}$. It can be easily seen that the last condition is equivalent to being closed under arbitrary intersections for $\mathcal{S}$. Hence, $\mathbf{2}$-closure systems coincide with closure systems, i.e. systems of sets closed under intersections (this is a well-known fact about ordinary closure operators).

(2) In general, being closed under arbitrary intersections is a weaker condition than being closed under S_K-intersections. Indeed, let $\mathcal{S}$ be closed under S_K-intersections. To show that $\mathcal{S}$ is closed under arbitrary intersections, it suffices to show that

$$\bigwedge_{j \in J} A_j(x) = \bigwedge_{i \in I, S(\bigcap_{j \in J} A_j, A_i) \in K} (S(\bigcap_{j \in J} A_j, A_i) \to A_i(x))$$

holds for any $J \subseteq I$. The inequality $\geq$ is clearly valid since for each $j' \in J$ we have $S(\bigcap_{j \in J} A_j, A_{j'}) \to A_{j'}(x) = 1 \to A_{j'}(x) = A_{j'}(x)$. The converse inequality holds iff

$$\bigwedge_{j \in J} A_j(x) \leq S(\bigcap_{j \in J} A_j, A_i) \to A_i(x)$$

for each $i \in I$ such that $S(\bigcap_{j\in J} A_j, A_i) \in K$ which is equivalent to

$$\bigwedge_{j\in J} A_j(x) \otimes S(\bigcap_{j\in J} A_j, A_i) \leq A_i(x),$$

i.e.

$$\bigwedge_{j\in J} A_j(x) \otimes (\bigwedge_{y\in X} (\bigwedge_{j\in J} A_j(y)) \to A_i(y)) \leq A_i(x)$$

which is true because

$$\bigwedge_{j\in J} A_j(x) \otimes (\bigwedge_{y\in X} (\bigwedge_{j\in J} A_j(y)) \to A_i(y)) \leq$$
$$\leq \bigwedge_{j\in J} (A_j(x) \otimes ((\bigwedge_{j\in J} A_j(x)) \to A_i(x))) \leq A_i(x).$$

On the other hand, put $X = \{x\}$, take the Łukasiewicz structure with $\mathbf{L} = \{0, \frac{1}{2}, 1\}$, $K = L$, $\mathcal{S} = \{\{{}^{0}/x\}, \{{}^{1}/x\}\}$, and $A = \{{}^{\frac{1}{2}}/x\}$. Then $\mathcal{S}$ is clearly closed under arbitrary intersections but not under S_K-intersections since $\bigcap_{i\in I, S(A,A_i)\in K} S(A, A_i) \to A_i = A \notin \mathcal{S}$.

Closedness under S_K-intersections is, however, equivalent to closedness under intersections of "K-shifted" $\mathbf{L}$-sets. Recall that for $a \in L$, $A \in L^X$, $a \to A$ denotes the $\mathbf{L}$-set defined by $(a \to A)(x) = a \to A(x)$.

Theorem 7.6 *$\mathcal{S}$ is an $\mathbf{L}_K$-closure system iff for any $a_i \in L$, $i \in I$, we have $\bigcap_{a_i\in K}(a_i \to A_i) \in \mathcal{S}$.*

Proof. If $\bigcap_{a_i\in K}(a_i \to A_i) \in \mathcal{S}$ for every $a_i \in L$, then taking $a_i = S(A, A_i)$ if $S(A, A_i) \in K$ and $a_i = 0$ otherwise for $A \in L^X$, it is easily seen that $\bigwedge_{i\in I, S(A,A_i)\in K} S(A, A_i) \to A_i(x) = \bigwedge_{a_i\in K}(a_i \to A_i(x))$, i.e. $\mathcal{S}$ is an $\mathbf{L}_K$-closure system.

Conversely, let $\mathcal{S}$ be an $\mathbf{L}_K$-closure system. Let $a_i \in L$ and put $A = \bigcap_{a_i\in K}(a_i \to A_i)$. We have to show $A \in \mathcal{S}$. Clearly, it is enough to show that $\bigcap_{i\in I, S(A,A_i)\in K}(S(A, A_i) \to A_i) = A$. The fact

$$\bigcap_{i\in I, S(A,A_i)\in K} (S(A, A_i) \to A_i) \supseteq A$$

is shown in Lemma 7.9. For the converse inclusion, observe first that if $a_j \in K$ then $S(A, A_j) \in K$. Indeed, by $\leq$-filter property of K it suffices to show that $a_j \leq S(A, A_j)$. This holds iff for each $x \in X$ we have

$$a_j \leq (\bigwedge_{a_i\in K} (a_i \to A_i(x)) \to A_j(x)),$$

i.e. $a_j \otimes (\bigwedge_{a_i \in K} (a_i \to A_i(x)) \leq A_j(x))$ which holds since $a_j \otimes (a_j \to A_j(x)) \leq A_j(x)$. Now, the converse inclusion to be proved holds true iff for each $x \in X$ we have

$$\bigwedge_{i \in I, S(A,A_i) \in K} S(A, A_i) \to A_i(x) \leq \bigwedge_{a_i \in K} a_i \to A_i(x)$$

which holds iff for each $a_j \in K$ it holds

$$a_j \otimes \bigwedge_{i \in I, S(A,A_i) \in K} S(A, A_i) \to A_i(x) \leq A_j(x)$$

which holds since by the above observation $a_j \leq S(A, A_j) \in K$, and therefore

$$a_j \otimes \bigwedge_{i \in I, S(A,A_i) \in K} S(A, A_i) \to A_i(x) \leq a_j \otimes (S(A, A_j) \to A_j(x)) \leq A_j(x).$$

The theorem is proved. □

Corollary 7.7 *A system $\mathcal{S}$ which is closed under arbitrary intersections is an $\mathbf{L}_K$-closure system iff for each $a \in K$ and $A \in \mathcal{S}$ it holds $a \to A \in \mathcal{S}$.*

The following theorem shows another way to obtain the closure in an $\mathbf{L}_K$-closure system.

Theorem 7.8 *Let $\mathcal{S} = \{A_i \in L^X \mid i \in I\}$ be an $\mathbf{L}_K$-closure system. Then for each $A \in L^X$ it holds*

$$\bigcap_{i \in I, S(A,A_i) \in K} S(A, A_i) \to A_i = \bigcap_{i \in I, A \subseteq A_i} A_i.$$

Proof. Clearly,

$$\bigcap_{i \in I, S(A,A_i) \in K} S(A, A_i) \to A_i \subseteq \bigcap_{i \in I, S(A,A_i) = 1} S(A, A_i) \to A_i = \bigcap_{i \in I, A \subseteq A_i} A_i.$$

On the other hand, it is easy to check that $A \subseteq \bigcap_{i \in I, S(A,A_i) \in K} S(A, A_i) \to A_i$. Since $\mathcal{S}$ is an $\mathbf{L}_K$-closure system, we have $\bigcap_{i \in I, S(A,A_i) \in K} S(A, A_i) \to A_i \in \mathcal{S}$ which immediately gives $\bigcap_{i \in I, A \subseteq A_i} A_i \subseteq \bigcap_{i \in I, S(A,A_i) \in K} S(A, A_i) \to A_i$. □

Lemma 7.9 *Let $\mathcal{S} = \{A_i \mid i \in I\}$ be an $\mathbf{L}_K$-closure system, K be a filter in $\mathbf{L}$. Then $\mathrm{C}_{\mathcal{S}} : L^X \to L^X$ defined by*

$$\mathrm{C}_{\mathcal{S}}(A)(x) = \bigwedge_{i \in I, S(A,A_i) \in K} (S(A, A_i) \to A_i(x))$$

is an $\mathbf{L}_K$-closure operator. Moreover, for $A \in L^X$ it holds $A \in \mathcal{S}$ iff $A = \mathrm{C}_{\mathcal{S}}(A)$.

Proof. We check (7.1)–(7.3).

(7.1): We have to show $A(x) \leq \mathrm{C}_{\mathcal{S}}(A)(x)$ for each $x \in X$ which holds true iff for each $i \in I$ such that $S(A, A_i) \in K$ we have $A(x) \leq S(A, A_i) \to A_i(x)$. This is, by adjointness, equivalent to $A(x) \otimes S(A, A_i) \leq A_i(x)$, i.e. $A(x) \otimes \bigwedge_{y \in X}(A(y) \to A_i(y)) \leq A_i(x)$ which holds because of

$$A(x) \otimes \bigwedge_{y \in X} (A(y) \to A_i(y)) \leq A(x) \otimes (A(x) \to A_i(x)) \leq A_i(x).$$

(7.2): Suppose $S(A_1, A_2) \in K$. We have to show

$$S(A_1, A_2) \leq S(\mathrm{C}_{\mathcal{S}}(A_1), \mathrm{C}_{\mathcal{S}}(A_2))$$

which is equivalent to the fact that for each $x \in X$ we have $S(A_1, A_2) \leq \mathrm{C}_{\mathcal{S}}(A_1)(x) \to \mathrm{C}_{\mathcal{S}}(A_2)(x)$, i.e. by adjointness,

$$\mathrm{C}_{\mathcal{S}}(A_1)(x) \otimes S(A_1, A_2) \leq \mathrm{C}_{\mathcal{S}}(A_2)(x) = \bigwedge_{i \in I, S(A_2,A_i) \in K} S(A_2, A_i) \to A_i(x)$$

which is true iff for each $j \in I$ with $S(A_2, A_j) \in K$ we have

$$\mathrm{C}_{\mathcal{S}}(A_1)(x) \otimes S(A_1, A_2) \otimes S(A_2, A_j) \leq A_j(x)$$

which is true. Indeed, since

$$S(A_1, A_2) \otimes S(A_2, A_j) \leq S(A_1, A_j),$$

$S(A_1, A_2) \in K$ and $S(A_2, A_j) \in K$, the filter property of K yields $S(A_1, A_j) \in K$, and we have

$$\begin{aligned} &\mathrm{C}_{\mathcal{S}}(A_1)(x) \otimes S(A_1, A_2) \otimes S(A_2, A_j) = \\ &= S(A_1, A_2) \otimes S(A_2, A_j) \otimes \bigwedge_{i \in I, S(A_1,A_i) \in K} (S(A_1, A_i) \to A_i(x)) \leq \\ &\leq S(A_1, A_j) \otimes (S(A_1, A_j) \to A_j(x)) \leq A_j(x). \end{aligned}$$

(7.3): Clearly, we only have to show $\mathrm{C}_{\mathcal{S}}(\mathrm{C}_{\mathcal{S}}(A)) \subseteq \mathrm{C}_{\mathcal{S}}(A)$. Since $\mathrm{C}_{\mathcal{S}}(A) \in \mathcal{S}$, there is some $j \in I$ such that $A_j = \mathrm{C}_{\mathcal{S}}(A)$. We therefore have

$$\begin{aligned} \mathrm{C}_{\mathcal{S}}(\mathrm{C}_{\mathcal{S}}(A))(x) &= \bigwedge_{i\in I, S(\mathrm{C}_{\mathcal{S}}(A),A_i)\in K} S(\mathrm{C}_{\mathcal{S}}(A), A_i) \to A_i(x) \leq \\ &\leq S(\mathrm{C}_{\mathcal{S}}(A), \mathrm{C}_{\mathcal{S}}(A)) \to (\mathrm{C}_{\mathcal{S}}(A))(x) = (\mathrm{C}_{\mathcal{S}}(A))(x). \end{aligned}$$

We now show that $A \in \mathcal{S}$ iff $A = \mathrm{C}_{\mathcal{S}}(A)$. Indeed, if $A = A_j \in \mathcal{S}$ then $A_j \subseteq \mathrm{C}_{\mathcal{S}}(A_j)$ as proved above.

Conversely,

$$\begin{aligned} \mathrm{C}_{\mathcal{S}}(A_j)(x) &= \bigwedge_{i\in I, S(A_j,A_i)\in K} (S(A_j, A_i) \to A_i(x)) \leq \\ &\leq S(A_j, A_j) \to A_j(x) = A_j(x), \end{aligned}$$

i.e. $\mathrm{C}_{\mathcal{S}}(A_j) \subseteq A_j$. If $A = \mathrm{C}_{\mathcal{S}}(A)$ then $A \in \mathcal{S}$ by the definition of the $\mathbf{L}_K$-closure system, completing the proof. □

Lemma 7.10 *Let* $\mathrm{C} : L^X \to L^X$ *be an* $\mathbf{L}_K$*-closure operator. Then* $\mathcal{S}_{\mathrm{C}} = \{A \in L^X \mid A = \mathrm{C}(A)\}$ *is an* $\mathbf{L}_K$*-closure system.*

Proof. Let I be such that $\mathcal{S}_{\mathrm{C}} = \{A_i \mid i \in I\}$. We have to show that for each $A \in L^X$ we have $\bigwedge_{i\in I, S(A,A_i)\in K}(S(A, A_i) \to A_i) \in \mathcal{S}_{\mathrm{C}}$. To this end it clearly suffices to show

$$\bigwedge_{i\in I, S(A,A_i)\in K} (S(A, A_i) \to A_i) = \mathrm{C}(A). \tag{7.5}$$

On the one hand, since $S(A, \mathrm{C}(A)) = 1 \in K$, we have

$$\bigwedge_{i\in I, S(A,A_i)\in K} (S(A, A_i) \to A_i(x)) \leq S(A, \mathrm{C}(A)) \to \mathrm{C}(A)(x) = \mathrm{C}(A)(x).$$

On the other hand,

$$\mathrm{C}(A)(x) \leq \bigwedge_{i\in I, S(A,A_i)} (S(A, A_i) \to A_i(x))$$

iff for each $i \in I$ such that $S(A, A_i) \in K$ it holds $\mathrm{C}(A)(x) \otimes S(A, A_i) \leq A_i(x)$. This is, indeed, true since

$$\begin{aligned} \mathrm{C}(A)(x) \otimes S(A, A_i) &\leq \mathrm{C}(A)(x) \otimes S(\mathrm{C}(A), \mathrm{C}(A_i)) \leq \\ &\leq \mathrm{C}(A)(x) \otimes (\mathrm{C}(A)(x) \to \mathrm{C}(A_i)(x)) \leq \mathrm{C}(A_i)(x) = A_i(x). \end{aligned}$$

To sum up, (7.5) is proved. □

Theorem 7.11 *Let* C *be an* $\mathbf{L}_K$*-closure operator on* X*,* $\mathcal{S}$ *be an* $\mathbf{L}_K$*-closure system on* X*,* K *be a filter in* $\mathbf{L}$*. Then* $\mathcal{S}_{\mathrm{C}}$ *is an* $\mathbf{L}_K$*-closure system on* X*,* $\mathrm{C}_{\mathcal{S}}$ *is an* $\mathbf{L}_K$*-closure operator on* X*, and we have* $\mathrm{C} = \mathrm{C}_{\mathcal{S}_{\mathrm{C}}}$ *and* $\mathcal{S} = \mathcal{S}_{\mathrm{C}_{\mathcal{S}}}$*, i.e. the mappings* $\mathrm{C} \mapsto \mathcal{S}_{\mathrm{C}}$ *and* $\mathcal{S} \mapsto \mathrm{C}_{\mathcal{S}}$ *are mutually inverse.*

Proof. By Lemmas 7.9 and 7.10 it remains to prove $\mathrm{C} = \mathrm{C}_{\mathcal{S}_{\mathrm{C}}}$, i.e. that for any $A \in L^X$, $x \in X$,

$$\mathrm{C}(A)(x) = \bigwedge_{A' \in L^X, A' = \mathrm{C}(A'), S(A,A') \in K} (S(A, A') \to A'(x)).$$

The inequality "$\leq$" holds iff for each $A' \in L^X$ such that $A' = \mathrm{C}(A')$ and $S(A, A') \in K$ we have $\mathrm{C}(A)(x) \otimes S(A, A') \leq A'(x)$ which is true since $\mathrm{C}(A)(x) \otimes S(A, A') \leq \mathrm{C}(A)(x) \otimes S(\mathrm{C}(A), \mathrm{C}(A')) \leq \mathrm{C}(A')(x) = A'(x)$. Conversely, putting $A' = \mathrm{C}(A)$ we get

$$S(A, \mathrm{C}(A)) \to \mathrm{C}(A)(x) = 1 \to \mathrm{C}(A)(x) = \mathrm{C}(A)(x),$$

i.e. "$\geq$" is also valid. □

$\mathbf{L}_K$-closure systems as systems of "almost closed" L-sets

A natural idea is to consider the property "to be closed (w.r.t. a given fuzzy closure operator C)" a graded property. An **L**-set A can be considered to be "almost closed w.r.t. C" iff "A almost equals $\mathrm{C}(A)$". This poses a question of whether fuzzy closure systems can be defined as systems of "almost closed" fuzzy sets.

Definition 7.12 An **L**-system $\mathbf{S} \in L^{L^X}$ is called an $\mathbf{L}_K$*-closure* **L**-*system in* X if for every $A, B \in L^X$ we have

$$\mathbf{S}\Big(\bigcap_{A_i \in L^X, S(A,A_i) \in K} (\mathbf{S}(A_i) \otimes S(A, A_i)) \to A_i\Big) = 1, \qquad (7.6)$$

$$\mathbf{S}(A) \otimes S(A, B) \otimes S(B, A) \leq \mathbf{S}(B) \qquad (7.7)$$
$$\text{whenever } S(B, A) \in K.$$

Remark 7.13 (1) The **L**-set $\bigcap_{A_i \in L^X, S(A,A_i) \in K} (\mathbf{S}(A_i) \otimes S(A, A_i)) \to A_i$ in X is defined by $(\bigcap_{A_i \in L^X, S(A,A_i) \in K} (\mathbf{S}(A_i) \otimes S(A, A_i)) \to A_i)(x) = \bigwedge_{A_i \in L^X, S(A,A_i) \in K} (\mathbf{S}(A_i) \otimes S(A, A_i)) \to A_i(x)$.

(2) An $\mathbf{L}_K$-closure $\mathbf{L}$-system is therefore an $\mathbf{L}$-set of $\mathbf{L}$-sets in X. We interpret $\mathbf{S}(A)$ as the degree to which $A \in L^X$ is closed. Condition (7.7) is naturally interpreted as the requirement that an $\mathbf{L}$-set that is both a subset and a superset of an "almost closed" $\mathbf{L}$-set is itself "almost closed".

We are going to investigate the relationship between $\mathbf{L}_K$-closure $\mathbf{L}$-systems, $\mathbf{L}_K$-closure operators, and $\mathbf{L}_K$-closure systems. To this end we define the following mappings.

For an $\mathbf{L}_K$-closure operator C in X and an $\mathbf{L}_K$-closure system $\mathcal{S}$ in X we define $\mathbf{L}$-sets $\mathbf{S}_{\mathrm{C}}$ and $\mathbf{S}_{\mathcal{S}}$ in L^X by

$$\mathbf{S}_{\mathrm{C}}(A) = E(A, \mathrm{C}(A)), \tag{7.8}$$

$$\mathbf{S}_{\mathcal{S}}(A) = E(A, \mathrm{C}_{\mathcal{S}}(A)). \tag{7.9}$$

Clearly, we have $\mathbf{S}_{\mathrm{C}}(A) = S(\mathrm{C}(A), A)$ and $\mathbf{S}_{\mathcal{S}}(A) = S(\mathrm{C}_{\mathcal{S}}(A), A)$.

For an $\mathbf{L}_K$-closure $\mathbf{L}$-system $\mathbf{S}$ in X we define a mapping $\mathrm{C}_{\mathbf{S}} : L^X \to L^X$ and a set $\mathcal{S}_{\mathbf{S}} \subseteq L^X$ by

$$(\mathrm{C}_{\mathbf{S}}(A))(x) = \bigwedge_{A_i \in L^X, S(A, A_i) \in K} (\mathbf{S}(A_i) \otimes S(A, A_i)) \to A_i(x) \tag{7.10}$$

$$\mathcal{S}_{\mathbf{S}} = \{A \in L^X \mid \mathbf{S}(A) = 1\}. \tag{7.11}$$

Lemma 7.14 *For an $\mathbf{L}_K$-closure operator* C *in* X *we have* $\mathrm{C}_{\mathbf{S}_{\mathrm{C}}} = \mathrm{C}_{\mathcal{S}_{\mathrm{C}}}$.

Proof. Take any $A \in L^X$, $x \in X$. We have to show $(\mathrm{C}_{\mathbf{S}_{\mathrm{C}}}(A))(x) = (\mathrm{C}_{\mathcal{S}_{\mathrm{C}}}(A))(x)$.

"$\leq$":

$$\begin{aligned}(\mathrm{C}_{\mathbf{S}_{\mathrm{C}}}(A))(x) &= \bigwedge_{A_i \in L^X, S(A, A_i) \in K} \mathbf{S}_{\mathrm{C}}(A_i) \otimes S(A, A_i) \to A_i(x) \leq \\ &\leq \bigwedge_{A_i \in L^X, S(A, A_i) \in K, \mathbf{S}_{\mathrm{C}}(A_i) = 1} \mathbf{S}_{\mathrm{C}}(A_i) \otimes S(A, A_i) \to A_i(x) = \\ &= \bigwedge_{A_i \in L^X, S(A, A_i) \in K, A_i = \mathrm{C}(A_i)} S(A, A_i) \to A_i(x) = (\mathrm{C}_{\mathcal{S}_{\mathrm{C}}}(A))(x).\end{aligned}$$

"$\geq$": By definitions, the inequality holds iff

$$\mathrm{C}_{\mathcal{S}_{\mathrm{C}}}(A)(x) \leq \mathbf{S}_{\mathrm{C}}(A_j) \otimes S(A, A_j) \to A_j(x)$$

for any j such that $S(A, A_j) \in K$. Since K is an $\leq$-filter in $\mathbf{L}$, $S(A, A_j) \in K$ and $S(A, A_j) \leq S(A, \mathrm{C}(A_j))$ imply $S(A, \mathrm{C}(A_j)) \in K$. Therefore,

$$\begin{aligned}\mathrm{C}_{\mathcal{S}_{\mathrm{C}}}(A)(x) &= \bigwedge_{A_i \in L^X, S(A, A_i) \in K, A_i = \mathrm{C}(A_i)} S(A, A_i) \to A_i(x) \leq \\ &\leq S(A, \mathrm{C}(A_j)) \to \mathrm{C}(A_j)(x).\end{aligned}$$

It follows that it is sufficient to show

$$S(A, \mathrm{C}(A_j)) \to \mathrm{C}(A_j)(x) \leq \mathbf{S}_{\mathrm{C}}(A_j) \otimes S(A, A_j) \to A_j(x).$$

The last inequality holds iff $\mathbf{S}_{\mathrm{C}}(A_j) \otimes S(A, A_j) \otimes (S(A, \mathrm{C}(A_j)) \to \mathrm{C}(A_j)(x)) \leq A_j(x)$ which is true. Indeed,

$$\begin{aligned}
& \mathbf{S}_{\mathrm{C}}(A_j) \otimes S(A, A_j) \otimes (S(A, \mathrm{C}(A_j)) \to \mathrm{C}(A_j)(x)) = \\
= \; & S(\mathrm{C}(A_j), A_j) \otimes S(A, A_j) \otimes (S(A, \mathrm{C}(A_j)) \to \mathrm{C}(A_j)(x)) \leq \\
\leq \; & S(\mathrm{C}(A_j), A_j) \otimes S(A, \mathrm{C}(A_j)) \otimes (S(A, \mathrm{C}(A_j)) \to \mathrm{C}(A_j)(x)) \leq \\
\leq \; & S(\mathrm{C}(A_j), A_j) \otimes \mathrm{C}(A_j)(x) \leq A_j(x).
\end{aligned}$$

□

Lemma 7.15 *For any $\mathbf{L}_K$-closure operator* C *in X, $\mathbf{S}_{\mathrm{C}}$ is an $\mathbf{L}_K$-closure* **L***-system in X.*

Proof. We verify (7.6) and (7.7).

(7.6): We have to show that for any $A \in L^X$ we have

$$\mathbf{S}_{\mathrm{C}}(\bigcap_{A_i \in L^X, S(A, A_i) \in K} (\mathbf{S}_{\mathrm{C}}(A_i) \otimes S(A, A_i)) \to A_i(x)) = 1,$$

i.e. $\mathbf{S}_{\mathrm{C}}(\mathrm{C}_{\mathbf{S}_{\mathrm{C}}}(A)) = 1$, i.e. $\mathrm{C}(\mathrm{C}_{\mathbf{S}_{\mathrm{C}}}(A)) = \mathrm{C}_{\mathbf{S}_{\mathrm{C}}}(A)$. The last equality, however, follows from idempotency of C by observing that $\mathrm{C}_{\mathbf{S}_{\mathrm{C}}} = \mathrm{C}_{\mathcal{S}_{\mathrm{C}}} = \mathrm{C}$ (Lemma 7.14).

(7.7): We have to show $\mathbf{S}(A) \otimes S(A, B) \otimes S(B, A) \leq \mathbf{S}(B)$, i.e.

$$S(\mathrm{C}(A), A) \otimes S(A, B) \otimes S(B, A) \leq S(\mathrm{C}(B), B)$$

which holds iff for each $x \in X$ we have $\mathrm{C}(B)(x) \otimes S(B, A) \otimes S(\mathrm{C}(A), A) \otimes S(A, B) \leq B(x)$. The last inequality is true:

$$\begin{aligned}
& \mathrm{C}(B)(x) \otimes S(B, A) \otimes S(\mathrm{C}(A), A) \otimes S(A, B) \leq \\
\leq \; & \mathrm{C}(B)(x) \otimes S(\mathrm{C}(B), \mathrm{C}(A)) \otimes S(\mathrm{C}(A), A) \otimes S(A, B) \leq \\
\leq \; & \mathrm{C}(B)(x) \otimes S(\mathrm{C}(B), B) \leq B(x).
\end{aligned}$$

□

Lemma 7.16 *Let K be a filter in* **L**. *For any $\mathbf{L}_K$-closure* **L***-system* **S** *in X, $\mathrm{C}_{\mathbf{S}}$ is an $\mathbf{L}_K$-closure operator in X.*

Proof. We verify (7.1)–(7.3).

(7.1): $A \subseteq \mathrm{C}_{\mathbf{S}}(A)$ holds iff $A(x) \otimes S(A, A_i) \otimes \mathbf{S}(A_i) \leq A_i(x)$ for any i such that $S(A, A_i) \in K$ which is evidently true since

$$A(x) \otimes S(A, A_i) \otimes \mathbf{S}(A_i) \leq A_i(x) \otimes \mathbf{S}(A_i) \leq A_i(x).$$

(7.2): Let $S(A, B) \in K$. $S(A, B) \leq S(\mathrm{C}_{\mathbf{S}}(A), \mathrm{C}_{\mathbf{S}}(B))$ is true iff for each $x \in X$ we have $S(A, B) \otimes \mathrm{C}_{\mathbf{S}}(A)(x) \leq \mathrm{C}_{\mathbf{S}}(B)(x)$ iff for any $A_j \in L^X$ such that $S(B, A_j) \in K$ we have

$$\begin{aligned}&\mathbf{S}(A_j) \otimes S(B, A_j) \otimes S(A, B) \otimes \\ &\otimes(\bigwedge_{A_i \in L^X, S(A, A_i) \in K} \mathbf{S}(A_i) \otimes S(A, A_i) \to A_i(x)) \leq A_j(x).\end{aligned}$$

The last inequality is true: since $S(A, B) \otimes S(B, A_j) \leq S(A, A_j)$, $S(A, B) \in K$ and $S(B, A_j) \in K$ yield $S(A, A_j) \in K$ (K is a filter). Therefore,

$$\begin{aligned}&\mathbf{S}(A_j) \otimes S(B, A_j) \otimes S(A, B) \otimes \\ &\otimes(\bigwedge_{A_i \in L^X, S(A, A_i) \in K} \mathbf{S}(A_i) \otimes S(A, A_i) \to A_i(x)) \leq \\ \leq\ &\mathbf{S}(A_j) \otimes S(A, A_j) \otimes (\mathbf{S}(A_j) \otimes S(A, A_j) \to A_j(x)) \leq A_j(x).\end{aligned}$$

(7.3):

$$\begin{aligned}&\mathrm{C}_{\mathbf{S}}(\mathrm{C}_{\mathbf{S}}(A))(x) = \\ =\ &\bigwedge_{A_i \in L^X, S(\mathrm{C}_{\mathbf{S}}(A), A_i) \in K} \mathbf{S}(A_i) \otimes S(\mathrm{C}_{\mathbf{S}}(A), A_i) \to A_i(x) \leq \\ \leq\ &\mathbf{S}(\mathrm{C}_{\mathbf{S}}(A)) \otimes S(\mathrm{C}_{\mathbf{S}}(A), \mathrm{C}_{\mathbf{S}}(A)) \to \mathrm{C}_{\mathbf{S}}(A)(x) = 1 \to \mathrm{C}_{\mathbf{S}}(A)(x) = \\ =\ &\mathrm{C}_{\mathbf{S}}(A)(x).\end{aligned}$$

□

The relationship between $\mathbf{L}_K$-closure operators, $\mathbf{L}_K$-closure systems, and $\mathbf{L}_K$-closure $\mathbf{L}$-systems is the subject of the following theorems.

Theorem 7.17 *Let* C *be an* $\mathbf{L}_K$*-closure operator in* X, $\mathbf{S}$ *be an* $\mathbf{L}_K$*-closure* $\mathbf{L}$*-system,* K *be a filter in* $\mathbf{L}$*. Then* $\mathbf{S}_{\mathrm{C}}$ *is an* $\mathbf{L}_K$*-closure* $\mathbf{L}$*-system in* X, $\mathrm{C}_{\mathbf{S}}$ *is an* $\mathbf{L}_K$*-closure operator in* X*, and* $\mathrm{C} = \mathrm{C}_{\mathbf{S}_{\mathrm{C}}}$ *and* $\mathbf{S} = \mathbf{S}_{\mathrm{C}_{\mathbf{S}}}$*, i.e. the mappings* $\mathrm{C} \mapsto \mathbf{S}_{\mathrm{C}}$ *and* $\mathbf{S} \mapsto \mathrm{C}_{\mathbf{S}}$ *are mutually inverse.*

Proof. By Lemmas 7.15 and 7.16, it remains to verify $\mathrm{C} = \mathrm{C}_{\mathbf{S}_{\mathrm{C}}}$ and $\mathbf{S} = \mathbf{S}_{\mathrm{C}_{\mathbf{S}}}$. By Lemma 7.14 and Theorem 7.11, $\mathrm{C}_{\mathbf{S}_{\mathrm{C}}} = \mathrm{C}_{\mathcal{S}_{\mathrm{C}}} = \mathrm{C}$. Since $\mathbf{S}_{\mathrm{C}_{\mathbf{S}}}(A) = S(\mathrm{C}_{\mathbf{S}}(A), A)$, it remains to prove $\mathbf{S}(A) = S(\mathrm{C}_{\mathbf{S}}(A), A)$:

On the one hand, $\mathbf{S}(A) \leq S(\mathrm{C}_{\mathbf{S}}(A), A)$ iff for each $x \in X$ we have $\mathbf{S}(A) \otimes \mathrm{C}_{\mathbf{S}}(A)(x) \leq A(x)$, i.e.

$$\mathbf{S}(A) \otimes \bigwedge_{A_i \in L^X, S(A,A_i) \in K} (\mathbf{S}(A_i) \otimes S(A, A_i) \to A_i(x)) \leq A(x)$$

which is true since

$$\begin{aligned}&\mathbf{S}(A) \otimes \bigwedge_{A_i \in L^X, S(A,A_i) \in K} (\mathbf{S}(A_i) \otimes S(A, A_i) \to A_i(x)) \leq \\ \leq\ & \mathbf{S}(A) \otimes (\mathbf{S}(A) \otimes S(A, A) \to A(x)) \leq A(x).\end{aligned}$$

On the other hand,

$$S(\mathrm{C}_{\mathbf{S}}(A), A) = \mathbf{S}(\mathrm{C}_{\mathbf{S}}(A)) \otimes S(\mathrm{C}_{\mathbf{S}}(A), A) \otimes S(A, \mathrm{C}_{\mathbf{S}}(A)) \leq \mathbf{S}(A),$$

by (7.7). □

Theorem 7.18 *Let $\mathcal{S}$ be an $\mathbf{L}_K$-closure system in X, $\mathbf{S}$ be an $\mathbf{L}_K$-closure $\mathbf{L}$-system, K be a filter in $\mathbf{L}$. Then $\mathbf{S}_{\mathcal{S}}$ is an $\mathbf{L}_K$-closure $\mathbf{L}$-system in X, $\mathcal{S}_{\mathbf{S}}$ is an $\mathbf{L}_K$-closure system in X, and $\mathcal{S} = \mathcal{S}_{\mathbf{S}_{\mathcal{S}}}$ and $\mathbf{S} = \mathbf{S}_{\mathcal{S}_{\mathbf{S}}}$, i.e. the mappings $\mathcal{S} \mapsto \mathbf{S}_{\mathcal{S}}$ and $\mathbf{S} \mapsto \mathcal{S}_{\mathbf{S}}$ are mutually inverse.*

Proof. By definition, $\mathbf{S}_{\mathcal{S}} = \mathbf{S}_{\mathrm{C}_{\mathcal{S}}}$, therefore, by Lemma 7.15, $\mathbf{S}_{\mathcal{S}}$ is an $\mathbf{L}_K$-closure $\mathbf{L}$-system. To see that $\mathcal{S}_{\mathbf{S}}$ is an $\mathbf{L}_K$-closure system it is, by Theorem 7.11, sufficient to show $\mathcal{S}_{\mathbf{S}} = \mathcal{S}_{\mathrm{C}_{\mathbf{S}}}$, i.e.

$$\{A \in L^X \mid \mathbf{S}(A) = 1\} = \{A \in L^X \mid A = \mathrm{C}_{\mathbf{S}}(A)\}.$$

On the one hand, $\mathbf{S}(A) = 1$ implies

$$\begin{aligned}\mathrm{C}_{\mathbf{S}}(A)(x) = & \bigwedge_{A_i \in L^X, S(A,A_i)} \mathbf{S}(A_i) \otimes S(A, A_i) \to A_i(x) \leq \\ \leq\ & \mathbf{S}(A) \otimes S(A, A) \to A(x) = A(x),\end{aligned}$$

i.e. $A = \mathrm{C}_{\mathbf{S}}(A)$. On the other hand, $A = \mathrm{C}_{\mathbf{S}}(A)$ implies (using (7.6)) $1 = \mathbf{S}(\mathrm{C}_{\mathbf{S}}(A)) = \mathbf{S}(A)$. Therefore, $\mathcal{S}_{\mathbf{S}} = \mathcal{S}_{\mathrm{C}_{\mathbf{S}}}$.

We show $\mathcal{S} = \mathcal{S}_{\mathbf{S}_{\mathcal{S}}}$: We have $A \in \mathcal{S}$ iff $A = \mathrm{C}_{\mathcal{S}}(A)$ iff $\mathbf{S}_{\mathrm{C}_{\mathcal{S}}}(A) = 1$ iff $\mathbf{S}_{\mathcal{S}}(A) = 1$ iff $A \in \mathcal{S}_{\mathbf{S}_{\mathcal{S}}}$. It remains to show $\mathbf{S}(A) = \mathbf{S}_{\mathcal{S}_{\mathbf{S}}}(A)$: We have $\mathrm{C}_{\mathbf{S}} = \mathrm{C}_{\mathcal{S}_{\mathrm{C}_{\mathbf{S}}}}$ and (by the above observation) $\mathcal{S}_{\mathrm{C}_{\mathbf{S}}} = \mathcal{S}_{\mathbf{S}}$. Therefore, $\mathrm{C}_{\mathbf{S}} = \mathrm{C}_{\mathcal{S}_{\mathbf{S}}}$. Using $\mathbf{S}(A) = S(\mathrm{C}_{\mathbf{S}}(A), A)$ (see the end of the proof of Theorem 7.17), we conclude $\mathbf{S}(A) = S(\mathrm{C}_{\mathbf{S}}(A), A) = S(\mathrm{C}_{\mathcal{S}_{\mathbf{S}}}(A), A) = \mathbf{S}_{\mathcal{S}_{\mathbf{S}}}(A)$ completing the proof. □

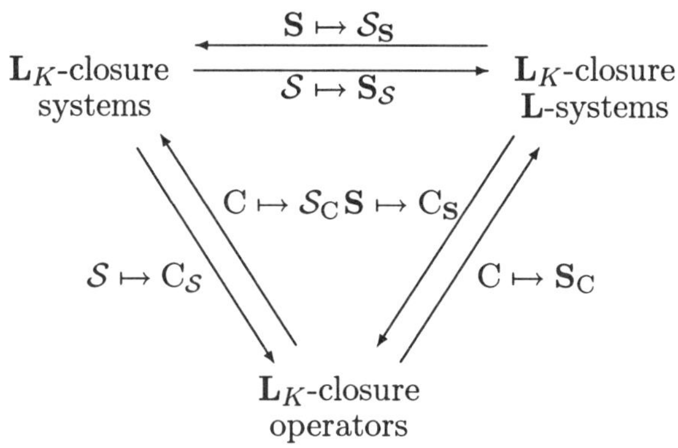

Figure 7.1. Commuting diagram of Corollary 7.19.

Corollary 7.19 *Under the above-introduced notation, the diagram in Fig. 7.1 commutes.*

Proof. Each oriented path in the diagram in Fig. 7.1 defines a mapping (that one composed of the mappings represented by the arrows). The assertion says that any two mappings corresponding to oriented paths with common starting and final node are equal. Using Theorems 7.11, 7.17, and 7.18, a moment's reflection shows that it is sufficient to show $\mathbf{S}_{\mathcal{S}} = \mathbf{S}_{\mathrm{C}_{\mathcal{S}}}$ and $\mathcal{S}_{\mathbf{S}} = \mathcal{S}_{\mathrm{C}_{\mathbf{S}}}$. However, both of these identities were established in the proof of Theorem 7.18. □

$\mathbf{L}_K$-Galois connections

We are going to investigate the relationship of fuzzy closure operators to fuzzy Galois connections. Recall that fuzzy Galois connections were investigated in Section 5.3.

Call two systems $\mathcal{S}_1 \subseteq L^X$ and $\mathcal{S}_2 \subseteq L^Y$ of $\mathbf{L}$-sets in X and $\mathbf{L}$-sets in Y, respectively, *S_K-dually isomorphic* iff there is a bijective mapping $\varphi : \mathcal{S}_1 \to \mathcal{S}_2$ such that for $A_1, A_2 \in \mathcal{S}_1$, $B_1 = \varphi(A_1), B_2 = \varphi(A_2) \in \mathcal{S}_2$ we have $S(A_1, A_2) = S(B_2, B_1)$ whenever $S(A_1, A_2) \in K$ or $S(B_2, B_1) \in K$.

Theorem 7.20 *Let $\langle {}^{\uparrow}, {}^{\downarrow} \rangle$ be an $\mathbf{L}_K$-Galois connection between X and Y, let C^X and C^Y be $\mathbf{L}_K$-closure operators on X and Y, respectively, such that $\mathcal{S}_{\mathrm{C}^X}$ and $\mathcal{S}_{\mathrm{C}^Y}$ are S_K-dually isomorphic with φ being the isomorphism. Put $\mathrm{C}^X_{\langle \uparrow, \downarrow \rangle} = {}^{\uparrow\downarrow}$ and $\mathrm{C}^Y_{\langle \uparrow, \downarrow \rangle} = {}^{\downarrow\uparrow}$ (the composite mappings), and let $(A)^{\uparrow_{\langle \mathrm{C}^X, \mathrm{C}^Y \rangle}} = \varphi(\mathrm{C}^X(A))$ and $(B)^{\downarrow_{\langle \mathrm{C}^X, \mathrm{C}^Y \rangle}} = \varphi^{-1}(\mathrm{C}^Y(B))$ for $A \in L^X$, $B \in L^Y$. Then the following is true.*

(1) $\mathrm{C}^X_{\langle\uparrow,\downarrow\rangle}$ *and* $\mathrm{C}^Y_{\langle\uparrow,\downarrow\rangle}$ *are* $\mathbf{L}_K$*-closure operators on* X *and* Y*, respectively, and* $\mathcal{S}_{\mathrm{C}^X_{\langle\uparrow,\downarrow\rangle}}$ *and* $\mathcal{S}_{\mathrm{C}^Y_{\langle\uparrow,\downarrow\rangle}}$ *are* S_K*-dually isomorphic.*

(2) $\langle {}^{\uparrow_{\langle \mathrm{C}^X,\mathrm{C}^Y\rangle}}, {}^{\downarrow_{\langle \mathrm{C}^X,\mathrm{C}^Y\rangle}}\rangle$ *is an* $\mathbf{L}_K$*-Galois connection.*

(3) The correspondences defined by $\langle {}^\uparrow, {}^\downarrow\rangle \mapsto \langle \mathrm{C}^X_{\langle\uparrow,\downarrow\rangle}, \mathrm{C}^Y_{\langle\uparrow,\downarrow\rangle}\rangle$ *and* $\langle \mathrm{C}^X, \mathrm{C}^Y\rangle \mapsto \langle {}^{\uparrow_{\langle \mathrm{C}^X,\mathrm{C}^Y\rangle}}, {}^{\downarrow_{\langle \mathrm{C}^X,\mathrm{C}^Y\rangle}}\rangle$ *are mutually inverse mappings.*

Proof. (1): (5.21) implies (7.1). If $S(A_1, A_2) \in K$ then, by (5.19), by the $\leq$-filter property of K, and by (5.20), we have

$$S(A_1, A_2) \leq S(A_2^\uparrow, A_1^\uparrow) \leq S(A_1^{\uparrow\downarrow}, A_2^{\uparrow\downarrow}),$$

i.e. (7.2) holds. The fact that $\langle {}^\uparrow, {}^\downarrow\rangle$ is a Galois connection between the lattices L^X and L^Y immediately gives (7.3) (cf. Chapter 1), i.e. $\mathrm{C}^X_{\langle\uparrow,\downarrow\rangle}$ is an $\mathbf{L}_K$-closure operator on X. The proof for $\mathrm{C}^Y_{\langle\downarrow,\uparrow\rangle}$ is completely analogous. The rest easily follows by taking $\varphi(A) = A^\uparrow$ observing that $A = A^{\uparrow\downarrow}$ and $B = B^{\downarrow\uparrow}$ for $A \in \mathcal{S}_{\mathrm{C}^X_{\langle\uparrow,\downarrow\rangle}}$ and $B \in \mathcal{S}_{\mathrm{C}^Y_{\langle\uparrow,\downarrow\rangle}}$.

(2): For simplicity, we write only $^\uparrow$ instead of $^{\uparrow_{\langle \mathrm{C}^X,\mathrm{C}^Y\rangle}}$; the same for $^\downarrow$. First, since

$$\mathrm{C}^X(A) = \varphi^{-1}(\varphi(\mathrm{C}^X(A))) = \varphi^{-1}(\mathrm{C}^Y(\varphi(\mathrm{C}^X(A)))) = A^{\uparrow\downarrow},$$

we have $\mathrm{C}^X = \mathrm{C}^X_{\langle\uparrow,\downarrow\rangle}$ whence (5.19) follows. If $S(A_1, A_2) \in K$ then by (7.2) and the assumption,

$$\begin{aligned} & S(A_1, A_2) \leq S(\mathrm{C}^X(A_1), \mathrm{C}^X(A_2)) = \\ = \; & S(\varphi(\mathrm{C}^X(A_2)), \varphi(\mathrm{C}^X(A_1))) = S(A_2^\uparrow, A_1^\uparrow), \end{aligned}$$

i.e. (5.20) holds. The rest of the statement can be proved analogously.

Part (3) now follows easily from the proof of (2) and the fact that $A^\uparrow = A^{\uparrow\downarrow\uparrow}$ and $B^\downarrow = B^{\downarrow\uparrow\downarrow}$ for $A \in L^X$, $B \in L^Y$. □

For $K = L$ the foregoing theorem can be strengthened.

Theorem 7.21 *Let* C *be an* $\mathbf{L}$*-closure operator (i.e. an* $\mathbf{L}_K$*-closure operator for* $K = L$*), and* $Y = \{\mathrm{C}(A) \mid A \in L^X\}$*. Then the pair of mappings* $^{\uparrow_\mathrm{C}} : L^X \to L^Y$*,* $^{\downarrow_\mathrm{C}} : L^Y \to L^X$ *defined for* $A \in L^X$*,* $B \in L^Y$ *and* $x \in X$*,* $A' \in Y$ *by*

$$\begin{aligned} A^{\uparrow_\mathrm{C}}(A') &= S(A, A'), \\ B^{\downarrow_\mathrm{C}}(x) &= \bigwedge_{A \in Y} B(A) \to (A)(x) \end{aligned}$$

forms an $\mathbf{L}$*-Galois connection such that* $\mathrm{C} = {}^{\uparrow_{\mathrm{C}}\downarrow_{\mathrm{C}}}$.

Proof. For brevity we write $^{\uparrow}$ and $^{\downarrow}$ instead of $^{\uparrow_{\mathrm{C}}}$ and $^{\downarrow_{\mathrm{C}}}$, respectively. We first verify (5.19)–(5.22).

(5.19): $S(A_1, A_2) \leq S(A_2^{\uparrow}, A_1^{\uparrow})$ holds true iff for each $A' \in Y$ we have $S(A_1, A_2) \leq S(A_2, A') \rightarrow S(A_1, A')$ which is true iff $S(A_1, A_2) \otimes S(A_2, A') \leq S(A_1, A')$ iff for each $x \in X$ we have $A_1(x) \otimes S(A_1, A_2) \otimes S(A_2, A') \leq A'(x)$ which is true since

$$\begin{aligned} & A_1(x) \otimes S(A_1, A_2) \otimes S(A_2, A') \leq \\ \leq\ & A_1(x) \otimes (A_1(x) \rightarrow A_2(x)) \otimes (A_2(x) \rightarrow A'(x)) \leq A'(x). \end{aligned}$$

(5.20): Let $B_1, B_2 \in L^Y$. We have to prove $S(B_1, B_2) \leq S(B_2^{\downarrow}, B_1^{\downarrow})$ which holds iff for each $x \in X$,

$$S(B_1, B_2) \otimes B_2^{\downarrow}(x) \leq B_1^{\downarrow}(x) = \bigwedge_{A \in Y} (B_1(A) \rightarrow A(x))$$

iff for each $A \in Y$ we have

$$B_1(A) \otimes S(B_1, B_2) \otimes B_2^{\downarrow}(x) \leq A(x)$$

which is valid since

$$\begin{aligned} & B_1(A) \otimes S(B_1, B_2) \otimes B_2^{\downarrow}(x) \leq \\ \leq\ & B_1(A) \otimes (B_1(A) \rightarrow B_2(A)) \otimes (B_2(A) \rightarrow A(x)) \leq A(x). \end{aligned}$$

To show (5.21), it suffices to show that $\mathrm{C} = {}^{\uparrow\downarrow}$. Let thus $A \in L^X$, $x \in X$. We show $\mathrm{C}(A)(x) = A^{\uparrow\downarrow}(x)$ by proving both of the inequalities.

"$\leq$": $\mathrm{C}(A)(x) \leq A^{\uparrow\downarrow}(x) = \bigwedge_{A' \in Y}(A^{\uparrow}(A') \rightarrow A'(x))$ holds iff for each $A' \in Y$ we have $A^{\uparrow}(A') \otimes \mathrm{C}(A)(x) \leq A'(x)$ which is true iff $A^{\uparrow}(A') \leq \mathrm{C}(A)(x) \rightarrow A'(x)$, i.e. $S(A, A') \leq \mathrm{C}(A)(x) \rightarrow A'(x)$ which holds true since

$$\begin{aligned} & S(A, A') \leq S(\mathrm{C}(A), \mathrm{C}(A')) \leq \\ \leq\ & \mathrm{C}(A)(x) \rightarrow \mathrm{C}(A')(x) = \mathrm{C}(A)(x) \rightarrow A'(x) \end{aligned}$$

as $\mathrm{C}(A') = A'$.

"$\geq$": $A^{\uparrow\downarrow}(x) \leq \mathrm{C}(A)(x)$, i.e. $\bigwedge_{A' \in Y}(S(A, A') \rightarrow A'(x)) \leq \mathrm{C}(A)(x)$ holds since for $A' = \mathrm{C}(A)$ we have

$$S(A, \mathrm{C}(A)) \rightarrow \mathrm{C}(A)(x) = 1 \rightarrow \mathrm{C}(A)(x) = \mathrm{C}(A)(x).$$

(5.22): We have $B(A) \leq B^{\downarrow\uparrow}(A)$ iff

$$B(A) \leq S(B^{\downarrow}, A) = \bigwedge_{x \in X} (B^{\downarrow}(x) \rightarrow A(x))$$

which holds iff for each $x \in X$ we have $B(A) \otimes B^{\downarrow}(x) \leq A(x)$, i.e.

$$B(A) \otimes \bigwedge_{A' \in Y} (B(A') \to A'(x)) \leq A(x)$$

which is true (putting $A' = A$ the inequality is evident). The proof is complete. □

Remark 7.22 In other words, Theorem 7.20 says (among others) that for an $\mathbf{L}_K$-Galois connection $\langle^{\uparrow}, ^{\downarrow}\rangle$, the composite mappings $^{\uparrow\downarrow}$ and $^{\downarrow\uparrow}$ are $\mathbf{L}_K$-closure operators. Moreover, Theorem 7.21 says that each $\mathbf{L}$-closure operator can be obtained this way (i.e. as a composition of suitable $^{\uparrow}$ and $^{\downarrow}$ which form an $\mathbf{L}$-Galois connection).

Induced relations: quasiorder and similarity

We will see that fuzzy closure operators induce in a natural way the following binary fuzzy relations that satisfy a generalized notion of transitivity: An $\mathbf{L}$-relation R on a set X is called

$$K\text{-transitive if} \quad R(x,y) \otimes R(y,z) \leq R(x,z) \\ \text{whenever } R(x,y) \in K \text{ and } R(y,z) \in K.$$

An $\mathbf{L}_K$-*quasiorder* on X is an $\mathbf{L}$-relation on X that is reflexive and K-transitive. An $\mathbf{L}_K$-*equivalence* (or $\mathbf{L}_K$-similarity) on X is an $\mathbf{L}$-relation on X that is reflexive, symmetric, and K-transitive.

Remark 7.23 (1) Clearly, putting $\mathbf{L} = \mathbf{2}$ we get the usual (bivalent) notion of a quasiorder and an equivalence relation (no matter what K). Thus, the notions of $\mathbf{L}_K$-quasiorder and $\mathbf{L}_K$-similarity are generalizations of the bivalent notions.

(2) If K is interpreted as the set of sufficiently high (designated) truth values, then K-transitivity means: "if the facts that $\langle x, y\rangle$ belongs to R and $\langle y, z\rangle$ belongs to R are sufficiently true, then $\langle x, z\rangle$ also belongs to R (and is sufficiently true in case K is a filter)". For $K = L$ (in which case we omit the subscript K) we get the usual notions of $\mathbf{L}$-quasiorder and $\mathbf{L}$-similarity (called usually fuzzy quasiorder and fuzzy similarity).

Theorem 7.24 *Let* C *be an* $\mathbf{L}_K$*-closure operator on* X*. Then the relation* Q_{C} *in* X *defined by*

$$Q_{\mathrm{C}}(x,y) = \mathrm{C}(\{\,^{1}/x\})(y)$$

is an $\mathbf{L}_K$*-quasiorder.*

Proof. Since $\{{}^1/x\} \subseteq \mathrm{C}(\{{}^1/x\})$, reflexivity follows by

$$Q_{\mathrm{C}}(x,x) = \mathrm{C}(\{{}^1/x\})(x) = 1.$$

K-transitivity: Let $Q_{\mathrm{C}}(x,y) \in K$, $Q_{\mathrm{C}}(y,z) \in K$. We have to show $Q_{\mathrm{C}}(x,y) \otimes Q_{\mathrm{C}}(y,z) \leq Q_{\mathrm{C}}(x,z)$. By adjointness and by definition of Q_{C} we thus have to show that

$$\mathrm{C}(\{{}^1/x\})(y) \leq \mathrm{C}(\{{}^1/y\})(z) \to \mathrm{C}(\{{}^1/x\})(z)$$

whenever $\mathrm{C}(\{{}^1/x\})(y)$, $\mathrm{C}(\{{}^1/y\})(z) \in K$. We have

$$S(\{{}^{\mathrm{C}(\{{}^1/x\})(y)}/y\}, \mathrm{C}(\{{}^1/x\})) = 1,$$

therefore, by (7.2) and using $\mathrm{C}(\mathrm{C}(\{{}^1/x\})) = \mathrm{C}(\{{}^1/x\})$, also

$$S(\mathrm{C}(\{{}^{\mathrm{C}(\{{}^1/x\})(y)}/y\}), \mathrm{C}(\{{}^1/x\})) = 1.$$

Furthermore, by $\mathrm{C}(\{{}^1/x\})(y) \in K$ and (7.2),

$$\begin{aligned} & \mathrm{C}(\{{}^1/x\})(y) = S(\{{}^1/y\}, \{{}^{\mathrm{C}(\{{}^1/x\})(y)}/y\}) \leq \\ \leq\ & S(\mathrm{C}(\{{}^1/y\}), \mathrm{C}(\{{}^{\mathrm{C}(\{{}^1/x\})(y)}/y\})). \end{aligned}$$

Therefore,

$$\begin{aligned} & \mathrm{C}(\{{}^1/x\})(y) \leq S(\mathrm{C}(\{{}^1/y\}), \mathrm{C}(\{{}^{\mathrm{C}(\{{}^1/x\})(y)}/y\})) = \\ =\ & S(\mathrm{C}(\{{}^1/y\}), \mathrm{C}(\{{}^{\mathrm{C}(\{{}^1/x\})(y)}/y\})) \otimes \\ & \otimes S(\mathrm{C}(\{{}^{\mathrm{C}(\{{}^1/x\})(y)}/y\}), \mathrm{C}(\{{}^1/x\})) \leq \\ \leq\ & S(\mathrm{C}(\{{}^1/y\}), \mathrm{C}(\{{}^1/x\})) \leq \mathrm{C}(\{{}^1/y\})(z) \to \mathrm{C}(\{{}^1/x\})(z). \end{aligned}$$

The proof is complete. □

Remark 7.25 Note that Q_{C} actually satisfies a stronger condition than K-transitivity. Namely, as it follows from the proof of Theorem 7.24, $Q_{\mathrm{C}}(x,y) \otimes Q_{\mathrm{C}}(y,z) \leq Q_{\mathrm{C}}(x,z)$ whenever $Q_{\mathrm{C}}(x,y) \in K$. This property is typical for Pavelka-style fuzzy logic (see [158] and also [105]): Take X to be the set of all formulas; put $K = \{1\}$; let C be the operator of syntactic consequence, i.e. for an **L**-set A of formulas and a formula $x \in X$, let $\mathrm{C}(A)(x)$ be the degree of provability of x from A. One easily verifies that C satisfies the above condition stronger than K-transitivity. On the other hand, C does not satisfy the in a sense symmetric condition, i.e. it is not true that if $Q_{\mathrm{C}}(y,z) \in K$ then $Q_{\mathrm{C}}(x,y) \otimes Q_{\mathrm{C}}(y,z) \leq Q_{\mathrm{C}}(x,z)$.

$Q_{\mathrm{C}}(x,y)$ is naturally interpreted as the truth degree to which y belongs to the closure of a singleton containing x. One might wonder what is the relationship between Q_{C} and $Q_{\mathcal{S}_{\mathrm{C}}}$ defined by

$$Q_{\mathcal{S}_{\mathrm{C}}}(x,y) = \bigwedge_{A\in\mathcal{S}_{\mathrm{C}},A(x)\in K} A(x) \to A(y),$$

i.e. the truth degree to which it holds that whenever it is sufficiently true that x belongs to some closure then y belongs to that closure as well.

Theorem 7.26 *For any $\mathbf{L}_K$-closure operator* C *we have $Q_{\mathrm{C}} = Q_{\mathcal{S}_{\mathrm{C}}}$. Therefore, $Q_{\mathcal{S}_{\mathrm{C}}}$ is an $\mathbf{L}_K$-quasiorder.*

Proof. On the one hand, $\mathrm{C}(\{{}^1/x\})(x) = 1 \in K$ yields

$$\begin{aligned} Q_{\mathcal{S}_{\mathrm{C}}}(x,y) = & \bigwedge_{A\in\mathcal{S}_{\mathrm{C}},A(x)\in K} A(x) \to A(y) \leq \\ \leq & \ \mathrm{C}(\{{}^1/x\})(x) \to \mathrm{C}(\{{}^1/x\})(y) = 1 \to \mathrm{C}(\{{}^1/x\})(y) = \\ = & \ \mathrm{C}(\{{}^1/x\})(y) = Q_{\mathrm{C}}(x,y). \end{aligned}$$

On the other hand, $Q_{\mathrm{C}}(x,y) \leq Q_{\mathcal{S}_{\mathrm{C}}}(x,y)$ is true iff for each $A \in \mathcal{S}_{\mathrm{C}}$ such that $A(x) \in K$ we have $\mathrm{C}(\{{}^1/x\})(y) \leq A(x) \to A(y)$. Applying adjointness twice, the last inequality is equivalent to $A(x) \leq \mathrm{C}(\{{}^1/x\})(y) \to A(y)$ which is true. Indeed, since $A(x) \in K$, (7.2) gives

$$\begin{aligned} & A(x) = 1 \to A(x) = \{{}^1/x\}(x) \to A(x) = S(\{{}^1/x\}, A) \leq \\ \leq \ & S(\mathrm{C}(\{{}^1/x\}), \mathrm{C}(A)) = S(\mathrm{C}(\{{}^1/x\}), A) \leq \mathrm{C}(\{{}^1/x\})(y) \to A(y). \end{aligned}$$

□

If K is, moreover, a filter in $\mathbf{L}$, the fact that Q_{C} is an $\mathbf{L}_K$-quasiorder follows from Theorem 7.26 and the following statement.

Lemma 7.27 *Let K be a filter in $\mathbf{L}$, $\mathcal{S} = \{A_i \in L^X \mid i \in I\}$ be a system of $\mathbf{L}$-sets. Then the $\mathbf{L}$-relation $Q_{\mathcal{S}}$ on X given by*

$$Q_{\mathcal{S}}(x,y) = \bigwedge_{i\in I,A_i(x)\in K} A_i(x) \to A_i(y)$$

is an $\mathbf{L}_K$-quasiorder.

Proof. Reflexivity follows from the fact that $A_i(x) \to A_i(x) = 1$ and from $\bigwedge\emptyset = 1$. K-transitivity: By definition, we have to show that if

$$\bigwedge_{i\in I,A_i(x)\in K} A_i(x) \to A_i(y) \in K$$

and

$$\bigwedge_{i\in I, A_i(x)\in K} A_i(y) \to A_i(z) \in K$$

then

$$(\bigwedge_{i\in I, A_i(x)\in K} A_i(x) \to A_i(y)) \otimes (\bigwedge_{i\in I, A_i(y)\in K} A_i(y) \to A_i(z)) \leq$$
$$\leq (\bigwedge_{i\in I, A_i(x)\in K} A_i(x) \to A_i(z)),$$

i.e. to show that for each $i \in I$ such that $A_i(x) \in K$ we have

$$A_i(x) \otimes (\bigwedge_{i\in I, A_i(x)\in K} A_i(x) \to A_i(y)) \otimes (\bigwedge_{i\in I, A_i(y)\in K} A_i(y) \to A_i(z)) \leq$$
$$\leq A_i(z).$$

This inequality is true. Indeed, $\bigwedge_{i\in I, A_i(x)\in K} A_i(x) \to A_i(y) \in K$ and $\bigwedge_{i\in I, A_i(x)\in K} A_i(x) \to A_i(y) \leq A_i(x) \to A_i(y)$ gives $A_i(x) \to A_i(y) \in K$. As also $A_i(x) \in K$, we have $A_i(x) \otimes (A_i(x) \to A_i(y)) \in K$. From $A_i(x) \otimes (A_i(x) \to A_i(y)) \leq A_i(y)$ we thus have $A_i(y) \in K$. Therefore, we conclude

$$A_i(x) \otimes (\bigwedge_{i\in I, A_i(x)\in K} A_i(x) \to A_i(y)) \otimes (\bigwedge_{i\in I, A_i(x)\in K} A_i(y) \to A_i(z)) \leq$$
$$\leq A_i(x) \otimes (A_i(x) \to A_i(y)) \otimes (A_i(y) \to A_i(z)) \leq A_i(z),$$

completing the proof. □

Indeed, putting $\mathcal{S} = \mathcal{S}_{\mathrm{C}}$, Lemma 7.27 yields that $Q_{\mathcal{S}_{\mathrm{C}}}$ is an $\mathbf{L}_K$-quasiorder. Theorem 7.26 then completes the argument.

Remark 7.28 (1) A closer look at the proof of Lemma 7.27 shows that like Q_{C}, $Q_{\mathcal{S}}$ satisfies a stronger form of K-transitivity: $Q_{\mathcal{S}}(x,y) \otimes Q_{\mathcal{S}}(y,z) \leq Q_{\mathcal{S}}(x,z)$ whenever $Q_{\mathcal{S}}(x,y) \in K$.

(2) The assumption of Lemma 7.27 that K be closed w.r.t. $\otimes$ is essential. As a counterexample, consider $I = \{i, j\}$, $X = \{x, y, z\}$, $L = [0,1]$ equipped with Łukasiewicz structure, $A_i(x) = 0.9$, $A_i(y) = 0.8$, $A_i(z) = 0.7$, $A_j(x) = A_j(y) = A_j(z) = 1$. Taking $K = [0.9, 1]$ which is an $\leq$-filter not closed w.r.t. $\otimes$, we have $Q_{\mathcal{S}}(x,y) \otimes Q_{\mathcal{S}}(y,z) = 0.9 \otimes 1 = 0.9 \not\leq 0.8 = Q_{\mathcal{S}}(x,z)$.

It can be easily seen that every $\mathbf{L}_K$-quasiorder on X induces an $\mathbf{L}_K$-equivalence E_Q on X by putting $E_Q(x,y) = Q(x,y) \wedge Q(y,x)$. Therefore, for any $\mathbf{L}_K$-closure operator C on X, the $\mathbf{L}$-relation E_{C} on X defined by

$$E_{\mathrm{C}}(x,y) = Q_{\mathrm{C}}(x,y) \wedge Q_{\mathrm{C}}(y,x)$$

is an $\mathbf{L}_K$-equivalence on X. Generalizing compatibility with $\mathbf{L}$-equivalence, we say that an $\mathbf{L}_K$-equivalence E on X is compatible with $A \in L^X$ if $A(x) \otimes E(x, y) \leq A(y)$ holds for any $x, y \in X$ such that $A(x) \in K$. The condition of compatibility translates verbally to "if it is sufficiently true that x belongs to A and if x and y are similar, then y belongs to A as well".

Theorem 7.29 *Let* C *be an* $\mathbf{L}_K$*-closure operator on* X*. Then* E_{C} *is the largest* $\mathbf{L}_K$*-equivalence on* X *that is compatible w.r.t. every* C*-closed* $\mathbf{L}$*-set (i.e. w.r.t. every* $A \in \mathcal{S}_{\mathrm{C}}$*).*

Proof. The fact that E_{C} is an $\mathbf{L}_K$-equivalence on X was established in the above paragraph. Let $x, y \in X$, $A \in \mathcal{S}_{\mathrm{C}}$, $A(x) \in K$. By Theorem 7.26,

$$E_{\mathrm{C}}(x, y) = Q_{\mathcal{S}_{\mathrm{C}}}(x, y) \wedge Q_{\mathcal{S}_{\mathrm{C}}}(y, x) \leq Q_{\mathcal{S}_{\mathrm{C}}}(x, y) =$$
$$= \bigwedge_{A \in \mathcal{S}_{\mathrm{C}}, A(x) \in K} A(x) \to A(y) \leq A(x) \to A(y),$$

i.e. $A(x) \otimes E_{\mathrm{C}}(x, y) \leq A(y)$ by adjointness. We proved that E_{C} is compatible w.r.t. any $A \in \mathcal{S}_{\mathrm{C}}$.

Let E be an $\mathbf{L}_K$-equivalence that is compatible w.r.t. any $A \in \mathcal{S}_{\mathrm{C}}$. Take $x, y \in X$ and an $A \in \mathcal{S}_{\mathrm{C}}$ such that $A(x) \in K$. Compatibility of E yields $A(x) \otimes E(x, y) \leq A(y)$, i.e. $E(x, y) \leq A(x) \to A(y)$. Since x, y, and A were chosen arbitrarily, we conclude

$$E(x, y) \leq \bigwedge_{A \in \mathcal{S}_{\mathrm{C}}, A(x) \in K} A(x) \to A(y) \leq Q_{\mathcal{S}_{\mathrm{C}}}(x, y).$$

Due to the symmetry of E we finally have

$$E(x, y) \leq Q_{\mathcal{S}_{\mathrm{C}}}(x, y) \wedge Q_{\mathcal{S}_{\mathrm{C}}}(y, x) = E_{\mathrm{C}}(x, y)$$

proving that E_{C} is the largest $\mathbf{L}_K$-equivalence compatible with all $A \in \mathcal{S}_{\mathrm{C}}$. □

Factorization: by similarity and by decrease of logical precision

For an $\mathbf{L}_K$-closure operator on X, $\mathcal{S}_{\mathrm{C}}$ (i.e. the set of fixed points of C) is a complete lattice w.r.t. $\subseteq$. This fact follows directly from the fact that $\mathcal{S}_{\mathrm{C}}$ is closed w.r.t. arbitrary intersections. For various reasons (e.g. for computational ones), it might not be desirable to distinguish the particular $\mathbf{L}$-sets in X. We are going to show that similarly as in the case of fuzzy concept lattices (which are in fact lattices of fixed points of fuzzy Galois connections), it is possible to factorize $\mathcal{S}_{\mathrm{C}}$ both by similarity and by decrease of logical precision. It will turn out that the corresponding factorization methods of fuzzy concept are special cases of the methods we are going to present.

Factorization of $\mathbf{L}_K$-closure systems by similarity. We deal with a situation similar to that of Section 5.2. We briefly recall our position: $E(A, B) = \bigwedge_x (A(x) \leftrightarrow B(x))$ defines an $\mathbf{L}$-equality on $\mathcal{S}_{\mathrm{C}}$; for each $a \in L$, ${}^aE$ is a tolerance relation on $\mathcal{S}_{\mathrm{C}}$. Provided ${}^aE$ is compatible with $\langle \mathcal{S}_{\mathrm{C}}, \subseteq \rangle$, we can apply Theorem 5.19 and obtain a factor lattice $\langle \mathcal{S}_{\mathrm{C}}, \subseteq \rangle / {}^aE$. For remarks on this construction we refer to Section 5.2. In the following we only aim to show that ${}^aE$ is a compatible tolerance on $\langle \mathcal{S}_{\mathrm{C}}, \subseteq \rangle$.

The first observation states that sufficiently high similarity between $\mathbf{L}$-sets is preserved by $\mathbf{L}_K$-closure operators.

Lemma 7.30 *For an $\mathbf{L}_K$-closure operator* C *on X and $A, B \in L^X$ we have $E(A, B) \leq E(\mathrm{C}(A), \mathrm{C}(B))$ whenever $E(A, B) \in K$.*

Proof. We have $E(A, B) = S(A, B) \wedge S(B, A)$. Therefore, $E(A, B) \in K$ yields $S(A, B) \in K$ and $S(B, A) \in K$. Applying (7.2) we get $S(A, B) \leq S(\mathrm{C}(A), \mathrm{C}(B))$ and $S(B, A) \leq S(\mathrm{C}(B), \mathrm{C}(A))$, and thus $E(A, B) = S(A, B) \wedge S(B, A) \leq S(\mathrm{C}(A), \mathrm{C}(B)) \wedge S(\mathrm{C}(B), \mathrm{C}(A)) = E(\mathrm{C}(A), \mathrm{C}(B))$. □

To show that the a-cuts ${}^aE$ can be used to factorize $\mathcal{S}_{\mathrm{C}}$ by the above-described procedure, we need to verify that ${}^aE$ is compatible w.r.t. $\subseteq$.

Lemma 7.31 *Let* C *be an $\mathbf{L}_K$-closure operator on X. For any $a \in K$, ${}^aE$ is a compatible tolerance relation on the complete lattice $\langle \mathcal{S}_{\mathrm{C}}, \subseteq \rangle$.*

Proof. We show that ${}^aE$ is compatible with both infima and suprema, i.e. we show that $\langle A_i, B_i \rangle \in {}^aE$ $(i \in I)$ implies both $\langle \bigwedge_{i\in I} A_i, \bigwedge_{i\in I} B_i \rangle \in {}^aE$ and $\langle \bigvee_{i\in I} A_i, \bigvee_{i\in I} B_i \rangle \in {}^aE$.

Infima: Suppose $\langle A_i, B_i \rangle \in {}^aE$, i.e. $a \leq \bigwedge_{x\in X} (A_i(x) \leftrightarrow B_i(x))$ $(i \in I)$. We have to show

$$a \leq \bigwedge_{x\in X} (\bigwedge_{i\in I} A_i(x) \leftrightarrow \bigwedge_{i\in I} B_i(x)),$$

i.e. to show that for each $i \in I$ we have

$$a \leq (\bigwedge_{i\in I} A_i(x) \leftrightarrow \bigwedge_{i\in I} B_i(x)).$$

The last inequality is true iff both $a \leq (\bigwedge_{i\in I} A_i(x) \to \bigwedge_{i\in I} B_i(x))$ and $a \leq (\bigwedge_{i\in I} B_i(x) \to \bigwedge_{i\in I} A_i(x))$ are valid. Due to symmetry we verify only $a \leq (\bigwedge_{i\in I} A_i(x) \to \bigwedge_{i\in I} B_i(x))$ which is equivalent to $a \otimes \bigwedge_{i\in I} A_i(x) \leq \bigwedge_{i\in I} B_i(x)$. This inequality is true. Indeed, by assumption, $\langle A_i, B_i\rangle \in {}^aE$, i.e. $a \leq \bigwedge_{x\in X}(A_i(x) \leftrightarrow B_i(x))$, from which it follows $a \otimes A_i(x) \leq B_i(x)$ for any $i \in I$, $x \in X$. We therefore have

$$a \otimes \bigwedge_{i\in I} A_i(x) \leq$$
$$\leq \bigwedge_{i\in I}(a \otimes A_i(x)) \leq \bigwedge_{i\in I} B_i(x).$$

Suprema: We have to show $a \leq E(\bigvee_{i\in I} A_i, \bigvee_{i\in I} B_i)$, i.e. $a \leq E(\mathrm{C}(\bigcup_{i\in I} A_i), \mathrm{C}(\bigcup_{i\in I} B_i))$. First, observe that $(*)$ $a \leq E(\bigcup_{i\in I} A_i, \bigcup_{i\in I} B_i)$: Indeed, the inequality is true iff both $a \leq S(\bigcup_{i\in I} A_i, \bigcup_{i\in I} B_i)$ and $a \leq S(\bigcup_{i\in I} B_i, \bigcup_{i\in I} A_i)$ hold. Because of symmetry, we verify only the former one: by definition, $a \leq S(\bigcup_{i\in I} A_i, \bigcup_{i\in I} B_i)$ is true iff $a \leq (\bigcup_{i\in I} A_i)(x) \to (\bigcup_{i\in I} B_i)(x)$, i.e. iff $a \otimes (\bigcup_{i\in I} A_i)(x) \leq (\bigcup_{i\in I} B_i)(x)$ holds for each $x \in X$. By assumption,

$$a \otimes (\bigcup_{i\in I} A_i)(x) = \bigvee_{i\in I}(a \otimes A_i(x)) \leq \bigvee_{i\in I} B_i(x) = (\bigcup_{i\in I} B_i)(x)$$

establishing $(*)$.

Now, $a \leq E(\bigcup_{i\in I} A_i, \bigcup_{i\in I} B_i)$ implies both $a \leq S(\bigcup_{i\in I} A_i, \bigcup_{i\in I} B_i)$ and $a \leq S(\bigcup_{i\in I} B_i, \bigcup_{i\in I} A_i)$. Since $a \in K$, (7.2) implies $a \leq S(\mathrm{C}(\bigcup_{i\in I} A_i), \mathrm{C}(\bigcup_{i\in I} B_i))$ and $a \leq S(\mathrm{C}(\bigcup_{i\in I} B_i), \mathrm{C}(\bigcup_{i\in I} A_i))$, i.e.

$$a \leq E(\mathrm{C}(\bigcup_{i\in I} A_i), \mathrm{C}(\bigcup_{i\in I} B_i)) = E(\bigvee_{i\in I} A_i, \bigvee_{i\in I} B_i)$$

completing the proof. □

Remark 7.32 (1) Factorization of $\mathcal{S}_{\mathrm{C}}$ by putting fixed points which are at least a-similar is thus feasible.

(2) Note that the tolerance relation ${}^aE$ used to factorize $\mathcal{S}_{\mathrm{C}}$ need not be supplied from the outside. It is determined by selecting an appropriate $a \in K$.

(3) Due to Theorem 7.20, Theorem 5.22 is a consequence of Lemma 7.31.

Factorization of $\mathbf{L}_K$-closure systems by decrease of logical precision. We now mention another way to factorize a system $\mathcal{S}_{\mathrm{C}}$ of an **L**-closure operator C; this time by a decrease of logical precision.

Theorem 7.33 *Let* C *be an* $\mathbf{L}_1$*-closure operator in* X, h *be a complete homomorphism of* $\mathbf{L}_1$ *onto* $\mathbf{L}_2$. *Then* $h(\mathrm{C})$, *defined for* $B \in L_2^X$ *by* $(h(\mathrm{C}))(B) = h(\mathrm{C}(A))$ *where* $B = h(A)$ *for* $A \in L_1^X$, *is an* $\mathbf{L}_2$*-closure operator in* X *and* $A \mapsto h(A)$ *is a complete lattice homomorphism of* $\mathcal{S}_{\mathrm{C}}$ *onto* $\mathcal{S}_{h(\mathrm{C})}$.

Proof. First, we show that $h(\mathrm{C})$ is defined correctly. In order to do it, take any $A_1, A_2 \in L_1^X$ such that $h(A_1) = h(A_2)$. We need to show $h(\mathrm{C}(A_1)) = h(\mathrm{C}(A_2))$. It is easy to show that $h(E(A_1, A_2)) = 1$. Since $E(A_1, A_2) \leq E(\mathrm{C}(A_1), \mathrm{C}(A_2))$, monotony of h gives $h(E(\mathrm{C}(A_1), \mathrm{C}(A_2)))$. As $h(E(\mathrm{C}(A_1), \mathrm{C}(A_2))) = E(h(\mathrm{C}(A_1)), h(\mathrm{C}(A_2)))$, we infer $E(h(\mathrm{C}(A_1)), h(\mathrm{C}(A_2))) = 1$, whence $h(\mathrm{C}(A_1)) = h(\mathrm{C}(A_2))$. Therefore, $h(\mathrm{C})$ is defined correctly.

Let $\langle {}^{\uparrow_{\mathrm{C}}}, {}^{\downarrow_{\mathrm{C}}} \rangle$ be the **L**-Galois connection between X and $\mathcal{S}_{\mathrm{C}}$ of Theorem 7.21. Consider an $\mathbf{L}_1$-relation I between X and $\mathcal{S}_{\mathrm{C}}$ defined for $x \in X$ and $A \in \mathcal{S}_{\mathrm{C}}$ by

$$I(x, A) = A(x).$$

Let $\langle {}^{\uparrow_I}, {}^{\downarrow_I} \rangle$ be the $\mathbf{L}_1$-Galois connection between X and $\mathcal{S}_{\mathrm{C}}$ according to Theorem 5.45. It is immediate that $\langle {}^{\uparrow_{\mathrm{C}}}, {}^{\downarrow_{\mathrm{C}}} \rangle = \langle {}^{\uparrow_I}, {}^{\downarrow_I} \rangle$. We will show that $h(\mathrm{C}) = {}^{\uparrow_{h(I)}\downarrow_{h(I)}}$, i.e. $h(\mathrm{C})$ is a composition of ${}^{\uparrow_{h(I)}}$ and ${}^{\downarrow_{h(I)}}$. To this end, take any $A \in L_1^X$. Using repeatedly the fact that h is a complete homomorphism, we get

$$h(A)^{\uparrow_{h(I)}\downarrow_{h(I)}} = \cdots = h(A^{\uparrow_I\downarrow_I}) = h(\mathrm{C}(A)) = (h(\mathrm{C}))(A)$$

proving $h(\mathrm{C}) = {}^{\uparrow_{h(I)}\downarrow_{h(I)}}$. By Theorem 7.21, $h(\mathrm{C})$ is an $\mathbf{L}_2$-closure operator in X.

Finally, by Theorem 5.30 and its proof, $\langle A, A^{\uparrow_I} \rangle \mapsto \langle h(A), h(A^{\uparrow_I}) \rangle$ is a complete lattice homomorphism of $\mathcal{B}_I = \{\langle A, A^{\uparrow_I} \rangle \mid A \in \mathcal{S}_{\mathrm{C}}\}$ onto $\mathcal{B}_{h(I)} = \{\langle A, A^{\uparrow_{h(I)}} \rangle \mid A \in \mathcal{S}_{h(\mathrm{C})}\}$ (with the lattice order $\leq$ on $\mathcal{B}_I$ given by $\langle A, A^{\uparrow_I} \rangle \leq \langle B, B^{\uparrow_I} \rangle$ iff $A \subseteq B$, similarly for $\mathcal{B}_{h(I)}$). The fact that $A \mapsto h(A)$ is a complete lattice homomorphism of $\mathcal{S}_{\mathrm{C}}$ onto $\mathcal{S}_{h(\mathrm{C})}$ now directly follows by observing that $\mathcal{B}_I$ is isomorphic to $\mathcal{S}_{\mathrm{C}}$, and $\mathcal{B}_{h(I)}$ is isomorphic to $\mathcal{S}_{h(\mathrm{C})}$. □

$\mathcal{S}_{h(\mathrm{C})}$ is therefore a homomorphic image of $\mathcal{S}_{\mathrm{C}}$, i.e. a factor lattice of $\mathcal{S}_{\mathrm{C}}$.

Remark 7.34 The fact that $h(\mathrm{C})$ is an $\mathbf{L}_2$-closure operator can be proved directly (without reference to Theorem 7.21) as follows: For any $A \in L_1^X$ we have $A \subseteq \mathrm{C}(A)$, whence $h(A) \subseteq h(\mathrm{C}(A)) = (h(\mathrm{C}))(A)$ by monotony of h, proving (7.1). Furthermore, $S(A_1, A_2) \leq$

$S(\mathrm{C}(A_1), \mathrm{C}(A_2))$ implies

$$\begin{aligned} & S(h(A_1), h(A_2)) = h(S(A_1, A_2)) \leq h(S(\mathrm{C}(A_1), \mathrm{C}(A_2))) = \\ = \; & S(h(\mathrm{C}(A_1)), h(\mathrm{C}(A_2))) = S((h(\mathrm{C}))(A_1), (h(\mathrm{C}))(A_2)) \end{aligned}$$

proving (7.2). Finally,

$$\begin{aligned} & (h(\mathrm{C}))(h(A)) = h(\mathrm{C}(A)) = h(\mathrm{C}(\mathrm{C}(A))) = \\ = \; & (h(\mathrm{C}))(h(\mathrm{C}(A))) = (h(\mathrm{C}))((h(\mathrm{C}))(h(A))) \end{aligned}$$

proving (7.3).

Representation by 2-closure operators

We show that there is a natural one-to-one correspondence between $\mathbf{L}_{\{1\}}$-closure operators on X and special closure operators on $X \times L$ (similarly as that for fuzzy Galois connections described in Section 5.3). To this end recall the representation of **L**-sets in X by representative subsets of $X \times L$.

Definition 7.35 A **2**-closure operator D on $X \times L$ is called *commutative w.r.t.* $\lfloor\lceil\ \rceil\rfloor$ if

$$\lfloor\lceil \mathrm{D}(A) \rceil\rfloor = \mathrm{D}(A) = \mathrm{D}(\lfloor\lceil A \rceil\rfloor) \tag{7.12}$$

holds for each $A \in X \times L$.

Remark 7.36 It is easy to verify that (7.12) holds iff both $\lfloor\lceil \mathrm{D}(A) \rceil\rfloor \subseteq \mathrm{D}(A)$ and $\mathrm{D}(\lfloor\lceil A \rceil\rfloor) \subseteq \mathrm{D}(A)$ hold.

For an operator $\mathrm{D} : X \times L \to X \times L$ define an operator $\mathrm{C}_\mathrm{D} : L^X \to L^X$ by

$$\mathrm{C}_\mathrm{D}(A) = \lceil \mathrm{D}(\lfloor \mathrm{A} \rfloor) \rceil$$

for $A \in L^X$. For an operator $\mathrm{C} : L^X \to L^X$ define an operator $\mathrm{D}_\mathrm{C} : X \times L \to X \times L$ by

$$\mathrm{D}_\mathrm{C}(A) = \lfloor \mathrm{C}(\lceil \mathrm{A} \rceil) \rfloor$$

for $A \in X \times L$.

Theorem 7.37 *Let* C *be an* $\mathbf{L}_{\{1\}}$*-closure operator on* X *and* D *be a* **2***-closure operator on* $X \times L$ *which is commutative w.r.t.* $\lfloor\lceil\ \rceil\rfloor$. *Then (1)* D_C *is a* **2***-closure operator on* X *which is commutative w.r.t.* $\lfloor\lceil\ \rceil\rfloor$, *(2)* C_D *is an* $\mathbf{L}_{\{1\}}$*-closure operator on* X, *and (3)* $\mathrm{C} = \mathrm{C}_{\mathrm{D}_\mathrm{C}}$ *and* $\mathrm{D} = \mathrm{D}_{\mathrm{C}_\mathrm{D}}$.

Proof. (1) Let $A, A_1, A_2 \subseteq X \times L$. We have

$$A \subseteq \lfloor\lceil A \rceil\rfloor \subseteq \lfloor \mathrm{C}(\lceil A \rceil) \rfloor = \mathrm{D}_\mathrm{C}(A),$$

proving extensionality of $\mathrm{D_C}$. If $A_1 \subseteq A_2$ then clearly $S(\lceil A_1\rceil, \lceil A_2\rceil) = 1$, hence $S(\mathrm{C}(\lceil A_1\rceil), \mathrm{C}(\lceil A_2\rceil)) = 1$, so

$$\mathrm{D_C}(A_1) = \lfloor \mathrm{C}(\lceil A_1\rceil)\rfloor \subseteq \lfloor \mathrm{C}(\lceil A_2\rceil)\rfloor = \mathrm{D_C}(A_2),$$

proving monotony of $\mathrm{D_C}$.

$$\mathrm{D_C}(\mathrm{D_C}(A)) = \lfloor \mathrm{C}(\lceil\lfloor \mathrm{C}(\lceil A\rceil)\rfloor\rceil)\rfloor = \lfloor \mathrm{C}(\mathrm{C}(\lceil A\rceil))\rfloor = \lfloor \mathrm{C}(\lceil A\rceil)\rfloor = \mathrm{D_C}(A),$$

proving idempotency. Finally,

$$\lfloor\lceil \mathrm{D_C}(A)\rceil\rfloor = \lfloor\lceil\lfloor \mathrm{C}(\lceil A\rceil)\rfloor\rceil\rfloor = \lfloor \mathrm{C}(\lceil A\rceil)\rfloor = \mathrm{D_C}(A),$$

and

$$\mathrm{D_C}(\lfloor\lceil A\rceil\rfloor) = \lfloor \mathrm{C}(\lceil\lfloor\lceil A\rceil\rfloor\rceil)\rfloor = \lfloor \mathrm{C}(\lceil A\rceil)\rfloor = \mathrm{D_C}(A),$$

verifying commutativity of $\mathrm{D_C}$.

(2) Let $A, A_1, A_2 \in L^X$. We have $A = \lceil\lfloor A\rfloor\rceil \subseteq \lceil \mathrm{D}(\lfloor A\rfloor)\rceil = \mathrm{C_D}(A)$, thus $\mathrm{C_D}$ is extensional. If $S(A_1, A_2) = 1$ then $\lfloor A_1\rfloor \subseteq \lfloor A_2\rfloor$, therefore $\mathrm{D}(\lfloor A_1\rfloor) \subseteq \mathrm{D}(\lfloor A_2\rfloor)$ and thus

$$\mathrm{C_D}(A_1) = \lceil \mathrm{D}(\lfloor A_1\rfloor)\rceil \subseteq \lceil \mathrm{D}(\lfloor A_2\rfloor)\rceil = \mathrm{C_D}(A_2),$$

monotony of $\mathrm{C_D}$. Using commutativity we further get

$$\mathrm{C_D}(\mathrm{C_D}(A)) = \lceil \mathrm{D}(\lfloor\lceil \mathrm{D}(\lfloor A\rfloor)\rceil\rfloor)\rceil = \lceil \mathrm{D}(\mathrm{D}(\lfloor A\rfloor))\rceil = \lceil \mathrm{D}(\lfloor A\rfloor)\rceil = \mathrm{C_D}(A),$$

idempotency of $\mathrm{C_D}$.

(3) For any $A \in L^X$ we have $\mathrm{C_{D_C}}(A) = \lceil\lfloor \mathrm{C}(\lceil\lfloor A\rfloor\rceil)\rfloor\rceil = \mathrm{C}(A)$. For any $A \subseteq X \times L$ we have by commutativity of D that $\mathrm{D_{C_D}}(A) = \lfloor\lceil \mathrm{D}(\lfloor\lceil A\rceil\rfloor)\rceil\rfloor = \mathrm{D}(A)$. □

Remark 7.38 Note that commutativity of D is essential in the foregoing proposition (a counterexample is easy to get).

Examples of fuzzy closure operators

In this section we introduce two further properties of $\mathbf{L}_K$-closure operators and show some examples. Call an **L**-set $A \in L^X$ *finite* if $\{x \in X \mid A(x) > 0\}$ is a finite set.

Definition 7.39 An $\mathbf{L}_K$ closure operator C on X is called *compact* (finitary, or algebraic) if

$$\mathrm{C}(A) = \bigcup\{\mathrm{C}(B) \mid B \in L^X, B \subseteq A, B \text{ is finite }\}$$

holds for each $A \in L^X$. An $\mathbf{L}_K$-closure operator C on X is called *topologic* if it satisfies

$$\mathrm{C}(A \cup B) = \mathrm{C}(A) \cup \mathrm{C}(B)$$

for any $A, B \in L^X$.

Remark 7.40 For $\mathbf{L} = \mathbf{2}$ we get compact closure operators. For $\mathbf{L} = \mathbf{2}$, topologic $\mathbf{L}_K$-closure operators are just closure operators of topologic spaces. The corresponding system $\mathcal{S}_\mathrm{C}$ consists of the closed sets of the topology.

Fuzzy subalgebras. Recall from Remark 3.135 the notion of a fuzzy subalgebra. Let $\mathbf{A} = \langle A, F\rangle$ be an algebra, i.e. A is a nonempty set and F is a system of operations on A. An **L**-set $B \in L^A$ is called an **L**-subalgebra of **A** if for each $f : A^n \to A$ of F and every $a_1, \dots, a_n \in A$, we have

$$B(a_1) \otimes \cdots \otimes B(a_n) \leq B(f(a_1, \dots, a_n)).$$

Denote by **L**-Sub **A** the set of all **L**-subalgebras of **A**.

Theorem 7.41 *For any algebra* $\mathbf{A} = \langle A, F\rangle$, **L**-Sub **A** *is an* $\mathbf{L}_{\{1\}}$-*closure system and the corresponding operator* $\mathrm{C} = \mathrm{C}_{\mathbf{L}\text{-Sub}\,\mathbf{A}}$ *is a compact* $\mathbf{L}_{\{1\}}$-*closure operator. Moreover, if* $\otimes = \wedge$ *(i.e.* **L** *is a Heyting algebra),* C *is an* **L**-*closure operator.*

Proof. It is easy to see that **L**-Sub **A** is closed under arbitrary intersections, hence **L**-Sub **A** is an $\mathbf{L}_{\{1\}}$-closure system and the corresponding C is an $\mathbf{L}_{\{1\}}$-closure operator. It remains to verify compactness of C. To this end, let $B \in L^A$ and put

$$\begin{aligned}[B](a) &= \bigvee\{B(a_1)^{|t|_{x_1}} \otimes \cdots \\ &\qquad \cdots \otimes B(a_n)^{|t|_{x_n}} \mid t \in T_n,\ a_i \in A, t(a_1, \dots, a_n) = a\}\end{aligned}$$

where T_n denotes the set of all n-ary terms of the type of **A**. It is a matter of routine to prove that $[B] = \mathrm{C}(B)$ (we have $B \subseteq [B]$, $[B] \subseteq \mathrm{C}(B)$, and $[B]$ is an **L**-subalgebra) which implies compactness of C.

Let $\otimes = \wedge$. We have to prove $S(B_1, B_2) \leq S([B_1], [B_2])$ for every $B_1, B_2 \in L^A$ which holds iff for each $a \in A$ we have $[B_1](a) \wedge S(B_1, B_2) \leq [B_2](a)$. Since $\wedge$ is idempotent we have

$$[B_1](a) \wedge S(B_1, B_2) = \tag{7.13}$$

$$= \bigvee_{t \in T_n, t(a_1, \dots, a_n) = a} (B_1(a_1) \wedge \cdots \wedge B_1(a_n)) \wedge S(B_1, B_2) = \tag{7.14}$$

$$= \bigvee_{t\in T_n, t(a_1,\dots,a_n)=a} (B_1(a_1) \wedge S(B_1, B_2) \wedge \cdots \tag{7.15}$$

$$\cdots \wedge B_1(a_n) \wedge S(B_1, B_2)) \leq \tag{7.16}$$

$$\leq \bigvee_{t\in T_n, t(a_1,\dots,a_n)=a} (B_2(a_1) \wedge \cdots \wedge B_2(a_n)) = [B_2](a). \tag{7.17}$$

□

Remark 7.42 Note that in general, $\mathrm{C} = \mathrm{C}_{\mathbf{L}\text{-Sub}\,\mathbf{A}}$ is not an **L**-closure system. As a counterexample consider $L = \{0, \frac{1}{2}, 1\}$ with a Łukasiewicz structure, and a four-element lattice with the support $A = \{a, b, c, d\}$ with the least element a, the greatest element d, and two mutually incomparable elements b and c, i.e. its Hasse diagram is a 45°-rotated square. Let $B_1, B_2 \in L^A$ be given by $B_1(a) = 0$, $B_1(b) = B_1(c) = B_1(d) = 1$, $B_2(a) = 0$, $B_2(b) = B_2(c) = B_2(d) = \frac{1}{2}$. Clearly, B_2 itself is an **L**-subalgebra of **A**. On the other hand, $[B_1](a) = 1$ since $B_1(b) \otimes B_1(c) = 1 \leq [B_1](b \wedge c) = [B_1](a)$. We therefore have $S(B_1, B_2) = \frac{1}{2} \not\leq 0 = [B_1](a) \to [B_2](a) = S([B_1], [B_2])$, i.e. $[\ ] = \mathrm{C}$ is not an **L**-closure operator.

Fuzzy relational closures. Our next example concerns fuzzy relational closures on X introduced in Chapter 4. We concentrate on $\mathrm{Cl}_{\mathrm{Ref}}$, $\mathrm{Cl}_{\mathrm{Sym}}$, and $\mathrm{Cl}_{\mathrm{Tra}}$; for simplicity we denote the closures $\mathrm{Cl}_{\mathrm{Ref}}(R)$, $\mathrm{Cl}_{\mathrm{Sym}}(R)$, and $\mathrm{Cl}_{\mathrm{Tra}}(R)$ of a binary **L**-relation R by R^r, R^s, and R^t, respectively. Recall furthermore that $R^r = R \cup_{x\in X} \{{}^1/\langle x, x\rangle\}$, $R^s = R \cup R^{-1}$, and $R^t = \bigcup_{i=1}^{\infty} R^i = R \cup R\circ R \cup R\circ R\circ R \cup \cdots$.

Since $R \subseteq R^r$, $R \subseteq R^s$, $S(R, S) \leq S(R^r, S^r)$, $S(R, S) \leq S(R^s, S^s)$ (both of the inequalities are easy to verify), and $R^r = R^{rr}$, $R^s = R^{ss}$, we conclude that both ^{r} and ^{s} are **L**-closure operators on $X \times X$. Moreover, by the above description of ^{r} and ^{s} we conclude that both of them are compact as well as topologic.

Consider ^{t}: We have $R \subseteq R^t$, $R^t = R^{tt}$, and $R \subseteq S$ implies $R^t \subseteq S^t$. Thus, ^{t} is an $\mathbf{L}_{\{1\}}$-closure operator in $X \times X$. In general, ^{t} is not an **L**-closure operator (consider $L = \{0, \frac{1}{2}, 1\}$ with Łukasiewicz structure, $X = \{a, b, c\}$, $R(a, b) = R(b, c) = 1$, $S(a, b) = S(b, c) = \frac{1}{2}$, and $R(x, y) = S(x, y) = 0$ otherwise). ^{t} is compact since

$$R^t(x, y) = \bigvee_{i=1}^{\infty} R^i(x, y) =$$

$$= \bigvee_{i=1}^{\infty} \bigvee_{x=z_1, z_2, \dots, z_{i+1}=y} R(z_1, z_2) \otimes \cdots \otimes R(z_i, z_{i+1}) =$$

$$= \bigvee_{i=1}^{\infty} \bigvee_{x=z_1,z_2,\ldots,z_{i+1}=y} \{ R(z_1, z_2)/\langle z_1, z_2\rangle, \cdots$$
$$\ldots, R(z_i, z_{i+1})/\langle z_i, z_{i+1}\rangle \}^t(x, y).$$

As is well-known from the classical case ($L = \{0, 1\}$), ^{t} is not topologic.

Remark 7.43 An easy inspection shows that if $\mathbf{L}$ is a Heyting algebra ($\otimes = \wedge$) then ^{t} is even an $\mathbf{L}$-closure operator on $X \times X$.

Fuzzy concept lattices. We only mention that due to previous results (Theorems 7.20 and 7.21), each $\mathbf{L}$-concept lattice is isomorphic to some $\langle \mathcal{S}_{\mathrm{C}}, \subseteq \rangle$ of some $\mathbf{L}$-closure operator and, conversely, each $\langle \mathcal{S}_{\mathrm{C}}, \subseteq \rangle$ of some $\mathbf{L}$-closure operator is isomorphic to some $\mathbf{L}$-concept lattice. Thus, in a sense, $\mathbf{L}$-concept lattices and $\mathbf{L}$-closure operators are the same.

Fuzzy closure operators and consequence relations. The concept of a closure operator is an important one from the point of view of logic. Typically, a closure operator C in a given logical calculus arises as follows: for a given set collection A of formulas, the closure $\mathrm{C}(A)$ is defined to be the collection of all formulas that follow (syntactically or semantically) from A. A detailed study of closure operators in the context of two-valued logic can be found in [208]. The situation is analogous in fuzzy logic. Namely, following the seminal work of Pavelka [158], the provability degree of a formula from a fuzzy set of formulas is defined in fuzzy logic. Then, a fuzzy closure operator is naturally induced by a fuzzy logical calculus as follows: for a given fuzzy set A of formulas and a given formula φ, the degree to which φ belongs to the closure $\mathrm{C}(A)$ of A is defined to be the provability degree of φ from A.

Our aim in this section is to present a general result on the relationship between fuzzy closure operators and fuzzy consequence relations. First, we show that each binary $\mathbf{L}$-relation between L^X (the set of all $\mathbf{L}$-sets in a given set X) and X induces in a natural way a fuzzy closure system (and the corresponding fuzzy closure operator). For an $\mathbf{L}$-relation R between L^X and X, and a subset $K \subseteq L$, we say that an $\mathbf{L}$-set $A \in L^X$ is R_K-closed if for any $B \in L^X$ and each $x \in X$ we have

$$S(B, A) \otimes R(B, x) \leq A(x)$$

whenever $S(B, A) \in K$.

Lemma 7.44 *For any $R \in L^{L^X \times X}$ and a filter $K \subseteq L$, the set $\mathcal{S}_R$ of all $\mathbf{L}$-sets in X that are R_K-closed forms an $\mathbf{L}_K$-closure system.*

Proof. By definition, we have to show that $\mathcal{S}_R$ is closed w.r.t. S_K-intersections, i.e. we have to show that for any $A \in L^X$, $(\bigcap_{B\in\mathcal{S}_R, S(A,B)\in K} S(A,B) \to B)$ is R_K-closed. Take any $C \in L^X$ such that $S(C, \bigcap_{B\in\mathcal{S}_R, S(A,B)\in K} S(A,B) \to B) \in K$. We have to show

$$\begin{aligned} & S(C, \bigcap_{B\in\mathcal{S}_R, S(A,B)\in K} S(A,B) \to B) \otimes R(C,x) \leq \\ \leq & \bigwedge_{B\in\mathcal{S}_R, S(A,B)\in K} S(A,B) \to B(x) \end{aligned}$$

is true for any $x \in X$. The last inequality holds iff for any $B \in L^X$ such that $S(A,B) \in K$ we have

$$S(A,B) \otimes S(C, \bigcap_{B\in\mathcal{S}_R, S(A,B)\in K} S(A,B) \to B) \otimes R(C,x) \leq B(x)$$

which is clearly true (since B is R_K-closed) provided both $S(A,B) \otimes S(C, \bigcap_{B\in\mathcal{S}_R, S(A,B)\in K} S(A,B) \to B) \leq S(C,B)$ and $S(C,B) \in K$ are valid. We have

$$\begin{aligned} & S(A,B) \otimes S(C, \bigcap_{B\in\mathcal{S}_R, S(A,B)\in K} S(A,B) \to B) \leq \\ \leq\ & S(A,B) \otimes S(C, S(A,B) \to B) = \\ =\ & S(A,B) \otimes \bigwedge_{y\in X} (C(y) \to (S(A,B) \to B(y))) = \\ =\ & S(A,B) \otimes \bigwedge_{y\in X} (S(A,B) \to (C(y) \to B(y))) \leq \\ \leq\ & \bigwedge_{y\in X} (S(A,B) \otimes (S(A,B) \to (C(y) \to B(y)))) \leq \\ \leq\ & \bigwedge_{y\in X} C(y) \to B(y) = S(C,B). \end{aligned}$$

Since $S(A,B) \in K$ and $S(C, \bigcap_{B\in\mathcal{S}_R, S(A,B)\in K} S(A,B) \to B) \in K$, we conclude $S(C,B) \in K$. □

In fact, each $\mathbf{L}_K$-closure system is induced in the way described in Lemma 7.44 by its corresponding $\mathbf{L}_K$-closure operator: Each $\mathbf{L}_K$-closure operator C on X induces an $\mathbf{L}$-relation $R_{\mathrm{C}} \in L^{L^X \times X}$ by

$$R_{\mathrm{C}}(A,x) = \mathrm{C}(A)(x).$$

Now, applying Lemma 7.44 to R_{C} we get the $\mathbf{L}_K$-closure system corresponding to C:

Lemma 7.45 *For an $\mathbf{L}_K$-closure operator* C *on* X *we have* $\mathcal{S}_{\mathrm{C}} = \mathcal{S}_{R_{\mathrm{C}}}$.

Proof. Let $A \in \mathcal{S}_{\mathrm{C}}$. We show that A is R_K-closed. Let $S(B, A) \in K$. Then, by (7.2), $S(B, A) \leq S(\mathrm{C}(B), \mathrm{C}(A))$, i.e. for each $x \in X$ we have

$$S(B, A) \otimes R_{\mathrm{C}}(B, x) = S(B, A) \otimes \mathrm{C}(B)(x) \leq \mathrm{C}(A)(x) = A(x),$$

whence A is R_K-closed. Conversely, if A is R_K-closed then for any B such that $S(B, A) \in K$ we have $S(B, A) \otimes R_{\mathrm{C}}(B, x) \leq A(x)$. Putting $B = A$ and considering $S(A, A) = 1 \in K$, we conclude

$$S(B, A) \otimes R_{\mathrm{C}}(B, x) = 1 \otimes R_{\mathrm{C}}(A, x) = \mathrm{C}(A)(x) \leq A(x),$$

thus $A \in \mathcal{S}_{\mathrm{C}}$. □

Recall (see e.g. [84]) that a (bivalent) relation $\vdash$ between 2^X and X is called a consequence relation in (a given set) X if (i) $X \vdash \varphi$ holds for each $\varphi \in X$; (ii) $X \vdash \varphi$ and $X \subseteq Y$ imply $Y \vdash \varphi$; and (iii) $X \cup Y \vdash \varphi$ and $X \vdash \psi$ for any $\psi \in Y$ imply $X \vdash \varphi$.

Definition 7.46 An $\mathbf{L}$-relation $\vdash$ between L^X and X is called an $\mathbf{L}_K$-consequence relation provided it satisfies

(i) $A(\varphi) \leq (A \vdash \varphi)$

(ii) $(A \vdash \varphi) \otimes S(A, B) \leq (B \vdash \varphi)$ whenever $(B \vdash \varphi) \in K$

(iii) $(\bigwedge_{\varphi \in X} B(\psi) \to (A \vdash \psi)) \otimes ((A \cup B) \vdash \varphi) \leq (A \vdash \varphi)$ whenever $(\bigwedge_{\varphi \in X} B(\psi) \to (A \vdash \psi)) \in K$

for any $A, B \in L^X$, $\varphi \in X$.

Remark 7.47 Note that condition (iii) may be equivalently replaced by (iii'):

$$(\bigwedge_{\varphi \in X} B(\psi) \to (A \vdash \psi)) \otimes (B \vdash \varphi) \leq (A \vdash \varphi)$$

whenever $(\bigwedge_{\varphi \in X} B(\psi) \to (A \vdash \psi)) \in K$. Indeed, $(B \vdash \varphi) \leq ((A \cup B) \vdash \varphi)$ by (ii), whence (iii) implies (iii'). Conversely, suppose (iii') and take $B' = A \cup B$. By $A(\psi) \leq (A \vdash \psi)$ we have

$$\begin{aligned} & \bigwedge_{\psi \in X} ((A \cup B)(\psi) \to (A \vdash \psi)) = \\ = \; & \bigwedge_{\psi \in X} (A(\psi) \to (A \vdash \psi)) \wedge \bigwedge_{\psi \in X} (B(\psi) \to (A \vdash \psi)) = \\ = \; & \bigwedge_{\psi \in X} (B(\psi) \to (A \vdash \psi)). \end{aligned}$$

Therefore,

$$
\begin{aligned}
&(\bigwedge_{\psi\in X}(B(\psi)\to(A\vdash\psi)))\otimes((A\cup B)\vdash\varphi)=\\
=\;&(\bigwedge_{\psi\in X}((A\cup B)(\psi)\to(A\vdash\psi)))\otimes((A\cup B)\vdash\varphi)=\\
=\;&(\bigwedge_{\psi\in X}(B'(\psi)\to(A\vdash\psi)))\otimes(B'\vdash\varphi)\leq(A\vdash\varphi)
\end{aligned}
$$

by (iii').

For a mapping $\mathrm{C}: L^X \to L^X$, define an $\mathbf{L}$-relation $\vdash_{\mathrm{C}}$ between L^X and X by

$$(A\vdash_{\mathrm{C}}\varphi)=\mathrm{C}(A)(\varphi).$$

For an $\mathbf{L}$-relation $\vdash$ between L^X and X define a mapping $\mathrm{C}_{\vdash}: L^X \to L^X$ by

$$\mathrm{C}_{\vdash}(A)(\varphi)=(A\vdash\varphi).$$

Theorem 7.48 *Let* $\mathrm{C}: L^X \to L^X$ *be a mapping,* $\vdash\in L^{L^X\times X}$ *be an* $\mathbf{L}$*-relation. Then (1)* C *is an* $\mathbf{L}_K$*-closure operator iff* $\vdash_{\mathrm{C}}$ *is an* $\mathbf{L}_K$*-consequence relation; (2)* $\vdash$ *is an* $\mathbf{L}_K$*-consequence relation iff* $\mathrm{C}_{\vdash}$ *is an* $\mathbf{L}_K$*-closure operator; (3)* $\mathrm{C}=\mathrm{C}_{\vdash_{\mathrm{C}}}$ *and* $\vdash=\vdash_{\mathrm{C}_{\vdash}}$.

Proof. Clearly, (3) is true. Therefore, it is sufficient to establish the "$\Rightarrow$"-parts of (1) and (2).

(1): We verify that $\vdash_{\mathrm{C}}$ is an $\mathbf{L}_K$-consequence operator. (i) and (ii) are direct consequences of (7.1) and (7.2). We now verify (iii'), a condition equivalent to (iii) (see Remark 7.47):

$$(\bigwedge_{\psi\in X}B(\psi)\to(A\vdash_{\mathrm{C}}\psi))\otimes(B\vdash_{\mathrm{C}}\varphi)\leq(A\vdash_{\mathrm{C}}\varphi)$$

is by definition equivalent to

$$(\bigwedge_{\psi\in X}B(\psi)\to\mathrm{C}(A)(\psi))\otimes\mathrm{C}(B)(\varphi)\leq\mathrm{C}(A)(\varphi)$$

which is true iff

$$\bigwedge_{\psi\in X}B(\psi)\to\mathrm{C}(A)(\psi)\leq\mathrm{C}(B)(\varphi)\to\mathrm{C}(A)(\varphi).$$

The last inequality is valid since $\bigwedge_{\psi\in X}B(\psi)\to\mathrm{C}(A)(\psi)\in K$ implies

$$
\begin{aligned}
&\bigwedge_{\psi\in X}B(\psi)\to\mathrm{C}(A)(\psi)=S(B,\mathrm{C}(A))\leq\\
\leq\;&S(\mathrm{C}(B),\mathrm{C}(\mathrm{C}(A)))\leq\mathrm{C}(B)(\varphi)\to\mathrm{C}(A)(\varphi).
\end{aligned}
$$

(2): We check that $\mathrm{C}_\vdash$ is an $\mathbf{L}_K$-closure operator. (7.1) and (7.2) are direct consequences of (i) and (ii). Putting $B = \mathrm{C}_\vdash(A)$, (iii') yields

$$(\mathrm{C}_\vdash(\mathrm{C}_\vdash(A)))(\varphi) = 1 \otimes (\mathrm{C}_\vdash(A) \vdash \varphi) =$$
$$= \bigwedge_{\psi \in X} (B(\psi) \to (A \vdash \psi)) \otimes (B \vdash \varphi) \leq (A \vdash \varphi) = \mathrm{C}_\vdash(A)(\varphi).$$

□

Remark 7.49 Theorem 7.48 thus establishes a one-to-one correspondence between fuzzy closure operators and fuzzy consequence relations. In [84], Gerla defines graded consequence relation in X as a fuzzy relation (he takes $L = [0,1]$ and $\otimes = \min$) between $\mathbf{2}^X$ (the power set of X) and X. Gerla then establishes a one-to-one correspondence between graded consequence relations and special fuzzy closure operators (Gerla deals, in our terms, with $\mathbf{L}_{\{1\}}$-closure operators), i.e. not all fuzzy closure operators. As can be easily seen, the difficulty is in condition (iii) of the definition of $\mathbf{L}_K$-consequence relation: particularly, the condition $(\bigwedge_{\varphi \in X} B(\psi) \to (A \vdash \psi)) \in K$ is missing in Gerla's definition. Instead, Gerla uses (iii"): $\bigwedge_{\psi \in B}(A \vdash \psi) \otimes ((A \cup B) \vdash \varphi) \leq (A \vdash \varphi)$ where $A, B \in 2^X$ (i.e. A, B are subsets of X). Clearly, for $A, B \in 2^X$, (iii") is equivalent to $(\bigwedge_{\varphi \in X} B(\psi) \to (A \vdash \psi)) \otimes ((A \cup B) \vdash \varphi) \leq (A \vdash \varphi)$. Now, it is not true (as Gerla observes) that (iii") is satisfied by $\vdash_{\mathrm{C}}$ for any $\mathbf{L}_{\{1\}}$-closure operator C, i.e. (iii") is too strong and the above natural relations do not establish a one-to-one correspondence. Theorem 7.48 shows a way to have a one-to-one correspondence, generalizing fully the bivalent case.

7.1.2 SIMILARITY SPACES

The aim of this section is to emphasize a concept that is in our view of crucial importance in fuzzy approach. Namely, we introduce the notion of a similarity space. A similarity space, as will be defined, is but a set equipped with a similarity relation (i.e. a fuzzy equivalence). In a sense, this is nothing new since we dealt with fuzzy equivalences throughout the whole book, and particularly in Sections 3.1, 3.3, and 4.2. Nevertheless, we still coin a special term (that of a similarity space), mainly because we are going to point out new aspects which we feel are important.

Definition 7.50 A *similarity space* is a set equipped with a fuzzy equivalence relation.

Remark 7.51 (1) We will denote similarity spaces by $\mathbf{X} = \langle X, \approx \rangle$; $\approx$ is a fuzzy equivalence on X. To make the structure $\mathbf{L}$ of truth values explicit, we say also $\mathbf{L}$-similarity space.

(2) If $\approx$ is an **L**-equality, $\langle X, \approx \rangle$ is called a strict similarity space.

One of the most important concepts in mathematics and its applications is that of a *metric space*. Our contention is that the notion of a similarity space has an analogous role in fuzzy approach to modeling of humanistic systems, i.e. systems where human judgment plays a pivotal role. Reason: Similarity is the main phenomenon that enables us to cope with the immense complexity and diversity of the outer world; therefore, if we are to build adequate models of humanlike (approximate) reasoning, a (formalized) notion of similarity plays a key role.

We start with some relationships between similarity and metric. Recall that a *metric space* is a pair $\langle X, \delta \rangle$ where X is a (nonempty) set and δ is a mapping (so-called metric) assigning to any $x, y \in X$ a nonnegative real $\delta(x, y)$ such that

$$\begin{aligned}
&\delta(x,y) = 0 \text{ iff } x = y,\\
&\delta(x,y) = \delta(y,x),\\
&\delta(x,z) \le \delta(x,y) + \delta(y,z).
\end{aligned}$$

If instead of the first condition we require only $\delta(x, x) = 0$, we get a more general notion of a pseudometric space. Sometimes, (pseudo)metric is used in a generalized sense allowing to assign also value ∞ (infinite distance).

Example 7.52 (1) Let **L** be the standard Łukasiewicz algebra on $[0, 1]$, let $\mathbf{X} = \langle X, \approx \rangle$ be a similarity space. Put $\delta_\approx(x, y) = \neg(x \approx y)$. Then $\delta_\approx(x, x) = 1 - (x \approx x) = 0$; $\delta_\approx(x, y) = \delta_\approx(y, x)$; $\delta_\approx(x, y) = \neg(x \approx y) \le \neg((x \approx z) \otimes (z \approx y)) = \neg(x \approx z) \oplus \neg(z \approx y) = \min(\neg(x \approx z) + \neg(z \approx y), 1) \le \neg(x \approx z) + \neg(z \approx y) = \delta_\approx(x, y) + \delta_\approx(y, z)$. Thus, $\delta_\approx$ is a pseudometric on X with $\delta_\approx(x, y) \in [0, 1]$. If **X** is moreover strict then $\delta_\approx$ is a metric.

Conversely, if δ is a (pseudo)metric on X with $\delta(x, y) \in [0, 1]$ then putting $(x \approx_\delta y) = 1 - \delta(x, y)$ we get that $\mathbf{X}_\delta = \langle X, \approx_\delta \rangle$ is a (strict) similarity space (transitivity: $(x \approx_\delta y) \otimes (y \approx_\delta z) = \neg\delta(x, y) \otimes \neg\delta(y, z) = \neg(\delta(x, y) \oplus \delta(y, z)) = \neg \min(\delta(x, y) + \delta(y, z), 1) \le \neg(\min(\delta(x, z), 1)) = \neg(\delta(x, z)) = (x \approx_\delta z)$).

(2) Let **L** be the standard product algebra on $[0, 1]$. For a similarity space $\mathbf{X} = \langle X, \approx \rangle$, ${}^{0+}{\approx} = \{\langle x, y \rangle \mid (x \approx y) > 0\}$ is an equivalence relation on X with equivalence classes, say, X_i $(i \in I)$. On each X_i, put $\delta_\approx(x, y) = -\log(x \approx y)$; for $x \in X_i$, $y \in X_j$ $(X_i \neq X_j)$ put $\delta(x, y) = \infty$. Then $\delta_\approx$ is a pseudometric on X (in a generalized sense since it may take also ∞ as its values). Indeed: $\delta_\approx(x, x) = -\log(1) = 0$; $\delta_\approx(x, y) = -\log(x \approx y) = -\log(y \approx x) = \delta_\approx(y, x)$; $\delta_\approx(x, z) =$

$-\log(x \approx z) \leq -\log((x \approx y) \cdot (y \approx z)) = (-\log(x \approx y)) + (-\log(y \approx z)) = \delta_{\approx}(x,y) + \delta_{\approx}(y,z)$. Moreover, $\delta_{\approx}$ is a metric iff $\mathbf{X}$ is strict.

(3) The foregoing two examples are special cases of the following general relationship between pseudometrics on X (in the generalized sense) and $\mathbf{L}$-equivalences on X where $\mathbf{L}$ is a residuated lattice on $[0,1]$ given by a continuous Archimedean t-norm (see Section 2.4). We need the following representation theorem for continuous Archimedean t-norms (for proof see e.g. [125]):

Theorem 7.53 (representation by additive generator) *A mapping $\otimes : [0,1]^2 \to [0,1]$ is a continuous Archimedean t-norm iff there is a continuous additive generator f such that*

$$a \otimes b = f^{(-1)}(f(x) + f(y)),$$

i.e. f is a strictly decreasing continuous mapping $f : [0,1] \to [0,\infty]$ with $f(1) = 0$ and $f^{(-1)}$ is the pseudoinverse of f defined by $f^{(-1)}(x) = f^{-1}(x)$ if $x \leq f(0)$ and $f^{(-1)}(x) = 0$ otherwise.

Łukasiewicz as well as product t-norms are both continuous and Archimedean. $f(x) = 1 - x$ and $f^{(-1)}(x) = \max(1-x, 0)$ are an additive generator and its pseudoinverse of the Łukasiewicz t-norm; $f(x) = -\log(x)$ and $f^{(-1)}(x) = e^{-x}$ are an additive generator and its pseudoinverse of the product t-norm. Now, we have the following result.

Theorem 7.54 (similarity spaces vs. (pseudo)metric spaces) *Let $\otimes$ be a continuous Archimedean t-norm with an additive generator f, $\mathbf{L}$ be a residuated lattice on $[0,1]$ given by $\otimes$, $\approx$ be an $\mathbf{L}$-equivalence on X, δ be a pseudometric on X in a generalized sense. Then (1) $\delta_{\approx} : [0,1]^2 \to [0,\infty]$ defined by*

$$\delta_{\approx}(x,y) = f(x \approx y)$$

is a pseudometric in a generalized sense which is a metric iff $\approx$ is an $\mathbf{L}$-equality; (2) $\approx_{\delta}: [0,1]^2 \to [0,1]$ defined by

$$(x \approx_{\delta} y) = f^{(-1)}(\delta(x,y))$$

is an $\mathbf{L}$-equivalence on X which is an $\mathbf{L}$-equality iff δ is a metric; (3) $\approx$ equals $\approx_{\delta_{\approx}}$ and if $\delta(X,X) \subseteq [0, f(0)]$ then δ equals $\delta_{\approx_{\delta}}$.

Proof. (1): $\delta_{\approx}(x,x) = 0$ and $\delta_{\approx}(x,y) = \delta_{\approx}(y,x)$ follow from $(x \approx x) = 1$, $(x \approx y) = (y \approx x)$, and $f(1) = 0$. Triangle inequality for $\delta_{\approx}$ can be obtained as follows: Transitivity of $\approx$ yields $f^{(-1)}(f(x \approx y) + f(y \approx$

$z)) \leq (x \approx z)$. Since f is decreasing, we have $f(x \approx z) \leq f(f^{(-1)}(f(x \approx y)+f(y \approx z)))$. Now, there are two possibilities: either $f(x \approx y)+f(y \approx z) > f(0)$ and then $f(x \approx z) \leq f(f^{(-1)}(f(x \approx y) + f(y \approx z))) = f(0) < f(x \approx y) + f(y \approx z)$, or $f(x \approx y) + f(y \approx z) \leq f(0)$ and then $f(x \approx z) \leq f(f^{(-1)}(f(x \approx y) + f(y \approx z))) = f(x \approx y) + f(y \approx z)$. In both of the cases we have $f(x \approx z) \leq f(x \approx y) + f(y \approx z)$ which means $\delta_{\approx}(x,z) \leq \delta_{\approx}(x,y) + \delta_{\approx}(y,z)$, i.e. the required triangle inequality.

That $\delta_{\approx}$ is a metric iff $\approx$ is an **L**-equality follows easily from the fact that f is strictly decreasing.

(2): $(x \approx_\delta x) = 1$ and $(x \approx_\delta y) = (y \approx_\delta x)$ follow from $\delta(x,x) = 0$, $\delta(x,y) = \delta(y,x)$, and $f^{(-1)}(0) = 1$. Transitivity of $\approx_\delta$: Note first that $f^{(-1)}(x) = f^{-1}(\min(f(0),x))$, $f(u \approx_\delta v) = \min(f(0), \delta(u,v))$, and that f^{-1} is decreasing. Using triangle inequality for δ, we have

$$\begin{aligned}(x \approx_\delta z) = f^{(-1)}(\delta(x,z)) &= f^{-1}(\min(f(0),\delta(x,z))) \geq \\ \geq\ & f^{-1}(\min(f(0),\delta(x,y)+\delta(y,z))) = \\ =\ & f^{-1}(\min(f(0), f(x \approx_\delta y) + f(y \approx_\delta z))) = (x \approx y) \otimes (y \approx z),\end{aligned}$$

verifying transitivity of $\approx_\delta$.

Since $f^{(-1)}$ is strictly decreasing, $\approx_\delta$ is an **L**-equality iff δ is a metric.

(3): $(x \approx_{\delta_{\approx}} y) = f^{(-1)}(\delta_{\approx}(x,y)) = f^{(-1)}(f(x \approx y)) = (x \approx y)$.

If $\delta(x,y) \leq f(0)$ then $\delta_{\approx_\delta}(x,y) = f(x \approx_\delta y) = f(f^{(-1)}(\delta(x,y))) = f(f^{-1}(\min(f(0),\delta(x,y)))) = f(f^{-1}(\delta(x,y))) = \delta(x,y)$. □

Morphisms, subsystems, and products of similarity spaces. Since a similarity space is a particular fuzzy structure (of type with $R = \{\approx\}$, $F = \emptyset$), the basic definitions and results are those of Section 3.4. Particularly in the case of similarity spaces, it is possible to investigate these basic structural notions further. However, we will not pursue this and skip to other issues.

Closure induced by a similarity space. It is well known that each metric space induces a topological closure operator (and thus a topological space). On the other hand, criteria have been found under which a given topological space is metrizable, i.e. induced by some metric. In the following, we pursue an analogous problem for similarity spaces.

Definition 7.55 Let $\mathbf{X} = \langle X, \approx \rangle$ be an **L**-similarity space. For **L**-sets $A, B \in L^X$, we put

$$\rho_{\mathbf{X}}(A,B) = \bigvee_{x,y \in X} (A(x) \otimes (x \approx y) \otimes B(y)).$$

Remark 7.56 (1) $\rho_{\mathbf{X}}(A,B)$ is naturally interpreted as the truth degree of the fact that there are some x in A and y in B which are similar.

(2) One can easily see that $\rho_{\mathbf{X}}$ extends $\approx$ in that $(x \approx y) = \rho_{\mathbf{X}}(\{{}^{1}/x\},\{{}^{1}/y\})$.

For an $\mathbf{L}$-similarity space $\mathbf{X} = \langle X, \approx\rangle$ denote by $\mathrm{C}_{\mathbf{X}}$ (or by $\mathrm{C}_{\approx}$) the operator $\mathrm{C}_{\mathbf{X}} : L^X \to L^X$ defined by

$$\mathrm{C}_{\mathbf{X}}(A)(x) = \rho_{\mathbf{X}}(\{{}^{1}/x\}, A). \tag{7.18}$$

Instead of $\mathrm{C}_{\mathbf{X}}$ we use also $\mathrm{C}_{\approx}$. We thus have $\mathrm{C}_{\approx}(A)(x) = \bigvee_y (x \approx y) \otimes A(y)$.

Lemma 7.57 *Let $\mathbf{X} = \langle X, \approx\rangle$ be an $\mathbf{L}$-similarity space. Then the mapping $\mathrm{C}_{\mathbf{X}}$ defined by (7.18) is an $\mathbf{L}$-closure operator satisfying, moreover,*

$$\mathrm{C}_{\mathbf{X}}(\bigcup_{i\in I} A_i) = \bigcup_{i\in I} \mathrm{C}_{\mathbf{X}}(A_i), \tag{7.19}$$

$$\mathrm{C}_{\mathbf{X}}(\{{}^{a}/x\}) = a \otimes \mathrm{C}_{\mathbf{X}}(\{{}^{1}/x\}), \tag{7.20}$$

$$\mathrm{C}_{\mathbf{X}}(\{{}^{1}/x\})(y) = \mathrm{C}_{\mathbf{X}}(\{{}^{1}/y\})(x) \tag{7.21}$$

for any $A_i \in L^X$ ($i \in I$), $x, y \in X$, $a \in L$.

Proof. We have $A(x) = A(x) \otimes (x \approx x) \leq \bigvee_{y\in X} A(x) \otimes (x \approx y) = \rho_{\mathbf{X}}(\{{}^{1}/x\}, A)$, thus $A \subseteq \mathrm{C}_{\mathbf{X}}(A)$, proving (7.1).

(7.3) is true iff for each $x \in X$ and every $A, B \in L^X$ we have $S(A,B) \leq \mathrm{C}_{\mathbf{X}}(A)(x) \to \mathrm{C}_{\mathbf{X}}(B)(x)$ which is equivalent to $\mathrm{C}_{\mathbf{X}}(A)(x) \otimes S(A,B) \leq \mathrm{C}_{\mathbf{X}}(B)(x)$. The last inequality is true. Indeed,

$$\begin{aligned} & \mathrm{C}_{\mathbf{X}}(A)(x) \otimes S(A,B) = \\ = \; & (\bigvee_{y\in X} A(y) \otimes (x \approx y)) \otimes (\bigwedge_{y\in X} A(y) \to B(y)) \leq \\ \leq \; & \bigvee_{y\in X} (A(y) \otimes (A(y) \to B(y)) \otimes (x \approx y)) \leq \\ \leq \; & \bigvee_{y\in X} (B(y) \otimes (x \approx y)) = \mathrm{C}_{\mathbf{X}}(B)(x) \end{aligned}$$

proving (7.3).

In order to show (7.2), we proceed as follows:

$$\begin{aligned} & (\mathrm{C}_{\mathbf{X}}(\mathrm{C}_{\mathbf{X}}(A)))(x) = \bigvee_{z\in X} \mathrm{C}_{\mathbf{X}}(A)(z) \otimes (x \approx z) = \\ = \; & \bigvee_{z\in X} (\bigvee_{y\in X} A(y) \otimes (z \approx y)) \otimes (x \approx z) = \end{aligned}$$

$$\begin{aligned} &= \bigvee_{z\in X}\bigvee_{y\in X}(A(y)\otimes(x\approx z)\otimes(z\approx y)) \le \\ &\le \bigvee_{z\in X}\bigvee_{y\in X}(A(y)\otimes(x\approx y)) = \\ &= \bigvee_{y\in X}(A(y)\otimes(x\approx y)) = \mathrm{C}_{\mathbf{X}}(A)(x) \end{aligned}$$

proving (7.2).

Furthermore, we have

$$\begin{aligned} \mathrm{C}_{\mathbf{X}}(\bigcup_{i\in I} A_i) &= \bigvee_{y\in Y}((\bigvee_{i\in I} A_i(y))\otimes(x\approx y)) = \\ &= \bigvee_{i\in I}(\bigvee_{y\in Y} A_i(y)\otimes(x\approx y)) = (\bigcup_{i\in I}\mathrm{C}_{\mathbf{X}}(A_i))(x) \end{aligned}$$

verifying (7.19).

(7.20) is true since

$$\begin{aligned} \mathrm{C}_{\mathbf{X}}(\{\,{}^{a}/x\})(y) &= \bigvee_{z\in X}(\{\,{}^{a}/x\}(z)\otimes(z\approx y)) = \\ &= a\otimes(x\approx y) = a\otimes\mathrm{C}_{\mathbf{X}}(\{\,{}^{1}/x\})(y). \end{aligned}$$

Finally, (7.21) follows directly from symmetry of $\approx$. □

Remark 7.58 **L**-closure operators induced by **L**-similarity spaces are therefore topologic. Moreover, note that (7.20) implies $\mathrm{C}_{\mathbf{X}}(\emptyset) = \emptyset$. Indeed, $\mathrm{C}_{\mathbf{X}}(\emptyset)(x) = \mathrm{C}_{\mathbf{X}}(\{\,{}^{0}/y\}) = 0\otimes\mathrm{C}_{\mathbf{X}}(\{\,{}^{1}/y\}) = 0$ for any $x, y \in X$.

Lemma 7.59 *Let* C *be an* **L**-*closure operator on* X *that satisfies (7.19)–(7.21). For* $x, y \in X$ *put*

$$(x \approx_{\mathrm{C}} y) = \mathrm{C}(\{\,{}^{1}/x\})(y).$$

Then $\mathbf{X}_{\mathrm{C}} = \langle X, \approx_{\mathrm{C}}\rangle$ *is an* **L**-*similarity space.*

Proof. We verify (3.23): $(x \approx_{\mathrm{C}} x) = \mathrm{C}(\{\,{}^{1}/x\})(x) \ge \{\,{}^{1}/x\}(x) = 1$, by (7.1).

By (7.21), $(x \approx_{\mathrm{C}} y) = \mathrm{C}(\{\,{}^{1}/x\})(y) = \mathrm{C}(\{\,{}^{1}/y\})(x) = (y \approx_{\mathrm{C}} x)$ proving (3.24).

Take $x, z \in X$ and put $A = \{\,{}^{1}/x\}$. We have $\mathrm{C}(A)(z) = (x \approx z)$ and by (7.19) and (7.20),

$$\mathrm{C}(\mathrm{C}(A))(z) = \mathrm{C}(\bigcup_{y\in X}\{\,{}^{\mathrm{C}(\{\,{}^{1}/x\})(y)}/y\})(z) =$$

$$= \bigvee_{y \in X} \mathrm{C}(\{ {}^{\mathrm{C}(\{ 1/x \})(y)}/y \})(z) =$$

$$= \bigvee_{y \in X} \mathrm{C}(\{ {}^{1}/x \})(y) \otimes \mathrm{C}(\{ {}^{1}/y \})(z) =$$

$$= \bigvee_{y \in X} (x \approx_{\mathrm{C}} y) \otimes (y \approx_{\mathrm{C}} z).$$

On account of (7.3) we have $\mathrm{C}(\mathrm{C}(A))(x) \leq \mathrm{C}(A)(x)$, i.e.

$$\bigvee_{y \in X} (x \approx_{\mathrm{C}} y) \otimes (y \approx_{\mathrm{C}} z) \leq (x \approx_{\mathrm{C}} z).$$

From this it follows that $(x \approx_{\mathrm{C}} y) \otimes (y \approx_{\mathrm{C}} z) \leq (x \approx_{\mathrm{C}} z)$ for any $y \in X$, establishing (3.25). □

Theorem 7.60 *The mappings sending* $\mathbf{X}$ *to* $\mathrm{C}_{\mathbf{X}}$*, and* C *to* $\mathbf{X}_{\mathrm{C}}$*, as defined in Lemmas 7.57 and 7.59, are mutually inverse mappings between the set of all* $\mathbf{L}$*-similarity spaces with support* X *and the set of all* $\mathbf{L}$*-closure operators on* X *satisfying (7.19)–(7.21).*

Proof. By Lemmas 7.57 and 7.59, we have to check that $\mathbf{X} = \mathbf{X}_{\mathrm{C}_{\mathbf{X}}}$ and $\mathrm{C} = \mathrm{C}_{\mathbf{X}_{\mathrm{C}}}$: We have

$$(x \approx_{\mathrm{C}_{\mathbf{X}}} y) = \mathrm{C}_{\mathbf{X}}(\{ {}^{1}/x \})(y) =$$

$$= \rho_{\mathbf{X}}(\{ {}^{1}/y \}, \{ {}^{1}/x \}) = (x \approx y).$$

Furthermore,

$$\mathrm{C}(A)(x) = \mathrm{C}(\bigcup_{y \in X} \{ {}^{A(y)}/y \})(x) =$$

$$= \bigvee_{y \in X} \mathrm{C}(\{ {}^{A(y)}/y \})(x) = \bigvee_{y \in X} A(y) \otimes \mathrm{C}(\{ {}^{1}/y \})(x) =$$

$$= \bigvee_{y \in X} A(y) \otimes (x \approx_{\mathrm{C}} y) = \mathrm{C}_{\mathbf{X}_{\mathrm{C}}}(A)(x)$$

completing the proof. □

We have established a bijection between similarity spaces and special fuzzy closure operators. We can thus speak of closed fuzzy sets of a similarity space $\mathbf{X}$ (these are the closed sets of $\mathrm{C}_{\mathbf{X}}$, i.e. the elements of $\mathcal{S}_{\mathrm{C}_{\mathbf{X}}}$).

Theorem 7.11 shows that there is a bijective correspondence between fuzzy closure operators and fuzzy closure systems. Fuzzy closure systems are axiomatized by Definition 7.4. In the following we find a suitable axiomatization of systems of closed elements of fuzzy closure operators which are induced by similarity spaces.

Lemma 7.61 *$A = \mathrm{C}_{\approx}(A)$ iff A is compatible with $\approx$. Thus, $\mathcal{S}_{\mathrm{C}_{\approx}} = L^{\langle X, \approx \rangle}$.*

Proof. Let $A = \mathrm{C}_{\approx}(A)$. Then $A(x) \otimes (x \approx y) = \mathrm{C}_{\approx}(A)(x) \otimes (x \approx y) = \bigvee_{x'}(A(x') \otimes (x' \approx x)) \otimes (x \approx y) = \bigvee_{x'}(A(x') \otimes (x' \approx x) \otimes (x \approx y)) \leq \bigvee_{x'}(A(x') \otimes (x' \approx y)) = \mathrm{C}_{\approx}(A)(y) = A(y)$.

Conversely, let A be compatible with $\approx$. Then $\mathrm{C}_{\approx}(A)(x) = \bigvee_{x'}(A(x') \otimes (x' \approx x)) \leq \bigvee_{x'} A(x) = A(x)$. □

Theorem 7.62 *A system S of **L**-sets in X is the system of closed sets of some similarity space iff it is an **L**-closure system satisfying $\bigcup_{i \in I} A_i \in \mathcal{S}$, $a \otimes A \in \mathcal{S}$, and $A \to a \in \mathcal{S}$ for each $A_i, A \in \mathcal{S}$, $a \in L$.*

Proof. Due to Theorems 7.60 and 7.11, we have to show that an **L**-closure system $\mathcal{S}$ satisfies (7.19)–(7.21) iff $\bigcup_{i \in I} A_i \in \mathcal{S}$, $a \otimes A \in \mathcal{S}$, and $A \to a \in \mathcal{S}$ for each $A_i, A \in \mathcal{S}$, $a \in L$. For simplicity, we write only C instead of $\mathrm{C}_{\mathcal{S}}$.

First, we show that (7.19) is equivalent to $\bigcup_{i \in I} A_i \in \mathcal{S}$. Assume (7.19); then for $A_i \in \mathcal{S}$ we have $\bigcup_i A_i = \bigcup \mathrm{C}(A_i) = \mathrm{C}(\bigcup_i A_i) \in \mathcal{S}$. Conversely, if $\bigcup_{i \in I} A_i \in \mathcal{S}$ for $A_i \in \mathcal{S}$, then from $\mathrm{C}(A_i) \in \mathcal{S}$ we get $\bigcup_i \mathrm{C}(A_i) \in \mathcal{S}$ and thus $\bigcup_i \mathrm{C}(A_i) = \mathrm{C}(\bigcup_i \mathrm{C}(A_i)) \supseteq \mathrm{C}(\bigcup_i A_i)$. Since we always have $\bigcup_i \mathrm{C}(A_i) \subseteq \mathrm{C}(\bigcup_i A_i)$, (7.19) follows.

Next, we show that (7.20) is equivalent to $a \otimes A \in \mathcal{S}$ (for $a \in L$, $A \in \mathcal{S}$). Assume (7.19) and (7.20). Then for $a \in L$ and $A \in \mathcal{S}$ we have

$$
\begin{aligned}
a \otimes A = a \otimes \mathrm{C}(A) &= a \otimes \mathrm{C}(\bigcup_{x \in X} \{ {}^{A(x)}/x \}) = \\
&= a \otimes \bigcup_{x \in X} \mathrm{C}(\{ {}^{A(x)}/x \}) = a \otimes \bigcup_{x \in X} A(x) \otimes \mathrm{C}(\{ {}^{1}/x \}) = \\
&= \bigcup_{x \in X} a \otimes A(x) \otimes \mathrm{C}(\{ {}^{1}/x \}) = \cdots = \bigcup_{x \in X} \mathrm{C}(\{ {}^{a \otimes A(x)}/x \}) = \\
&= \mathrm{C}(\bigcup_{x \in X} \{ {}^{a \otimes A(x)}/x \}) = \mathrm{C}(a \otimes A).
\end{aligned}
$$

Conversely, assume that for $a \in L$ and $A \in \mathcal{S}$ we have $a \otimes A \in \mathcal{S}$. For each **L**-closure system we have $a = S(\{ {}^{1}/x \}, \{ {}^{a}/x \}) \leq S(\mathrm{C}(\{ {}^{1}/x \}), \mathrm{C}(\{ {}^{a}/x \}))$ and thus $a \otimes \mathrm{C}(\{ {}^{1}/x \}) \subseteq \mathrm{C}(\{ {}^{a}/x \})$. On the other hand, $\{ {}^{a}/x \} \subseteq a \otimes \mathrm{C}(\{ {}^{1}/x \})$ and thus $\mathrm{C}(\{ {}^{a}/x \}) \subseteq$

$\mathrm{C}(a \otimes \mathrm{C}(\{{}^{1}/x\})) = a \otimes \mathrm{C}(\{{}^{1}/x\})$ (as $a \otimes \mathrm{C}(\{{}^{1}/x\})$ is C-closed) establishing (7.20).

Finally, let $A \to a \in \mathcal{S}$ for $A \in \mathcal{S}$, $a \in L$. By Lemma 7.9 we have

$$\begin{aligned} \mathrm{C}(\{{}^{1}/x\})(y) &= \bigwedge_{A \in \mathcal{S}} S(\{{}^{1}/x\}, A) \to A(y) = \\ &= \bigwedge_{A \in \mathcal{S}} A(x) \to A(y) \leq \bigwedge_{A \in \mathcal{S}, a \in L} (A(x) \to a) \to (A(y) \to a) \leq \\ &\leq (\mathrm{C}(\{{}^{1}/y\})(x) \to \mathrm{C}(\{{}^{1}/y\})(x)) \to (\mathrm{C}(\{{}^{1}/y\})(y) \to \mathrm{C}(\{{}^{1}/y\})(x)) = \\ &= 1 \to (1 \to \mathrm{C}(\{{}^{1}/y\})(x)) = \mathrm{C}(\{{}^{1}/y\})(x) \end{aligned}$$

(we put $A = \mathrm{C}(\{{}^{1}/y\})$ and $a = \mathrm{C}(\{{}^{1}/y\})(x)$). Symmetrically, $\mathrm{C}(\{{}^{1}/y\})(x) \leq \mathrm{C}(\{{}^{1}/x\})(y)$ establishing (7.21). Conversely, if C satisfies (7.19)–(7.21) then by Theorem 7.60, C is induced by some similarity $\approx$ on X and thus, by Lemma 7.61, $A \in \mathcal{S}$ means that A is compatible with $\approx$. In order to show $A \to a \in \mathcal{S}$ we thus need to show that $A \to a$ is compatible with $\approx$. We have $(A \to a)(x) \otimes (x \approx y) \leq (A \to a)(y)$ iff $A(y) \otimes (y \approx x) \otimes (A(x) \to a) \leq a$ which is true. The proof is complete. □

Corollary 7.63 *A system $\mathcal{S}$ of* **L***-sets in X is the system of all* **L***-sets in X compatible with some* **L***-equivalence $\approx$ on X (i.e. $\mathcal{S} = \langle X, \approx \rangle$ for some $\approx$) iff $\mathcal{S}$ satisfies*

$$\bigcap_{i \in I} A_i \in \mathcal{S}, \ \bigcup_{i \in I} A_i \in \mathcal{S}$$

$$a \otimes A \in \mathcal{S}, \ a \to A \in \mathcal{S}, \ A \to a \in \mathcal{S}.$$

for each $A_i, A \in \mathcal{S}$ and $a \in L$.

Proof. Directly by Theorem 7.62 and Corollary 7.7. □

If $\mathcal{S}$ satisfies the five conditions of Corollary 7.63, i.e. $\mathcal{S} = L^{\langle X, \approx \rangle}$, then $\approx$ may be obtained from $\mathcal{S}$ by the Leibniz rule:

Corollary 7.64 *If $\mathcal{S}$ satisfies the five conditions of Corollary 7.63 then $\mathcal{S} = L^{\langle X, \approx_{\mathcal{S}} \rangle}$.*

Proof. By Corollary 7.63, $\mathcal{S} = L^{\langle X, \approx \rangle}$ for some **L**-equivalence $\approx$. Since $[x]_{\approx} \in \mathcal{S}$, we have $(x \approx_{\mathcal{S}} y) \leq [x]_{\approx}(x) \leftrightarrow [x]_{\approx}(y) = (x \approx y)$. On the other hand, for each $A \in \mathcal{S}$ we clearly have $(x \approx y) \leq A(x) \leftrightarrow A(y)$ (by compatibility of A) whence $(x \approx y) \leq (x \approx_{\mathcal{S}} y)$. □

Corollary 7.65 *$\langle L^{\langle X, \approx \rangle}, \subseteq \rangle$ is a complete sublattice of $\langle L^X, \subseteq \rangle$.*

Some further issues. We only mention two issues that seem to be interesting. First, *convergence in similarity.* A sequence $x_1, x_2, \ldots$ of elements $x_i \in X$ of a similarity space $\langle X, \approx\rangle$ may be said to converge in similarity to $x \in X$ if for each truth degree $a < 1$ there is n_0 such that for each $n > n_0$ we have that the similarity of x_n to x is at least (greater than) a.

Second, *degree of continuity of a space* and related problems. Adopting Menger [144], people consider a set X a continuum (continuous), if for each two $x, y \in X$ there is a chain $x = x_1, \ldots, x_n = y$ of elements $x_i \in X$ such that x_i is indistinguishable from x_{i+1}. Now, if we assume that X is equipped with a similarity $\approx$ and if we replace (the degree of) indistinguishability by (the degree of) similarity, we may define for $\mathbf{X} = \langle X, \approx\rangle$ the degree

$$\mathrm{cont}(\mathbf{X}) = \bigwedge_{x,y\in X} \bigvee_{x=x_1,\ldots,x_n=y\in X} (x_1 \approx x_2) \wedge \cdots \wedge (x_{n-1} \approx x_n)$$

to which $\mathbf{X}$ is continuous (w.r.t. $\approx$).

Example 7.66 (1) Let $\mathbf{L}$ be the standard Łukasiewicz algebra on $[0, 1]$. Put $X = \mathbf{R}$ (reals) and $(x \approx y) = \min(1, |x - y|)$. Then $\mathrm{cont}(\mathbf{X}) = 1$.

(2) Let $L = \{0, 1/n, \ldots, 1\}$ be equipped with Łukasiewicz structure. Then we have $\mathrm{cont}(L^X, \approx) = 1/n$.

Remark 7.67 We dealt with the closure $\mathrm{C}_\approx(A)$ of an $\mathbf{L}$-set A in X w.r.t. an $\mathbf{L}$-equivalence relation $\approx$. From the preceding results we know that $\mathrm{C}_\approx(A)$ is the least $\mathbf{L}$-set compatible with $\approx$ that contains A. In a similar way, one could ask for the greatest $\mathbf{L}$-set $\mathrm{I}_\approx(A)$ in X compatible with $\approx$ which is contained in A, i.e. the interior of A w.r.t. $\approx$. As one can verify we have $(\mathrm{I}_\approx(A))(x) = \bigwedge_{y\in X}(x \approx y) \to A(y)$. Furthermore, results obtained for $\mathrm{C}_\approx$ can be obtained for $\mathrm{I}_\approx$ in an analogous form.

7.2 APPLICATIONS OF SELECTED RESULTS TO SOME PROBLEMS AND MODELS

7.2.1 PRESERVATION OF SIMILARITY AS A TOOL FOR SENSITIVITY ANALYSIS

The aim of this section is to illustrate how results on preservation of similarity (Section 3.3) can contribute to sensitivity analysis of fuzzy relational models. Note that previous chapters (namely, Chapters 4 and 5) contain several applications of results from Section 3.3. Recall that the basic question addressed in Section 3.3 is the following. Given fuzzy relations $R_i, \ldots$ considered as inputs to some computational scheme c

that yields $c(R_i, \ldots)$ as output, we are interested to what extent changes in inputs lead to changes in output. That is, to what extent it is true that if we take $R'_i, \ldots$ that is similar to $R_i, \ldots$ (for instance, if R_i and R'_i are binary, then being similar means that the statement "$\langle x, y \rangle$ belongs to R_1 iff it belongs to R_2" has a big truth value; that is, from a logical point of view, $R_i(x, y)$ is not too much different from $R'_i(x, y)$ for each x, y), then $c(R_i, \ldots)$ and $c(R'_i, \ldots)$ are similar. That is, we ask a very important question:

(To what extent) Do the exact truth degrees really matter?

Extension principle. The extension principle was formulated by Zadeh [220] for the following purpose. Let $g : X \to Y$ be an (ordinary) function from a set X to a set Y. It can happen that instead of exact values from X, we know only their approximate linguistic descriptions. For instance, we may know that the input is "approximately x", "around the freezing point", etc. These verbally described inputs in X are naturally modeled by fuzzy sets in X. Now, if we still want to "apply g" to these approximate inputs we need a principle for extending g (which assigns elements from Y to elements from X) to $\overline{g}$ which assigns fuzzy sets in Y to fuzzy sets in X in an appropriate way. The basic requirement is that if applied to points (singletons), $\overline{g}$ and g coincide ($\overline{g}(\{\,{}^1/x\}) = \{\,{}^1/g(x)\}$). According to Zadeh's extension principle,[1] $\overline{g} : L^X \to L^Y$ is defined by

$$(\overline{g}(A))(y) = \sup\{A(x) \mid g(x) = y\}$$

for $A \in L^X$ and $y \in Y$ (this is a generalization of so-called interval methods). Since the $A \in L^X$ are basically thought of as approximations of elements of X, the question of to what extent the exact "shape of A" matters is particularly appealing. Intuitively, one expects that if the extension principle is well-proposed then $\overline{g}(A_1)$ is similar to $\overline{g}(A_2)$ whenever A_1 is similar to A_2, i.e. similar inputs are mapped to similar outputs. We shall see that this is indeed the case; we elaborate details.

Let us go back to Example 3.71 where we showed (for $X = Y$) that the function $\overline{g}$ obtained by the extension principle is in fact defined by logical formula. We recall the situation; however, in order to be able to cope with $X \neq Y$ we need to modify the language from Example 3.71 for the many-sorted case. Let thus $S = \{\mathcal{X}, \mathcal{Y}\}$ and consider an S-sorted language $\mathcal{J}_{\mathrm{EP}}$ with a unary relation symbol r_A, binary relation symbols $\approx_{\mathcal{X}}$ and $\approx_{\mathcal{Y}}$, and a unary function symbol f such that $\sigma(r_A) = \mathcal{X}$, $\sigma(\approx_{\mathcal{X}}) = \mathcal{X}\mathcal{X}$, $\sigma(\approx_{\mathcal{Y}}) = \mathcal{Y}\mathcal{Y}$, $\sigma(f) = \mathcal{X}\mathcal{Y}$. An $\mathbf{L}$-structure $\mathbf{M}$ for $\mathcal{J}_{\mathrm{EP}}$

[1]Note that Zadeh considers a special case for $L = [0, 1]$.

thus consists of a set $M_{\mathcal{X}}$, a set $M_{\mathcal{Y}}$, equivalence relations $\approx_{\mathcal{X}}^{\mathbf{M}}$ on $M_{\mathcal{X}}$ and $\approx_{\mathcal{Y}}^{\mathbf{M}}$ on $M_{\mathcal{Y}}$, and a function $f^{\mathbf{M}} : M_{\mathcal{X}} \to M_{\mathcal{Y}}$ which is compatible with $\approx_{\mathcal{X}}^{\mathbf{M}}$ and $\approx_{\mathcal{Y}}^{\mathbf{M}}$. Let EP$(f)$ be the formula

$$(\exists x)(r_A(x) \otimes (f(x) \approx_{\mathcal{Y}} y))$$

where x and y are variables of sort $\mathcal{X}$ and $\mathcal{Y}$, respectively. Let now $\mathbf{M}_1$ and $\mathbf{M}_2$ be two $\mathbf{L}$-structures for $\mathcal{J}_{\text{EP}}$ with the same $\mathbf{L}$-equivalence relations. As $|\text{EP}(f)|_{r_A} = 1$ and $|\text{EP}(f)|_f = 1$, (many-sorted version of) Theorem 3.87 yields that

$$(r_A^{\mathbf{M}_1} \approx r_A^{\mathbf{M}_2}) \otimes (f^{\mathbf{M}_1} \approx f^{\mathbf{M}_2}) \leq \|\text{EP}(f)\|_{\mathbf{M}_1,v} \leftrightarrow \|\text{EP}(f)\|_{\mathbf{M}_2,v}. \quad (7.22)$$

Now, EP(f) induces a mapping assigning to an $\mathbf{L}$-set A in X, a function $g : X \to Y$, and an $\mathbf{L}$-equivalence E on Y the $\mathbf{L}$-set EP$(f)^{(\mathbf{L},M)}$ in Y which is given by

$$(\text{EP}(f)^{(\mathbf{L},M)})(y) = \|\text{EP}(f)\|_{\mathbf{M},v}$$

where $y \in Y$, $\mathbf{M}$ is an $\mathbf{L}$-structure for $\mathcal{J}_{\text{EP}}$ such that $M = \{M_{\mathcal{X}}, M_{\mathcal{Y}}\}$, $M_{\mathcal{X}} = X$, $M_{\mathcal{Y}} = Y$, $f^{\mathbf{M}} = g$, $\approx_{\mathcal{Y}}^{\mathbf{M}} = E$, and $r_A^{\mathbf{M}} = A$, and v is a valuation such that $v(y) = y$. Recall that

$$\|\text{EP}(f)\|_{\mathbf{M},v} = \bigvee_{v' =_{x} v} r_A^{\mathbf{M}}(v'(x)) \otimes (f^{\mathbf{M}}(v'(x)) \approx_{\mathcal{Y}}^{\mathbf{M}} v'(y)) =$$

$$= \bigvee_{x \in X} A(x) \otimes E(g(x), y).$$

If E is the identity on Y ($E(y_1, y_2) = 1$ for $y_1 = y_2$; $E(y_1, y_2) = 0$ otherwise), then we get

$$(\text{EP}(f)^{(\mathbf{L},M)})(y) = \bigvee\{A(x) \mid g(x) = y\} = (\overline{g}(A))(y), \quad (7.23)$$

i.e. EP$(f)^{(\mathbf{L},M)}$ is exactly $\overline{g}(A)$—thus, $\overline{g}$ is in the above sense defined by formula EP(f). Coming back to our problem of whether $\overline{g}$ preserves similarity, consider (7.22) and take $\mathbf{M}_1$ and $\mathbf{M}_2$ such that $M_i = \{M_{\mathcal{X}}, M_{\mathcal{Y}}\}$, $M_{\mathcal{X}} = X$, $M_{\mathcal{Y}} = Y$, $f^{\mathbf{M}_i} = g$, $\approx_{\mathcal{Y}}^{\mathbf{M}_i}$ be the identity on Y, $r_A^{\mathbf{M}_i} = A_i$ ($i = 1, 2$). (7.22) and (7.23) give

$$(A_1 \approx A_2) \leq (\overline{g}(A_1))(y) \leftrightarrow (\overline{g}(A_2))(y)$$

for each $y \in Y$ whence we get

Theorem 7.68 *For a function $g : X \to Y$ and $A_1, A_2 \in L^X$ we have*

$$(A_1 \approx A_2) \leq (\overline{g}(A_1) \approx \overline{g}(A_2)).$$

Proof. Directly from the above considerations recalling $(\overline{g}(A_1) \approx \overline{g}(A_2)) = \bigwedge_{y\in Y}(\overline{g}(A_1))(y) \leftrightarrow (\overline{g}(A_2))(y)$. □

answering affirmatively the question of whether $\overline{g}$ preserves similarity. In a sense, this result justifies the following claim:

> When using the extension principle, the exact shape of fuzzy sets representing approximate inputs does not matter!

Note, however, that we have to be careful when interpreting this claim: $\overline{g}(A_1)$ and $\overline{g}(A_2)$ can be subject to further processing p; clearly, it may happen that in general p does not preserve similarity and thus $p(\overline{g}(A_1))$ and $p(\overline{g}(A_2))$ may not be similar at all even if $\overline{g}(A_1)$ and $\overline{g}(A_2)$ are.

The extension principle is sometimes formulated to extend a function $g : X_1 \times \cdots \times X_n \to Y$ to a function from $L^{X_1} \times \cdots \times L^{X_n}$ to L^Y. This is done basically in two ways: either one defines $\overline{g}^{\otimes}$ by

$$\overline{g}^{\otimes}(A_1,\ldots,A_n)(y) = \bigvee_{x_1\in X_1,\ldots,x_n\in X_n, f(x_1,\ldots,x_n)=y} A_1(x_1)\otimes\cdots\otimes A_n(x_n) \tag{7.24}$$

or $\overline{g}^{\wedge}$ by

$$\overline{g}^{\wedge}(A_1,\ldots,A_n)(y) = \bigvee_{x_1\in X_1,\ldots,x_n\in X_n, f(x_1,\ldots,x_n)=y} A_1(x_1)\wedge\cdots\wedge A_n(x_n) \tag{7.25}$$

for $A_i \in L^{X_i}$. For instance, this is the case for fuzzy arithmetic where the arithmetic operations on fuzzy numbers are defined using the extension principle. For this case we have the following assertion.

Theorem 7.69 *For the functions $\overline{g}^{\otimes}$ and $\overline{g}^{\wedge}$ defined by (7.24) and (7.25) we have*

$$(A_1 \approx B_1)\otimes\cdots\otimes(A_n \approx B_n) \leq (\overline{g}^{\otimes}(A_1,\ldots,A_n) \approx \overline{g}^{\otimes}(B_1,\ldots,B_n))$$

and

$$(A_1 \approx B_1)\wedge\cdots\wedge(A_n \approx B_n) \leq (\overline{g}^{\wedge}(A_1,\ldots,A_n) \approx \overline{g}^{\wedge}(B_1,\ldots,B_n)).$$

Proof. The assertion is an easy consequence of Theorem 7.68. Namely, putting $X = X_1 \times \cdots \times X_n$ and defining A and B by $A(x_1,\ldots,x_n) = A_1(x_1)\otimes\cdots\otimes A_n(x_n)$, $B(x_1,\ldots,x_n) = B_1(x_1)\otimes\cdots\otimes B_n(x_n)$ for $\overline{g}^{\otimes}$ and $A = A_1(x_1)\wedge\cdots\wedge A_n(x_n)$, $B = B_1(x_1)\wedge\cdots\wedge B_n(x_n)$ for $\overline{g}^{\wedge}$ Theorem 7.68 yields

$$(A \approx B) \leq (\overline{g}^{\otimes}(A_1,\ldots,A_n) \approx \overline{g}^{\otimes}(B_1,\ldots,B_n))$$

and

$$(A \approx B) \leq (\overline{g}^{\wedge}(A_1, \ldots, A_n) \approx \overline{g}^{\wedge}(B_1, \ldots, B_n)).$$

To prove the assertion, it is therefore sufficient to show

$$(A_1 \approx B_1) \otimes \cdots \otimes (A_n \approx B_n) \leq (A \approx B)$$

and

$$(A_1 \approx B_1) \wedge \cdots \wedge (A_n \approx B_n) \leq (A \approx B).$$

The first inequality is true iff for every $\mathbf{x} = \langle x_1, \ldots, x_n \rangle \in X_1 \times \cdots \times X_n$, $\mathbf{y} = \langle y_1, \ldots, y_n \rangle \in Y_1 \times \cdots \times Y_n$, we have

$$(A_1 \approx B_1) \otimes \cdots \otimes (A_n \approx B_n) \leq A(\mathbf{x}) \to B(\mathbf{x}) \wedge B(\mathbf{y}) \to A(\mathbf{y}).$$

To check this inequality we verify

$$(A_1 \approx B_1) \otimes \cdots \otimes (A_n \approx B_n) \leq A(\mathbf{x}) \to B(\mathbf{x})$$

(due to symmetry we omit checking the second inequality to verify) which is equivalent to

$$A(\mathbf{x}) \otimes (A_1 \approx B_1) \otimes \cdots \otimes (A_n \approx B_n) \leq B(\mathbf{x}).$$

This is indeed true since

$$\begin{aligned} A(\mathbf{x}) \otimes (A_1 \approx B_1) \otimes \cdots \otimes (A_n \approx B_n) = \\ = \ & A_1(x_1) \otimes (A_1 \approx B_1) \otimes \cdots \otimes A_n(x_n) \otimes (A_n \approx B_n) \leq \\ \leq \ & A_1(x_1) \otimes (A_1(x_1) \to B_1(x_1)) \otimes \cdots \otimes A_n(x_n) \otimes (A_n(x_n) \to B_n(x_n)) \leq \\ \leq \ & B_1(x_1) \otimes \cdots \otimes B_n(x_n) = B(\mathbf{x}). \end{aligned}$$

The second part for $\overline{g}^{\wedge}$ is similar. □

Fuzzy automata. The notion of a fuzzy automaton generalizes that of a nondeterministic automaton. The main aim is to have a device for recognizing fuzzy languages, i.e. languages to which words belong to truth degrees not necessarily equal to 0 or 1. Such languages are very natural: a language consisting of words that are being used only seldom, a language consisting of words that are difficult to pronounce, a language consisting of sequences of notes of music which represent a nice sound, etc.

Recall [72] that a nondeterministic automaton over an alphabet Σ (a set of elementary symbols) is given by a finite set Q of so-called states, subsets $Q_I \subseteq Q$ of so-called initial states and $Q_F \subset Q$ of so-called final states, and a relation $\delta \subseteq Q \times \Sigma \times Q$. δ is called a transition relation; the meaning is the following: if the actual state of the automaton is $q \in Q$ and the actual symbol on input is $s \in \Sigma$, then the automaton can proceed so that its next actual state will be any $q' \in Q$ such that $\langle q, s, q' \rangle \in \delta$. An input word $s_1 \dots s_n$ ($s_i \in \Sigma$) is accepted by the automaton iff there is a sequence $q_1, \dots, q_{n+1}$ of states such that q_1 is an initial state, q_{n+1} is a final state, and $\langle q_i, s_1, q_{i+1} \rangle \in \delta$ (transition). The set of all words accepted by the automaton is called the language recognized by the automaton.

In fuzzy setting, we have the following. An $\mathbf{L}$-*automaton* $\mathcal{M}$ over a finite alphabet Σ is given by a finite set Q of states, an $\mathbf{L}$-set Q_I in Q (for $q \in Q$, $Q_I(q)$ is the degree to which q is an initial state); an $\mathbf{L}$-set Q_F in Q (for $q \in Q$, $Q_F(q)$ is the degree to which q is a final state); an $\mathbf{L}$-relation δ between Q, Σ, and Q (for $q, q' \in Q$ and $s \in \Sigma$, $\delta(q, s, q')$ is the degree to which the $\mathbf{L}$-automaton can transfer from q to q' if the actual input symbol is s). Then, for an input word $s_1 \dots s_n$ we define the degree $(\mathcal{L}(\mathcal{M}))(s_1 \dots s_n)$ to which $\mathcal{M}$ accepts $s_1 \dots s_n$ by

$$(\mathcal{L}(\mathcal{M}))(s_1 \dots s_n) = \bigvee_{q_1, \dots, q_{n+1} \in Q} Q_I(q_1) \wedge \delta(q_1, s_1, q_2) \wedge \dots \wedge \delta(q_n, s_n, q_{n+1}) \wedge Q_F(q_{n+1}).$$

The thus defined $\mathbf{L}$-set $\mathcal{L}(\mathcal{M})$ is called the $\mathbf{L}$-language recognized by $\mathcal{M}$. We can easily see that for $\mathbf{L} = \mathbf{2}$ we get the above notion of a nondeterministic automaton and the corresponding recognized language.

An $\mathbf{L}$-automaton can be viewed as an $\mathbf{L}$-relational system: Consider an S-sorted language $\mathcal{J}_{\text{Aut}}$ where $S = \{\mathcal{Q}, \mathcal{S}\}$, $R = \{r_\delta, r_{Q_I}, r_{Q_F}\}$, $F = \emptyset$. Then an $\mathbf{L}$-automaton $\mathcal{M}$ can be viewed as an $\mathbf{L}$-structure $\mathbf{M}$ for $\mathcal{J}_{\text{Aut}}$ where $r_\delta^{\mathbf{M}} = \delta$, $r_{Q_I}^{\mathbf{M}} = Q_I$, and $r_{Q_F}^{\mathbf{M}} = Q_F$. Using the rules for evaluating truth degrees of formulas we easily see that the degree $(\mathcal{L}(\mathcal{M}))(s_1 \dots s_n)$ to which $\mathcal{M}$ accepts $s_1 \dots s_n$ equals the truth degree of a formula (x_i are variables of sort $\mathcal{Q}$, y_i are variables of sort $\mathcal{S}$)

$$(\exists x_1, \dots x_{n+1})[r_{Q_I}(x_1) \mathbin{\&} r_\delta(x_1, y_1, x_2) \mathbin{\&} \cdots \mathbin{\&} r_\delta(x_n, y_n, x_{n+1}) \mathbin{\&} r_{Q_F}(x_1)]$$

for a valuation v such that $v(y_i) = s_i$. Denoting this formula accept$_n$ we thus have

$$(\mathcal{L}(\mathcal{M}))(s_1 \dots s_n) = \|\text{accept}_n\|_{\mathbf{M},v}.$$

This enables us to apply (many-sorted version of) Lemma 3.85 to answer the question

> Do fuzzy automata with pairwise similar fuzzy sets of initial states, final states, and transition relations recognize similar languages?

The answer to our question is given by the following theorem.

Theorem 7.70 *Let $\mathcal{M}_1$ and $\mathcal{M}_2$ be* $\mathbf{L}$*-automata over an alphabet* Σ*; then we have*

$$(Q_{I1} \approx Q_{I2}) \otimes (\delta_1 \approx \delta_2) \otimes (Q_{F1} \approx Q_{F2}) \leq (\mathcal{L}(\mathcal{M}_1) \approx \mathcal{L}(\mathcal{M}_2)).$$

Proof. Let $\mathbf{M}_1$ and $\mathbf{M}_2$ be the $\mathbf{L}$-structures for $\mathcal{J}_{\mathrm{Aut}}$ corresponding to $\mathcal{M}_1$ and $\mathcal{M}_2$ (i.e. $r_{Q_I}^{\mathbf{M}_i} = Q_{Ii}$, $r_\delta^{\mathbf{M}_i} = \delta_i$, and $r_{Q_F}^{\mathbf{M}_i} = Q_{Fi}$, $i = 1, 2$). Since $|\mathrm{accept}_n|_{r_{Q_I}} = 1$, $|\mathrm{accept}_n|_{r_\delta} = 1$, and $|\mathrm{accept}_n|_{r_{Q_F}} = 1$, (many-sorted version of) Lemma 3.85 yields

$$\begin{aligned} &(r_{Q_I}^{\mathbf{M}_1} \approx r_{Q_I}^{\mathbf{M}_2}) \otimes (r_\delta^{\mathbf{M}_1} \approx r_\delta^{\mathbf{M}_2}) \otimes (r_{Q_F}^{\mathbf{M}_1} \approx r_{Q_F}^{\mathbf{M}_2}) \leq \\ \leq \;& \|\mathrm{accept}_n\|_{\mathbf{M}_1,v} \leftrightarrow \|\mathrm{accept}_n\|_{\mathbf{M}_2,v}. \end{aligned}$$

As for v such that $v(\mathfrak{y}_i) = s_i$ ($i = 1, \ldots, n$) we have $\|\mathrm{accept}_n\|_{\mathbf{M}_i,v} = (\mathcal{L}(\mathcal{M}_i))(s_1 \ldots s_n)$, recalling that

$$\begin{aligned} &(\mathcal{L}(\mathcal{M}_1) \approx \mathcal{L}(\mathcal{M}_2)) = \\ = &\bigwedge_{n \in \mathbf{N}, s_1, \ldots, s_n \in \Sigma} (\mathcal{L}(\mathcal{M}_1))(s_1 \ldots s_n) \leftrightarrow (\mathcal{L}(\mathcal{M}_2))(s_1 \ldots s_n), \end{aligned}$$

the assertion follows. □

7.2.2 CUTWORTHY PROPERTIES AND REPRESENTATION OF FUZZY NOTIONS BY CRISP NOTIONS

Some concepts and properties in fuzzy setting are in a cutlike manner related to their bivalent counterparts. For instance, a fuzzy set A is the intersection of fuzzy sets B and C iff each a-cut of A is the intersection of the a-cut of B and the a-cut of C, i.e. ${}^aB \cap C = {}^aB \cap {}^aC$. Indeed, $x \in {}^aB \cap C$ iff $a \leq (B \cap C)(x)$ iff $a \leq B(x) \wedge C(x)$ iff $a \leq B(x)$ and $a \leq C(x)$ iff $x \in {}^aB$ and $x \in {}^aC$ iff $x \in {}^aB \cap {}^aC$.

Klir and Yuan [127, p. 23] write: "Any property generalized from classical set theory into the domain of fuzzy set theory that is preserved in all α-cuts for $\alpha \in (0, 1]$ in the classical sense is called a *cutworthy property* ...". Later on (p. 36): "... the standard fuzzy intersection and fuzzy union are both cutworthy ...". We shall see that although Klir and Yuan speak of cutworthy properties and cutworthy operations, the problem has a common core.

The aim of this section is to show how results of Section 3.5 can contribute to shed light on "cutworthiness". The point is whether the fact that a given property applies to a fuzzy relation or a collection of fuzzy relations (in general, to a given fuzzy structure) can be "translated" into saying that the property applies to (suitable) a-cuts of the fuzzy relation or a collection of fuzzy relations. We shall show that as far as properties expressible by logical formulas are concerned, cutworthiness can be approached from a unifying point of view: what plays a role is a logical formula φ and the question of whether ${}^a\|\varphi\|_{\mathbf{M},v}$ can be expressed by $\|\varphi\|_{{}^b\mathbf{M},v}$ (for some b's); moreover, we show that Section 3.5 provides a complete solution to the problem. First, we propose a formal definition of cutworthiness. We take several examples which have been discussed in the literature in connection with cutworthiness and discuss them in detail; the main emphasis is on showing how (many-sorted versions of) Lemmas 3.139, 3.140, and Theorem 3.141 apply. Although some conclusions we will obtain can be inferred directly and more easily (and our justification is thus somewhat intricate), we still prefer to use results of Section 3.5. On the one hand, it provides a general method; on the other hand, we illustrate applications of the general results.

We start with a definition of cutworthiness (of a formula); the definition will be illustrated in subsequent examples.

Definition 7.71 A formula φ is *cutworthy* for $\mathbf{L}$ and $K \subseteq L$ if for each $a \in K$ there exists $C_a \subseteq L$ such that

$$
{}^a\|\varphi\|_{\mathbf{M},v} = 1 \quad \text{iff} \quad \text{for each } b \in C: \ \|\varphi\|_{{}^b\mathbf{M},v} = 1
$$

for each $\mathbf{M}$ and v.

We say that φ is cutworthy for $\mathbf{L}$ if it is cutworthy for $\mathbf{L}$ and $K = L$; if φ is cutworthy for each $\mathbf{L}$ we say that it is cutworthy.

Cutworthiness and operations with fuzzy sets. First, consider intersection. Cutworthiness of intersection was discussed directly (not using Section 3.5) above. Let A, B be $\mathbf{L}$-sets in a universe X; consider a language $\mathcal{J}$ with unary relation symbols r_A and r_B, and an $\mathbf{L}$-structure $\mathbf{M}$ for $\mathcal{J}$ such that $r_A^{\mathbf{M}} = A$ and $r_B^{\mathbf{M}} = B$. Then the intersection $A \cap B$ "is defined" (cf. Section 3.2) by formula $r_A(\mathsf{x}) \wedge r_B(\mathsf{x})$: for a valuation v such that $v(\mathsf{x}) = x$ we have $(A \cap B)(x) = \|r(\mathsf{x}) \wedge s(\mathsf{x})\|_{\mathbf{M},v}$. Now, applying Lemma 3.139 we get ${}^a(A \cap B)(v(\mathsf{x})) = {}^a\|r(\mathsf{x}) \wedge s(\mathsf{x})\|_{\mathbf{M},v} = \|r(\mathsf{x}) \wedge s(\mathsf{x})\|^a_{\mathcal{C}_{\mathbf{M}},v} = \|r(\mathsf{x})\|^a_{\mathcal{C}_{\mathbf{M}},v} \wedge^{\mathbf{2}} \|s(\mathsf{x})\|^a_{\mathcal{C}_{\mathbf{M}},v} = {}^aA(v(\mathsf{x})) \wedge^{\mathbf{2}} {}^aB(v(\mathsf{x}))$, i.e. ${}^a(A \cap B) = {}^aA \cap {}^aB$ showing that the formula $r(\mathsf{x}) \wedge r(\mathsf{x})$ which represents intersection is cutworthy (on has $C_a = \{a\}$). Note that cutworthiness of intersection as understood in [127] means exactly cutworthiness of $r(\mathsf{x}) \wedge s(\mathsf{x})$.

For union, the situation is in general different. Namely, if the structure $\mathbf{L}$ of truth values is not linearly ordered, we do not have ${}^a(A \cup B) =$

${}^{a}A \cup {}^{a}B$. Indeed, let $a, b \in L$ be noncomparable and let A and B be **L**-sets in X such that for some $x \in X$ we have $A(x) = a$ and $B(x) = b$. Then we have $(A \cup B)(x) = a \vee b$ and thus $x \in {}^{a \vee b}A \cup B$; on the other hand, $x \notin {}^{a \vee b}A$ and $x \notin {}^{a \vee b}B$. However, if **L** is linearly ordered, union of fuzzy sets is cutworthy, i.e. ${}^{a}A \cup B = {}^{a}A \cup {}^{a}B$. This can be proved as above for intersection using Lemma 3.140.

Consider now negation. Again, we do not have ${}^{a}\neg A = \neg {}^{a}A$ (where $\neg {}^{a}A$ is the set-theoretical complement of ${}^{a}A$); the reader can easily find a counterexample. However, Lemma 3.139 yields immediately how ${}^{a}\neg A$ can be expressed in terms of b-cuts of B. Consider an **L**-structure **M** for a language $\mathcal{J}$ containing a unary relation symbol r_A and suppose $r_A^{\mathbf{M}} = A$. Then for a valuation v such that $v(x) = x$ we have by Lemma 3.139 ${}^{a}\neg A(x) = {}^{a}\|\neg r_A(x)\|_{\mathbf{M},v} = \|\neg r_A(x)\|^{a}_{\mathcal{C}_{\mathbf{M},v}} = \|r_A(x) \Rightarrow 0\|^{a}_{\mathcal{C}_{\mathbf{M},v}} = \bigwedge^{\mathbf{2}}_{b \in L} \|r_A(x)\|^{b}_{\mathcal{C}_{\mathbf{M},v}} \Rightarrow^{\mathbf{2}} \|0\|^{a \otimes b}_{\mathcal{C}_{\mathbf{M},v}} = \bigwedge^{\mathbf{2}}_{b \in L} {}^{b}A(x) \Rightarrow^{\mathbf{2}} {}^{a \otimes b}0(x)$ (recall that ${}^{a \otimes b}0(x) = 1$ iff $a \otimes b = 0$ iff $b \leq \neg a$). We thus get that $x \in {}^{a}\neg A$ iff for each $b \in L$ we have that if $x \in {}^{b}A$ then $b \leq \neg a$. This can be simplified: it is easy to see that it is sufficient to test the latter condition only for $b = A(x)$; this yields $x \in {}^{a}\neg A$ iff $A(x) \leq \neg a$.

We thus have

Observation 7.72 *Intersection of* **L***-sets is cutworthy for* **L** *(for $C_a = \{a\}$); union of* **L***-sets is cutworthy for* **L** *(for $C_a = \{a\}$) iff* **L** *is linearly ordered; negation is not cutworthy.*

Cutworthiness and reflexivity, symmetry, and transitivity. Let R be a binary **L**-relation on a set X; let **M** be an **L**-structure for a language $\mathcal{J}$ containing a binary relation symbol r such that $R = r^{\mathbf{M}}$. Recall that reflexivity, symmetry, and transitivity of R are defined using formulas $(\forall x)r(x, x)$, $(\forall x, y)(r(x, y) \Rightarrow r(y, x))$, and $(\forall x, y, z)((r(x, y) \otimes r(y, z)) \Rightarrow r(x, z))$, abbreviated by Ref, Sym, and Tra, respectively. R is called a-reflexive if $a \leq \|\text{Ref}\|_{\mathbf{M}}$; a-symmetric if $a \leq \|\text{Sym}\|_{\mathbf{M}}$; a-transitive if $a \leq \|\text{Tra}\|_{\mathbf{M}}$. For $a = 1$ we omit the prefix a- and speak of reflexivity instead of 1-reflexivity etc.

Consider first reflexivity. Using Lemma 3.139 we have ${}^{a}\|\text{Ref}\|_{\mathbf{M},v} = \|\text{Ref}\|^{a}_{\mathcal{C}_{\mathbf{M},v}} = \bigwedge^{\mathbf{2}}_{v' =_x v} \|r(x, x)\|^{a}_{\mathcal{C}_{\mathbf{M},v}} = \bigwedge^{\mathbf{2}}_{x \in X} {}^{a}r^{\mathbf{M}}(x, x)$ (note that v can be taken arbitrarily since $\|\text{Ref}\|_{\mathbf{M},v} = \|\text{Ref}\|_{\mathbf{M}}$). As R is a-reflexive iff ${}^{a}\|\text{Ref}\|_{\mathbf{M},v} = 1$, observing that $\bigwedge^{\mathbf{2}}_{x \in X} {}^{a}r^{\mathbf{M}}(x, x) = 1$ is equivalent to a-reflexivity of ${}^{a}r^{\mathbf{M}}$, we conclude that R is a-reflexive iff ${}^{a}R$ is reflexive (as a bivalent relation). Particularly, R is reflexive iff ${}^{1}R$ is a reflexive relation. Note also that since ${}^{a}R \subseteq {}^{1}R$, R is reflexive iff each ${}^{a}R$ $(a \in L)$ is reflexive.

Symmetry: Lemma 3.139 gives ${}^a\|\mathrm{Sym}\|_{\mathbf{M},v} = \|\mathrm{Sym}\|^a_{\mathcal{C}_\mathbf{M},v} = \bigwedge^{\mathbf{2}}_{v'=_{x,y}v} \|r(x,y) \Rightarrow r(y,x)\|^a_{\mathcal{C}_\mathbf{M},v} = \bigwedge^{\mathbf{2}}_{v'=_{x,y}v} \bigwedge^{\mathbf{2}}_{b\in L} \|r(x,y)\|^b_{\mathcal{C}_\mathbf{M},v} \to^{\mathbf{2}} \|r(y,x)\|^{a\otimes b}_{\mathcal{C}_\mathbf{M},v} = \bigwedge^{\mathbf{2}}_{x,y\in X} \bigwedge^{\mathbf{2}}_{b\in L} {}^b r^\mathbf{M}(x,y) \to^{\mathbf{2}} {}^{a\otimes b} r^\mathbf{M}(y,x)$. R is a-symmetric iff ${}^a\|\mathrm{Sym}\|_{\mathbf{M},v} = 1$. Therefore, R is a-symmetric iff for every $x, y \in X$ and for each $b \in L$ we have that whenever $\langle x, y\rangle$ belongs to b-cut of R then $\langle y, x\rangle$ belongs to $(a\otimes b)$-cut of R. That is, we have a condition equivalent to a-symmetry of R which uses only c-cuts of R. Particularly, for symmetry of R ($a = 1$) we get that R is symmetric iff for each $b \in L$, if $\langle x, y\rangle$ belongs to b-cut of R then $\langle y, x\rangle$ belongs to $(1\otimes b)$-cut, i.e. to b-cut of R. That is, R is symmetric iff each a-cut of R is symmetric.

Transitivity: By Lemma 3.139, ${}^a\|\mathrm{Tra}\|_{\mathbf{M},v} = \|\mathrm{Tra}\|^a_{\mathcal{C}_\mathbf{M},v} = \bigwedge^{\mathbf{2}}_{v'=_{x,y,z}v} \|(r(x,y) \otimes r(y,z)) \Rightarrow r(x,z)\|^a_{\mathcal{C}_\mathbf{M},v} = \bigwedge^{\mathbf{2}}_{v'=_{x,y,z}v} \bigwedge^{\mathbf{2}}_{b\in L} (\bigvee^{\mathbf{2}}_{b\le c\otimes d} \|r(x,y)\|^c_{\mathcal{C}_\mathbf{M},v} \wedge^{\mathbf{2}} \|r(y,z)\|^d_{\mathcal{C}_\mathbf{M},v}) \to^{\mathbf{2}} \|r(x,z)\|^{a\otimes b}_{\mathcal{C}_\mathbf{M},v} = \bigwedge^{\mathbf{2}}_{x,y,z} \bigwedge^{\mathbf{2}}_{b\in L} (\bigvee^{\mathbf{2}}_{b\le c\otimes d} {}^c r^\mathbf{M}(x,y) \wedge^{\mathbf{2}} {}^d r^\mathbf{M}(y,z)) \to^{\mathbf{2}} {}^{a\otimes b} r^\mathbf{M}(x,z)$. As a-transitivity of R is equivalent to ${}^a\|\mathrm{Tra}\|_{\mathbf{M},v} = 1$, we get that R is a-transitive iff for each $x, y, z \in X$ and each $b \in L$ we have that if $\langle x, y\rangle \in {}^cR$ and $\langle y, z\rangle \in {}^dR$ for some $c\otimes b \ge b$ then $\langle x, z\rangle \in {}^{a\otimes b}R$. This is how a-transitivity of R is expressed using b-cuts of R and bivalent logical operations. Particularly, consider transitivity and $\mathbf{L}$ satisfying $x\otimes y = x \wedge y$ (i.e. $\mathbf{L}$ is a Heyting algebra). Then from $\|r(x,y) \otimes r(y,z)\|^b_{\mathcal{C}_\mathbf{M},v} = \|r(x,y)\|^b_{\mathcal{C}_\mathbf{M},v} \otimes^{\mathbf{2}} \|r(y,z)\|^b_{\mathcal{C}_\mathbf{M},v}$ we get that R is transitive iff for every $x, y, z \in X$ we have that for each $b \in L$, if $\langle x, y\rangle \in {}^bR$ and $\langle y, z\rangle \in {}^bR$, then $\langle x, z\rangle \in {}^bR$; that is, R is transitive iff each b-cut of R is transitive.

To sum up, we have

Observation 7.73 *reflexivity is cutworthy for any* $\mathbf{L}$ *(for* $C_a = \{a\}$*); symmetry is cutworthy for any* $\mathbf{L}$ *and* $K = \{1\}$ *(for* $C_a = L$*); transitivity is cutworthy for* $\mathbf{L}$ *and* $K = \{1\}$ *(for* $C_a = \{a\}$*) whenever* $\mathbf{L}$ *is a Heyting algebra.*

Recall that we applied results from Section 3.5 to show that fuzzy Galois connections can be viewed as $\mathbf{L}$-nested systems of Galois connections in Section 5.3.

Another case: fuzzy automata as nested systems of (nondeterministic) automata. We are interested in how the notions of a fuzzy automaton and that of a nondeterministic automaton are related; particularly, we shall show that fuzzy automata can be thought of as nested systems of ordinary nondeterministic automata. Recall from Section 7.2.1 that a fuzzy automaton $\mathcal{M}$ can be seen as an $\mathbf{L}$-structure for a two-sorted language $\mathcal{J}_{\text{Aut}}$ and that the degree to which a word $s_1 \ldots s_n$ is accepted by $\mathcal{M}$ is the truth degree of a formula accept$_n$.

For an S-sorted **L**-structure **M** which represents an **L**-automaton $\mathcal{M}$, ${}^a\mathbf{M}$ is a bivalent S-sorted structure which represents in a natural way a crisp nondeterministic automaton ${}^a\mathcal{M}$: the alphabet and the set of states of ${}^a\mathcal{M}$ are Σ and Q, respectively; the transition relation is ${}^a\delta$, the set of initial states is ${}^aQ_I$, the set of final states is ${}^aQ_F$. The collection of all ${}^a\mathbf{M}$ ($a \in L$) is **L**-nested; we can thus call **L**-nested the collection of all ${}^a\mathcal{M}$ ($a \in L$).

Using Lemmas 3.139 and 3.140 we get that for $a \in L$ we have

$$\begin{aligned}\|\text{accept}_n\|^a_{\mathcal{C}\mathbf{M},v} = \bigvee^{\mathbf{2}}_{v'=_{x_1,\ldots,x_{n+1}}v} & \|r_{Q_I}(x_1)\|^a_{\mathcal{C}\mathbf{M},v'} \wedge^{\mathbf{2}} \\ & \wedge^{\mathbf{2}} \|r_\delta(x_1,y_1,x_2)\|^a_{\mathcal{C}\mathbf{M},v'} \wedge^{\mathbf{2}} \cdots \\ & \cdots \wedge^{\mathbf{2}} \|r_\delta(x_n,y_n,x_{n+1})\|^a_{\mathcal{C}\mathbf{M},v'} \wedge^{\mathbf{2}} \|r_{Q_F}(x_{n+1})\|_{\mathcal{C}\mathbf{M},v'}.\end{aligned}$$

Recalling $\|r_{Q_I}(x_1)\|^a_{\mathcal{C}\mathbf{M},v'} = {}^ar^{\mathbf{M}}_{Q_I}(v'(x_1))$, $\|r_\delta(x_i,y_i,x_{i+1})\|^a_{\mathcal{C}\mathbf{M},v'} = {}^ar^{\mathbf{M}}_\delta(v'(x_i),v'(y_i),v'(x_{i+1}))$, and $\|r_{Q_F}(x_{n+1})\|^a_{\mathcal{C}\mathbf{M},v'} = {}^ar^{\mathbf{M}}_{Q_F}(v'(x_{n+1}))$, we obtain

$$\|\text{accept}_n\|^a_{\mathcal{C}\mathbf{M},v} = \|\text{accept}_n\|_{{}^a\mathbf{M},v}.$$

This means that for a word $s_1 \ldots s_n$ ($s_i \in \Sigma$), the degree to which $s_1 \ldots s_n$ is accepted by an **L**-automaton $\mathcal{M}$ is at least a if and only if $s_1 \ldots s_n$ is accepted by the nondeterministic automaton ${}^a\mathcal{M}$. Taking into account (many-sorted version of) Theorem 3.141, we can summarize the observed results:

Theorem 7.74 *The mapping sending an* **L**-*automaton* $\mathcal{M}$ *to a system* $\{{}^a\mathcal{M} \mid a \in L\}$ *is a bijective correspondence between* **L**-*automata over an alphabet* Σ *and* **L**-*nested systems of nondeterministic automata over* Σ. *Moreover, we have*

$${}^a\mathcal{L}(\mathcal{M}) = \mathcal{L}({}^a\mathcal{M})$$

and thus

$$(\mathcal{L}(\mathcal{M}))(s_1 \ldots s_n) = \bigvee\{a \mid s_1 \ldots s_n \in \mathcal{L}({}^a\mathcal{M})\}$$

for each word $s_1 \ldots s_n$ *of symbols from* Σ.

Remark 7.75 Note that Theorem 7.74 is the main result obtained in [145] (in [145], however, this is written in terms of category theory).

Cutworthiness and extension principle: computing $\overline{g}$ "cut-by-cut". In the bivalent case (only 0 and 1 as truth values), the extension principle yields the following: for a function $g : X \to Y$, $\overline{g}$ is a function assigning subsets of Y to subsets of X by

$$\overline{g}(A) = \{g(x) \mid x \in A\}.$$

For now, we will denote the function obtained from g by the extension principle in the bivalent case by g^* (and not by $\overline{g}$). A natural question arises as to whether it is possible to reduce the general case (**L** contains also other truth values than 0 and 1) to the bivalent one; particularly, whether $\overline{g}(A)$ can be computed cut by cut, i.e. whether we have

$$^a\overline{g}(A) = g^*(^aA) \tag{7.26}$$

for each $a \in L$. We shall see that if **L** is linearly ordered and if X and Y are finite, then this is indeed the case.

To this end, let us go back to the previous subsection where we showed that $\overline{g}$ is in fact defined by EP(f) which is a formula $(\exists \mathbb{x})(r_A(\mathbb{x}) \otimes (f(\mathbb{x}) \approx_{\mathcal{Y}} \mathbb{y}))$ of two-sorted language $\mathcal{J}_{\mathrm{EP}}$. It is easy to see that all of our arguments remain valid if we take $(\exists \mathbb{x})(r_A(\mathbb{x}) \wedge (f(\mathbb{x}) \approx_{\mathcal{Y}} \mathbb{y}))$ instead of EP(f). Recalling (see above) that

$$(\overline{g}(A))(v(\mathbb{y})) = \|(\exists \mathbb{x})(r_A(\mathbb{x}) \wedge (f(\mathbb{x}) \approx_{\mathcal{Y}} \mathbb{y}))\|_{\mathbf{M},v},$$

(many-sorted version of) Lemma 3.139 yields

$$\begin{aligned}
^a\overline{g}(A) &= {}^a\|(\exists \mathbb{x})(r_A(\mathbb{x}) \wedge (f(\mathbb{x}) \approx_{\mathcal{Y}} \mathbb{y}))\|_{\mathbf{M},v} = \\
&= \|(\exists \mathbb{x})(r_A(\mathbb{x}) \wedge (f(\mathbb{x}) \approx_{\mathcal{Y}} \mathbb{y}))\|^a_{\mathcal{C}_{\mathbf{M}},v} = \\
&= \bigvee^{\mathbf{2}}_{v' =_{\mathbb{x}} v} \|r_A(\mathbb{x})\|^a_{\mathcal{C}_{\mathbf{M}},v'} \wedge^{\mathbf{2}} \|f(\mathbb{x}) \approx_{\mathcal{Y}} \mathbb{y}\|_{\mathcal{C}_{\mathbf{M}},v'} = \\
&= \bigvee^{\mathbf{2}}_{v' =_{\mathbb{x}} v} r_A^{^a\mathbf{M}}(v'(\mathbb{x})) \wedge^{\mathbf{2}} (g(v'(\mathbb{x})) \approx_{\mathcal{Y}}^{^a\mathbf{M}} v'(\mathbb{y})) = \\
&= \|(\exists \mathbb{x})(r_A(\mathbb{x}) \wedge (f(\mathbb{x}) \approx_{\mathcal{Y}} \mathbb{y}))\|_{^a\mathbf{M},v} = \\
&= g^*(^aA)
\end{aligned}$$

verifying (7.26).

7.3 FUZZY LOGIC WITH TRUTH DEGREES IN RESIDUATED LATTICES

So far we were not interested in deductive aspects of fuzzy logic (the exception being fuzzy equational logic in Section 3.6). The aim of this section is to present (with only a sketch of proof) a logical calculus with a formalized notion of provability of formulas such that provable formulas are those which are always (i.e. in each fuzzy structure) true in degree 1. This calculus is called *monoidal (predicate) logic* (MPL) and is due to Höhle [115, 117].

Monoidal predicate logic is in fact our first-order fuzzy logic with $\mathbb{L} = \{\mathbb{0}\}$ where $\mathbb{1}$ is an abbreviation for $\mathbb{0} \Rightarrow \mathbb{0}$ (cf. Remark 3.49). The axioms of monoidal predicate logic are

$$
\begin{array}{l}
(\varphi \Rightarrow \psi) \Rightarrow ((\psi \Rightarrow \chi) \Rightarrow (\varphi \Rightarrow \chi)) \\
(\varphi \otimes \psi) \Rightarrow \varphi \\
(\varphi \otimes \psi) \Rightarrow (\psi \otimes \varphi) \\
(\varphi \Rightarrow (\psi \Rightarrow \chi)) \Rightarrow ((\varphi \otimes \psi) \Rightarrow \chi) \\
((\varphi \otimes \psi) \Rightarrow \chi) \Rightarrow (\varphi \Rightarrow (\psi \Rightarrow \chi)) \\
(\varphi \wedge \psi) \Rightarrow \varphi \\
(\varphi \wedge \psi) \Rightarrow (\psi \wedge \varphi) \\
(\varphi \otimes (\varphi \Rightarrow \psi)) \Rightarrow (\varphi \wedge \psi) \\
\mathbb{0} \Rightarrow \varphi \\
\varphi \Rightarrow (\varphi \vee \psi) \\
\psi \Rightarrow (\varphi \vee \psi) \\
(\varphi \Rightarrow \chi) \Rightarrow ((\psi \Rightarrow \chi) \Rightarrow ((\varphi \vee \psi) \Rightarrow \chi)) \\
(\forall x)\varphi \Rightarrow \varphi(x/t) \\
\qquad \text{whenever } \varphi(x/t) \text{ is a correct substitution} \\
\varphi(x/t) \Rightarrow (\exists x)\varphi \\
\qquad \text{whenever } \varphi(x/t) \text{ is a correct substitution} \\
(\forall x)(\varphi \Rightarrow \psi) \Rightarrow (\varphi \Rightarrow (\forall x)\psi) \\
\qquad \text{provided } x \text{ is not free in } \psi \\
(\forall x)(\varphi \Rightarrow \psi) \Rightarrow ((\exists x)\varphi \Rightarrow \psi) \\
\qquad \text{provided } x \text{ is not free in } \varphi
\end{array}
$$

for all formulas φ, ψ, χ, term t, and a variable x.

The inference rules are *modus ponens* which says

$$\text{from } \varphi \text{ and } \varphi \Rightarrow \psi \text{ infer } \psi,$$

and *generalization* which says

$$\text{from } \varphi \text{ infer } (\forall x)\varphi.$$

A *proof* in MPL is a sequence $\varphi_1, \ldots, \varphi_n$ such that each φ_i is either an axiom, or follows from some preceding φ_j, φ_k $(j, k < i)$ by *modus ponens* or from some preceding φ_j $(j < i)$ by generalization. A formula φ is *provable* in MPL, denoted by $\vdash\varphi$, if φ is the last member of some proof. If for each complete residuated lattice **L**, each **L**-structure **M**, and each **M**-valuation v we have $\|\varphi\|_{\mathbf{M},v} = 1$, i.e. the truth degree of φ is always 1, we say that φ is a *tautology* and denote this fact by $\models\varphi$. We have the following result.

Theorem 7.76 (completeness of monoidal predicate logic) *For each formula φ of monoidal predicate logic we have*

$$\vdash\varphi \quad \textit{iff} \quad \models\varphi,$$

i.e. φ is provable iff its truth degree is always 1.

Proof. "$\vdash\varphi$ implies $\models\varphi$" (so-called soundness): It is easy to verify that each axiom is a tautology. Furthermore, if φ and $\varphi \Rightarrow \psi$ are tautologies then so is ψ, and if φ is a tautology then $(\forall x)\varphi$ is a tautology; therefore, being a tautology is preserved under performing inference by *modus ponens* and generalization. It follows by definition of a proof that each provable formula is a tautology.

"$\models\varphi$ implies $\vdash\varphi$": We give only a sketch. The first step is to show that one can construct a so-called Lindenbaum algebra for MPL and that this algebra is a residuated lattice. Lindenbaum algebra of MPL is the algebra of provably equivalent formulas: Consider binary relation $\sim$ on the set $\mathcal{F}$ of formulas of MPL defined by

$$\varphi \sim \psi \quad \text{iff} \quad \vdash\varphi \Leftrightarrow \psi,$$

i.e. $\varphi \sim \psi$ holds iff formula $\varphi \Leftrightarrow \psi$ is provable. One can show that $\sim$ is an equivalence relation on $\mathcal{F}$ (this needs to show that certain formulas are provable in MPL). Now, $\mathcal{F}$ may be turned into an algebra of the same type as residuated lattices, i.e. with binary operations $\wedge^{\mathcal{F}}, \vee^{\mathcal{F}}, \otimes^{\mathcal{F}}, \to^{\mathcal{F}}$ and nullary operations $0^{\mathcal{F}}, 1^{\mathcal{F}}$ defined by

$$\begin{aligned}
\varphi \wedge^{\mathcal{F}} \psi &= \varphi \mathbin{\&\!\wedge} \psi, \\
\varphi \vee^{\mathcal{F}} \psi &= \varphi \mathbin{\vee\!\!\vee} \psi, \\
\varphi \overset{\mathcal{F}}{\otimes} \psi &= \varphi \circledast \psi, \\
\varphi \to^{\mathcal{F}} \psi &= \varphi \Rightarrow \psi, \\
0^{\mathcal{F}} &= \mathbb{0}, \\
1^{\mathcal{F}} &= \mathbb{1}.
\end{aligned}$$

It turns out that $\sim$ is a congruence on $\mathcal{F}$ and that $\mathcal{F}/\sim$ (which is called a Lindenbaum algebra) is in fact a residuated lattice (again, this needs to show that some formulas are provable).

The second step is to show that each residuated lattice $\mathbf{K}$ can be embedded into a complete residuated lattice, i.e. that there is an injective morphism of $\mathbf{K}$ into some complete residuated lattice $\mathbf{K}^*$. One can take here the well-known Dedekind–MacNeille completion for $\mathcal{F}/\sim^*$ (introduced in [136], cf. Section 4.3). Particularly, this yields that $\mathcal{F}/\sim$ can be embedded into some complete residuated lattice $\mathcal{F}/\sim^*$; denote the embedding by h. Now, one can consider an $\mathcal{F}/\sim^*$-structure $\mathbf{M}$ defined as follows: The universe M is the set of all terms. Functions are defined by $f^{\mathbf{M}}(t_1,\ldots,t_n) = f(t_1,\ldots,t_n)$ for each n-ary $f \in F$ and terms $t_1,\ldots,t_n$. For each n-ary $r \in R$, the $\mathcal{F}/\sim^*$-relation $r^{\mathbf{M}}$ is defined by

$$r^{\mathbf{M}}(t_1,\ldots,t_n) = h([r(t_1,\ldots,t_n)]_\sim).$$

For each formula φ without free variables we have that $\|\varphi\|_{\mathbf{M}} = h([\varphi]_\sim)$. Moreover each formula φ with free variables among $x_1,\ldots,x_n$ is provably equivalent to $(\forall x_1,\ldots,x_n)\varphi$, i.e. $[\varphi]_\sim = [(\forall x_1,\ldots,x_n)\varphi]_\sim$. Now, if some formula φ is not provable in MPL, we have that $[\varphi]_\sim \neq [\mathbb{1}]_\sim$ and since h is an embedding we have $\|\varphi\|_{\mathbf{M}} = h([\varphi]_\sim) \neq h([\mathbb{1}]_\sim) = 1^{\mathcal{F}/\sim^*}$, and thus φ is not a tautology. □

This proof can easily be extended to a language with $\approx$: the set of axioms is extended by

$$x \approx x$$
$$(x \approx y) \Rightarrow (y \approx x)$$
$$(x \approx y) \otimes (y \approx z) \Rightarrow (x \approx z)$$
$$(x_1 \approx y_1) \otimes \cdots \otimes (x_{\sigma(f)} \approx y_{\sigma(f)}) \Rightarrow$$
$$\Rightarrow (f(x_1,\ldots,x_{\sigma(f)}) \approx f(y_1,\ldots,y_{\sigma(f)}))$$
$$(x_1 \approx y_1) \otimes \cdots \otimes (x_{\sigma(r)} \approx y_{\sigma(r)}) \otimes r(x_1,\ldots,x_{\sigma(r)}) \Rightarrow r(y_1,\ldots,y_{\sigma(r)})$$

for all $f \in F$ and $r \in R$. It is easily seen that these *axioms of equality* are true in degree 1 for each valuation v in a given $\mathbf{L}$-structure $\mathbf{M}$ iff $\approx^{\mathbf{M}}$ is an $\mathbf{L}$-equivalence compatible with all $f^{\mathbf{M}}$ and $r^{\mathbf{M}}$. Then each formula is provable from this extended list of axioms iff it is a tautology w.r.t. all $\mathbf{L}$-structures where $\approx^{\mathbf{M}}$ is an $\mathbf{L}$-equivalence compatible with functions and $\mathbf{L}$-relations of $\mathbf{M}$.

7.4 BIBLIOGRAPHICAL REMARKS

Section 7.1: The notion of a fuzzy closure operator in a set X was used in Pavelka [158] to denote what we call $\mathbf{L}_{\{1\}}$-closure operator in X. Fuzzy closure spaces were investigated in [139]. Further interest in fuzzy closure operators was motivated by interpreting them as consequence operators of fuzzy logic, see e.g. [50, 52, 53, 84] and also recent monograph [85] where these topics are discussed. Among other works on fuzzy closure operators we recall [37, 38, 148] and also [85]. It seems that up to [14, 29] the monotony condition was formulated by "$A \subseteq B$ implies $\mathrm{C}(A) \subseteq \mathrm{C}(B)$"; moreover, usually only $L = [0,1]$ is taken into account, i.e. only $\mathbf{L}_{\{1\}}$-closure operators usually with $L = [0,1]$ have been considered. Interesting to note is that in this case, residuum $\rightarrow$ plays no role and the essential structure on L is that of a complete lattice only. [173] defines a so-called implicative closure operator which is in fact our $\mathbf{L}$-closure operator.

Results on similarity spaces are based on [35]. The relationship between pseudometrics and fuzzy equivalences which forms parts (1) and (2) of Theorem 7.54 is basically due to [63]. Corollary 7.63 is present in [123].

Section 7.2: The first part concerning the extension principle is based on [22] (the proof there is a direct one, not using results from Section 3.3). The extension principle was formulated already in [216]; for the role of the extension principle in fuzzy sets and their applications see [127]. Cutworthiness appeared explicitly in [11] and is discussed in [127]; several particular forms of cutworthiness appear very frequently in the literature on fuzzy sets. Our definition of cutworthy formula makes precise the informal definition of [11]. That being a subgroup is cutworthy appears in [174]. A pioneering work on fuzzy automata is [182]; fuzzy automata and their representation of nested systems of finite automata appear (with $L = [0,1]$) in [145].

Section 7.3: Our presentation of monoidal logic follows Gottwald's [99, Chapter 14] which is in fact a simplified version of original Höhle's treatment [115, 117]. One can consider only particular residuated lattices as suitable structures of truth values (e.g. prelinear residuated lattices, BL-algebras, MV-algebras, G-algebras, Π-algebras) and can have the corresponding logical calculi (monoidal t-norm logic, basic logic, Łukasiewicz logic, Gödel logic, product logic, respectively; see [99, 105]). Note that all these calculi use syntax in the classical sense. On the other hand, Pavelka [158] applied fuzzy approach also to syntax and initiated a series of very interesting results. For instance, Pavelka considers evaluated rules of inference (enabling us to infer lower estimations of validity of formulas from estimations of lower validity of other formulas), fuzzy set of axioms (enabling us to postulate assumptions like "x is big in degree 0.9" etc.), the notion of provability in a certain degree, etc. Our fuzzy equational logic presented in Section 3.6 is in this "Pavelka style". Evaluated syntax is pursued by Novák, see [154].

References

[1] *Arbeiten der Forschungsgruppe Begriffsanalyse bis 1989. Arbeiten der Forschungsgruppe Begriffsanalyse ab 1990.*

[2] Arbib M. (Ed.): *The Handbook of Brain Theory and Neural Networks.* MIT Press, Cambridge, MA, 1995.

[3] Arnauld A., Nicole P.: *La logique ou l'art de penser.* 1662.

[4] Audi R. (Ed.): *The Cambridge Dictionary of Philosophy* 2nd ed. Cambridge University Press, Cambridge, 1999.

[5] Banaschewski B.: Hüllensysteme und Erweiterungen von Quasiordnungen. *Z. Math. Logic Grundl. Math.* **2**(1956), 117–130.

[6] Bandler W., Kohout L. J.: Fuzzy relational products as a tool for analysis and synthesis of the behaviour of complex natural and artificial systems. In: P. P. Wang and S. K. Chang (Eds.): *Fuzzy Sets: Theory and Applications to Policy Analysis and Information Systems.* Plenum Press, New York, 1980, pp. 341–367.

[7] Bandler W., Kohout L. J.: Semantics of implication operators and fuzzy relational products. *Int. J. Man-Machine Studies* **12**(1980), 89–116.

[8] Bandler W., Kohout L. J.: Mathematical relations. In: M. G. Singh *et al.* (Eds.): *International Encyclopedia of Systems and Control.* Pergamon Press, Oxford, 1985, pp. 4000–4008.

[9] Bandler W., Kohout L. J.: On the general theory of relational morphisms. *Int. J. Gen. Syst.* **13**(1986), 47–68.

[10] Bandler W., Kohout L.: Special properties, closures and interiors of crisp and fuzzy relations. *Fuzzy Sets Syst.* **26**(3)(1988), 317–331.

[11] Bandler W., Kohout L.: Cuts commute with closures (preprint).

[12] Barwise J.: *Handbook of Mathematical Logic.* North–Holland, Amsterdam, 1977.

[13] Bělohlávek R.: *Networks Processing Indeterminacy.* PhD thesis, xviii+198 pp., Technical University of Ostrava, 1998.

[14] Bělohlávek R.: Fuzzy Galois connections. *Math. Logic Quarterly* **45**(4)(1999), 497–504.

[15] Bělohlávek R.: Lattices of fixed points of fuzzy Galois connections. *Math. Logic Quarterly* **47**(1)(2001), 111–116.

[16] Bělohlávek R.: Galois connections and concept lattices with global-valued equalities. *Multiple-Valued Logic* (to appear).

[17] Bělohlávek R.: Similarity relations in concept lattices. *J. Logic Computation* **10**(6)(2000), 823–845.

[18] Bělohlávek R.: On the regularity of MV-algebras and Wajsberg hoops. *Algebra Universalis* **44**(2000), 375–377.

[19] Bělohlávek R.: Concept representation by bidirectional associative memories. *Neural Computation* **12**(10)(2000), 2279–2290.

[20] Bělohlávek R.: Similarity relations and BK-relation products. *Inf. Sci.* **126**(1–4)(2000), 287–295.

[21] Bělohlávek R.: Fuzzy logical bidirectional associative memory *Inf. Sci.* **128**(1–2)(2000), 91–103.

[22] Bělohlávek R.: A note on the extension principle. *J. Math. Anal. Appl.* **248**(2000), 678–682.

[23] Bělohlávek R.: Reduction and a simple proof of characterization of fuzzy concept lattices. *Fundamenta Informaticae* **46**(4)(2001), 277-285.

[24] Bělohlávek R.: Logical precision in concept lattices. *J. Logic Computation* (to appear).

[25] Bělohlávek R.: Some properties of residuated lattices. *Czech. Math. J.* (to appear).

[26] Bělohlávek R.: Fuzzy equational logic. *Arch. Math. Logic* (to appear).

[27] Bělohlávek R.: Concept lattices and order in fuzzy logic. *Ann. Pure Appl. Logic* (to appear).

[28] Bělohlávek R.: Concept equations (submitted).

[29] Bělohlávek R.: Fuzzy closure operators. *J. Math. Anal. Appl.* **262**(2001), 473-489.

[30] Bělohlávek R.: Fuzzy closure operators II. *Soft Computing* (to appear).

[31] Bělohlávek R.: Cutlike semantics for fuzzy logic and its applications (submitted).

[32] Bělohlávek R.: Birkhoff's variety theorem and fuzzy logic (submitted).

[33] Bělohlávek R.: Model theory for fuzzy logic in the foundations of soft computing (submitted).

[34] Bělohlávek R.: Algebras with fuzzy equalities (submitted).

[35] Bělohlávek R.: Similarity spaces (submitted).

[36] Bělohlávek R.: *Fuzzy Concept Lattices and Related Structures.* Habilitation thesis, Ostrava, 2000, ix+130 pp. (available on request).

[37] Biacino L., Gerla G.: Closure systems and L-subalgebras. *Inf. Sci.* **33**(1984), 181–195.

[38] Biacino L., Gerla G.: An extension principle for closure operators. *J. Math. Anal. Appl.* **198**(1996), 1–24.

[39] Biacino L., Gerla G., Ying M.: Approximate reasoning based on similarity. *Math. Logic Quarterly* **46**(2000), 77–86.

[40] Birkhoff G.: On the structure of abstract algebras. *Proc. Cambridge Philos. Soc.* **31**(1935), 433–454.

[41] Birkhoff G.: *Lattice Theory* (3rd ed.). Am. Math. Soc. Colloq. Publ., Vol. XXV, Providence, Rhode Island, 1967.

[42] Black M.: Vagueness: an exercise in logical analysis. *Philosophy of Science* **4**(1937), 427–455 (reprinted in: *Int. J. Gen. Syst.* **17**(1990), 107–128).

[43] Blyth T. S., Janowitz M. F.: *Residuation Theory.* Pergamon Press, Oxford, 1972.

[44] Bodenhofer U.: *A Similarity-Based Generalization of Fuzzy Orderings.* PhD thesis, Vol. 26 of *Schriftenreihe der Johannes-Kepler-Universität Linz.* Universitätsverlag Rudolf Trauner, 1999.

[45] Bodenhofer U.: A similarity-based generalization of fuzzy orderings preserving the classical axioms. *Int. J. Uncertainty, Fuzziness and Knowledge-Based Systems* **8**(5)(2000), 593–610.

[46] Bohm D.: *Quantum Theory.* Dover Publ., New York, 1979.

[47] Burusco A., Fuentes-Gonzáles R.: The study of the L-fuzzy concept lattice. *Mathware & Soft Computing* **3**(1994), 209–218.

[48] Burusco A., Fuentes-Gonzáles R.: Concept lattices associated with L-fuzzy W-contexts. *Mathware & Soft Computing* **3**(1996), 321–327.

[49] Burusco A., Fuentes-Gonzáles R.: Concept lattices defined from implication operators. *Fuzzy Sets Syst.* (to appear).

[50] Castro J. L.: Fuzzy logics as families of bivaluated logics. *Fuzzy Sets Syst.* **64**(1994), 321–332.

[51] Chajda I.: *Algebraic Theory of Tolerance Relations.* Palacký University Press, Olomouc, 1991.

[52] Chakraborty M. K.: Use of fuzzy set theory in introducing graded consequence in multiple valued logic. In: M. M. Gupta, T. Yamakawa: *Fuzzy Logic in Knowledge-Based Systems, Decision and Control.* North–Holland, Amsterdam, 1988, pp. 247–257.

[53] Chakraborty M. K.: Graded consequence: further studies. *J. Appl. Non-classical Logics* **5**(1994), 227–237.

[54] Chakraborty M. K., Das M.: On fuzzy equivalence I, II. *Fuzzy Sets Syst.* **11**(1983), 185–193, 299–307.

[55] Chakraborty M. K., Sarkar S.: Fuzzy antisymmetry and order. *Fuzzy Sets Syst.* **21**(1987), 169–182.

[56] Chang C. C.: Algebraic analysis of many-valued logics. *Trans. Amer. Math. Soc.* **88**(1958), 467–490.

[57] Chang C. C.: A new proof of the completeness of the Łukasiewicz's axioms. *Trans. Amer. Math. Soc.* **93**(1959), 74–80.

[58] Chang C. C., Keisler H. J.: *Continuous Model Theory.* Princeton University Press, Princeton, NJ, 1966.

[59] Chang C. C., Keisler H. J.: *Model Theory.* North–Holland, Amsterdam, 1973.

[60] Cignoli R., D'Ottaviano I.M.L., Mundici D.: *Algebraic Foundations of many-valued Reasoning.* Kluwer, Dordrecht, 2000.

[61] Cignoli R., Esteva F., Godo L., Torrens A.: Basic fuzzy logic is the logic of continuous t-norms and their residua. *Soft Computing* **4**(2000), 106–112.

[62] Czédli G.: Factor lattices by tolerances. *Acta Sci. Math. (Szeged)* **44**(1982), 35–42.

[63] De Baets B., Mesiar R.: Pseudo-metrics and T-equivalences. *The Journal of Fuzzy Mathematics* Vol. **5**, No. 2(1997), 471–481.

[64] Di Nola A., Gerla G.: Fuzzy models of first order languages. *Z. Math. Logic Grundl. Math.* **32**(1986), 331–340.

[65] Di Nola A., Gerla G.: Lattice valued algebras. *Stochastica* **11**(1987), 137–150.

[66] Di Nola A., Sessa S., Pedrycz W., Sanchez E.: *Fuzzy Relation Equations and Their Applications to Knowledge Engineering.* Kluwer, Dordrecht, 1989.

[67] Dubois D., Prade H.: *Fuzzy Sets and Systems: Theory and Applications.* Academic Press, New York, 1980.

[68] Dubois D., Prade H.: Soft computing, fuzzy logic, and artificial intelligence. *Soft Computing* **2**(1998), 7–11.

[69] Dubois D., Prade H. (Eds.): *Fundamentals of Fuzzy Sets.* Kluwer, Boston, 2000.

[70] Dubois D., Prade H., Esteva F., Garcia P., Godo L.: A logical approach to interpolation based on similarity relations. *Int. J. Approx. Reasoning* **17**(1997), 1–36.

[71] Ebbinghaus H. D., Flum J., Thomas W.: *Mathematical Logic.* Springer-Verlag, New York, 1984.

[72] Eilenberg S.: *Automata, Languages and Machines,* Vol. A, Academic Press, New York, 1974.

[73] Elloumi S., Jaoua A.: Implication based fuzzy Galois connection: application to the conceptual classification and information retrieval (preprint, 2000).

[74] Erceg M. A.: Functions, equivalence relations, quotient spaces and subsets in fuzzy set theory. *Fuzzy Sets Syst.* **3**(1980), 75–92.

[75] Fan J., Xie W., Pei J.: Subsethood measure: new definitions. *Fuzzy Sets Syst.* **106**(1999), 201–209.

[76] Faucett W. M.: Compact semigroups irreducibly connected between two idempotents. *Proc. Amer. Math. Soc.* **6**(1955), 741–747.

[77] Fodor J.: On contrapositive symmetry of implications in fuzzy logic. In: H. J. Zimmermann (Ed.): *Proc. EUFIT'93,* Aachen, 1993. Verlag Augustinus Buchhandlung, Aachen, pp. 1342–1348.

[78] Font J., Rodriguez A., Torrens A.: Wajsberg algebras. *Stochastica* **8**(1984), 5–31.

[79] Gaines B. R.: Precise past–fuzzy future. *Int. J. Man-Machine Studies* **19**(1983), 117–134.

[80] Ganter B.: *Lattice theory and formal concept analysis—a subjective introduction.* Preprint MATH-AL-2-1994, Technical University Dresden, 1994.

[81] Ganter B., Wille R.: *Formale Begriffsanalyse.* Springer-Verlag, Berlin, 1996 (English translation: *Formal Concept Analysis,* Springer-Verlag, 1999).

[82] Ganter B., Wille R.: Applied lattice theory: formal concept analysis. In [102], pp. 591–605.

[83] Ganter B., Wille R., Wolff K. E.: *Beiträge zur Begriffsanalyse.* B. I. Wissenschaftsverlag, Mannheim, 1987.

[84] Gerla G.: Graded consequence relations and fuzzy closure operators. *J. Appl. Non-Classical Logics* **6**(1996), 369–379.

[85] Gerla G.: *Fuzzy Logic. Mathematical Tools for Approximate Reasoning.* Kluwer, Dordrecht, 2001.

[86] Giles R.: The concept of grade of membership. *Fuzzy Sets Syst.* **25**(1988), 297–323.

[87] Girard J.Y.: Linear logic. *Theor. Comput. Sci.* **50**(1987), 1–102.

[88] Glivenko V.: Sur quelques points de la logique de M. Brouwer. *Bull. Acad. Sci. Belg.* **15**(1929), 183–188.

[89] Gödel K.: Zum intuitionistischen Aussagenkalkül. *Akad. Wiss. Wien, Math.-Naturwiss. Kl.* **69**(1932), 65–66.

[90] Goguen J. A.: L-fuzzy sets. *J. Math. Anal. Appl.* **18**(1967), 145–174.

[91] Goguen J. A.: The logic of inexact concepts. *Synthese* **19**(1968–69), 325–373.

[92] Gottwald S.: Fuzzy uniqueness of fuzzy mappings. *Fuzzy Sets Syst.* **3**(1980), 49–74.

[93] Gottwald S.: On the existence of solutions of systems of fuzzy equations. *Fuzzy Sets Syst.* **12**(1984), 301–302.

[94] Gottwald S.: Generalized solvability criteria for fuzzy equations. *Fuzzy Sets Syst.* **17**(1985), 285–296.

[95] Gottwald S.: *Mehrwertige Logik.* Akademie-Verlag, Berlin, 1988.

[96] Gottwald S.: *Fuzzy Sets and Fuzzy Logic. Foundations of Applications—from a Mathematical Ponit of View.* Vieweg, Wiesbaden, 1993.

[97] Gottwald S.: Approximately solving fuzzy relation equations: some mathematical results and some heuristic proposals. *Fuzzy Sets Syst.* **66**(1994), 175–193.

[98] Gottwald S.: Approximate solutions of fuzzy relational equations and characterization of t-norms that define metrics for fuzzy sets. *Fuzzy Sets Syst.* **75**(1995), 189–201.

[99] Gottwald S.: *A Treatise on Many-Valued Logics.* Research Studies Press, Baldock, Hertfordshire, England, 2001.

[100] Gottwald S., Pedrycz W.: Solvability of fuzzy relational equations and manipulation of fuzzy data. *Fuzzy Sets Syst.* **18**(1986), 45–65.

[101] Grätzer G.: *Universal Algebra.* Springer-Verlag, New York, 1968.

[102] Grätzer G.: *General Lattice Theory.* Birkhäuser Verlag, Basel, 1998.

[103] Gupta M. M., Rao D. H.: On the principles of fuzzy neural networks. *Fuzzy Sets Syst.* **61**(1994), 1–18.

[104] Hájek P.: Basic fuzzy logic and BL-algebras. *Soft Computing* **2**(1998), 124–128.

[105] Hájek P.: *Metamathematics of Fuzzy Logic.* Kluwer, Dordrecht, 1998.

[106] Hájek P.: Ten questions and one problem on fuzzy logic. *Ann. Pure Appl. Logic* **96**(1999), 157–163.

[107] Hájek P.: Function symbols in fuzzy predicate logic. *East-West Fuzzy Logic Days*, Zittau, 2000.

[108] Hájek P., Godo L., Esteva F.: A complete many-valued logic with product conjunction. *Arch. Math. Logic* **35**(1996), 191–208.

[109] Heyting A.: Die formalen Regeln der intuitionistischen Logik. *Sitzungsber. Preuss. Akad. Wiss. Berlin, Phys.-Math. Kl.* II(1930), 42–56.

[110] Higashi M., Klir G. J.: Resolution of finite fuzzy relation equations. *Fuzzy Sets Syst.* **13**(1984), 65–82.

[111] Hisdal E.: The philosophical issues raised by fuzzy set theory. *Fuzzy Sets Syst.* **25**(1988).

[112] Hisdal E.: Are grades of membership probabilities? *Fuzzy Sets Syst.* **25**(3)(1988), 349–356.

[113] Höfler A.: *Grundlehren der Logik und Psychologie.* G. Freytag, Leipzig, 1906.

[114] Höhle U.: Quotients with respect to similarity relations. *Fuzzy Sets Syst.* **27**(1988), 31–44.

[115] Höhle U.: Monoidal logic. In: R. Kruse, J. Gebhardt, R. Palm (Eds.): *Fuzzy Systems in Computer Science.* Vieweg, Wiesbaden, 233–243.

[116] Höhle U.: Commutative, residuated l-monoids. In: U. Höhle, E.P. Klement (Eds.): *Non-classical logics and their applications to fuzzy subsets.* Kluwer, Dordrecht, 1995.

[117] Höhle U.: On the fundamentals of fuzzy set theory. *J. Math. Anal. Appl.* **201**(1996), 786–826.

[118] Höhle U.: Many-valued equalities, singletons and fuzzy partitions. *Soft Computing* **2**(1998), 134–140.

[119] Jacas J.: Similarity relations – the calculation of minimal generating families. *Fuzzy Sets Syst.* **35**(1990), 151–162.

[120] Kaufmann A.: *Introduction to the Theory of Fuzzy Subsets.* Academic Press, New York, 1975.

[121] Klaua D.: Grunbegriffe einer mehrwertigen Mengenlehre. *Monatsber. Dtsch. Akad. Wiss. Berlin* **12**(1966), 749–758.

[122] Klawonn F.: Fuzzy points, fuzzy relations and fuzzy functions. In: Novák V., Perfileva I. (eds.): *Discovering the World with Fuzzy Logic.* Physica-Verlag (Springer-Verlag Company), Heidelberg, New York, 2000.

[123] Klawonn F., Castro J. L.: Similarity in fuzzy reasoning. *Mathware & Soft Computing* **2**(1995), 197–228.

[124] Klawonn F., Kruse R.: Fuzzy control as interpolation on the basis of equality relations. *Proc. IEEE Int. Conf. on Fuzzy Systems*, San Francisco, March 1993, pp. 8.

[125] Klement E. P., Mesiar R., Pap E.: *Triangular Norms.* Kluwer, Dordrecht, 2000.

[126] Klir G. J.: *Fuzzy Sets: An Overview of Fundamentals, Applications, and Personal Views.* Beijing Normal University Press, Beijing, 2000.

[127] Klir G. J., Yuan B.: *Fuzzy Sets and Fuzzy Logic. Theory and Applications.* Prentice–Hall, Upper Saddle River, NJ, 1995.

[128] Klir G. J., Yuan B. (Eds.): *Fuzzy Sets, Fuzzy Logic, and Fuzzy Systems. Selected Papers by Lotfi A. Zadeh.* World Scientific, Singapore, 1996.

[129] Kosko B.: *Fuzzy Thinking. The New Science of Fuzzy Logic.* Flamingo, 1994.

[130] Kruse R., Gebhardt J., Klawonn F.: *Fuzzy-Systeme.* B. G. Teubner, Stuttgart, 1995.

[131] Kuhn T. S.: *The Structure of Scientific Revolutions.* University of Chicago Press, Chicago, 1962.

[132] Lee S. C., Lee E. T.: Fuzzy neural networks. *Math. Biosci.* **23**(1975), 151–177.

[133] Lewis H. W.: *The Foundations of Fuzzy Control.* Plenum Press, New York, 1997.

[134] Łukasiewicz J.: O logice trówartosciowej (On three-valued logic). *Ruch filozoficzny* **5**(1920), 170–171.

[135] Łukasiewicz J.: *Selected Works.* PWN—Polish Scientific Publishers, Warsaw, and North–Holland, Amsterdam, 1970.

[136] MacNeille H. M.: Partially ordered sets. *Trans. Amer. Math. Soc.* **42**(1937), 416–460.

[137] Mal'cev A. I.: *Algebraic Systems* (in Russian). Nauka, Moscow, 1970.

[138] Mamdani E. H., Assilian S.: An experiment in linguistic synthesis with a fuzzy logic controller. *Int. J. Man-Machine Studies* **7**(1)(1975), 1–13.

[139] Mashour A. S., Ghanim M. H.: Fuzzy closure spaces. *J. Math. Anal. Appl.* **106**(1985), 154–170.

[140] McNeill D., Freiberger P.: *Fuzzy Logic. The Discovery of a Revolutionary Computer Technology—and How It Is Changing Our World.* Simon & Schuster, New York, 1993.

[141] Menger K.: Statistical metrics. *Proc. Natl. Acad. Sci. U.S.A.* **28**(1942), 535–537.

[142] Menger K.: Ensembles flous et fonctions aleatoires. *C. R. Acad. Sci.* **232**(1951), 2001–2003.

[143] Menger K.: Probabilistic theories of relations. *Proc. Natl. Acad. Sci. U.S.A.* **37**(1951), 178–180.

[144] Menger K.: *Selected Papers in Logic and Foundations, Didactics, Economics.* Reidel, Dordrecht, 1979.

[145] Močkoř J.: Fuzzy and non-deterministic automata. *Research report No. 8*, Institute for Research and Applications of Fuzzy Modeling, University of Ostrava, 1997, pp. 11.

[146] Monk J. D.: *Mathematical Logic.* Springer-Verlag, New York, 1976.

[147] Mostert P. S., Shields A. L.: On the structure of semigroups on a compact manifold with boundary. *Ann. Math.* **65**(1957), 117–143.

[148] Murali V.: Lattice of fuzzy subalgebras and closure systems in I^X. *Fuzzy Sets Syst.* **41**(1991), 101–111.

[149] Negoita C. V., Ralescu A.: *Applications of Fuzzy Sets to Systems Analysis.* J. Wiley, New York, 1975.

[150] Novák V.: *Fuzzy Sets and Their Applications.* Adam–Hilger, Bristol, 1989.

[151] Novák V.: On the syntactico-sematical completeness of first-order fuzzy logic. Part I — Syntactical aspects; Part II — Main results. *Kybernetika* **26**(1990), 47–66; 134–154.

[152] Novák V.: Ultraproduct theorem and recursive properties of fuzzy logic. In: U. Höhle, E.P. Klement (Eds.): *Non-classical logics and their applications to fuzzy subsets.* Kluwer, Dordrecht, 1995.

[153] Novák V.: Paradigm, formal properties and limits of fuzzy logic. *Int. J. Gen. Syst.* **24**(4)(1996), 377–405.

[154] Novák V., Perfilieva I., Močkoř J.: *Mathematical Principles of Fuzzy Logic.* Kluwer, Dordrecht, 1999.

[155] Ore O.: *Galois connexions. Trans. Amer. Math. Soc.* **55**(1944), 493–513.

[156] Ovchinikov S. V.: Structure of fuzzy binary relations. *Fuzzy Sets Syst.* **6**(2)(1981), 169–195.

[157] Ovchinikov S. V.: Similarity relations, fuzzy partitions, and fuzzy orderings. *Fuzzy Sets Syst.* **40**(1)(1991), 107–126.

[158] Pavelka J.: On fuzzy logic I, II, III. *Z. Math. Logik Grundl. Math.* **25**(1979), 45–52, 119-139, 447–464.

[159] Pawlak Z.: *Rough Sets: Theoretical Aspects of Reasoning about Data.* Kluwer, Boston, 1991.

[160] Pedrycz W.: Fuzzy relational equations with generalized connectives and their applications. *Fuzzy Sets Syst.* **10**(1983), 185–201.

[161] Pedrycz W.: Numerical and applicational aspects of fuzzy relational equations. *Fuzzy Sets Syst.* **11**(1983), 1–18.

[162] Pedrycz W.: Approximate solutions of fuzzy relational equations. *Fuzzy Sets Syst.* **28**(1988), 183–202.

[163] Pedrycz W.: *Fuzzy Sets Engineering.* CRC Press, Boca Raton, 1995.

[164] Pedrycz W.: Fuzzy neural networks and neurocomputations. *Fuzzy Sets Syst.* **56**(1993), 1–28.

[165] Perfilieva I., Tonis A.: Compatibility of systems of fuzzy relation equations. *Int. J. Gen. Syst.* **29**(4)(2000), 511-528.

[166] Poincaré H.: *The Foundations of Science.* University Press of America, Washington, DC, 1982 (contains *Science and Hypothesis, The Value of Science, Science and Method*).

[167] Pollandt S.: *Fuzzy Begriffe.* Springer-Verlag, Berlin, 1997.

[168] Pultr A.: Fuzziness and fuzzy equality. In [190], pp. 119–135.

[169] Quine W. V.: What price bivalence? *J. Philos.* **78**(1981), 90–95.

[170] *Random House Webster's College Dictionary.* Random House, New York, 1991.

[171] Rasiowa H.: *An Algebraic Approach to Non-classical Logics.* PWN—Polish Scientific Publishers, Warsaw, and North–Holland, Amsterdam, 1974.

[172] Riguet J.: Relations binaires, fermetures, correspondances de Galois. *Bull. Soc. Math. Fr.* **76**(1948), 114–155.

[173] Rodriguez R. O., Esteva F., Garcia P., Godo L.: On implicative closure operators in approximate reasoning. *Proc. FUZZ-IEEE'00*, San Antonio, Texas, 2000.

[174] Rosenfeld A.: Fuzzy groups. *J. Math. Anal. Appl.* **35**(3)(1971), 512–517.

[175] Rosser J. B., Turquette A. R.: *Many-valued Logics.* North–Holland, Amsterdam, 1952.

[176] Russell B.: Vagueness. *Austral. J. Philos.* **1**(1923), 84–92.

[177] Samhan M. A.: Fuzzy quotient algebras and fuzzy factor congruences. *Fuzzy Sets Syst.* **73**(1995), 269–277.

[178] Sanchez E.: *Equations de relations floues.* Thèse de Doctorat, Faculté de Médicine de Marseille, 1974.

[179] Sanchez E.: Resolution of composite fuzzy relation equations. *Inf. Control* **30**(1976), 38–48.

[180] Sanchez E.: Solutions in composite fuzzy relation equations: application to medical diagnosis in Brouwerian logic. In: M. M. Gupta *et al.* (Eds.): *Fuzzy Automata and Decision Processes.* North–Holland, Amsterdam, 1977, pp. 221–234.

[181] Sanchez E.: Resolution of eigen fuzzy sets equations. *Fuzzy Sets Syst.* **1**(1978), 69–74.

[182] Santos E. S.: Maximin automata. *Inf. Control* **13**(1968), 363–377.

[183] Schmidt J.: Zur Kennzeichnung der Dedekind-MacNeillschen Hülle einer geordneten Menge. *Arch. Math.* **7**(1956), 241–249.

[184] Schröder E.: *Algebra der Logik I, II, III.* Leipzig, 1890, 1891, 1895.

[185] Schreider Y. A.: *Equality, Resemblance, and Order.* Mir Publishers, Moscow, 1975.

[186] Schweizer B., Sklar A.: Statistical metrics. *Pac. J. Math.* **10**(1960), 313–334.

[187] Schweizer B., Sklar A.: *Probabilistic Metric Spaces.* North–Holland, Amsterdam, 1983.

[188] Shoenfield J. R.: *Mathematical Logic.* Addison-Wesley, Reading, MA, 1967.

[189] Sikorski R.: *Boolean Algebras.* Springer-Verlag, Berlin, 1964.

[190] Skala H. J., Termini S., Trillas E.: *Aspects of Vagueness.* Reidel, Dordrecht, 1984.

[191] Sugeno M. *et al.*: Helicopter flight control based on fuzzy logic. *Proc. IFES'91*, Yokohama, Japan, 1991, pp. 1120–1121.

[192] Takeuti G., Titani S.: Fuzzy logic and fuzzy set theory. *Arch. Math. Logic* **32**(1992), 1–32.

[193] Tarski A.: *Introduction to Logic and to the Methodology of Deductive Sciences.* Oxford University Press, New York, 1965.

[194] Trillas E., Valverde L.: An inquiry into indistinguishability operators. In [190], pp. 231–256.

[195] Tugendhat E., Wolf U.: *Logisch-semantische Propädautik.* Philipp Reclam jun., Stuttgart, 1986.

[196] Turunen E.: *Mathematics Behind Fuzzy Logic.* Physica Verlag, Heidelberg, 1999.

[197] Valverde L.: On the structure of F-indistinguishability operators. *Fuzzy Sets Syst.* **17**(1985), 313–328.

[198] Venugopalan P.: Fuzzy ordered sets. *Fuzzy Sets Syst.* **46**(1992), 221–226.

[199] Vopěnka P.: *Mathematics in the Alternative Set Theory.* Teubner, Teubner Texte zur Mathematik, Leipzig, 1979.

[200] Vopěnka P.: *Úvod do matematiky v alternatívnej teórii množín* (Introduction to mathematics in the alternative set theory, in Slovak). Alfa, Bratislava, 1989.

[201] Vopěnka P.: The philosophical foundations of alternative set theory. *Int. J. Gen. Syst.* **20**(1991), 115–126.

[202] Ward M., Dilworth R. P.: Residuated lattices. *Trans. Amer. Math. Soc.* **45**(1939), 335–354.

[203] Wille R.: Restructuring lattice theory: an approach based on hierarchies of concepts. In: I. Rival (Ed.): *Ordered Sets.* Reidel, Dordrecht, 1982, pp. 445–470.

[204] Wille R.: Complete tolerance relations of concept lattices. In: G. Eigenthaler *et al.*: *Contributions to General Algebra*, Vol. 3. Hölder-Pichler-Tempsky, Wien, 1985, pp. 397–415.

[205] Wille R.: Concept lattices and conceptual knowledge systems. *Comp. Math. Appl.* **23**(1992), 493-515.

[206] Williamson T.: *Vagueness.* Routledge, London and New York, 1994.

[207] Winston P. H.: *Artificial Intelligence*, 3rd ed. Addison-Wesley, Reading, MA, 1992.

[208] Wójcicki R.: *Theory of Logical Calculi. Basic Theory of Consequence Operations.* Kluwer (Synthese Library, Vol. 199), Dordrecht, 1988.

[209] Wu W. M.: Fuzzy reasoning and fuzzy relational equations. *Fuzzy Sets Syst.* **20**(1986), 67–78.

[210] Wygarlak M.: Fuzzy inclusion and fuzzy equality of two fuzzy subsets, fuzzy operations for fuzzy subsets. *Fuzzy Sets Syst.* **10**(1983), 157–168.

[211] Yager R. R., Ovchinikov S., Tong M. M., Nguyen H. T.: *Fuzzy Sets and Applications: Selected Papers by L. A. Zadeh.* John Wiley, New York, 1987.

[212] Yeh R. T., Bang Y.: Fuzzy relations, fuzzy graphs, and their applications. In: L. Zadeh *et al.* (Eds.): *Fuzzy Sets and their Applications to Cognitive and Decision Processes.* Academic Press, New York, 1975, pp. 125–149.

[213] Ying M.-S.: The fundamental theorem of ultraproduct in Pavelka's fuzzy logic. *Z. Math. Logik Grundl. Math.* **38**(1992), 197–201.

[214] Ying M.-S.: Compactness, Löwenheim-Skolem property and the direct product of lattices of truth values. *Z. Math. Logik Grundl. Math.* **38**(1992), 521–524.

[215] Zadeh L. A.: From circuit theory to system theory. *Proc. Inst. Radio Eng.* **50**(1962), 856–865.

[216] Zadeh L. A.: Fuzzy sets. *Inf. Control* **8**(3)(1965), 338–353.

[217] Zadeh L. A.: Towards a theory of fuzzy systems. In: R. E. Kalman, R. N. De Clairis (Eds.): *Aspects of Networks and Systems Theory.* Holt, Rinehart & Winston, New York, 1971, pp. 469–490.

[218] Zadeh L. A.: Similarity relations and fuzzy orderings. *Inf. Sci.* **3**(1971), 159–176.

[219] Zadeh L. A.: Outline of a new approach to the analysis of complex systems and decision processes. *IEEE Trans. Syst. Man Cybernetics* **1**(1)(1972), 28–44.

[220] Zadeh L. A.: The concept of a linguistic variableand its application to approximate reasoning I, II, III. *Inf. Sci.* **8**(3)(1975), 199–251, 301–357, **9**(1975), 43–80.

[221] Zadeh, L. A.: Towards the theory of fuzzy information granulation and its centrality in human reasoning and fuzzy logic. *Fuzzy Sets Syst.* **90**(2)(1997), 111–127.

[222] Zadeh L. A.: Some reflections on soft computing, granular computing and their roles in the conception, design and utilization of information/intelligent systems. *Soft Computing* **2**(1998), 23–25.

Index